Artificial Intelligence and Other Innovative Computer Applications in the Nuclear Industry

Artificial Intelligence and Other Innovative Computer Applications in the Nuclear Industry

Edited by

M. Catherine Majumdar
Westinghouse Idaho Nuclear Corporation
Idaho Falls, Idaho

Debu Majumdar
U.S. Department of Energy
Idaho Operations Office
Idaho Falls, Idaho

and

John I. Sackett
Argonne National Laboratory West
Idaho Falls, Idaho

Springer Science+Business Media, LLC

Library of Congress Cataloging in Publication Data

American Nuclear Society Topical Meeting on Artificial Intelligence and Other Innovative Computer Applications (1987: Snowbird, Utah)
   Artificial intelligence and other innovative computer applications in the nuclear industry / edited by M. Catherine Majumdar, Debu Majumdar, and John I. Sackett.
      p.   cm.
   "Proceedings of the American Nuclear Society Topical Meeting on Artificial Intelligence and Other Innovative Computer Applications, held August 31–September 2, 1987, in Snowbird, Utah"—T.p. verso.
   Includes bibliographies and indexes.
   ISBN 978-1-4612-8290-7   ISBN 978-1-4613-1009-9 (eBook)
   DOI 10.1007/ 978-1-4613-1009-9
   1. Nuclear industry—Data processing—Congresses. 2. Artificial intelligence—Industrial applications—Congresses. I. Majumdar, M. Catherine. II. Majumdar, D. (Debu) III. Sackett, John I. IV. Title.
TK9006.A48   1987   88-9396
621.48′3′028563—dc19   CIP

Proceedings of the American Nuclear Society Topical Meeting on Artificial
Intelligence and Other Innovative Computer Applications,
held August 31–September 2, 1987, in Snowbird, Utah

© 1988 Springer Science+Business Media New York
   Originally published by Plenum Press, New York in 1988
Softcover reprint of the hardcover 1st edition 1988

This conference brought together experts from 15 countries to
discuss application of Artificial Intelligence (AI) techniques
to the nuclear industry. It was apparent from the meeting that
even those active in the field were surprised at the extent of
work and the progress made. There was a strong impression that
application of this technology to nuclear power plants is
inevitable. The benefits to improved operation, design, and
safety are simply too significant to be ignored.

This is a much different conclusion than might have been
reached a few years ago when the technology was new and people
were struggling to understand its significance. We believe
that this meeting reflects a major turning point for the
technology. It has moved from being a topic understood only by
specialists to a situation where users are the most active
people in the field.

A broad array of innovative work is described from all of the
participating countries. The activity in the U.S. is large and
diverse. Although there is no nationally focussed policy for
AI research in the U.S., many of these activities are reported
here. Japan and France have a strong drive to integrate AI
technology into their nuclear plants, and this is reflected in
these proceedings.

There is now a sense of excitement about what can be achieved,
a refreshing departure from the often defensive posture taken
in nuclear power development these days. The excitement stems
from recognition that the technology has the potential to
greatly expand our capabilities as designers and operators. To
take but one example, great strides have been made in
developing systems that can diagnose the condition of the plant
and provide options for response. Artificial Intelligence work
has not yet led to extensive automation, but it will.

There was much discussion at the meeting about the fact that
Artificial Intelligence is a poor title for this work.
Knowledge engineering comes closer to what is being achieved.
We hope that you will find these proceedings both interesting
and useful. This is an emerging field that will affect us all.

J. Curtis Haire,
General Chairman

Debu Majumdar and John I. Sackett,
Technical Program Co-Chairmen

MEETING OFFICIALS

General Chairman   J. Curtis Haire
                          EG&G Idaho, Inc.

Program Co-Chairman   Debu Majumdar
                          U.S. Department of Energy
                          Idaho Operations Office

Program Co-Chairman   John I. Sackett
                          Argonne National Laboratory-West

Arrangements   Richard W. Lindsay
                          Argonne National Laboratory-West

Finance   Orville R. Meyer
                          EG&G Idaho, Inc.

Registration   Carl H. Cooper
                          EG&G Idaho, Inc.

Exhibits   Glen A. Mortensen
                          EG&G Idaho, Inc.

Publications   M. Catherine Majumdar
                          Westinghouse Idaho Nuclear Co.

Public Information   George A. Freund
                          Science Applications, Inc.

Guest Program   Karen Sackett

Registration   Betty Haire and Marge Lindsay
Desk

CO-SPONSORS

Idaho Section of the American Nuclear Society
European Nuclear Society
ANS Human Factors Division
ANS Remote Systems Technology Division

PROGRAM COMMITTEE

Technical Program Co-Chairmen - Debu Majumdar, U.S. Department
of Energy - Idaho Operations Office and John I. Sackett,
Argonne National Laboratory-West

A.D. Alley.............. General Electric Company
Hakan Andersson........ Studsvik Energiteknik
Bill Bertch............ General Dynamics, Inc.
Paul Blanch............ Northeast Utilities Service Co.
William V. Botts....... EI International, Inc.
Michael Bray........... EG&G Idaho, Inc.
John S. Brtis.......... Sargent and Lundy
David Cain............. EPRI
Allen Christie......... Westinghouse Electric Corp.
Robert Colley.......... EPRI
Gail A. Cordes......... Intermountain Technologies, Inc.
Brent Dixon............ EG&G Idaho, Inc.
James Dukelow, Jr...... Boeing Computer Service
Robert Engelmore....... Stanford University
Jerry D. Griffith...... U.S. Department of Energy
Dennis Harrison........ U.S. Department of Energy
James P. Jenkins....... U.S. Nuclear Regulatory Commission
C. Edward Johnson...... Westinghouse Idaho Nuclear Co.
Harry Julian........... Volian Enterprises, Inc.
Jan G. Kretzschmar..... SCK/CEN Informatics
Bob Lang............... Middle South Services Company
Henry Makowitz......... EG&G Idaho, Inc.
William Nelson......... EG&G Idaho, Inc.
Pedro J. Otaduy........ Oak Ridge National Laboratory
Howard Rohm............ U.S. Department of Energy
Gary Sandquist......... University of Utah
Donald Schurman........ EG&G Idaho, Inc.
Devin Smith............ S.U.N.Y., Stony Brook
Bob Stiger............. Delian Corporation
Al Sudduth............. Duke Power Company
Gilles Zwingelstein.... Electricite de France

PAPER REVIEWERS

John Sackett........... Argonne National Laboratory-West
Debu Majumdar.......... U.S. Department of Energy
Jan Kretzschmar........ SCK/CEN Informatics
Hakan Andersson........ Studsvik Energiteknik
Paul Blanch............ Northeast Utility Service Co.
John Brtis............. Sargent and Lundy
Henry Makowitz......... EG&G Idaho, Inc.
Mike Bray.............. EG&G Idaho, Inc.
Al Sudduth............. Duke Power Company
Gary Sandquist......... University of Utah
Jim Jenkins............ U.S. Nuclear Regulatory Commission
Brent Dixon............ EG&G Idaho, Inc.
Rich Mark.............. Volian Enterprises
Jim Dukelow............ Boeing Computer Service
Gail A. Cordes......... Intermountain Technologies, Inc.
Pedro J. Otaduy........ Oak Ridge National Laboratory
C. Edward Johnson...... Westinghouse Idaho Nuclear Co.
Bill Nelson............ EG&G Idaho, Inc.
Howard Rohm............ U.S. Department of Energy
Bob Stiger............. Delian Corporation

SESSION CHAIRMEN

A.D. Alley
General Electric Co.

H. Andersson
Studsvik Energiteknik

T. Bjorlo
Institutt for Energiteknikk

P. Blanch
Northeast Utilities Service Co.

W.V. Botts
EI International, Inc.

M.A. Bray
EG&G Idaho, Inc.

J. Brtis
Sargent and Lundy

P.C. Cacciabue
Joint Research Centre Ispra

D. Cain
Electric Power Research
   Institute

A. Christie
Westinghouse Electric Corp.

G.A. Cordes
Intermountain Technologies, Inc.

B. Dixon
EG&G Idaho, Inc.

J. Dukelow, Jr.
Boeing Computer Service

L. Ho
Institute of Nuclear Energy
   Research

E. Hollnagel
Computer Resources International

A. Jaeschke
Nuclear Research Center

J. Jenkins
U.S. Nuclear Regulatory
   Commission

C.W. Johnson
Westinghouse Idaho Nuclear
   Co.

H. Julian
Volian Enterprises, Inc.

T. Kiguchi
Hitachi Energy Research
   Laboratory

J. Kvetzchmar
SCK/CEN Informatics

R.W. Lindsay
Argonne National Laboratory
   West

H. Makowitz
EG&G Idaho, Inc.

P. Malvache
CEA, France

K. Monta
Nippon Atomic Industry Group

J. Mott
EI International, Inc.

W. Nelson
EG&G Idaho, Inc.

J. Panoussian
Framatome

Y. Shinohara
Japan Atomic Energy Research
   Institute

D. Smith
SUNY at Stony Brook

Y. Souchet
CEA, France

A. Sudduth
Duke Power Company

J.B. Thomas
CEA, France

G. Zwingelstein
Electricite de France

CONTENTS

CHAPTER 3
ALARM AND SIGNAL VALIDATION

CHAPTER 4
EMERGENCY RESPONSE

CHAPTER 5
PROCESS DIAGNOSTICS AND TRANSIENT ADVISOR

CHAPTER 10
OPERATION ANALYSIS AIDS

CHAPTER 1

INTERNATIONAL OVERVIEW

KNOWLEDGE BASED SYSTEMS FOR NUCLEAR APPLICATIONS IN GERMANY

F. Schmidt

IKE, University of Stuttgart, Pfaffenwaldring 31
D-7000 Stuttgart 80, West Germany, Tel. 0711/685-2116

Abstract

Several national and international research programs which are dealing with artificial intelligence and other innovative computer applications are in progress in Germany. However in contrast to the development of computer applications in the past, the new research programs are not very much determined from needs of the nuclear industry. Thus, applications of AI techniques in German nuclear industry are not very innovative in the sense of artificial intelligence. They may be divided into two categories:

1. projects which are aimed to explore the new technologies,

2. projects which are aimed to open new areas of work.

This situation changes due to the fact that supercomputers with large memory, workstations with cheap disc devices and fast networks are becoming available. These hardware devices allow the connection of locally available knowledge and data bases with powerful central computer capacity. Using such hardware tools new applications can be developed in nuclear engineering using even existing software tools. These new applications may be characterized as integrated systems. The Integral Planning and Simulation System IPSS which is under development at the University of Stuttgart is such a system. The basic concept of the system and its various connections to knowledge based techniques will be described in the second part of the paper.

1 Artificial Intelligence in Germany - National and International Programs

The use of methods developed in Artificial Intelligence allows new applications for all kinds of computers. Also existing applications can be improved by the new techniques. To explore this new potential various national and international research programs were established in the FRG.

Probably the most known program is the ESPRIT program (European Strategic Program for Research in Information Technology) of the European Community. Phase 1 was established in 1984. Its total budget for a 5 years' period is approximately 2.4 billion US $. Phase 2 of the program is planned to be started in 1988. Its budget and program are not yet decided. Phase 1 concentrated on advanced microelectronics, software technology, advanced information processing, office systems and computer integrated manufacture.Results are published in the ESPRIT Status Reports [1].

Techniques and methods developed in the European High Technology Program ESPRIT will be applied in the frame of various EUREKA projects. Among them are such projects as the European Software Factory or the PROMETHEUS (Program for an European Traffic with Highest Efficiency and Unprecendented Safety) project. The goal of these programs it to initiate cooperations between industry and universities on a European wide basis. By the synergetic effects of such cooperations one hopes to get a pay-off greater than the coordination efforts necessary to organize such cooperations.

The international programs are supported by national efforts. The most known and most powerful program was initiated by the German Ministry of Research and Technology (BMFT). Its budget is about 1.6 billion US $. Coordinated research projects are supported in the areas of computer aided design of computers and software, new computer structures and expert systems. First results were reported at a recent conference [2].

Smaller projects were initiated from various other institutions. Probably most noteworthy are several special research programs initiated by the Deutsche Forschungsgemeinschaft. They include a "Sonderforschungsbereich" on artificial intelligence.

Also the industry is continuously becoming aware of the importance of the new techniques. Thus Daimler-Benz, the manufacturer of Mercedes cars and one of the partners in the PROMETHEUS project is planning to install a special research center on knowledge based systems at the University of Ulm.

2 Artificial Intelligence in the German Nuclear Industry

Artificial intelligence techniques still play a minor role in nuclear applications. In my opinion this is due to several reasons. The most important one is the ongoing debate on the possible role of nuclear energy. Even if there are some tendencies to postpone the final decision on nuclear energy this debate resulted in a decrease of the money available for the development of new projects. Also it seems to be difficult to participate in the funding of the newly established programs with nuclear applications. This is due to the fact that the nuclear community took away in the past quite a big portion of the research funds generally available.

As a result applications of artificial intelligence techniques in nuclear industry can be divided into two categories:

- projects which are aimed to explore the new technologies,

- projects which are aimed to open new areas of work outside the nuclear business.

Both types of projects have to be considered as conventional types of application of artificial intelligence techniques. They concentrate to utilize knowledge based techniques for planning, monitoring, analysis and diagnosis [3]. However it is easy to foresee that this situation will change. There are two developments which require the intensive use of artificial intelligence techniques and especially knowledge based methods. Both developments were caused by the rapid changes in computer technologies, which resulted in an increase by at least an order of magnitude of both measured and calculated data which are available to describe the actual status of a power plant.

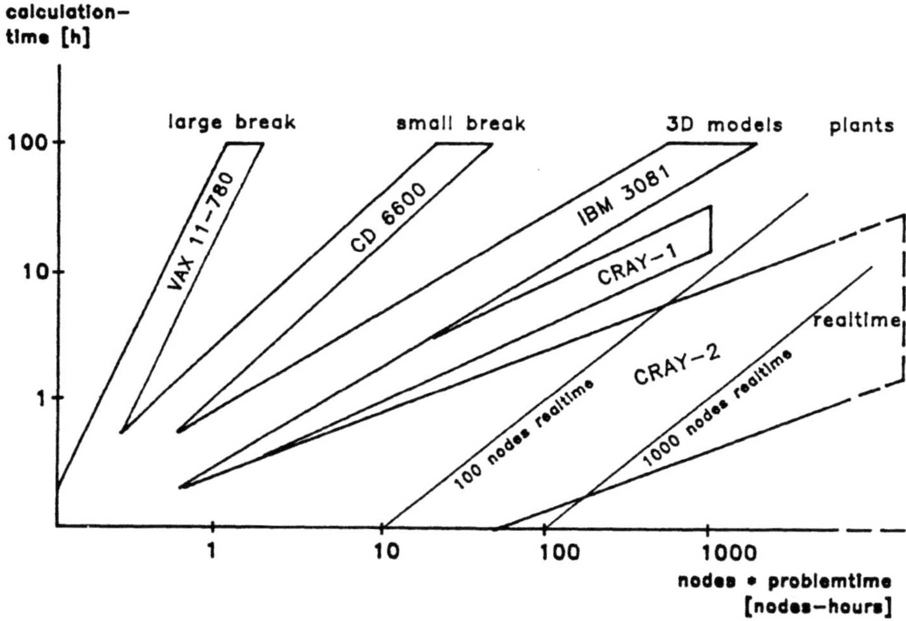

Figure 1: Classes of thermal hydraulic problems solvable on various computers

The increase in measured data is due to the newly available process control systems. The increase in calculated data is a result of the availability of supercomputers as for instance the CRAY-2. CRAY-2006 was installed at the University of Stuttgart at the end of 1986. Due to its large memory, its high computational speed and its low computational costs this computer allows a new quality of engineering models. In Fig. 1 this is demonstrated by showing typical applications of thermal hydraulic codes on various past and present computers.

In this figure the applications are characterized by the number of nodes used for discretisation times the duration (in hours) of the transient simulated. The calculation time necessary to run a problem depends in a first approximation linearly from this quantity. Of course it also is a function of the computer used. The band widths reflect our experiences with various codes like RELAP [4], TRAC [5] or ATHLET [6] on the computers available at the University of Stuttgart. Only with the CRAY-2 and very advanced versions of the codes like RELAP5 Mod2 it will become feasible to simulate accidents in real time or to include all the main components of a plant into a simulation.

Thus with the new computers available very complex analyses are becoming feasible. Such analyses allow further optimisations of operation, preventive maintenance to reduce failures or realistic simulations of accident situations to train operators in accident management procedures. However to realize such complex analyses poses quite a lot of new problems to the engineer. Instead of treating special subsystems or even components he has to treat complex systems. To perform consistent analyses requires from the view of simulation that one has to have consistent data to describe the plant, consistent tools (programs) to do the simulation and consistent knowledge on the behaviour of both the plant and the simulation system.

Everybody who knows how difficult it is to perform a meaningful thermal hydraulic analyses with one of the system codes mentioned before can imagine the problems connected with even more complex analyses. The engineer who has to perform such analyses can hardly specialize to a single application or a certain set of codes. Instead he has to develop an understanding of the plant behaviour which takes into account quite different effects and tries to integrate them. It is our belief that this kind of understanding has to be supported by the introduction of highly integrated systems.

Such systems have to integrate data, tools and knowledge. They may be considered either as knowledge based systems or simulation systems dependend on the point of view one is taking [7]. The Integrated Planning and Simulation System (IPSS) [8] which presently is under development at the University of Stuttgart is a first trial to develop specifications for such systems.

3 Integration of Tools, Data and Knowledge — the Integrated Planning and Simulation System IPSS

The main components of the Integrated Planning and Simulation System are shown in Fig. 2. They are from the knowledge based system and in Fig. 3 the simulation system point of view.

- the IPSS data and knowledge bases,

- the IPSS data and knowledge acquisition (planning) subsystem,

- the IPSS simulation subsystem.

The heart of the IPSS data and knowledge bases is the plant data bank. As our final goal we intend to have all the data necessary to describe the actual status of a plant in such a data bank. This requires both data to describe the components of the plant and data to describe the actual status. It is also intended to get the descriptive data directly from the planning and maintenance processes. Thus an automatic updating of the plant description may become possible. The data which describe the plant status may be provided either by measurements or if this is not feasible by simulations. Usually they will have to be aggregated over space and time. The plant data bank has to be supported by a planning data bank and a simulation data bank. The planning data bank contains information on manufacture data, design data and data on the actual components of the plant. These data are necessary to generate the descriptive data of the plant data bank. The simulation data bank contains primarily knowledge on how to use the information in the plant data bank. This knowledge may be formulated in programs (procedural description), program sequences (operational description) or rules (declarative description).

Even from this very sparse and incomplete description of the content of the various data banks it becomes clear that the realisation of such data banks requires complex data models with complex data structures. At present we plan to structure the information on a hierarchical basis. The strucutre of the data will reflect the structure of the plant. It is necessary however also to be able to formulate relations between different plant components. This requires new data models. The RSYST data base system [9] which is under development at the University of Stuttgart supports such models.

The data model is based on abstract data types. However due to the complexity of the information necessary to describe a component of a plant it is necessary to collect data of different type in data classes [10]. Data classes are tools to describe and structure complex data objects on a formal basis and therefore support consistency.

The main task of the IPSS planning subsystem is to support data and knowledge acquisitation. This has to be done primarily in an automatic way. We therefore intend to couple the acquisita-

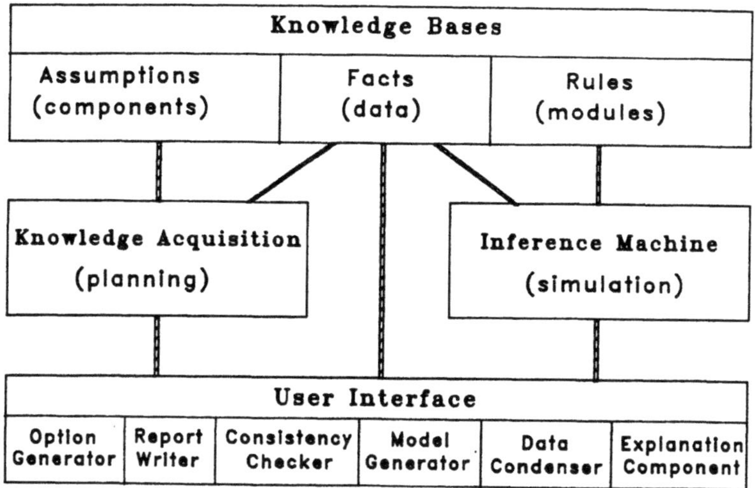

Figure 2: The main components of the Integrated Planning and Simulation System IPSS of IKE from the knowledge based system point of view

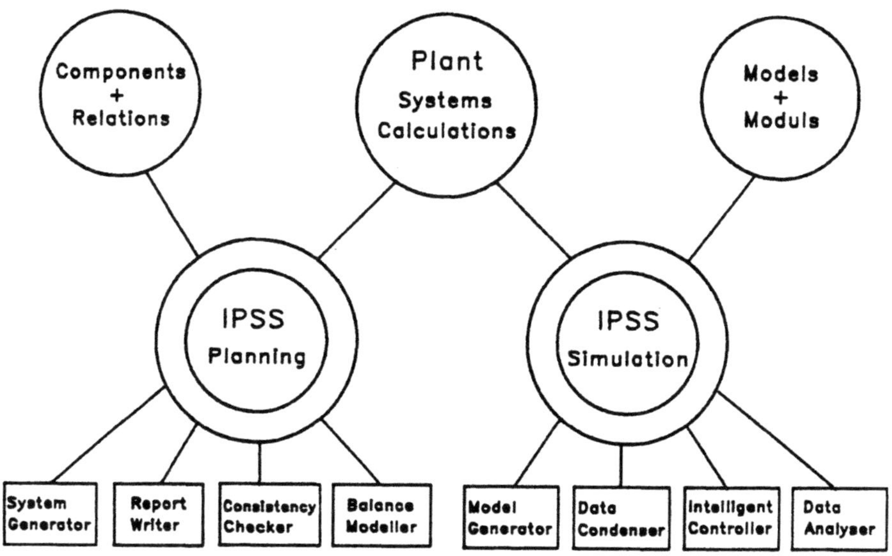

Figure 3: The main components of the Integrated Planning and Simulation System IPSS of IKE from the simulation point of view

tion process with planning and maintenance tasks. To encourage the engineer to work with the IPSS during planning and maintenance we intend to provide tools which are powerful enough to make the new system attractive. Several possible tools are listed in Fig. 3. They include:

- Computer aided design systems on graphical workstations which have access to the planning data bank and provide software to construct subsystems and combine them to plants.

- A report generator which is able to document systems or subsystems in various ways. Automatic generation of drawings, lizencing documents or even 3D plots seem to be possible even with available software and hardware.

- A consistyency checker seems to be a quite useful tool especially during modification processes. Rules to check consistency of connections, flow rates or material combinations are available and can be applied very often on a routinely basis.

- To get a first impression from the quality of a design engineers frequently consider stationary behaviour. To do this they perform balance calculations by using quite simple models or even rules of thumb. These models could be the base of a balance modeller which has access to data in both the plant data bank and the planning data bank.

If it becomes possible by these or similar tools to picture the actual status of a plant in the plant data bank one gets a quite powerful base for various kinds of simulations. Such simulations may be performed by numerical methods (for example thermal hydraulics or structural mechanics programs) or by the newly developed methods of qualitative reasoning [11]. Eventually they may support each other in a sense that qualitative reasoning will provide procedures to interpret numerical results on a routine basis and numerical calculations will provide numbers to adjust the qualitative arguments.

In our first design of the IPSS we have included four tool packages to support simulations. They are

- a computer aided design package to support the construction of an actual plant model on a graphical workstation. The actual plant model (components of the plant included in a simulation — usually specified through input) includes only those parts of the plant which should be treated in the desired simulation and allows to specify the accuracy of the simulation model (equations and correlations to describe the behaviour of a component — usually given by programs) for each component of the plant model.

- If one has constructed a plant model one has to provide the simulation codes with input data according to this model. These data should be derived automatically from the plant data bank by a data condenser. The data condenser will be a collection of abstract data type moduls which operate on the plant data bank and produce code specific input data.

- Automatically derived input data need special attention in performing actual simulations. This should be supported by an intelligent controller. Our first steps in the direction of the development of such a controller are described in [12].

- Finally results of complex simulations have to be analysed on a more routinely basis. This includes the automatic generation of standard plots as well as the automatic detection of unexpected results and the conditions which caused them. These and similar tasks have to be supported by a data analyser.

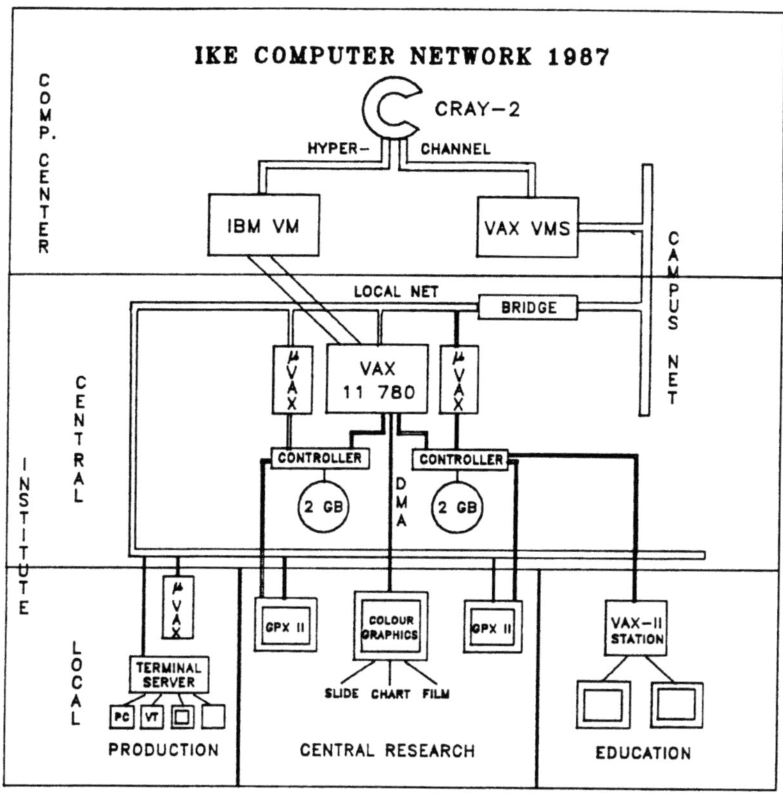

Figure 4: Computer network of IKE to realize the IPSS

4 Hardware Requirements of the IPSS

The software system has to be supplemented by a corresponding hardware concept. At our institute modelling and analysis work of the engineer is supported by local computer power provided by workstations. They are clustered in a local net (Fig. 4) which allows each engineer to share our common data and knowledge bases (central discs). Of course our concept relies also on the CPU power provided by the CRAY-2, on fast networks for data transmission and on extensive colour graphics facilities.

5 Conclusions

New devices for both performing measurements and calculations allow the development of a new approach to engineering solutions which may be characterized by new dimensions in complexity and integration. To realize this approach requires innovative computer applications including methods developed in artificial intelligence. The realisation of this new approach is fascinating in many aspects. In my opinion one of the most important ones is that engineers become able again to reflect the complexity of the real world on a rational base.

References

[1] ESPRIT '84 Status Report on Ongoing Work,
ESPRIT '85 Status Report on Ongoing Work,
ESPRIT '86 Status Report on Ongoing Work,
all published by North/Holland, Amsterdam 1985 to 1987

[2] Brauer, W., and Radig, B., ed., 1985, Wissensbasierte Systeme, GI-Kongress 1985, Springer Verlag, Berlin, Heidelberg

[3] Künstliche Intelligenz: Praktische Bedeutung und konkrete Anwendungen in der Industrie, 1986, Interatom GmbH, Bergisch Gladbach

[4] Ransom, V.H., et al., 1985, RELAP5-MOD2 Code Manual, Idaho National Engineering Laboratory

[5] Liles, D.R., et al., 1984, TRAC-PF1/MOD1 an Advanced Best-/Estimated Computer Program for Pressurized Water Reactor Thermal Hydraulic Analyses, Los Alamos National Laboratory

[6] Lerchel, G., Nov. 1986, Systemcode ATHLET, GRS (private communication)

[7] Cain, D.G., and Schmidt, F., April 1987, Expert Systems — Basic Principles and Possible Applications in Nuclear Energy, Int. Top. Meeting on Adv. in Reactor Physics, Mathematics and Computation, Paris

[8] Schmidt, F., Scheuermann, W., and Schatz, A., Einsatz von Supercomputern zur Lösung von Problemen der Kerntechnik (Problem Solving in Nuclear Engineering Using Supercomputers), to appear in Atomkernenergie-Kerntechnik in Dec. 1987

[9] Rühle, R., 1975, RSYST I—III, Experience and Further Development, Atomkernenergie 26,3

[10] Löffler, K., and Rühle, R., 1985, Ein Klassenkonzept für die Datenobjekte eines wissenschaftlich-technischen Anwendungssystems (RSYST), Stuttgart, IKE 4-120

[11] Burger, B., and Schmidt, F., THEXSYST — An Intelligent Monitor for Reactor Safety Calculations, this conference paper 4C/1

[12] Bobrow, D.G., ed., 1985, Qualitative Reasoning about Physical Sciences, MIT Press

APPLICATIONS OF ARTIFICIAL INTELLIGENCE IN THE U. S. NUCLEAR INDUSTRY

Robert E. Uhrig

Instrumentation and Control Division, Oak Ridge National
Laboratory*, Oak Ridge, TN 37831 and
Department of Nuclear Engineering, University of Tennessee
Knoxville, TN 37996

ABSTRACT

In the United States, the introduction of artificial intelligence (AI)
into use in the nuclear power field is being carried out by a wide spectrum
of organizations (i.e., nuclear equipment vendors, architect-engineer firms,
universities, national laboratories, federal agencies, the electric utility
industry, and small entrepreneurial groups). The most coherent of these
efforts is an Electric Power Research Institute program to demonstrate the
usefulness of AI in nuclear power plants (including augmenting plant
automation) and an agreement with NASA to transfer the technology of their
multi-year AI "Core Technology in Systems Autonomy" to the nuclear power
industry. A few vendors are offering commercial AI products that reduce the
burden on reactor operators during both normal and abnormal operation.
Several AI programs at universities and national laboratories have automation as their primary focus, and individual AI projects have been
initiated under the Small Business Innovative Research Program. The
fundamental and synergistic relationship between training and expert systems
supports the use of AI in the training of nuclear personnel. In the long
run, the most significant contribution of AI may well be the introduction of
AI programming techniques. With multi-million line computer codes
contemplated for use in automated nuclear plants, the ability to readily
modify programs and to utilize verified and validated "building blocks" of
code would offer extraordinary advantages.

Artificial intelligence may be defined as:

"A computerized process that attempts to emulate the human thought
processes associated with activities that require the use of intelligence".

Artificial intelligence burst on the scientific scene about thirty
years ago with much fanfare and promise. Recognition that computer symbols
could represent characteristics of the real world and that computer programs
could relate these features provided the means by which computers could be
used to simulate certain important aspects of intelligence and provided an

---

* Operated by Martin Marietta Energy Systems, Inc., for the U.S. Department
of Energy under Contract No. DE-AC05-84OR21400.

information-processing model of the human mind. The subsequent frustrations
and limited success of researchers in trying to use this model to gain a
better understanding of the working of the mind, in even the most elementary
situation, is well documented by H. and S. Dreyfus in their book _Mind Over
Machine_[1], which has a subtitle "The Power of Human Intuition and Expertise
in the Era of the Computer." It is ironic that as progress in the use of AI
to enhance the intellectual understanding of the workings of the human mind
floundered, certain practical applications of this information-processing
model spawned whole new technologies that promise to revolutionize the way
both business and industrial organizations operate.

Robotics and expert systems are two particularly successful products of
AI research. Although both have important applications in the nuclear
industry, the primary focus of this paper will be on the use of expert
systems to enhance the engineering, management, and operation of nuclear
power plants in the United States. As a direct derivative of the definition
of AI, expert systems are defined as

"Computerized processes or programs that attempt to emulate the human
thought processes associated with the application of expertise to problems."

As expert systems have evolved over the past decade, they have
typically consisted of two separate components, an "inference engine" (i.e.
an information processor) and a "knowledge base." The inference engine
gathers the information needed, guides the search process in accordance with
the strategy programmed into it, uses rules of logic to draw inferences
about the processes involved, and presents conclusions (when warranted)
along with an explanation of the bases for the conclusions. The inference
engine may use either "forward chaining" (i.e. forward reasoning), in which
it starts with the given data and proceeds toward a solution, or "backward
chaining," in which it assumes a conclusion and then looks for evidence to
support that conclusion. Since the inference engine and the knowledge base
are entirely separate, changes in the knowledge base can be made easily
without any influence on the inference engine.

Generally, the knowledge base of the expert systems relies on the
expertise of experts or expert knowledge that has been codified in
publications, books, or regulations to provide advice under a wide variety
of conditions. When the data and/or information in the knowledge base are
specific and precise, expert systems give results that are unambiguous.
However, when the needed information is imprecise or "fuzzy," incomplete,
missing, or even conflicting, expert systems can still reach a rational
conclusion or solution through the use of confidence factors or Bayesian
probabilities. Under these conditions, an expert system will give the "most
probable" solution or the "best" solution, but not necessarily the correct
solution. For this reason expert systems usually identify alternative or
less probable solutions along with the associated probabilities or
confidence factors. This characteristic of expert systems is one of their
greatest advantages, although it is of concern to regulatory authorities
when these systems are installed in nuclear power plants.

In the operation of a nuclear power plant, great quantities of numeric,
symbolic, and quantitative information are handled by the reactor operators
even during routine operation. The sheer magnitude of the number of process
parameters and systems interactions poses difficulties to the operators,
particularly during abnormal or emergency situations. Recovery from an
upset situation depends upon the facility with which available raw data can
be converted to and assimilated as meaningful knowledge. In operating a
nuclear power plant people are sometimes affected by stress and emotion that
may have varying degrees of influence on their performance. Expert systems
can take some of the uncertainty and guesswork out of their decisions by

providing expert advice and rapid access to a large information base. The
application of AI technologies, particularly expert systems, to the control
room activities in a nuclear power plant can reduce operator error and
enhance plant safety and reliability. Furthermore, a large number of
nonoperating activities (e.g. testing, routine maintenance, outage planning,
equipment diagnostics, fuel management, etc.) exist where expert systems can
increase the efficiency and effectiveness of overall plant operation.

Expert systems can utilize almost any type of computer (including
mainframe machines) for off-line applications. However, for on-line
applications, especially those with the computers located in the control
room, operating personnel seem to prefer units with which they are familiar.
Generally, those are personal computers or micro/minicomputers similar to
those already installed in the plant. Special AI computers, such as LISP
machines and engineering workstations, can be used for development of expert
systems that can then be ported to the plant computers.

In the United States, the development of expert systems in the nuclear
power field is being carried out by a wide spectrum of groups (i.e., nuclear
equipment vendors, architect-engineer firms, universities, national
laboratories, federal agencies, the electric power utility industry, and
small entrepreneurial groups). Most coherent of these efforts is the
program undertaken in 1983 by the Electric Power Research Institute (EPRI)
to demonstrate the usefulness of AI in a number of areas, including
augmenting plant automation activities. Special emphasis was given to fault
recognition and diagnosis, fault recovery, task planning and replanning,
intelligent operator interfaces, and intelligent systems control. EPRI also
has a program to transfer the technology of NASA's multi-year "AI Core
Technology in Systems Automation" to the nuclear power industry.

The recent EPRI Seminar, "Expert Systems Applications in Power Plants"[2]
gave ample evidence that expert systems for use in power plants are being
developed extensively, not only in the United States but also in Europe,
Canada, and Japan. The introduction of expert systems into plant operations
is proceeding more rapidly in fossil plants, even though the anticipated
benefits are greater in nuclear plants. This is probably due to the
reluctance of utilities to introduce to regulatory review a new technology
that involves dealing with uncertainty until they are convinced that
benefits gained warrant the effort involved. Perhaps the principal concern
of the regulators with expert systems is the ability to encode expertise
properly, particularly the fine nuances and shades of meaning, into the
knowledge base of an expert system so that it can emulate human expertise
with fidelity. Another major concern is the narrow scope of the expertise
and the associated limited area of applicability of expert systems. One of
the consequences of these limitations is the inability of an expert system
to exhibit common sense and its limited ability to recognize when it is
operating outside its field of knowledge. Researchers have sought to
minimize the impact of these limitations by building "robustness" into
expert systems (i.e. the ability to fail gradually and predictably when it
gets outside its operating regime). These limitations, as well as the
lower precision associated with answers when data are missing or have low
confidence factors, are of major concern to regulators in dealing with
proposed expert systems in nuclear power plants.

The examples of applications of expert systems to the nuclear power
field cited here, which are typical of those being developed in the United
States, constitute only a small fraction of those being developed today.
However, only a few systems are actually in use in nuclear plants today.

One of the first EPRI projects in expert systems was REALM (Reactor
Emergency Alarm Level Monitor) which was developed by Technology

Applications, Inc.[3] The NRC has about 20 pages of guidance on classifying
an emergency as an unusual event, an alert, a site area emergency or a
general emergency. Each level of emergency has a specific set of responses
that the utility must undertake. The decision as to the level of the
emergency has to be made rapidly and sometimes in a time frame in which the
true nature of the event is not yet clear. While there are many sensory and
manual observations available, certain needed data may be missing, ambiguous, or even conflicting. Then judgment is required for proper interpretation. REALM is designed to operate in a real-time process environment. It
incorporates what might be called a "first-level" diagnostic system that
readily identifies the cause of the emergency on the basis of a comparison
of the symptoms that are observed and the events that are possible in a
nuclear power plant. In addition, REALM provides a rationale as to why it
recommends a particular classification. It then carries out a "vulnerability analysis" that tells the operators which events would lead to a
higher emergency level. It also tells them what needs to be done to get to
the next lower level. REALM was developed for Indian Point-2 in cooperation
with Consolidated Edison of New York, and it performed well when operated in
parallel with normal plant operations during their two most recent emergency
drills.

EPRI is also developing a computerized tracking system for Emergency
Operating Procedures.[4] This expert system is co-resident on the Safety
Parameter Display System computer and is presently being tested on the
Kuosheng Nuclear Power Plant, a BWR-6 nuclear reactor in Taiwan that has
been operating since 1981. The Emergency Operating Procedures are written
in about 250 rules that can be evaluated in less than one second.
Conclusions as to the steps that should be taken are available within
seconds after a parameter change. Its inference engine looks for pattern
matches between the rule premises and the operating conditions that then
lead to the recommendation of action to be taken. It is an on-line system
that requires no input from the operators. Explanations for the conclusions
are available to the operators.

Westinghouse Hanford Company has developed two expert systems that are
"clones" of experts at the Hanford Engineering Development Laboratory (HEDL)
and the FFTF (Fast Flux Test Reactor).[5] CLEO (Clone of Leo, an expert on
refueling the FFTF) is an expert system that is able to generate a list of
necessary refueling moves in less than 30 seconds, given the present and
future core configuration of the FFTF. It has a "front end" that allows the
selection of the strategy needed for the particular fuel loading. CRAW is a
clone of an expert in diagnosing fuel cladding failures in FFTF. Indications of fuel failure (i. e. tag gas detection) requires interpretation and
diagnosis by experts within a short period of time, 24 hours a day. This
expert system is an effective substitute when the resident expert is not
available. Westinghouse Hanford Company also has developed a prototype on-
line diagnostic system that can diagnose faults in both instruments and
equipment as well as multiple, concurrent faults in complicated process
systems such as the FFTF. It uses the Luenberger observer diagnostic
system, a fault detection methodology based on analytic redundancy. A
separate PROLOG program is used to provide an explanation of the conclusions
of this system.

Oak Ridge National Laboratory (ORNL) is developing and implementing an
on-line operator advisor for their 100 MWt High Flux Isotope Reactor.[6] An
expert advisor maps procedural and operational knowledge onto process and
historical data to provide the operator with plant status and operational
advice. The system's main features are recognition and tracking of plant
status, intelligent display generation, alarm filtering, and explanatory
capabilities of the logic behind its action. This prototype monitoring

system integrates a multi-lingual set of distributed expert systems and plant system models written in OPS-5, OPS-83, LISP, PROLOG, C, and FORTRAN.

Idaho National Engineering Laboratory (INEL) is developing the Reactor Safety Assessment Expert System[7] to aid an NRC reactor safety team to maintain a "big picture" of a transient-in-progress, such as the TMI-2 accident. The system would monitor and project core conditions, containment conditions, and status of fission product barriers. It has been designed to fit existing operations centers and to be extendable to many plants. Approximately 800 rules are organized in frame-like structures, and the system uses forward chaining to reach conclusions. Multiple diagnoses are allowed.

Middle South Utilities has developed TRIBES (Trip Buffer Expert System).[8] TRIBES analyzes trips caused by the core protection calculator and the control element assembly calculator, which form a group of six digital computers that monitors nuclear power plant parameters and control element assembly positions. These core protection systems will initiate a trip to prevent violation of fuel design limits (i.e., kilowatts per foot, DNB limits, rate of power increase, etc.). An analysis of the computer output is required to establish the cause of the trip before the plant can be restarted. Since this analysis is a narrow and well-defined specialty in the nuclear power field and the activity occurs after the plant has been shut down, it is an appropriate problem for an expert system to undertake.

Stone and Webster have developed an expert system to analyze the limiting conditions of operation (LCOs) and technical specifications in a nuclear power plant.[9] These limitations are imposed by regulation, and violations can result in regulatory action that may include civil penalties as well as shutdown of the plant. One of the uses of this system is to assess the effect of both operational changes and the removal of equipment from service to determine whether either of these activities will lead to a violation of LCOs or technical specifications. The program has a "what if" mode that allows the operators to determine the impact of the proposed maintenance actions and operational changes before they are authorized. This mode is used to detect the subtle interactions that might otherwise go undetected and cause a trip of the plant or a violation of the technical specifications or limiting conditions of operation.

Southern California Edison has developed TAGS (Tagout Administration and Generation System) for their San Onofre Nuclear Power Plant.[10] TAGS is a conventional computer program that administers the safety tagout process. It has been integrated with an expert system in the form of an intelligent work station that uses PLEXSYS (plant expert system) which has been developed by the EPRI. PLEXSYS will present piping and instrumentation drawings (P&IDs) and electrical one-line schematics for the systems of interest. When the components to be tested are selected, PLEXSYS and TAGS recommend a "safety tagout boundary" that allows the maintenance to be performed without danger of tripping the plant.

Texas Utilities and Westinghouse jointly have developed GenAID[TM], an on-line generator diagnostic system,[11] to diagnose 15 conditions with damage potential to the generator and to recommend corrective action for each condition. The diagnostic system utilizes special monitors attached to the generators located in Texas that are coupled to computer terminals and are continuously linked via phone lines to Westinghouse's Diagnostics Center in Orlando, Florida. The goals of using this program are to maximize the plant availability and to reduce the forced outage rates. The diagnoses and recommendations are based on the knowledge of the best experts (designers, service engineers, field engineers, operators, etc.). GenAID is now in operation and has proven to be an effective tool in reducing the risks of error in human judgment, thereby improving plant productivity and availability.

Westinghouse is also using their Intelligent Eddy Current Data Analysis (IEDA) system[12] to analyze the eddy current data of the 45 miles of tubing in a typical nuclear plant steam generator tube bundle. The analysis typically requires 60,000 judgments, some extremely difficult. IEDA is based on a set of highly defined rules (developed from an "expert model data analyst") to which the eddy current data are compared. Incorporated into the system is a versatile and user-friendly operating mode that allows manual evaluations of those signals the computer cannot categorize properly.

Combustion Engineering has developed a Generic Diagnostic System (GDS) software shell that it is applying to power plant diagnostics.[13] The GDS is a generic framework system that provides a skeletal software system to facilitate utility personnel building "problem specific" diagnostic expert systems for complex plant and process systems. The user supplies the knowledge base information in the form of functional rules and a fault-tree description of the problem domain. The system diagnoses malfunctions at two levels, a functional level and a root cause level, and is capable of efficiently handling multiple failures in complex situations. Additional features include an intelligent request-for-information module, an automated instrument validation system, a process for filtering information noise, a success path recovery module, and a learning module.

Ohio State University is developing an expert system to diagnose operational problems in nuclear power plants[14] even in the presence of some incorrect and/or conflicting data. It uses classification techniques that can diagnose a large percentage of problems that are commonly found in mechanical systems. Validation routines include: limit checking on trends and absolute values, auctioneering among like sensors, and parity space checking. The evaluation of malfunction appropriateness is based on knowledge embedded in each malfunction node of the hierarchy of malfunction hypotheses. Each node is capable of performing a simple kind of problem solving that evaluates whether the malfunction at this node exists, given the presently available data. This process was implemented in the design of an expert system for diagnosing coolant system malfunctions for a General Electric BWR-6 boiling water nuclear power plant.

The Department of Energy (DOE) has supported a number of AI projects for application to nuclear power plants through its Small Business Innovation Research (SBIR) program. Earlier this year DOE announced 113 initial awards, including several that dealt with "Computer Applications to Nuclear Power Plants."[15] These awards, averaging $50,000 for six months, are to determine the feasibility of the proposed concept. Included were the following projects in which AI is used:

"An Expert System Operator Aid for Nuclear Power Plant Maneuvers," Applied Research Associates, Inc.

"An Expert System-Based Decision Aid for Reactor Trip Reduction and Post Trip Analysis," Expert-EASE Systems, Inc.

"Signal Validation by Combining Model-Based and Evidential Reasoning Approaches," Expert-EASE Systems, Inc.

"The Design of a Reliable Fuzzy Fault-Tolerant Automatic Control," Technology International, Inc.

At least two SBIR projects involving AI have reached the second stage for which the awards, averaging $460,000, are to support the development work that is expected to lead to commercialization. These include:

"A Bayesian Diagnostic System: An Expert System to Aid Reactor Operations," Pickard, Lowe and Garrick, Inc.[16]

"Residual Heat Removal Advisor," Odetics, Inc.[17]

Other reported applications of expert systems in various stages of development include outage planning, heat rate improvement, alarm filtering, sequencing and suppression, diagnostics for instruments and equipment, welding rod selection advisor, generating welder procedure specifications that comply with regulatory codes, signal validation, disturbance analyses, condensate feedwater monitor, radwaste processing system advisor, bypass-inoperable status indicator system, sequencing BWR control rods after maneuvering, pressure-temperature control during start-up (to avoid pressurized thermal shock problems), real-time emergency evacuation planning, and real-time radiation exposure management.

The fundamental and synergistic relationship between training and expert systems offers a unique opportunity to improve the training of nuclear power plant personnel. One of the features that makes an expert system so compatible with diagnostics in nuclear plants is its ability to explain its reasoning and conclusions for postulated or real conditions given to it. All supporting evidence for machine opinions about systems or events can be cited for final evaluation and decision by human operators. As the operators work with an expert system, there is constant exposure to the bases, limits and nature of system interrelations. Recent work at The University of Tennessee[18] has dealt with the symbiotic relationship between diagnostics and training. Indeed, the understanding gained in developing and encoding the knowledge base on the workings of a nuclear power plant into an expert system may be as important, if not more important, as the use of that system in actual plant operation. This effort further enhances the quality of the training of nuclear personnel.

Demands by the safety and environmental regulatory authorities for increased safety margins and lower environmental impacts and those by the economic regulatory authorities and the financial community for increased efficiency in operation (e.g. fewer trips, higher availability, plant investment protection, etc.) inevitably lead to more sophisticated plants with additional systems that must be controlled and/or automated. Digital systems inevitably will totally dominate the control systems of the next generation of nuclear power plants, unless they are specifically forbidden by regulatory authorities. Indeed, the integration of expert systems into the safety, control, and management systems of power plants is a logical next step in the automation process.

The high reliability and desirable operating characteristics of the digital systems are very attractive, but the costs of writing, validating, verifying, checking out, and certifying the software under the quality assurance requirements associated with nuclear power plants must be addressed. The critical issue in introducing digital systems is the elimination of those errors that show up only under rare and unique circumstances. Such was the case with the THERAC-25 in which several people undergoing radiation treatments for cancer were severely overexposed, two of them fatally.[19] These unfortunate incidents with the THERAC-25 emphasize the absolute necessity for the thorough validation, verification, check-out, and certification of software prior to its installation in the primary control system of an operating nuclear power plant. It is incumbent upon those who are introducing digital technology into nuclear power plants for the benefits it offers to take whatever steps are necessary to see that those costs do not include accidents induced by software "bugs."

The development of software instructions to read and analyze the plant data and to take appropriate control action is a major undertaking. However, use of the concepts of computer-aided design and engineering has led to programming tools, generally called computer aided software engineering (CASE),[20] that aid in the development of the complex logic flows, manage the vast amount of data involved, and develop the detailed lines of code based on more macroscopic instructions. Japanese experience with this type of tools has shown that advanced control software costs can be reduced by a factor of about 15.

Once the need for automated digital systems is established, there must be one or more places where the software required for operation of digital systems in nuclear plants can be prepared, tested, verified, validated, checked out, and certified. Estimates of the software required to automate the control and monitoring of a single nuclear plant may exceed one-million lines of code. Use of the traditional error rate of one error per thousand lines of code gives at least one-thousand errors that must be found and corrected before the software is suitable for use in a nuclear power plant.

If utilities are to accept automation, they will need a facility like like the Advanced Control Test Operation (ACTO) facility currently being developed at ORNL[21] to check out and demonstrate the operation of their system. The validity of their digital system must be demonstrated to the Nuclear Regulatory Commission. ACTO is the only facility presently envisioned within the national nuclear arena that can satisfy the needs delineated above. It will be linked to on-going DOE programs in instrumentation and controls, robotics, and AI. Indeed, ACTO can and must play a multi-function role in introducing digital control and automation into nuclear power plants. Simultaneously it can be a test vehicle for demonstrating the ability of CASE to reduce the cost per line of certified software, while providing the engineering simulation of nuclear power plants needed to develop digital control and automation systems.

In the long run, the most significant contribution of AI may well be the introduction of AI programming techniques, sometimes called "modern computer science."[22] With multi-million line computer codes that are contemplated for some future multi-unit nuclear power plants, the use of programs that can be modified readily and that utilize "building blocks" of code offers extraordinary advantages. The separation of the knowledge base from the processing portion of a computer program and the use of powerful AI oriented computer languages (e. g. LISP, FORTH, Smalltalk, PROLOG, etc.) make this possible. Instead of telling the computer precisely what to do at every point (as in a conventional program), the computer is told what to know. Then the computer uses its knowledge base and reasons as to what to do when new situations arise. The ability to cope with uncertainty in situations that cannot be predicted and the ability to use "rapid proto-typing" are essential to the wide-scale use of automation.

In summary, a nuclear power plant is too complex a system to be managed or operated by anyone's "gut feeling." An expert system can be the ever alert, knowledgeable assistant to the operators as well as a valuable tool for plant management. Demands for increased safety margins, lower environmental impacts, increased performance, and greater investment protection will inevitably lead to automation of most functions of nuclear power plants. In turn, automation will be paced by the ability to develop efficiently the needed software through the use of modern computer science brought about by AI programming techniques. The regulators and the public must be assured that these plants are properly designed, properly built, properly operated, and properly maintained. Artificial intelligence and expert systems can and must play a major role in providing this assurance.

REFERENCES

1. H. Dreyfus and S. Dreyfus, "Mind Over Machine: The Power of Human Intuition and Expertise in the Era of the Computer," The Free Press, (1986).
2. Seminar Notebook of Seminar: Expert systems Applications in Power Plants (Prepared by Expert-EASE Systems, Inc., for the Electric Power Research Institute, Palo Alto, CA), Boston, MA (May 27-29, 1987).
3. R. Touchton, A. Gunter, and D. Cain, Reactor Emergency Action Monitor; An Expert System for Classifying Emergencies, included in Ref. 2.
4. W. Petrick, C. Stewart, and K. Ng, A Production System for Computerized Emergency Procedures Tracking, included in Ref. 2.
5. S. E. Seeman, Private Communication, Application of Artificial Intelligence at Hanford, DOE/ANL Training Course on "The Potential Safety Impact of New and Emerging Technologies on the Operation of DOE Nuclear Facilities", Knoxville, TN (August 18-20, 1987).
6. D. K. Wehe and P. Otaduy, "A Summary of the Artificial Intelligence Applications at the HFIR," Proceedings of the Sixth Power Plant Dynamics, Control and Testing Symposium, Knoxville, TN (April 14-16, 1986).
7. D. Sebo, M. A. Bray, and M. A. King, "An Expert System for USNRC Response," IEEE Second Expert Systems in Government Symposium, McLean, VA (October 20-24, 1986).
8. R. Lang, TRIBES - A CPC/CEAC Trip Buffer Expert System, included in Ref. 2.
9. G. Finn, F. Whittum, and R. Bone, An Expert System for Technical Specifications, included in Ref. 2.
10. J. Munchausen and K. Glazer, An Expert System Technology for Work Authorization Information System, included in Ref. 2.
11. J. Carson and M. Coffman, TU Electric Experience with On-Line Generator Monitoring and Diagnostics, included in Ref. 2.
12. J. L. Gallagher, Westinghouse Nuclear Technology Systems Division, Private communication (August 1987).
13. C. Neuschaefer, P. Rzasa, E. Filshtein, R. Burrington, and R. Donais, Application of C-E's Generic Diagnostic System to Power Plant Diagnostics, included in Ref. 2.
14. S. Hashemi, W. F. Punch III, and B. K. Hajek, CSRL Application to Nuclear Power Plant Diagnosis and Sensor Data Validation, included in Ref. 2.
15. DOENEWS, Energy Department Announces Awards to 113 Small Businesses, Office of the Press Secretary, Washington, DC (March 27, 1987).
16. S. Kaplan, M. V. Frank, and D. C. Bley, COPILOT: An Expert System to Aid Reactor Operations--Conceptual Framework and Sample Applications of the Bayesian Diagnostic Module, PLG-0532, Pickard, Lowe and Garrick, Inc., Newport Beach, CA (1987).
17. E. Kurrasch, Expert system for Reactor Plant Safety and Control, Odetics, Inc., Final Report, USDOE Contract No. DE-AC03-86ER80432 (February 25, 1987).
18. R. E. Uhrig and M. T. Buenaflor, Artificial Intelligence and Training of Nuclear Reactor Personnel, International OECD-CNSI Specialist Meeting on Training of Nuclear Reactor Personnel, Orlando, FL, (April 21-24, 1987).
19. E. Joyce, Software Bugs: A Matter of Life and Liability, Datamation (May 15, 1987).
20. A. R. Stubberud, "A Hard Look at Software," IEEE Control Systems Magazine, pp. 9-10 (February 1985).
21. R. L. Shepard and S.J. Ball, Private communication, "ACTO: How and Why," Oak Ridge National Laboratory, Oak Ridge, TN (June 26, 1986).
22. M. M. Waldrop, Artificial Intelligence Moves into Mainstream, Science, Vol. 237 (July 31, 1987).

NKA/INF - ADVANCED INFORMATION TECHNOLOGY:

INFORMATION TECHNOLOGY FOR ACCIDENT AND EMERGENCY MANAGEMENT

V. Andersen and K. Møllenbach
Risø National Laboratory, Box 49
DK-4000 Roskilde, Denmark

R. Heinonen and S. Jakobsson
Technical Research Centre of Finland, VTT
Otakaari 7 B, SF-02150 ESPOO, Finland

T. Kukko
Imatran Voima Oy, PL 138, SF-00100 Helsinki, Finland

Ø. Berg, J.S. Larsen, and T. Westgård
The Institute for Energy Technology , IFE, Box 173
N-1751 Halden, Norway

B. Magnusson
The Swedish Nuclear Power Inspectorate, SKI
Box 27106, S-102 52 Stockholm, Sweden

H. Andersson and C. Holmström
Studsvik Energiteknik AB, S-611 82 Nyköbing, Sweden

B. Brehmer and R. Allard
The Psychological Institute, Uppsala University
Box 227, S-751 04 Uppsala, Sweden

ABSTRACT

There is an increasing potential for severe accidents as the industrial development tends towards large, centralized production units. In several industries this has led to the formation of large organizations which are prepared for fighting accidents and for emergency management. The functioning of these organizations critically depends upon efficient decision making and exchange of information.

This project is aimed at securing and possibly improving the functionality and efficiency of the accident and emergency management by verifying, demonstrating, and validating the possible use of advanced information technology in the organisations mentioned above. With the nuclear industry in focus the project consists of five main activities:

1) The study and detailed analysis of accident and emergency scenarios based on records from incidents and drills in nuclear installations.

2) Development of a conceptual understanding of accident and emergency management with emphasis on distributed decision making, information flow, and control structures that are involved.

3) Development of a general experimental methodology for evaluating the effects of different kinds of decision aids and forms of organisation for emergency management systems with distributed decision making.

4) Development and test of a prototype system for a limited part of an accident and emergency organisation to demonstrate the potential use of computer and communication systems, data-base and knowledge base technology, and applications of expert systems and methods from artificial intelligence.

5) Production of guidelines for the introduction of advanced information technology in the organizations based on evaluation and validation of the prototype system.

INTRODUCTION

The basis for the study of the potential use of advanced information technology for accident and emergency management was established in a pilot project undertaken in 1985. The subjects addressed in this project led to a preliminary description of accident and emergency scenarios, a state-of-the-art review of models and methods available for construction of a conceptual system, and a review of available tools from Artificial Intelligence, e.g. expert systems.

The pilot project suggested three main activities in the first phase of the main project:

1) Detailed descriptions and analysis of accident and emergency   scenarios

Some information is available from detailed records from incidents and accidents and emergency drills in the nuclear industry. The analysis must lead to identification of observable weaknesses in the organisations.

2) Comprehensive conceptual work

A general framework for a functional analysis of the accident and emergency management organisations has been suggested, but not thoroughly tested. In particular it is important to verify the applicability of this problem domain description model with respect to an understanding of the distributed and hierarchical decision processes, the mental strategies and the cognitive control domains.

3) Development of a prototype system

As the project aims at an evaluation of the potential use of advanced information technology in the application domain chosen, it is very important to develop a technological test bed for main features, e.g. in data-base and knowledge-base technology, communication systems, and tools from advanced information processing.

In an early stage of the project a limited target area must be defined. Based on the scenario descriptions in 1), a "vertical slice" is identified dependent primarily on two criteria: it must (a) be able to display the major features of the conceptual system, and (b) be limited to the extent where the prototype development is possible using the available resources.

In the later phases of the project the scenario descriptions will gradually change to data and knowledge acquisition, the conceptual work will be followed by development of a general experimental methodology, and by experimental work using the prototype as test bed. The prototype system will experience a dynamical development throughout the major part of the project. The keyword for the project is system studies with emphasis on system integration. This will be reflected in the recommendations and guidelines developed in the final phase of the project.

As the project described here receives limited funding it is of the utmost importance to choose a realistic scope and milestones to be reached in the project.

STATUS OF THE PROGRAMME

Conceptual Work

The general point of departure for the conceptual work has been to design a framework for analyzing different kinds of emergencies.

In the first stage, we have been concerned with the problems of hierarchical command and control systems in emergency management. Such systems were found to be of limited use in this context because

- all kinds of emergencies cannot be foreseen, and this may create a need for a more flexible structure with the capacity to reconfigure itself;

- information delays would make it hard to exercise control by means of a hierarchical system that would be too slow;

- some aspects of emergency management cannot be modelled hierarchically but require a different control structure; and

- hierarchical command and control systems are not needed for all kind of emergencies.

In the second stage, we have tried to create a general framework for analyzing emergency management based on the view of emergency management as a control system. This

- provides a clear specification of the goals of an emergency management system;

- provides a specification of what the components of such a system should be;

- specifies the information needs; and

- specifies what can, and what cannot, be controlled in emergency management.

Further work is now directed towards solving two problems:

1. To develop a conceptual framework for those aspects of emergency management that cannot be controlled hierarchically. The problems here are those of coordination in a system characterized by distributed decision making.

2. Using the time-area diagrams developed as part of the analysis of emergency management as a control system to analyze a variety of emergencies. This is done in an attempt to test the general usefulness of these diagrams as an analytical tool for analyzing information needs in emergency management.

In addition, some first thoughts on how the decision support system should be evaluated have been looked into. Here a distinction between two forms of evaluation has been discussed: analytical evaluation and empirical evaluation. It is recommended that an analytical evaluation be performed first. This comprises two steps:

- mapping the decision support system on to a set of general decision tasks, and

- assessing the extent to which these tasks are supported by analyzing (a) the nature of the situation, (b) the kind of displays that are provided, and (c) the knowledge required for understanding these displays.

It is also recommended that the empirical evaluation be directed towards limited and well-defined functions of the decision support system. DESSY-D, a general interactive program for simulating dynamic systems, is being developed for this purpose in Uppsala.

The methodological problems in using this system for the evaluation of a decision support system are now being analyzed.

Data Acquisition and Specification of Data and Knowledge Base

The analysis of information requirements both on-site and off-site is almost finished. A detailed description of the information flow has been completed for four different Emergency Organization Centres (EOC): the County EOC, the EOC at the National Institute of Radiation Protection, the EOC at the Swedish Nuclear Power Inspectorate (SKI), and at the Technical Support Centre within the on-site EOC. Furthermore, the analysis will concentrate on the Plant Emergency Manager, the Radiation Protection Manager and the Plant Operation Manager. Experience from previous incidents and preparedness drills has revealed that many difficulties can be traced back to information management problems. This has been addressed in the emergency preparedness planning in setting-up of formal status departments, that are key departments in the information management activities. A detailed list has been completed identifying the basic information types needed within the status departments of the different EOCs.

Prototype Implementation

The work has concentrated on choosing a suitable implementation strategy for the prototype system. A common data structure has been defined for process data, rules and message passing. Frame structures and object-oriented programming techniques Flavors on Symbolic computers are utilised.

The implementation of a program that interprets the on-site classification rules for emergency situations is in progress. One can edit new rules, and then apply the rules on process data. The user can interact with the system when additional information is needed or if explanations are requested. The intention is to create a general inference mechanism that can be applied in other parts of the system where the rules have the same structure.

A basic message structure has been established supporting information exchange in the organisation. Functions for manipulating various kinds of messages have been specified. The implementation must handle both fixed and free format messages and the possibilities of automatic distribution of information.

A first proposal for the off-site man-machine interface has been implemented. This interface is based on the complete item list from the SKI status department. The further development will be based on this framework.

The work on the user-interface for the on-site part has been continued. More specificially it is intended for use by the Plant Emergency Manager. The basic window-frame with its panes has been established and the technique for menu-handling as well as mouse-tracking has been utilized.

The trend- and bar-chart displays have been implemented. The functions for information concerning the manning of the preparedness organization have been developed and are working satisfactorily. Also some basic functions for displaying and producing maps have been implemented.

The work with the items described above will continue to establish a prototype where the functions are implemented incrementally for both on-site and off-site parts of the organisation.

FUTURE PLANS

Having the first proposal of a prototype system, a test scenario for a limited part of an emergency situation will be designed to make experiments to evaluate empirically the combined effect of the database, the implementation strategy, and the conceptual model for the developed part of the expert system. Through these experiments the prototype system will be optimized by an iterative procedure, and the final product will be used for recommendations and guidelines for a possible further development into an expert system capable of coping with emergency situations in risky industrial plants.

REPORTS PUBLISHED UNTIL MAY 1987

1. A descriptive analysis of the management of nuclear power plant emergencies. Roger Johansson, Håkan Andersson, Conny Holmström, Studsvik Technical Note NI-86/7, 86-09-01.
2. Potential applications of advanced information technology in emergency management. Håkan Andersson, Studsvik Technical Note NI-86/2, 86-02-03.
3. Delstudie rörande tekniskt stöd och utbildning inom havariorganisation för kärnkraftverk, samt kort genomgång av tidigare genomförda studier och informationstillgång. Conny Holmström, Studsvik Technical Note NI-86/15, 86-09-01.
4. Identification and analysis of information needs within nuclear power plants' preparedness organizations by use of emergency scenarios. Conny Holmström, Studsvik Technical Note NI-87/Draft, 87-04-30.
5. Scenario description of an emergency and accident situation at a nuclear power plant. Rauno Heinonen and Stefan Jakobsson, Technical Research Centre of Finland, Electrical Engineering Laboratory, 1985.
6. On-site emergency operations management conceptual analysis. Rauno Heinonen, Stefan Jakobsson, Technical Research Centre of Finland, Electrical Engineering Laboratory, 1986.
7. Description of a multifunctional decision support system for on-site and off-site emergency management. Rauno Heinonen, Stefan Jakobsson, Technical Research Centre of

Finland, Electrical Engineering Laboratory, and Tommi Kukko, Imatran Voima, 1987.

8. Review of expert system techniques and relevance to computerised support systems in emergency management. Ø. Berg and M. Yokobayashi, The Institute for Energy Technolocy, INF-630(85)1.

9. Overall system specification for a multifunctional decision support system in emergency operation of a nuclear power plant using expert system technology. Ø. Berg, J.S. Larsen and M. Yokobayashi, The Institute for Energy Technology, INF-630(86)1.

10. Description of a multifunctional decision support system in emergency management, applying expert system techniques. Ø. Berg, J.S. Larsen, T. Westgaard, C. Holmström, H. Andersson, S. Jakobsson, R. Heinonen, T. Kukko, The Institute for Energy Technology, HWR-193, 1987.

11. A preliminary specification for a multifunctional support system in emergency management, applying expert system technology. Ø. Berg, J.S. Larsen, T. Westgaard, C. Holmström, H. Andersson, S. Jakobsson, R. Heinonen, T. Kukko, The Institute for Energy Technology, HWR-205, 1987.

12. Framework for modelling decision making contexts. Jens Rasmussen, Risø National Laboratory, N-6-86, March 1986.

13. A cognitive engineering approach to the modelling of decision making and its organization. Jens Rasmussen, Risø-M-2589, 1986.

14. Evaluation of the use of advanced information technology (expert systems) for data base system development and emergency management in non-nuclear industries. Jens Rasmussen, O.M. Pedersen and C.D. Grønberg, Risø-M-2639, 1987.

15. System design and psychology of complex systems. Berndt Brehmer, Uppsala University, Unpublished Note.

16. Organization for decision making in complex systems. Berndt Brehmer, Uppsala University, Unpublished Note.

A SURVEY OF FRAMATOME'S EXPERT SYSTEMS ACTIVITY

Didier Delaigue and Michel Grundstein

Framatome, Corporate Strategy - Planning - Organization
Expert Systems Development Unit
Tour Fiat, Cedex 16, 92Ø84 Paris La Défense, France

INTRODUCTION

The French multinational nuclear energy world leader, Framatome, has designed and installed more than 40000 MWe of power in both France and abroad using Pressurized Water Reactor (PWR) technology. The French nuclear program ranks as one of the most succesful in the world. In 1983, Framatome entered the Applied Artificial Intelligence (A.I.) field by setting up FRAMENTEC S.A., a joint venture with TEKNOWLEDGE Inc. Today, Framentec is a wholly-owned subsidiary of Framatome and is among the leading european companies specializing in Applied Artificial Intelligence. Framatome now has a 7.5% stake in Teknowledge Inc.

A.I. can be applied to a great variety of areas in both the nuclear engineering and plant management sectors, ranging from computer aided design (CAD) to process control, diagnosis and planning. Safety, in all such areas is of the utmost importance. The contribution of fault diagnosis expert systems, integrated into costly equipment or production resources designed for long service lives, cannot be overstressed. The main applications in the nuclear industry can be summarized as follows :

- quality assurance
- design of systems subject to extreme operating conditions
- maintenance of complex systems
- control of complex phenomenon producing high velocity transients
- expert advice in multiple fields
- compliance with complex regulations
- high-skill personnel requirements
- heavy financial investments

Adapting A.I. technologies to knowledge based systems implies working on partially unknown objects. The innovative features, scale and complexity of problems involved mean that novel approaches to work within companies need to be explored.

The aim of the expert system organization is to remain in the forefront of progress and to disseminate A.I. techniques and methodologies within FRAMATOME and subsidiaries :

. The "System Expert Applications" unit is a task force concerned with methodology. It is under the responsibility of Corporate Strategy - Planning - Organization Director. It coordinates work on methodology, initiates and stimulates action and monitors project status.

. A study and technical exchange group monitors progress on novel approaches, disseminates results and carrries out experiments on inter-group exchanges and performs rigourous studies on methods and techniques. Group members support and extend the system expert applications Unit activity within their respective departments.

. The data Processing division supplies the appropriate hardware and software.

. FRAMENTEC markets three tools -M1, K1, S1- for the development of expert systems, delivers industrial expert systems, develops generic and semi-generic products -MAINTEX-. It organizes training courses and seminars on A.I. technology and is committed to Research & Development on a large scale.

MAIN NUCLEAR EXPERT SYSTEM APPLICATIONS

Various objects are as follows:

. Research and Demonstration grade systems based on a fraction of expertise for test project feasibilty and knowledge base architecture evaluation;

. Prototypes and advanced prototypes displaying most of the functionalities (characteristics) of the final knowledge engineering system. They can be integrated into conventional hard and software. Their domain of action is limited but their structure is designed to easily integrate functional extensions required for global application; . Operational systems which are extensions of prototypes to the total application domain and are integrated into the environment.

Parameters of a specific system are described as : XXX (X, Y, Z, T) XXX = name of the expert system , X = owner, Y = prime contractor, Z = expert system development tool, T = hardware

Research and Demonstration Grade Systems

Design-optimization
* OPACS (FRA, FTC, S1, Xerox 1108) :a system assisting the locating of pipe support components in a nuclear plant. The main functions of the system are :
    . to simulate the general description of the supporting
    . to help in the choice of the initial solution for the supporting
    . to reduce repetitions between calculation and installation
    . to minimize supports number
    . to minimize expensive or nonstandardized solutions

* PREPIANO and PIANO (NOVATOME, M1, IBM PC's) : these are two complementary systems which assist in the design of electromagnetic pumps. PREPIANO searches through a catalog of existing pumps for the most appropriate model. It then establishes a proposal for the design of the required pump. PIANO takes the proposal and recommends detailed design modifications to apply to the model selected.

Training-diagnosis
* EXPERT-GV (FRA, FTC, S1, VAX) : it aims to determine the causes of cracking in steam generator tubing based on the leak's location and the history of the tube. EXPERT-GV is currently used for staff training.

Diagnosis
* ADTC (FRA, M1, V286 AT) : it aims to appraise the acceptability of surface defects appearing in pressure vessel fasteners. The knowledge base connects to a video defect recognition model.

Diagnosis-regulation
* RCCM (FRA, FTC, S1, VAX) : an expert system for assisting weld qualification following RCC-M requirements (The French equivalent of the ASME code). The final system will be connected to a computer aided production management tool.

Monitoring
* PMS-1 (FRA, FTC, AGE, Interlisp-D, Xerox 1108) : it aims to apply A.I. to build a real-time advisor for the control of a PWR. A sub-system has been developed focusing on the control of boron concentration in the PWR primary system.

* ACCINAL (FRA, FTC, S1, Xerox 1108, VAX) : analysis of PWR primary system parameters following safety injection. The kind of accident or incident can thereby be determined. The system provides appropriate driving guidelines as a function of the situation detected.

Planning-optimization
* SYREP (FRA, M1, IBM PC's) : Selection, planning and optimization of refueling configurations. The system is semi-graphic and uses a window modul.

Prototypes (P) and Advanced Prototypes (AP)

Optimization
* CLEF (FRA, S1, IBM PC's, AP) : Helps in classification of mechanical and electrical equipment installed in nuclear plants. Classification is based on RCC-M safety and quality assurance (QA) classes and categories governing the ability of mechanical systems to withstand earthquakes.

Diagnosis-advising
* MACHA (FRA, FTC, MAINTEX, PC's, P) : establishes corrective diagnosis on a revolving cylinder manipulator crane used for PWR refueling. This system is at the experimental testing grade at PALUEL. Using such a diagnosis assistance system implies :
    . a decreasing of the outage time of the equipments by easy diagnosis of the failure
    . an increasing of diagnosis reliability
    . a permanent memorization of the knowledge and data of the equipment

Operational Systems

. inform and turn out the different hierarchy levels on the essentials concepts of knowledge based systems and give them an opportunity to manipulate demonstration grade systems
. stimulate ideas on fields of application
. make participants aware of their contribution to an operation

Site Awareness and Problem Selection

It consists in a conventionnal impact and opportunity analysis enhanced with the psycho-sociological factors related to the concerned actors. These factors are partially explicitated during a one day "progress session" which helps too to define the actors' expectations and to specify the application.

The three next stages are quite similar to the knowledge acquisition process as described by Hayes-Roth et al., 1983.

Problem Identification and Project Evaluation

The main objectives are :
. to characterize the system and specify its functionnalities;
. to determine and evaluate the techniques able to support these functions;
. to check the technical, economical and organizationnal feasability of the proposed solution;
. to plan a development program.

From a knowledge engineering standpoint, it consists of :
. identifying the problem and the associated knowledge;
. specifying the knowledge representation and then selecting an adequate tool;
. estimating the knowledge base size and some of the future difficulties;
. preparing and enforcing the cooperation between the expert(s) and the knowledge engineer.

The issue of the session are a technico-economical (or commercial) report and, possibly, a demonstration system.

Prototype's Conception and Realization

With the prototype the knowledge architecture is tested on a part of the expertise domain; the process can be iterative until stabilization During this phase, the knowledge acquisition process must be constructive and not only extractive. At the end of this stage the object has to be accepted by the experts of the domain and most of the functionalities of the final system must be simulated.

Final System Realization

Based on the prototype's architecture, the system is enhanced and lineary filled with the knowledge; the different integrative aspects are also developed. Contrary to the former phase which was built on the basis of a dialog between the expert(s) and the knowledge engineer during the realization process, all the participants must be present to prepare the integration phase. It is time for involving the final users' interface wishes, of thinking about training people who will complete the system as well as those who will use it and

preparing the organizational structure of the receipt group. During this
stage a true transfer of technology operates.

Diagnosis-advising
* TIG (FRA, FTC, M1, IBM PC's) : a valuable aid in resolving malfunctions
and failures encountered   using a new weld cladding process (Gas
Tungsten Arc Welding). A   further   purpose of this system is to
ensure   transmission   of   comprehensive   process-related   expert
opinions   which   occurs   in   parallel to the actual transfer of the
process from its development   environment to the production line (de
Bonnières et al.,1986).

* SIRACUS (FRA, FTC, M1, PC's) :A non-destructive testing (NDT) advisor
for ultrasonic testing of   welds   and   classifies located flaws
into linear and volumic   defects.

Diagnosis-optimization-planning
* CERBERE (FRA, K1, Sun 3-52, P) : Computes   optimum   refueling
configurations and establishes related   documents. The system has been
used since april 1987. Its main functionnal characteristics are :
     . automated configuration of the input files for the neutronics codes
used in these studies
     . automatic connection through FRAMATOME's computer network
     . automatic treatment of the outputs
     . computation of intermediate results during the study
     . analysis of errors and of safety criteria violations
     . automated edition of a study report and of the safety report
     . optimized management of jobs' submission
     . creation of a data-base for the entire set of power stations in
operation

* FOCUS (FRA, M1, PC's AT) : detects similarities between two nuclear
fuel load configurations   for   a   new   fuel   load. The system
identifies   the   most similar   previously installed load. Computation
time can thus be shortened   and   reliability enhanced. This is a
small scope system and a new   extended version known as CERBERE is now
at the advanced prototype   stage.

Many   others   applications   are   under   development and are at
different stages of development.   Not all are described here owing
either to confidentiality or to the fact they do not apply to the nuclear
field.

METHODOLOGICAL FEATURES : LIFE CYCLE OF A KNOWLEDGE BASED SYSTEM

The   development   process of the operational system TIG is based
on close cooperation between the various project participants in the same
manner   as   for   conventional computer systems. Observation of TIG's
history makes visible a specific life cycle which situates the
operation in a space-time referential and highlights the contribution of
each   participant   in   their   synergical   set-up   using successive
impulses. The main features of this life cycle are now described
(Grundstein and de Bonnières, 1987).

Fertilization-Pollinization of the Sociological-Professionnal Environment

The objectives of this first stage are to :
     . promote specifics of expert systems with regard to conventional
systems

<u>Qualification and Sociological-Professionnal Integration</u>

Experimental and operational integration are two sub-phases associated with the completion of the transfer of technology process: the user and improvement groups take over the system.

<u>Exploitation and Evaluation of the System</u>

This is the validation phase of the operation regarding to the initial purpose of such an object. Observation of the life of the expert system in its environment must be carried on during a long period to detect the problems in maintaining and extending it.

The following scheme (fig.1) will summarize the life cycle development of a knowledge-based system project. It points out through the different loops the recursion and regulation mechanisms (von Bertalanffy, 1968; Morin, 1977) inherent to the building process.

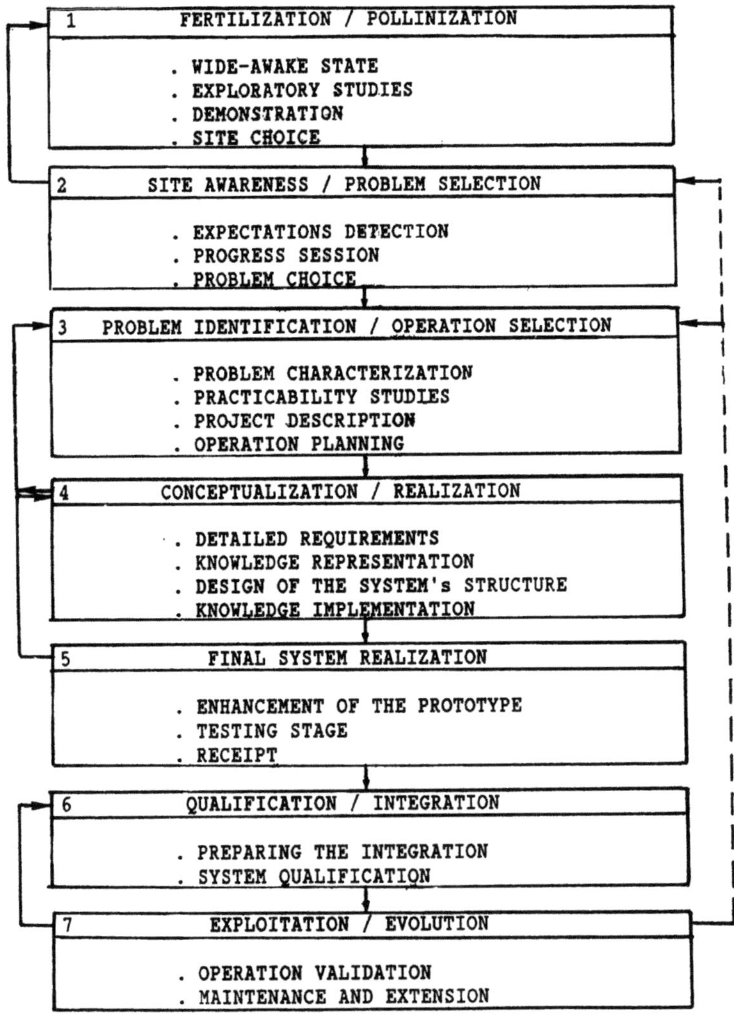

Fig. 1 Life cycle of a knowledge-based system.

CONCLUSION

All these experiments help to progress in the direction of
building expert systems and of understanding the complex process of
their integration in a socialogical-professionnal environment.

From a technical point of view, a base of appropriate
selection criteria have emerged and priority has been given to
middle size systems; The TIG application provided FRAMENTEC with an
opportunity of conceiving a generic maintenance product MAINTEX which is
used now as a generator for many applications.

From a methodological point of view, a sort of life cycle
of knowledge-base systems with specific features in each development
stage assists project management.

From the socialogical-professionnal integration point of view,
the organizational device as well as the underlying proceeding
methods seem adequate at the present phase of introducing a new
technology in a company.

<u>References</u>

de Bonnières P., Boutes J.L., Calas M.A., Para S.,1986, A
Knowledge-based diagnosis System for Welding Machine Problem-solving,
International Conference on Computer Technology in Welding, London.
Grundstein M., de Bonnières P., 1987, De la Fertilisation à
l'Insertion : vers une Méthodologie de Développement des Systèmes à Base
de Connaissances, Revue d'Intelligence Artificielle, Ed. Hermes.
Hayes-Roth F., Waterman D.A., Lenat D.B., 1983. "Building Expert
Systems". Addison Wesley,
Morin E., 1977. "La Methode. 1. La Nature de la Nature". Seuil,
Paris.
von Bertalanffy L., 1968. "General Systems Theory. Essays on its
Foundation and Development". Braziller, New-York.

ARTIFICIAL INTELLIGENCE APPLICATIONS FOR OPERATION AND MAINTENANCE :

JAPANESE INDUSTRY CASES

Mutsumi Itoh and Ichiro Tai
Nuclear Energy Group, Toshiba Corporation, Tokyo, Japan

Kazuo Monta and Koichi Sekimizu
Nippon Atomic Industry Group Co., Ltd., Kawasaki, Japan

ABSTRACT

A nuclear power plant as a typical man-machine system of the modern industry needs an efficient human window through which operators can observe every necessary details of the plant for its safe and reliable operation. Much efforts have been devoted to the development of the computerized operator support systems (COSS).

Recent development of artificial intelligence (AI) seems to offer new possibility to strengthen the performance of the COSS such as more powerful diagnosis and procedure synthesis and user friendly man-machine interfaces. From this point of view, a national project of "Advanced Man-Machine System Development for Nuclear Power Plants" has been carried out since 1984. This eight years project consists of Phase I (Conceptual Design, 1984-1986) and Phase II (Detailed Design and Implementation, 1987-1991).

Artificial intelligence application to nuclear power plant operation and maintenance is also selected as a major theme for the promotion of research and development on frontiers in the recently revised long term national program for development and utilization of nuclear energy in JAPAN.

In addition to the aboves, various studies are being performed for exploring the benefit of AI applications for nuclear power plant operation and maintenance, since practical and short term studies are necessary and helpful to foster AI applications including training necessary personnel.

INTRODUCTION

Based on the lessons learned from TMI-2 accident, development of Computerized Operator Support Systems (COSS) were undertaken supported by the Japanese Ministry of International Trade and Industry (MITI) during 1980- 1984.[1],[2],[3] The major functions of the systems such as the standby safety system management, the disturbance analysis and the post trip operational guidance were specified based on the operator's role.[1] Although without being explicitly noticed, these functions were realized by the various knowledge on plant behavior and operational procedures.

Meanwhile the Japanese fifth generation computer project was started in 1982 aiming to provide the basis for a revolutionary new kind of knowledge information processing by the early

1990's. These studies seek to provide higher-level functions for new computers by using artificial intelligence techniques, and new software and hardware (architecture) technologies. So this project will furnish fundamentals in general and provide various tools when they fit in particular applications in nuclear industry.

Near the end of the COSS project, the applications of artificial intelligence to nuclear power plant operation were tried by several researchers and some earlier works were reported.[4],[5],[6]

Recent development of artificial intelligence (AI) seems to offer new possibility to strengthen the performance of the COSS such as more powerful diagnosis and procedure synthesis and user friendly man-machine interfaces. From this point of view, a national project of "Advanced Man-Machine System Development for Nuclear Power Plants" (MMS-NPP) has been carried out since 1984 also supported by MITI. This eight years project consists of Phase I (Conceptual Design, 1984-1986) and Phase II (Detailed Design and Implementation, 1987-1991). The detail of this project is described in section 2 of this paper.

The long term national program for development and utilization of nuclear energy was recently revised by Japanese Atomic Energy Commission (JAEC)[7]. As one of the major goals of the program, JAEC first selected the promotion of research and development on frontiers which will contribute to the common base technology for future nuclear energy. As such, artificial intelligence application to nuclear plant control, plant anomaly diagnosis, teleoperation in nuclear radwaste disposal facilities et cetera is selected along with new materials, laser technology and radiation reduction and assessment. The government related research organizations, i.e., Japan Atomic Energy Research Institute (JAERI) and Power Reactor and Nuclear Fuel Development Corporation (PNC) will pursue this goal progressively. Current activities of PNC in this area was recently published.[8]

Electric utility companies and vendors of the nuclear steam supply systems are also eager to develop application of artificial intelligence. The Central Research Institute of Electric Power Industry (CRIEPI) is now developing the expert system for abnormal event recurrence prevention.[9] Development of nuclear plant maintenance and equipment diagnosis support systems is undertaken by various utility companies and NSSS vendors.[10] An example of this application is described later in this paper.

ADVANCED MAN-MACHINE SYSTEM DEVELOPMENT FOR NUCLEAR POWER PLANT (MMS-NPP)

The participants of the project are six private companies* engaged in NSSS supply in Japan and contribute to the development of MMS-NPP on the division of work basis according to the reactor type and the proposed functions described later.

Goals and functions

The goals of the project are as follows :

(1) Support of the flexible planning of operation during plant start up, shutdown and load following operation,

(2) Support to efficient and reliable plant operation and maintenance management,

(3) Support of detection of abnormalities in the plant, identification of their causes and provision of appropriate operational countermeasures,

(4) An efficient, reliable and interactive man-machine interface which offers the above information to the operators.

---

*   Hitachi Ltd., Mitsubishi Heavy Industries, Ltd., Mitsubishi Electric Corporation, Mitsubishi Atomic Power Industries, Inc., Toshiba Corporation and Nippon Atomic Industry Group Co.,Ltd.

The major functions of MMS-NPP are as follows :

(1) Support of operational planning and supervisory control
Reactor re-startup planning during unplanned shutdown requires the search of the plant status for the optimal startup sequence. Also, the flexible response of the plant to the demand of electric power grid requires the quick evaluation of several operational constraints. These planning activities require ad-hoc processing of large amount of plant operational data and constraints in short time which can benefit from the AI application. Also the supervisory control during both processes can benefit from AI when abnormalities happen and some countermeasures are needed.

(2) Support of operational management
Operators of nuclear power plant are required to manage a large quantity of equipment. Standby safety systems must be in order to respond the demand at all times. So they should be monitored for their proper configurations and should be dynamically tested at appropriate intervals. Redundant equipment should be rotated for periodic maintenance and economic usage. These clerical tasks often require skilled and reliable workers and then can benefit from AI. AI technique is suitable for application to more advanced automated systems, such as an automatic load following system, which are inevitably complex and difficult to manage when nonstandard situations occur.

(3) Support of maintenance management
In order to maintain equipment, it is necessary to set up appropriate maintenance work condition and to provide procedures to the workers. Also it is necessary to supervise the maintenance work. Using vast amount of plant data and relevant knowledge base, AI technique can support these activities.

(4) Support for abnormalities and accidents
In the former COSS project, this function was developed mainly for anticipated transients and accidents. AI technique seems to promise more flexible and more adaptable diagnosis and procedure synthesis due to its more fundamental and more comprehensive knowledge base and versatile inference mechanisms. In order to cope with unanticipated transients and accidents, cooperation of man and computer is so inevitable that one of the knowledge bases of the system should be compatible with operator's mental model of the plant.

(5) Intelligent man-machine interfaces
All the above support functions are available to the operator through the man-machine interface. So the information should be offered in a fittest form for the operator in abstraction level, object coverage and timing. Also the interface should provide interactive capability for the operator as close as his colleagues.

These functions will be integrated with the existing COSS functions as shown in Fig. 1 in future man-machine interfaces. The COSS functions are mainly support of operator's rule based behavior such as are represented by plant symptom-action pairs. So in general, they can be rather quickly derived and presented to the operators. However, they are more or less based on the anticipated scenarios. So in case of an unanticipated abnormal plant condition beyond the anticipated scenarios, the operators behavior must shift to the knowledge based behavior and the support functions also must shift to its support. MMS-NPP functions principally aim at support of this knowledge based behavior. MMS-NPP intelligent man-machine interface function is expected to replace the existing interface due to its human friendliness.

Developmental Approaches for Support for Plant Anomaly and Intelligent Man-Machine Interface

To realize MMS-NPP functions, especially the support for abnormalities and accidents and the intelligent man-machine interface, the following approaches are considered.

From the functions allocated to MMS-NPP relative to COSS shown in Fig.1, the topographic search strategy [11] seems quite appropriate for the support for abnormalities and accidents.

To diagnose plant anomaly, plant input signals are mapped into the domain (i.e. nuclear power plants) model, and good/bad judgments are made in the domain. Then these judgments are interpreted based on the domain knowledge base to identify plant situation and to formulate current plant goal. Then to achieve the goal, the domain knowledge base is investigated to synthesize the appropriate procedures.

The domain knowledge-base used here is a hierarchical abstract functional model based on mass, energy and information flow in the plant.[12] In this model, the two ultimate goals of the nuclear power plants, i.e. electricity production and safety are set at the top level. These two goals are represented by the respective flow structures. The production goal is represented by the energy flow from nuclear energy source to electric power grid and the safety goal is represented and maintained by many barriers between fission products in the fuel and the environment. These top level flow structures are then decomposed into substructures which constitute or support them. This process is repeated successively to represent the plant functional structure hierarchically and in sufficient detail.

This hierarchical connection can serve as the road map for diagnostic search and every flow structure at the node in this connection can possess the knowledge for diagnosis and dynamic procedure synthesis. This knowledge is related to the internal structure of the flow structure which is represented by several flow functions such as storage, transport et cetera, as well as to its functional goal and task.

To supplement the hierarchical abstract functional model, the actual physical representation of the plant is described by using an object model. This object model corresponds to the lowest level of the hierarchical abstract functional model and is useful for detailed diagnosis, detailed planning and evaluation of operational procedure synthesis.

In order to realize intelligent man-machine interfaces, following two approaches[13] are adopted.

One is to incorporate an operator model in the system to attain conformity between operator's cognitive process and the information which MMS-NPP will provide for the operator. In anomalous plant conditions, the operator imagines some scripts in mind to recover the conditions, based on the plant knowledge which he has acquired through training and operational experience. The system infers operator's cognitive process, corresponding to the above script, or operator's focus of attention by use of the operator model. Guidance information for the operator's decision making is arranged on the basis of the operator's focus of attention. The domain knowledge base described above seems to be usable as the operator model.

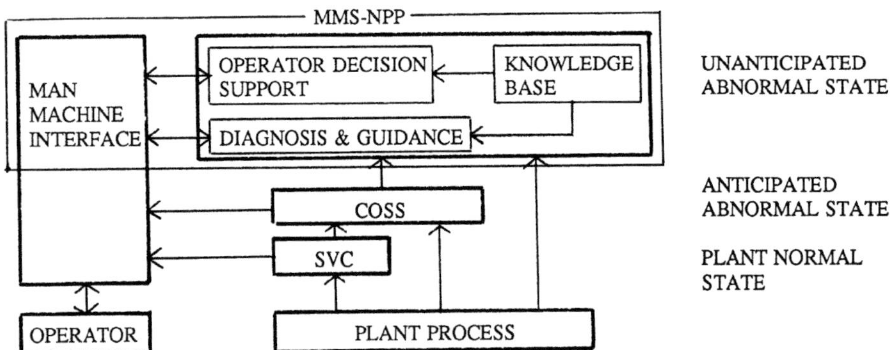

Fig.1 *Man Machine System Integration*
COSS : Computerized Operator Support System
SVC : Supervisory Control System

The other is to develop advanced communication environment for smooth dialogue between the operator and the system. Advanced man-machine devices considered for use in the system includes a voice recognition system, audio response unit, high speed graphic display units with touch screen and a large format display by which more than one operators can access to the same information at the same time. The voice recognition system which permits the operator to use about 1,000 words vocabulary and to speak simple sentence continuously is the current goal. Furthermore, the high speed graphic display system allows us to use understandable icons in pictures and to easily present information with complex hierarchy through window management technique.

ARTIFICIAL INTELLIGENCE APPLICATION AS RESEARCH AND DEVELOPMENT ON FRONTIERS

Research and development for addition of intelligence to nuclear power facilities aims to ultimately develop an autonomous system with fully automatic decision, control and maintenance through the interim target of manual supervisory control and management supported by artificial intelligence technique applied to plant control, plant anomaly diagnosis and corrective action, maintenance work, fuel management, radioactive waste management et cetera.

The present targets are the development of the inspection and maintenance robot with capability of sophisticated decision making and manipulation while working narrow, highly radioactive environments and the development of plant supervision and control system with excellent man-machine interface capability.

PNC, for instance, from its twenty years experience on nuclear fuel cycle activities and development of advanced power reactors, has accumulated various knowledge and by acquiring appropriate knowledge bases from it, is developing various expert systems for operator training, equipment management, failure diagnosis and preventive maintenance, plant operational support and various engineering support.[8] PNC claims that the expert system technology is the most appropriate means for utilizing the accumulated knowledge and knowhow as well as design and operational information which an organization possesses for its future activities.

For examples, concerning the experimental fast breeder reactor JOYO, an operator support system, a maintenance support system and a fuel management planning support system are now being developed. The JOYO operator support system aims to provide the operators with plant anomaly diagnosis and operational procedures to cope with. This system is planned to be fully operational for JOYO's major subsystems in 1990.

ACTIVITIES OF ELECTRIC UTILITIES AND NSSS VENDORS

Japanese electric utilities engage research and development for artificial intelligence application by themselves, by groups such as the PWR owners group and by their cofunding research organization, CRIEPI.

Development of CSPAR

CRIEPI is developing an expert system for abnormal event recurrence prevention,[9] CSPAR (Consultation System for Prevention of Abnormal Event Recurrence). This system intends to abstract, organize and store the valuable information for abnormal event recurrence prevention from world wide or national abnormal event report systems such as INPO's Nuclear Network and on demand to provide it for users in appropriately processed forms for their need.

CSPAR identifies the abnormal events which need some measures to prevent their recurrence and supports users to synthesize those which may involve changes in the plant design, construction, operation and maintenance.

CSPAR's fundamental functions are as follows :

1. Cause identification

2. Consequence prediction

3. Similar event prediction

4. Importance evaluation

5. Preventive measure synthesis

The functional framework of CSPAR has been completed in March 1987 and its knowledge base will be enriched in a few years for its practical use.

Development of a Maintenance Work Support System

Since the availability of the nuclear power plants has stably reached to the level of about 75%, Japanese utilities gradually turn their attention to the efficiency improvement of the mainte-nance and repair work. Maintenance work of nuclear power plants in Japan has been supported by plural companies and their experiences are distributed in individuals in those companies. It is preferable for those experiences and acquired knowledge to be rearranged and synthesized to be utilized for the design and maintenance work for the future plants as well as the present ones. Specialised and divided work and know-how should be rearranged and kept in a proper manner and transfered to the next generation smoothly.

From above mentioned circumstances, Toshiba made a trial to build up a maintenance sup-port system which aims to detect a certain anomaly, identify the cause of the anomaly and give a guidance for the inspection and repair work, based on the expert system technique[14].

This system has following scope:

(a) Plant diagnosis support by

   Plant dynamics diagnosis

   Containment atmosphere diagnosis

(b) Plant subsystem maintenance support by

   Control rod and drive system maintenance support

   Reactor core instrumentation maintenance support

   Feedwater system maintenance support[15] .

(c) Plant equipment maintenance support by

   Recirculation flow pump maintenance support

   Control system electronics maintenance support

   Mainsteam isolation valve maintenance support

The detection of an anomaly is, in many cases, done through the conventional monitoring systems installed in the control room but the detection of some particular status changes requires the analysis of process signals by an optional instruments. As a support system as the latter case, this maintenance support system has a subsystem to detect the anomaly. The identification of the kind of the anomaly and location is realized by a series of the check sequence such as alarm sequence analysis and process parameter check sequence stored in the computer as a knowledge data base.

42

The guidance for the inspection and repair work is realized based on the standard evaluation technique of the system and equipment, which is established by the specialists. The decision making on the treatment regarding to the anomaly is done through the evaluation of all data facts acquired through the dialogue with an engineer who is consulting with the system on the anomaly.

This maintenance support system is composed of plural expert systems each of which has a inference engine and knowledge bases, respectively on a hardware such as TOSBAC super mini-computer G8050. The maintenance work such as inspection and repair needs various kinds of knowledge so that the computerization of the knowledge is done in plural expressions such as production rule, frame, list, trend curve, graphs and so on. In case of the necessity, numerical simulation also assists the diagnosis function and the future trend estimation.

CONCLUSION

The major activities of Japanese nuclear industry on artificial intelligence application are surveyed. There also are many other studies such as application to robotics[16] and studies by academic which can not be mentioned in a limited scope of the paper.

From the survey it seems that one of the major motivations for AI applications is to materialize the past knowledge and knowhow accumulated during the course of nuclear power development. Behind this motivation there is an alteration of generation of operators and an accumulation of various engineering assets during these two or three decades. Another motivation is to improve the man-machine interface and to effectively utilize operational experience to further improve the operational safety of nuclear power plant. From the same motivation, application to operator training is expected.

From the technical point of view, following the spread of applications, various paradigms of knowledge engineering are sought according to the applications and the knowledge acquisition is one of the most important theme regarding the above motivations.

Japanese activities on AI applications could be classified into the long term projects sponsored by the government and the practical and short term studies by the nuclear industries themselves. The former provide the opportunities to pursue the revolutionary targets which AI seems to promise for future technological era, while the latter are necessary and helpful to foster AI applications including training necessary personnel.

REFERENCES

[1] K.Monta, S. Fukutomi, M. Itoh and I. Tai, Development of a Computerized Operator Support System for Boiling Water Reactor Power plants, International Topical Meeting on Computer Applications for Nuclear Power Plant Operation and Control, Pasco, Washington, Sept. (1985)

[2] T. Masui, M. Tani and Y. Okamoto, The Development and Evaluation of Pressurized Water Reactor Advanced Control Room Concepts in Japan, Part II : Computerized Operator Support System (COSS), ibid.

[3] Y. Higashikawa, F. Murata, S. Hashimoto and T. Kiguchi, A Computerized Operator Support System for Boiling Water Reactor Abnormal Conditions, ibid.

[4] K. Yoshida, M. Yokobayashi, T. Aoyagi and Y. Shinohara, Development and Verification of an Accident Diagnosis System for Nuclear Power Plant by Using a Simulator, ibid.

[5] T. Kiguchi, H. Motoda, N. Yamada and Y. Yoshida, A knowledge Based System for Plant Diagnosis, ibid.

[6] T. Washio, M. Kitamura, K. Katajima and K. Sugiyama, Semantic Network Approach to Automated Failure Diagnosis in Nuclear Power Plant, ibid.

[7] Atoms in Japan, June 1987, Japan Atomic Industrial Forum

[8] A. Ishido, T. Asano, K. Sato, H. Aoki, K. Setoguchi and T. Okada, Status of Expert Systems Development at PNC, Genshiryoku Kogyo, 33,4(1987) (In Japanese)

[9] T. Nishiyama and Y. Shinohara, Development of Expert System for Abnormal- Event Recurrence Prevention, J. of Atomic Energy Society of Japan, 29,4(1987) (In Japanese)

[10] H. Ujita, T. Kiguchi, K. Onodera and M. Komata, Development of Maintenance Support System for Nuclear Power Plants, ibid. 29,6(1987) (In Japanese)

[11] J. Rasmussen, Strategies for State Identification and Diagnosis in Supervisory Control Tasks and Design of Computer-based Support System, in Advances in Man-Machine Systems Research, Vol.1, W. Rouse (ed), JAI Press Inc.(1984)

[12] M. Lind, Multilevel Flow Modelling of Process Plant for Diagnosis and Control, International Meeting on Thermal Nuclear Reactor Safety, Chicago, Sept.(1982), NUREG/CP-0027, pp.1653

[13] T. Ogino, Y. Nishizawa, T. Morioka, N. Naitoh, M.Tani and Y. Fujita, Intelligent Decision Support Systems for NPP in Japan, International Post-SMIRT 9 Seminor, Accident Sequence Modeling : Human Actions, System Response, Intelligent Decision Support, Munich, FRG., Aug.(1987)

[14] Y. Sasaki, I. Tai and Y. Andoh, Maintenance Support Expert System for Nuclear power plant, Toshiba Review, 42,5(1987) (In Japanese)

[15] R. Meguro, Expert System for Nuclear Power Plant Feedwater System Diagnosis, This Topical Meeting, Session 6-A, Equipment Diagnostics,(1987)

[16] T. Arai, H. Yoshikawa, M. Takano, S. Ozono, G. Odawara, T. Miyoshi, K. Shimo and T. Mikami, A Stair-Climbing Robot for Maintenance : "A MOOTY", IAEA Conference on Robotics and Remote Handling in Nuclear Industry, Oxford, UK, Sept.(1984)

DEVELOPMENT IN THE APPLICATION OF KNOWLEDGE BASE SYSTEMS

IN THE CANADIAN NUCLEAR INDUSTRY

J.W.D. Anderson, A. Natalizio, and J.E.S. Stevens

Atomic Energy of Canada Limited, CANDU Operations
Sheridan Park Research Community
Mississauga, Ontario. L5K 1B2

ABSTRACT

Atomic Energy of Canada Limited (AECL) is a leader in the area of computer monitoring and control of nuclear power plants. The high level of plant computerization in CANDU stations has not only provided benefits in terms of more reliable plant operation but also in terms of improved man-machine interface.

Today, new opportunities for improvements in safety and performance include distributed control systems, data highways and improved operator aids. The most exciting innovation is the development of expert systems aimed at assisting the operator in areas such as: fault management and diagnosis, identification of limiting conditions of operation, assessment of heat sink availability, optimization of on-power fuelling schemes, and computerization of procedures and communication.

The "Operator Companion", an expert system intended to diagnose plant faults and to advise the operator on appropriate restoring and corrective actions, is a major undertaking which is receiving support within the research and engineering groups of AECL and from McMaster University in Hamilton, Ontario.

INTRODUCTION

The nuclear industry is presently faced with two major challenges: the need for cost reduction to be competitive with coal in the early years of operation; and the need to improve public confidence in the safety of nuclear power plants. Atomic Energy of Canada Limited (AECL) is meeting these challenges now because it believes in the long term future of and opportunities for nuclear power.

The program developed by AECL to meet these challenges has the following major goals:

1.   The achievement of greater plant automation and computerization to reduce capital as well as operating costs, particularly through greater plant availability;

2.   The enhancement of plant safety through significant improvements
        in the man-machine interface; and

3.   The achievement of greater productivity and high quality in the
        process of engineering and constructing nuclear power plants.

AECL recognizes the importance of knowledge base system technology
in the achievement of these objectives and has thus identified this as a
strategic technology to be fully exploited in the next generation of CANDU
PHW reactors.

The nuclear research laboratory at Chalk River, Ontario, (CRNL) works
closely with the Engineering organization in Mississauga, Ontario in all
aspects of nuclear power plant development including knowledge base systems.
In this particular field, cooperative programs have also been established
with the Computer Science and Systems Department and Engineering Science
Department at McMaster University, which are doing research on knowledge
base systems. Activity on artificial intelligence and knowledge base systems
within AECL is being coordinated by a steering committee.

THE BENEFITS OF KNOWLEDGE BASE SYSTEMS

AECL sees several potential benefits from the application of knowledge
base systems. First, there are benefits to CANDU reactor operations. It is
anticipated that the application of this technology to various aspects of
operation will result in lower system maintenance requirements, better
planning of maintenance outages, higher plant availability and lower occupational exposure to station personnel. In addition, this technology will
improve productivity and enhance efficiency throughout the company. Reduc-
tions in design, project management and operating station support service
costs are anticipated.

CHALLENGES

The introduction of any new technology brings with it challenges as
well as opportunities. The challenges that need to be met in exploiting this
technology are both technical and organizational. The first challenge is to
establish an education and awareness program to build support. This has
been a lengthy process as the words artificial intelligence, expert systems
and knowledge base systems mean different things to different people and
everyone has their own perceptions and expectations of this developing
technology. However, this particular challenge has been dealt with.

Having established some credibility within the organization, the
next problem is to obtain funds at a time when the nuclear industry as a
whole is shrinking and new reactor sales are virtually non-existent.

After establishing a modest level of financial support, the next
question is "How do you spend it?". The challenge is to obtain a balance
between the purchase of hardware and software tools, training to build up
a core group of technically competent people, and the early development of
products to demonstrate delivery capability. AECL's approach is to develop
a few simple systems to obtain hands-on experience with artificial intelligence techniques and to concentrate on three early delivery products: a
shell and tube heat exchanger design expert, a demonstration of an operator
aid system, and a fuel defect detection and evaluation system. Spending
on tools and training is limited to what is essential to complete these
projects.

The course that AECL is following is to establish a group of experts in this area that can provide services to the corporation. The long term goal is to turn this core group into a centre of excellence for all AI related work. This group will be responsible for the evaluation of methods and technology and, eventually, for the development of new technology. This will be accomplished through promotion of the technology within the organization, through direct involvement and assistance on ongoing initiatives and through development training courses and seminars for the rest of the company. In addition to actively promoting the advantages of knowledge base systems, there is a need to overcome scepticism about the value of the technology and its applicability within the organization. It is important to demonstrate that the first applications are useful and cost effective and that they can be produced using a technology that is now mature.

There are also anticipated technical challenges in adopting and implementing knowledge base systems. One of these will be to develop a large system for reactor operation support. This system will require significant effort to implement effectively and efficiently because of its size. It will probably have to be implemented as a set of separate tasks running on a distributed hardware system. The task of implementing this distributed system and coordinating the communications of the subsystems will be a major undertaking. In addition, the delivery hardware and implementation language will need to be specified.

ORGANIZATION OF THE AECL PROGRAM

AECL has committed financial support for the formation and development of a Knowledge Systems Engineering Group (KSEG). This group will initially be supported with development funds with the expectation that over a five year period it will become a self-sustaining business unit.

Training for the staff of the KSEG has largely been done through attendance at topical courses and at software vendors' training courses. The rest of the training has been achieved by hands on experience with various software packages. This has allowed the learning period to be used to actually produce some running applications. Future training will focus on professional and technical staff who may be involved in knowledge base system development work. Much of this training will be done as an extra-curricular activity due to the enthusiastic staff response, as was also the case for staff training in CADDS.

In addition to the above group, AECL is supporting a small research project at McMaster University in Hamilton, Ontario. Cooperative efforts between AECL and McMaster are being explored and it is anticipated that they will grow in scope.

APPLICATION EVALUATION

Identification of areas in which expert systems could provide potential benefits was relatively easy compared to the evaluation task. A company wide search for potential applications was conducted. Literally hundreds of possible applications were suggested. The applications were separated into three major groups according to the potential benefits that they offer to the company: enhancement of the CANDU product line, engineering productivity improvement, and the development of new commercial opportunities.

The applications were then evaluated by a task force consisting of
representatives from the research and engineering organizations to identify
those with the best combinations of potential benefit and feasability of
implementation. Benefit was evaluated in terms of the strategic fit of the
application into AECL's corporate goals, economic pay-back, and the value
to the company of the expertise being put into the system. Feasability
was evaluated in terms of appropriateness of the application for a knowledge base system, appropriateness of the task for implementation, and
availability of required resources. The applications which were selected
were required to exceed a threshold score in terms of both benefit and
feasibility. The existing company screening procedures and criteria were
used, with minor modifications, in the evaluation process.

SELECTED APPLICATIONS

The following are some of the applications that were evaluated and
passed the screening criteria. They are also those that are currently
being pursued.

1.   Operator Companion

The intent of this system is to advise the operator of equipment and/
or system failures, suggest appropriate corrective action and allow
the operator to interact with the system by providing an explanation
for all advice given. Operator Companion will contain the knowledge
of designers, safety engineers and plant shift supervisors.

A demonstration program for the central module of this system has
been written.

2.   Fuel Defect Detection and Evaluation

Fuel defects, while rare, can lead to increased radiation exposure
for station staff if they are not removed. CANDU reactors are continuously refuelled on power so it is possible to remove defected
fuel bundles without shutting down the reactor. If, however, fuel
bundles are removed when it is not really necessary, the decision
could be costly. An expert system to advise on the location and
condition of fuel defects would aid in plant operations and in the
development of methodology for fuel defect analysis.

The first phase of this system has been developed and is about to
undergo testing. Pending a successful outcome, the system will be
expanded in breadth and depth to operate as a true expert.

3.   Shell and Tube Heat Exchanger Advisor

This system advises designers on the selection of appropriate design
configurations and parameters of shell and tube heat exchangers. This
work is being done for the Heat Transfer and Fluid Flow Service (HTFS)
of which CRNL is the North American sponsor.

This system has been implemented and is about to become a commercial
system.

4.   Automatic Fault Tree Generator

The intent of this system is to use failure conditions associated
with basic components and the upstream and downstream connections of
the components and expert knowledge of the system to generate a

computer resident fault tree which can then be manipulated by the
computer to determine minimal cut sets and merged with other fault
trees.

Work is presently underway to define the scope and technical specification of this program.

COOPERATION WITH OTHER ORGANIZATIONS

To foster the growth of the knowledge base systems technology in the
small Canadian nuclear industry it was abundantly clear that strong cooperation with other industry groups is essential. Exchange of information with
other organizations has been established at two levels: informal and formal.

At the informal level, AECL is a founding member of an AI applications
user group. This group consists of private and publicly owned companies and
government departments. The objectives of the group are to promote expert
system technology and to share experiences in using available tools and
producing useful applications.

At the formal level, AECL is a partner in PRECARN (Pre Competitive
Applied Research Network) Associates Inc., a consortium which promotes the
adoption of knowledge base technology in Canada and coordinates contract
research for its member organizations.

Within the nuclear industry, AECL maintains close cooperation with
nuclear utilities operating CANDU reactors. This contact allows AECL to
identify ways in which the operation of future reactors could be enhanced,
as well as assisting in enhancing the operability of current plants. On a
major project, such as Operator Companion, utility input and participation
will be required. AECL has strong ties with the Canadian nuclear industry
and these will assist in forging the cooperation required for the exploita-
tion of this new technology.

The nuclear utilities have established AI programs of their own and
have produced some interesting applications. Ontario Hydro has produced,
among other systems, an equipment surface coating advisor, a boiler tube
eddy current inspection expert system, and a system for diagnosing faults
in the emergency core coolant injection system. Ontario Hydro and SRI have
also received a contract from EPRI to produce a generator monitoring system
called GEMS.

The New Brunswick Electric Power Commission has identified a large
number of useful applications including the monitoring and display of
regional over power trip margins, a fuelling schedule advisor, and an
abnormal plant operating procedure annunciation and display system.

EXPERIENCE

Obtaining experience in a new field is painful because it requires
skills to navigate in dense fog. To get started in a short period of time,
it was necessary to assess and become familiar with several commercial AI
development tools. These included a number of expert system shells, AI
programming environments and programming languages covering a large range
of price, capability and ease of use. It also requires the assessment of
various types of hardware systems that were being marketed very vigorously.
The projects which have been started so far have been implemented using
expert system shells and Prolog on personal computers. To gain a better
understanding of this technology, some simple expert system shells have

been developed using conventional programming languages.

AECL has made no specific commitment to obtaining hands-on experience
with dedicated AI hardware and "high-end" software. The marketplace for
these products is changing very rapidly and it is the company's belief that
it is unwise to make a commitment at this time. In addition, the software
packages for these machines require a long time to learn to use effectively.
Since a primary objective of the program is to produce as part of the learning process, simple tools were used to start with. Another consideration
is the ability to deliver applications to a user base which relies primarily
on personal computers. The use of "high-end" tools will continue to be
evaluated.

It is encouraging that significant and useful applications were
developed while still learning the technology. This indicates the power
of the tools for prototyping and developing applications. It has also been
observed that program modification and maintenance are significantly
improved compared to conventional programming techniques.

CONCLUSION

In summary, while we are still only scratching the surface of AI
technology, our experience with the application of knowledge base systems
indicates that they will contribute to solving many of the challenges
facing the nuclear industry. AECL firmly believes in the incorporation
of such systems into CANDU power plants (present and future) and that
significant productivity gains can be made by the application of this
technology in the engineering process.

CHAPTER 2

ROBOTICS AND MAINTENANCE

BEYOND HUMAN PERFORMANCE

Floyd E. Gelhaus

Electric Power Research Institute
Palo Alto, U.S.A.

HUMANS AND ROBOTS

Number Five <u>is</u> alive! In fact, Eric Allard and his
talented All Effects crew have several new robots ready for
the filming of a sequel to <u>Short circuit</u>. As I watched the
movie, I was struck by a personal association with that
robot, and have since concluded that its anthropomorphic
form and its near-human role promoted those feelings.
However, this response probably should be considered
typical, since it has been deeply conditioned. Starting
with Capek's 1920s play, "Rossums Universal Robots," we have
continually been taught by science fiction authors that
robots are devices that look like humans and perform human
tasks.[1]

Furthered by Webster's dictionary definition ("2. Any
automatic apparatus or device that performs functions
normally ascribed to human beings, or operates with what
appears to be almost human intelligence"), this human-robot
association has subliminally gained position as the result
of the AI focus that began in the 1950's with a grant
application to the Rockefeller Foundation which was [2] "...to
proceed on the basis of the conjecture that every aspect of
learning or any other feature of intelligence can in
principle be so precisely described that a machine can be
made to simulate it." The machine being referred to was a
computer, the heart of every robot system. Thus, the AI
community of activities further promotes the Capek-Webster
point of view regarding the humanistic characteristics of a
robot. In fact, those promoting "strong" AI believe that
the appropriately programmed computer <u>is</u> mind.

The computing power required for a robot to think like a
human is considerable. As suggested by Sheil,[3] "we cannot
capture all of the knowledge used by a human decision
maker... bulk alone will make acquiring and encoding the
knowledge required to cover... any... subject a major
problem." However in, the nuclear power industry (and

elsewhere) we have several good examples of how narrowly-
focused expert systems can successfully be used as
assistants to human specialists. Although this is not
intelligence, such a program can offer one additional
perception that a human (e.g., a robot operator) can
consider when making a decision.

There is considerable and growing evidence in every
industry that human error has both initiated and promulgated
major accident events. While it may prove true that an AI
expert system would help decrease the probability of
misjudgement because of its collective knowledge base it
could be that, by attempting to emulate human processes,
unnecessary detail clutter is created that can increase the
chance of an error.

The robot's computer must be programmed based on
thinking like a robot, promoting and supporting those
capabilities and attributes that robots, not humans, can
have. When specifying and designing robot hardware and
software, our embedded tendency to "make it humanlike" must
be consciously addressed. It is the purpose of this invited
paper to comment on the possible ways that this mind set can
influence how we establish the specifications for and
ultimately use robots.

ROBOTS AND HUMANS

Robots can be advantageously employed to aid and/or
substitute for humans in work activities that are hazardous
to health. However, in many robot designs, especially in
those touting some level of autonomous behavior, the focus
has been on emulating a human _process_ rather than on
achieving the _results_ that are the goal-purpose of the
system.

Homo sapiens definitely are capable entities. In the
broad spectrum called Nature, however, they take a back seat
to many other members in aspects that are fundamental in
robot design. Speed, dexterity, vision, mobility, and
strength-to-weight ratio are examples of key robot design
parameters where man's capabilities should be examined, but
not automatically emulated. Replacing an effort with a
robot that only performs those tasks can definitely improve
the human's lot, but represents only marginal use of
potential robot capability.

Robot Specifications

Our end user, the electricity generating industry and
specifically EPRI's member utilities, must provide the key
data that determine the specifications for robot design.
The process of modifying existing robot systems for power
plant use has met with very marginal success.

Setting the Goal

The adage that 'well begun is half done' definitely
applies. The focus must be on the results desired from the
mission, with no a priori judgements as to what form the

robotic solution will take. If this goal focus is not
firmly established and maintained, experience would indicate
that arriving at an optimum robot specification is highly
unlikely.

Task Analysis

Familiar to all human factors practitioners, task
analysis is a systematic method for obtaining qualitative
descriptions of the activities that must be performed to
accomplish a task required for a system goal[4]. Since its
development, task analysis has been a primary means for the
design, development, test, and evaluation of _human_
components of complex systems. Task sequence charts and
task data forms are useful tools that help give structure
and completeness to this detailed process.

Here, too, new focus must be applied. Task analysis is
a necessary step in determining the desired robot's
specifications, but its successful application demands that
the practitioner _think like a robot_, not like a human. For
example, to reach a valve twelve feet above the floor, the
human might subscribe to a task like 'ladder selection' or,
perhaps, 'scaffold erection', whereas the robot may choose
'manipulator extension'. The robot thought path would then
lead to subsequent task statements that define end effector
movements such as continuous rotation for 'unbolt flange'.

During this process there is a tendency for the mind to
slide back into the more familiar human-based approach
whenever a task occurs that seems to demand unfamiliar
capability. Thought discipline is obviously required, not
to force every task into the robotic arena, but to relegate
to humans only those activities at which they are best.

Robot Characteristics

Having thought like a robot, a task sequence still
results that contains an intermingling of human and robot
efforts. A breakdown of the robot tasks into subtasks (and
these even further into subtask elements) is now required to
formulate the robot specifications. If this process is
followed for several different missions, another key benefit
will ensue. The matrices of characteristics for the
missions can be compared, revealing similarities that would
support shared modularization rather than redundancy between
robot systems.

When discussing these specifications with the robot
manufacturer, mentally reject the Capet-Webster definition
of a robot and look to all of Nature for your design
details. Be encouraged by Her examples that attest to
performance that is well beyond human capabilities.

ROBOT AND ROBOTS

Adherence to this methodology does not preclude the
decision that human attributes are the best ones to
emulate. Its application will, however, expand the design
process towards robots that go beyond human performance.

Today's expert robots represent a state of methodology compromise[5], since the full optimization of robot application will require some modification of both the system hardware acted upon and the procedures that permit/define how these actions may occur.

How, then, do humans use these extra human machines?[6] Can we follow established procedures to develop AI and/or expert systems? Perhaps not, unless the knowledge engineer is fluent in machine language - in the absolute sense. Our tendency is to apply our human thought and logic so that machines emulate us. This definitely will not work with robots with non-human characteristics and super-human capabilities. Short of robots being able to interview robots, the knowledge engineer will have to transpose human "expert" perceptions into rules based upon the way the robot 'sees' the mission's tasks. A robot with beyond-human capabilities will not be well served if its thinking computer is loaded with what is now referred to as artificial intelligence.

A computer (robot) has a difficult time recognizing a tree, a more difficult time distinguishing between elm and oak. I think that this is an example of what will continue to occur if we try, almost by brute force, to impose on a robot system, the human's way of doing something. We should approach robot usage differently, beginning with fundamental questions such as, "how would a robot define seeing?" To augment its superb mechanical attributes, a new-thinking group of technologists is required to embed intelligence in the computer of this robot that is based on its, not their, view of the world. Meeting these challenges, will create a set of robot tools that complete tasks in a manner that goes beyond human performance.

REFERENCES

1. I. Asimov and K. A. Frenkel, Robots. Machines in Man's Image, Harmony Books, NY (1985).

2. H. Gardner, The Mind's New Science, Basic Books, Inc. New York (1985).

3. B. Sheil, "Thinking About Artificial Intelligence," Harvard Business Review, pp. 91-97, July-August (1987).

4. E. L. Shriver, S. E. Zach, and J. P. Foley, Jr., Section 2, "Task Analysis and Job Performance Aids: Development of a Technology", in EPRI Report NP-2676 (1982), and

    R. W. Pack, et al, Chapter II-A, Section 3, "Analytic Techniques", in EPRI Report NP-4350 (1985).

5. J. S. Byrd, et al, "Expert Robots in Nuclear Plants," Session II-A, This Conference (1987).

6. H. McIlvaine Parsons, "An Overview of Human Factors in
    Robotics," pp. 27-31, <u>Robotics Today</u>, February (1987),
    and

    "Robot Vehicles: A Review", pg. 29, <u>Human Factors in
    Organizational Design and Management-II</u>, O. Brown, Jr.
    and H. W. Hendrick (Editors), Elsevier Science
    Publishers B. V. (North-Holland) (1986).

EXPERT ROBOTS IN NUCLEAR PLANTS

J. S. Byrd, J. J. Fisher, K. R. DeVries, and T. P. Martin*

E. I. du Pont de Nemours and Company
Savannah River Laboratory
Aiken, SC 29808
*Victoria Plant, Victoria, TX 77901

ABSTRACT

Expert robots will enhance safety and operations in nuclear
plants. E. I. du Pont de Nemours and Company, Savannah River
Laboratory, is developing expert mobile robots for deployment in
nuclear applications at the Savannah River Plant. Knowledge-based
expert systems are being evaluated to simplify operator control, to
assist in navigation and manipulation functions, and to analyze
sensory information. Development work using two research vehicles
is underway to demonstrate semiautonomous, intelligent, expert
robot system operation in process areas. A description of the
mechanical equipment, control systems, and operating modes is
presented, including the integration of onboard sensors. A control
hierarchy that uses modest computational methods is being used to
allow mobile robots to autonomously navigate and perform tasks in
known environments without the need for large computer systems.

INTRODUCTION

E. I. du Pont de Nemours and Company, Savannah River Laboratory
(SRL), is developing expert robots for deployment in nuclear applications
at the Savannah River Plant (SRP). A control hierarchy is being developed
to allow robots to autonomously navigate and perform simple tasks in known
environments. Knowledge-based expert systems are being evaluated to
assist in navigation and manipulation functions, to analyze sensory information, and to simplify operator control. Development work using two
research vehicles is underway to demonstrate semiautonomous operations in
process areas. A description of the mechanical equipment, control
systems, and operating modes of these vehicles is presented in the paper.

Nuclear process operations provide many incentives for applying
remote mobile robot technologies. Human operations in certain areas are
restricted to short durations due to radiation levels. In addition,
operating personnel may be required to wear nuclear contamination protective clothing, which limits their productivity. Traditionally, inspection
and monitoring tasks in inaccessible areas have been difficult to perform,
since operating personnel have only had access to fixed mounted cameras,
viewing ports, or periscopes. Remote maintenance tasks can require costly

process shutdowns before human intervention is allowed. Mobile vehicles and robots provide solutions to many of these problems.

We are developing and employing a variety of robots at SRL and SRP, including application areas such as nuclear waste bagout,[1] laboratory automation,[2] shielded cells automation,[3] and telerobotics.[4] Applications in nuclear waste cleanup and disassembly of contaminated equipment have employed nonprogrammable, teleoperated mobile vehicles that have limited capabilities but are quite useful. Since people are required to guide the vehicles to the worksites, sophisticated camera systems must be used to locate obstacles (i.e., narrow passages and overhead pipes) that might impede motion. When wall-mounted process cameras or windows are not present, human operators have difficulty keeping track of the vehicle position and orientation in the work area. Another drawback is that important sensory and diagnostic information relating to onboard equipment and sensors is presented to the system operator as raw data that must be interpreted and acted upon, thereby increasing the control burden. Autonomous and semiautonomous vehicle research that addresses some of their drawbacks is underway.

DEVELOPMENTAL SYSTEMS

In order to develop semiautonomous systems we have two mobile telerobots and a variety of computers and software. The telerobots vary greatly in their characteristics and performance, allowing us to examine and compare the advantages and disadvantages of each system for different applications. One vehicle is a three-wheeled unit from Cybermation, Inc., and the other is a six-legged walking robot from Odetics, Inc. A primary goal in developing the higher level controls is to implement flexible, general control techniques that are not highly hardware-dependent. Software tools for navigation, mapping, and operator interaction must be adaptable to different expert robot systems. Therefore, development on both of our mobile robots is tightly coupled to ensure that flexibility is maintained. Much of our efforts has been integrating the individual systems together and developing the interfaces that allow the higher levels of control to treat both vehicles nearly identically.

Cybermation K2A -- ALVIN

ALVIN (Autonomous Lab Vehicle with Intelligent Navigation) (Fig. 1) consists of a Cybermation (Roanoke, VA) K2A mobile platform, ultrasonic range sensors, a docking transponder, onboard computers with data acquisition, and a Cybermation host computer. ALVIN is a three-wheeled (2° of freedom) telerobotic system. All three wheels are driven and turned in synchronization at the same speed without slipping or skidding. This gives the vehicle excellent dead reckoning, typically less than 0.5% error over the length of the path on most floor surfaces, even during zero-radius turns. The vehicle is approximately 33 in. wide, 42 in. tall, and weighs 300 lb. It can carry a continuous payload of 250 lb at its maximum speed of 2.5 ft/sec. The vehicle's onboard computers, without offboard assistance, are programmable to follow a path and avoid obstacles using the integrated sonar system. Also onboard is an infrared transponder that is used to dock the robot and reorient it relative to a fixed docking station. The transponder is also used to transfer data between the robot and host computer. When the robot is away from the docking station, communication is over an RF radio link. SRL added a camera system with a microwave link back to the control station, an onboard computer, a data acquisition system, radiation monitors, and a speech synthesizer. The Cybermation host computer provides menu-driven graphic displays of the status of the robot.

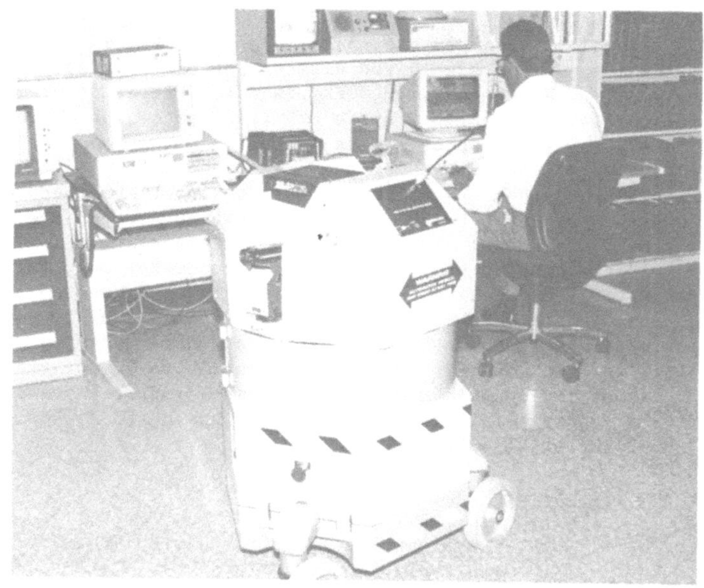

FIG. 1. ALVIN negotiating a navigational plan in the
SRL Robotics Laboratory

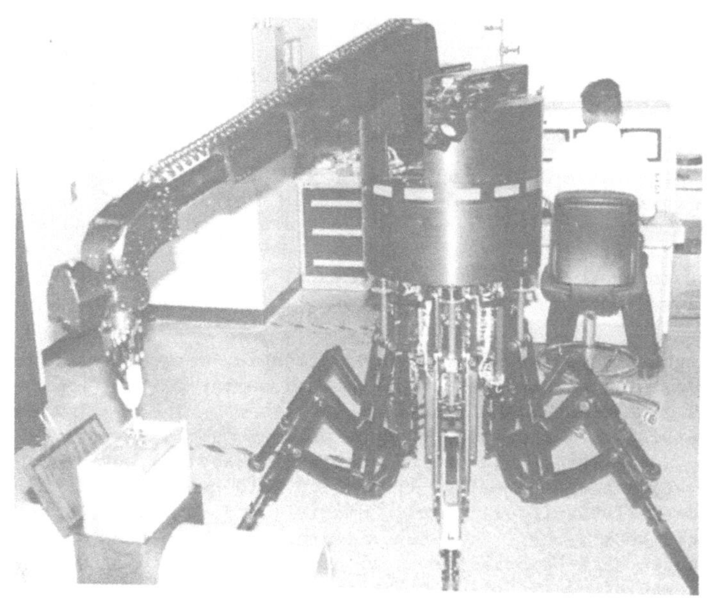

FIG. 2. ROBIN performing a simulated filter change in the
SRL Robotics Laboratory

<u>**Odetics Odex -- ROBIN**</u>

ROBIN (<u>Ro</u>botic <u>In</u>sect) is the third model of the ODEX series
(Odetics, Anaheim, CA) of six-legged walking robots (Fig. 2). SRL- specified enhancements over the earlier ODEXs include a telescoping manipulator
arm, a tension-controlled, fiber-optic umbilical cable, an onboard data
acquisition system, and a three-camera TV system. The robot has an
onboard data acquisition module for additional equipment and sensors. The
arm can lift 30 lb at the end of its 7-ft extension. The robot can climb
an incline with a slope up to 30°, climb stairs, or step onto a platform
or over a pipe 18 in. high. These operations are performed semiautono-
mously by multiple onboard computers, with the operator only entering
initial parameters, overseeing the operation, giving directions, and
commanding the operation to continue. The operator's console contains
three joysticks, a keypad for menu selection and data entry, a status and
menu monitor, and two monitors that are switchable to any of the onboard
cameras. Most of the console operations (control and monitor) can be
performed by a supervisory computer (currently an IBM PC/AT) over a
standard RS-232C serial interface.

<u>**Computers and Peripherals**</u>

Our primary computer systems are IBM PC/AT's with Vectrix
(Greensboro, NC) VX/PC high resolution color graphics systems, Stargate
Technologies (Eastlake, OH) eight-port serial processors, and Digitronics
(Clifton Park, NY) Sixnet distributed I/O systems. The Sixnet systems are
used to read sensors and control devices that are fixed in the work area.
We are using Turbo Pascal (Borland International, Scotts Valley, CA) to
code the realtime I/O routines and to do other "conventional programming
language" work. Turbo Pascal is easy to use and can easily interface to
assembly language routines. We also have an AI VAXstation II/GPX (Digital
Equipment Corp., Maynard, MA). This is an AI workstation with high
resolution color graphics, well suited for the higher level control tasks,
which require large amounts of memory and computation.

<u>**Insight Expert System**</u>

The ability to rapidly produce small-scale, working prototype expert
systems for ROBIN, with a minimal initial investment in designated hardware, makes the IBM PC/AT a very appealing, first-try development tool.
These initial prototypes are being written using the expert system shell
Insight 2+, Version D (Insight) (Level Five Research, Indialantic, FL), an
enhanced version of the popular Insight 2+, Version 1.2. The enhancements, written specifically for Du Pont, deal with direct database access,
graphics displays, and "help" features. One particular addition that is
important to our work is the LOOP command, which allows a rule to repeti-
tively "fire" until a condition fails.

Insight offers a relatively user-friendly, menu-driven development
environment with a straightforward, though semantically sensitive, method
of entering knowledge in the form of heuristics. The heuristics are
incorporated into a knowledge base by writing fairly structured
IF...AND...OR...THEN...ELSE rules, as required by Insight's knowledge
representation language called Production Rule Language (PRL). PRL can
handle simple facts (true/false), numeric facts, string facts (text), and
object-attribute facts. Confidence factors may be placed on any rule in
the knowledge base by both the knowledge engineer and the user, and confidence thresholds may be used to set the minimum confidence (0-100) at
which a particular rule will fire. Once compiled, the rules are manipulated by a primarily backward-chaining, goal-driven inference engine. This

fits our needs due to a large number of conditions that lead to a relatively small number of ultimate goals. Within Insight, knowledge bases may be chained with as much parameter passing as desired. This feature allows for a hierarchical breakdown of the knowledge into many small manageable segments, which may be tested independently before being integrated by a master knowledge base.

As is the case with expert shells in general, Insight is designed for static applications. It does, however, offer three features that allow us to simulate a quasi-realtime advisor. First, it has the ability to call external routines--Pascal routines in particular. It can access DBPascal, DBASEII, AND DBASEIII directly through PRL commands. The majority of our calls goes to programs written in Turbo Pascal. The second feature is the LOOP function, mentioned previously. This allows for the continuous re-execution of a rule until a condition fails. Going one step further brings out the final key feature: the CYCLE command. With it, we maintain knowledge-base continuum, rather than the traditional termination of the session once a goal has been reached. This is made possible by the ability to dynamically choose which conditions and rules to forget, and under which circumstances it is desired to forget them.

Gensym G2 Realtime Expert System

We have recently procured a software license from Gensym Corp. (Cambridge, MA) to use a new software product, the G2 Realtime Expert System. Operating on the DEC AI VAXstation II/GPX, we will use G2 to develop realtime expert systems for our mobile robots, ALVIN and ROBIN. An expert system to be used in control of active robots needs to have not only the complex knowledge of the robot's surrounding environment, assigned tasks, operational history, and constraints, but also the ability to diagnose alarm conditions, monitor and quickly respond to environmental conditions, monitor sensory systems, and execute operations in a true realtime sense. G2 provides the software tools for developing these systems.

The Gensym G2 Realtime Expert System accomplishes intelligent advising, diagnosing, and controlling in situations in which there are variables that change in time. Like most expert systems, G2 has an inference engine that is a fixed component of the system and a knowledge base that is separate and created by a user to represent the expertise of a particular domain of knowledge and a particular application; but it goes beyond other expert systems in several areas. It is a complete system designed for the professional who is typically neither a programmer nor a knowledge engineer. It does not require LISP programming and allows for easy operation by the organization of knowledge bases using schematics, tables, and structured natural language statements, as well as interactive editing facilities.

G2 is an interrupt-driven, realtime system using time-dependent variables that can be deployed as an online monitor and controller of events that are measured by sensors, rather than primarily as a dialog system. It requests data selectively according to its goals, monitoring different sensors at different rates, and can focus on areas of interest to its various goals, examining some sensors only when there is an indication of an unusual condition. It can initiate actions ranging from issuing warning messages to direct online control. The system is organized around the concept of realtime variables whose values must be kept current. Time- based properties of variables, such as rate of change and variance, are well developed. Realtime procedures give G2 a unique capability to concurrently execute many different plans and strategies over extended periods of time, giving an attention span beyond the present moment.

Knowledge engineering development and maintenance can be performed in
the field, either while G2 is connected to a real process or with a simulated process. Knowledge editing facilities can be used online while
causing only short lapses of attention to online activities. Simulation
is an integral part of the system. It is a way of representing deep
knowledge about the domain, including prediction of consequences and
estimation of variables that cannot be measured, as well as testing normal
operating conditions.

EXPERT ROBOTS AND CONTROLLERS

Our definition of an expert robot is an intelligent machine controlled
by computers running knowledge-based expert systems. Obviously, machine
intelligence must be a relative measurement, since present machines have
very low intelligence if defined as, or compared to, human intelligence.

A totally autonomous robot system operates in an unknown, unstructured environment of very large and complex work arenas and task domains,
using cognitive planning and decision-making during critical situations.
Self-learning algorithms are coupled with fast response knowledge bases
and realtime sensory systems. A semiautonomous robot system on a specific
mission uses cognitive planning and decision-making, but focuses on a
narrow problem. It operates in a small neighborhood and a semistructured
environment (i.e., an occasional cluttering of the norm). Autonomous
supervisory action to perform a task (redefining its location, monitoring
the environment, manipulating an object, etc.) is provided by a host
computer that obtains assistance from a human companion when necessary.
Realtime sensing and reflexive actions are used to perform the given task
autonomously.

Our philosophy is that deploying intelligent mobile robots in nuclear
environments is most practically performed using a task semiautonomous
approach. In contrast to the fully autonomous operations, a human
controller remains in the loop to plan complex tasks and evaluate preplex-
ing, system-generated conclusions. The robot autonomously executes many
of the detailed operations, including navigation planning, obstacle avoidance, and data collection/interpretation. Obviously, provisions are made
for the system to interact with the human controller at all levels to
respond to out-of-bound problems or interpret unpredictable situations.

Real intelligence and artificial intelligence are realized at all
levels of the control hierarchy. At the highest level, a cognitive expert
system will assist the human in diagnosing tasks, recalling past historical events, and initiating tasks at the lower levels. At the middle
level, expert systems will interpret patterns of diagnostic responses and
sensory data to match known conditions in a process or problems with
equipment on the robot. Navigational expert systems will assist in route
selection and vehicle orientation for obtaining required sensory data.
Expert systems will also assist in the operation of manipulators. At the
lowest level, for example in legged mobility, an expert system will deduce
optimal foot placement to surmount obstacles and uneven terrain.

ALVIN's Navigational Control

Mobile robots require area maps for both route planning and for
navigational purposes. Area maps can be created dynamically by the robot
from sensory information or be created by a human operator for the robot
to retrieve as necessary. Dynamic mapping in nuclear process areas is
usually impractical due to congested piping, irregularly shaped equipment,

and grated floors, which easily confound navigation systems based on ultrasonic and vision systems. Using predetermined maps provides a very satisfactory alternative to dynamic mapping since nuclear work environments are highly structured and rarely changed. Modifications to stored maps, due to blocked corridors or obstructions along a path, can be easily accommodated by declaring certain areas or pathways on the map out-of-bounds.

In our control system, skeletal maps are generated with a data type called a "Location," which contains the name of the Location, its X and Y coordinates, and its adjacent neighboring Locations. Data representing the type of floor surface (necessary to calculate dead reckoning errors) and the pathwidth to neighbor Locations are also included. Since all the Locations are referenced relative to each other, locating paths between two points is performed using linked lists and tree search techniques. A route selection algorithm, written in Pascal, locates all the possible paths on an area map to a desired location, then evaluates the paths based on floor surfaces, distance, and speed to locate the optimum route of travel. This algorithm is normally run on an offboard computer, and the computed path trajectory is sent to the robot over a radio or infrared communications link. If the robot is unable to complete its commanded route due to a blocked passageway, the algorithm can be run onboard the robot to locate alternative routes. This redundant capability is also used if the robot looses communications with its offboard host, since it can still carry out its last command and return to its docking station.

Position referencing and calibration are necessary to avoid large cumulative odometric errors that would cause the robot to miss a desired target location or wander off its programmed path. ALVIN uses a docking transponder manufactured by Cybermation to reorient itself to a known location in its work area. Other position recalibration techniques using optical position sensors (Op-Eye Corp.) and scanning lasers (LaserNet Corp.) have also been successfully lab-tested. Before ALVIN executes a path, the navigation algorithm computes the theoretical position error (based on the path distance and floor type) and compares it with a predeclared value to determine if reorientation is necessary. If so, its path is automatically altered to include a docking maneuver.

In order for ALVIN to perform high level functions, a programming environment was developed to allow complex task operations to be defined in terms of subtasks and logical expressions. The result is a hierarchical control system where the operator can interact with the system on a task level with commands like, "perform radiation survey of valve corridor." This expression is decomposed into subtasks that the robot can implement such as planning paths, executing path segments, and performing radiation measuring functions. The details of the subtasks (i.e., if docking were necessary or if obstacle avoidance were required) is kept transparent to the operator unless the required task cannot be performed.

Software is currently being developed on the VAXstation to enable ALVIN to autonomously interact in a simulated process application. A data acquisition system, linked via DECnet to the VAXstation, will control and monitor all the process variables such as flow rates, humidity, pressure, and the radiation field. In case of a simulated emergency condition, the VAXstation will automatically deploy ALVIN to take corrective action or to take further process data. The expert system environment offered by G2, once employed, will offer excellent control flexibility since the robot can be dispatched within IF-THEN-ELSE rules. Integrating "ALVIN-like" vehicles into computer-controlled processes could significantly reduce the expense of fixed mounted instrumentation and could improve the emergency response time in certain conditions.

We are employing Insight as the tool for ROBIN's expert controller by building a master-slave knowledge base chaining structure. The main thrust of the design has been to keep as much of the decision-making, operational sequencing, and inquiry about the situation at hand as transparent to the operator as possible. Typical tasks fall mainly in the category of routine maintenance that integrates both mobility intelligence as well as tedious manipulator maneuverability at close range. Our ultimate goal is to allow the operator to choose a desired task to be performed and be interrupted only if an unusual or dangerous situation is encountered by the robot. In the event of such an interruption, the operator will normally be given the option of telling ROBIN to continue or switching to joystick control to remedy the circumstance. Sensors both on the robot and in the workspace are looked to as the primary means of perception. When their data are inadequate to confidently justify a positive response, we have what was previously referred to as an unusual or dangerous situation.

The master knowledge base contains the highest level of choices from the operator as well as the most comprehensive task descriptions to the robot. A rule on this level might read the equivalent of:

```
RULE for changing locations
IF ASK the desired destination
AND ROBIN goes to the destination
THEN ROBIN has arrived at the destination
```

if translated to a human. Rather than facts in words, however, a condition would actually be a chain to a lower level knowledge base or an activated external program, dedicated to a more specific task. For instance,

```
AND ROBIN goes to the destination
```

might actually be:

```
AND CHAIN movement
```

where "movement" is the title of another knowledge base and the desired destination is passed to "movement" as a shared string fact. This lower level knowledge base has less complex conditions. A rule at this level might read the equivalent of:

```
RULE for determining path
IF where is ROBIN presently
AND where is the desired destination
AND determine the best path
THEN begin to move there
```

again, if translated in English. Of course, this set of conditions would actually be calling on the sensors via an external Turbo Pascal routine to determine the present position; using the shared string fact from the master stating the destination; and calling the same optimum path determination routine as used for ALVIN, also written in Turbo Pascal, before going on to a movement rule of which "begin to move there" in the initial condition. For example,

```
AND determine the best path
```

would actually read:

```
        .
        .

        .
    AND ACTIVATE C:/krd/path.com
    DISK C:/krd/pathdata.txt
    SEND present location
    SEND desired destination
    RETURN path string
    RETURN closest known point to present location
    RETURN closest known point to destination
    THEN begin to move there.
```

This hierarchical breakdown continues until the movement routine commands are at their lowest, most generic level.

The most difficult portion of this problem, once the limitations imposed by Insight are realized, is the reduction of all predictable situations to a generic set of movement and decision modules. It is said that the limitations of any technology can be traced to the scope of its designer's foresight. Our intentions for this system are to give it the capability to react properly under a fairly structured, yet not entirely predictable, set of environmental and positional circumstances.

The key to this design is its modularity. Tasks can be added and new knowledge bases chained with no need to change any of the existing rules. This can be done at very little additional cost to an already inexpensive system.

Depending upon the success of the above completed system, our next step will be the addition of a second IBM PC/AT into the loop as the true executor and monitor of the sequences desired by the Insight routines running on the host AT. This second machine would keep a log of all instructions sent to it by the host and run them in the order in which they were received. This serves the purpose of "freeing" the master Insight expert system for use as an advisor, which could then be used simultaneously with the controller. As it stands presently, Insight is inaccessible while any routine is being run unless input is requested of the operator by the system itself. The second machine will also display a graphical picture of both ROBIN's position in the workspace as well as the position of the arm and turret.

FUTURE SYSTEMS

Future robots will apply more sophisticated hardware and AI in knowledge-based systems at each control level. These more intelligent systems will evolve from independent efforts that are presently being made in this research endeavor and from knowledge bases that result from the present applications in industry. Many production problems in nuclear and other industrial plants can be solved with the effective employment of truly intelligent robots. However, today's robots are relatively unintelligent, and more intelligent systems will require much laboratory research and development before they are turned loose in the real work environment without close supervision.

The robot evolutionary process begins with the systems dedicated to specific tasks, as we are developing in our present work. As knowledge is acquired during the performance of these tasks, the task domain will gradually expand. It will be many years before independent mobile robots will approach the intelligence and autonomy of a simple insect. However, in many instances this may not be necessary or even desirable. A network of generic mobile telerobots, multipurpose modular arms, and toolboxes of special tools (or end-effectors), connected to a hierarchical computer system made up of a large intelligent data-base and expert knowledge modules, appears to be an effective automation approach. That system would be capable of recognizing a process problem, taking direct action through a standard process control network, defining and planning the task, and employing the appropriate robot that would be on standby at robot workbenches located at strategic locations in the plant. When the robot was called on to work, instructions and knowledge about the task (cause, location, previous history, possible fixes, etc.) would be provided in an expert knowledge module. The robot would be instantly trained as an expert to perform the task. It would maintain good communications with its supervisor (the main computer) and would acquire additional knowledge at the jobsite through its sensory system. If the robot's capability and cognition capacity were exceeded, a human could be called in to help. Relating all this to the human world, it would be like having the best, most highly trained person on each task standing by to do the job on demand. Generic robots could be interchanged between jobs and areas without any effect on job performance. Our research and development programs will evolve toward these objectives that lead to the totally autonomous, cognitive, expert robot system.

ACKNOWLEDGMENT

The authors acknowledge the work of E. E. Gillenwater for the opertion and tooling for ROBIN and physical body modifications for ALVIN.

The information contained in this paper was developed during the course of work under Contract No. DE-AC09-76SR00001 with the U.S. Department of Energy.

REFERENCES

1. J. J. Fisher, "Shielded Cells Transfer Automation," _ANS National Topical Meeting on Robotics and Remote Handling_, Gatlinburg, TN (April 1984).

2. G. M. Dyches and S. D. Burkett, "Laboratory Robotics Systems at the SRL," _Eastern Analytical Symposium_, New York, NY (November 1983).

3. W. I. Lewis, and R. M. Taylor, "A Flexible Computer-Integrated Robotic Transfer System," _Robots 11_, Chicago, IL (April 1987).

4. T. P. Martin, J. S. Byrd, and J. J. Fisher, "Mobile Robots in Research and Development Programs at the Savannah River Site," _ANS International Topical Meeting on Remote Handling and Robotics_, Pasco, WA (March 1987).

5. A. Wolf, "An Easier Way to Build a Realtime Expert System," _Electronics_, pp. 71-73, (March 1987).

RELIABILITY ANALYSIS OF THE INSPECTION BY A MAN AND A ROBOT

FOR THE MAINTENENCE OF A NUCLEAR POWER PLANT

Yoshiaki Oka, Hideki Araki and Yasumasa Togo

Department of Nuclear Engineering
   University of Tokyo
Hongo, Bunkyo-ku, Tokyo 113, Japan

INTRODUCTION

The tour inspection or the walk-arround inspection is very important for safe and reliable operation of a nuclear power plant. The inspection has been mainly carried out by a man. The reliability analysis of the inspectors has, however, not been well studied compared with that of the reactor operators. The inspection has relatively passive nature in contrast with the operation. Also the probability of finding anything wrong is very low due to the low failure rate of the inspected apparatus.

In the present paper, the reliability of a tour inspector is analysed using probabilistic network models. Robots are being developed for the inspection. The reliability of the inspection by a robot, a man and their combination is also analysed.

THE TOUR INSPECTION

The inspection consists of a scheduled tour of a specified area in the nuclear power plant. The inspector reports anything unusual or any deviant condition of apparatus. There are several inspection items for each apparatus such as a pump, a control panel, a motor-generator set etc. For example, as for a pump, the inspector checks the abnormal sound and smell, leakage from the ground-seal, leakage from valves and pipes, and the position of valves. Generally the inspection items are summarized as follows :

(1) valve, switch and lamp
(2) monitor display, liquid level
(3) sound, smell and vibration
(4) leakage and spill

The inspection is entirely visual. Any deviation is fairly obvious for the inspector. He generally knows the plant well, and can recognize deviant conditions if he notices them. Furthermore the normal state of the component, such as valve position is usually indicated at the place where it is installed. A patrol checklist is usually carried by the in-

spector. It indicates the order of the inspection. The frequency of the
inspection depends on the importance of the apparatus. Important apparatus are inspected three times a day, while others are once a day.

MODEL

     Probabilistic network models are proposed for the reliability
analysis of the inspection. The basic model is the one-inspector model,
which is shown in Fig.1. The nodes which have two output branches
express the probabilistic nodes. The model describes omission error,
commission error and recovery of the error by confirmation. The component is assumed to be in failure state or under deviant conditions
when it is inspected. The probability of omission and commission errors
is the basic input data of the model. The values are shown in Table I.
The probabilities were determind by referring to the Swain's
handbook(1980). Some data could not be found in the handbook. They were
determined by comparing with the other data: The abnormal sound, smell
and vibration are easily detected and the omission probability is low.
The leakage is rather defficult to find because the location is not always identical. Then the high omission probability is assigned. The commission error probability for sound, smell and vibration is high, because they require subjective recognition.

     The success probability of the confirmation process depends on the
probability of the commission error. The conditional probability was
calculated by using the dependence equations of the Swain's
handbook(1980). Medium dependence (MD) was assumed. The success and
failure probabilities of the confirmation process for 5 % commission error rate are shown in Table II.

     The probability of initiating the confirmation is also required as
the input data. The confirmation is not always carried out. When the
inspector succeeds to detect the deviant condition, the confirmation is
done almost all times. When the inspector fail to detect, the confirmation is omitted in many cases. The probabilities of 95 % and 10 % are
assigned to the initiation of the confirmation for the two cases respectively.

RESULT

<u>One-inspector model</u>

     The failure probabilities of detecting a deviation are depicted in
Fig.2. The failure probabilities are rather high due to the passive nature of inspection coupled with the inspector's low expectation of finding anything wrong. Since the failure probability of the inspected component itself is generally below 0.1 %, the failure probability of
detecting a deviation for one inspection becomes below 0.01 %. The ratio
of the omission and the commission errors in Fig.2 for each inspected
item is understood from the ratio of the input data of Table I.

     The results of Fig.2 have been derived under the following condi-
tions :

                         a. good visual accessibility
                         b. optimum stress level
                         c. use of a checklist
                         d. once a day inspection
                         e. one inspector

The effect of these parameters was studied.

Table I. Commission and omission error probabilities

	valve, switch,	monitor display	water level	sound	smell	vibration	leakage
commission error (%)	5	7	4	10	10	10	1
omission error (%)	5	5	10	3	1	1	15

Table II. Probability of the recovery after the commission error

	Commission	
	success	error
Recovery success	95.7 %	71.4 %
error	4.3 %	18.6 %

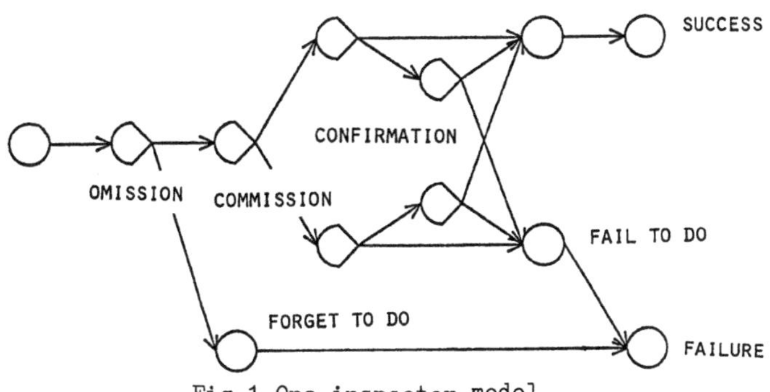

Fig.1 One-inspector model

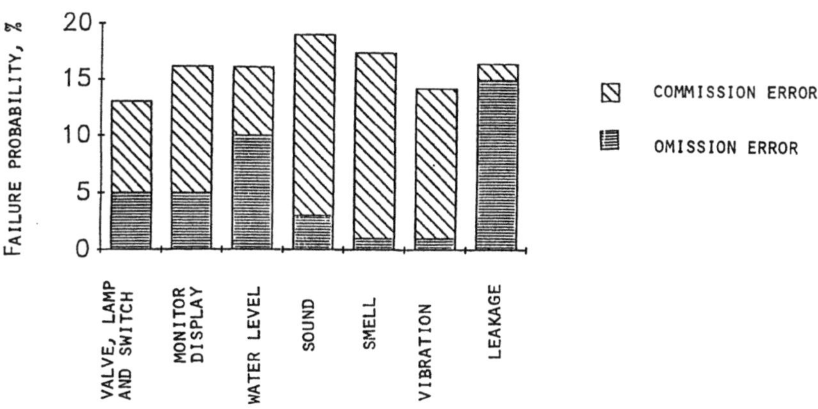

Fig.2 Failure probability of the inspection (one-inspector model)

Visual accessibility . The visual accessibility changes the performance. The inspection of a valve was tried to be evaluated as an example. The visual accessibility changes with the valve location and the position indicator of the valve. If the location of the valve is not clearly seen, the omission error increases. If the position indicator of the valve is difficult to be seen, the commission error may increase. The omission and the commission probabilities were determined by referring to the Swain's handbook. The result is shown in Fig.3.

Stress level . The performance effectiveness becomes maximum at the optimum stress level. Both low and high stress decrease the effectiveness. The tour of inspection is very monotonous work. The stress level is lower than the optimum. This increases the failure probability. As an example, the result of analysis is shown in Fig.4 for the inspection of sound, leakage and valve.

Use of a patrol checklist . A checklist is used for a tour inspection. It indicates the sequence of inspections. The proper way to use a checklist is to read each item in turn, inspect the item and check-off that item on the checklist. With increased familiarity, people take shortcuts : an inspector check several items all at once on the checklist. The incorrect procedure increases the probability of errors. The probability of improper use of the checklist is assumed to be about 50 % (Swain and Guttman, 1980). The result for the case is depicted in Fig.5.

Frequency of inspection . The frequency of the inspection depends on the importance and failure rate of the apparatus. Once a day-inspection is carried out normally, but important apparatus are inspected three times a day, every 8 hours. The increase in the frequency may increase the dependence on the previous inspection. The failure probability increases with the degree of dependence, but decreases with the number of inspection. The effect of the dependence on the failure probability was analysed for the three-times-a day-inspection. The result is compared with that of once-a day inspection in Fig.6. The medium-dependence (MD) case of the three-times-a day-inspection still shows higher probability after 48 hours than the zero-dependence (ZD) case of the once-a day-inspection. The decrease of the dependence is important for increasing the reliability of inspection.

Two-inspectors, a supervisor model

When the inspection is performed by two-persons, the skillful inspector becomes a supervisor. The supervisor checks the result of the first inspection. The supervisor model shown in Fig.7 was proposed and used for the analysis. Recovery by the supervisor is considered in the model. The recovery changes with the degree of his dependence to the first inspector. The dependence equation of the Swain's handbook was used for calculating the probability. The failure probability is compared with that of the one-man-inspection in Fig.8. High dependence (HD) of the supervisor was assumed and the results of the two-inspector model gives rather conservative value. The failure probability changed with the dependence. The result is depicted in Fig.9 for the inspection of valve and monitor display.

INSPECTION BY A MAN AND A ROBOT

Robots are being developed for reducing the occupational exposure and man-power for the inspection. There are two types of robots.

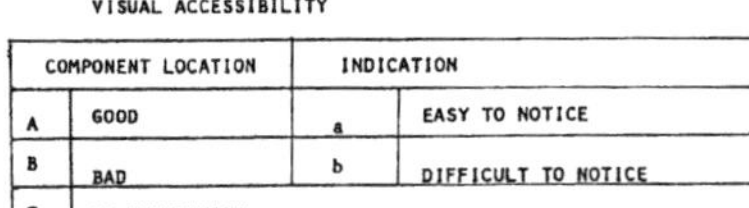
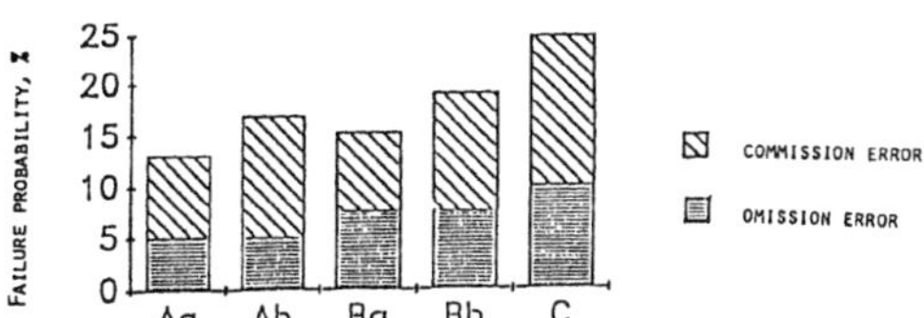

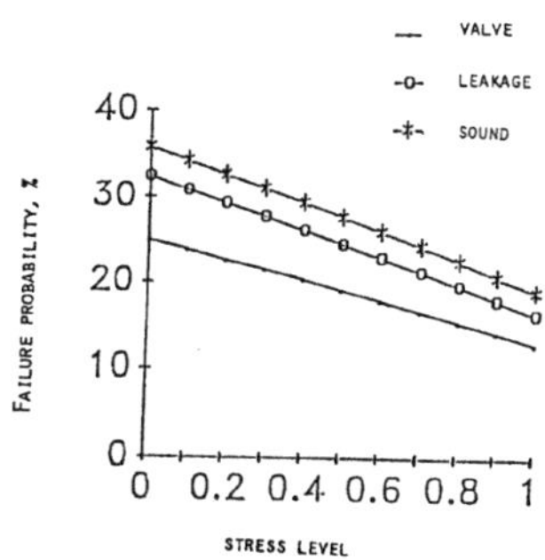

Fig.3 Effect of visual accessibility

Fig.4 Effect of stress level

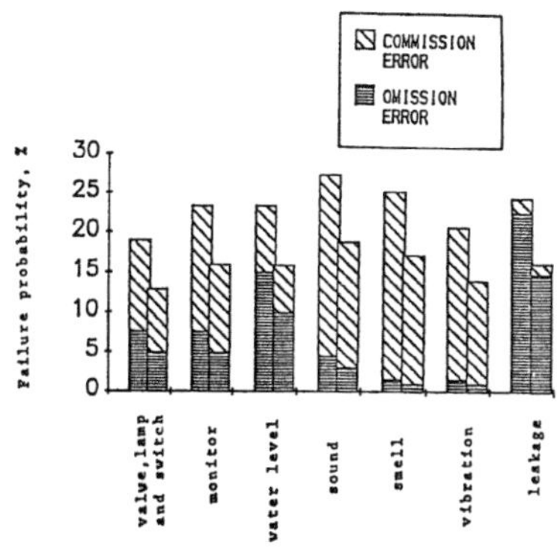

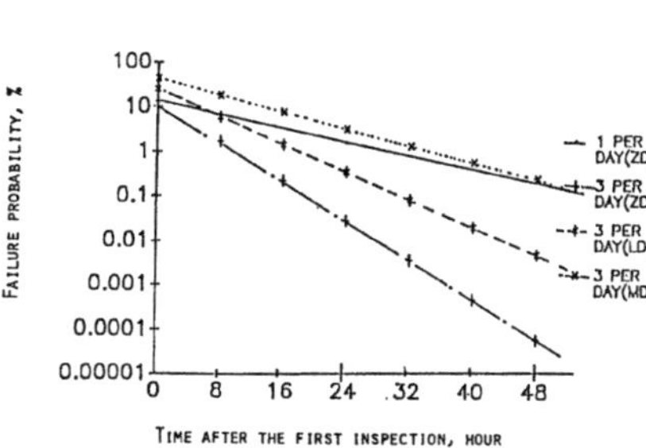

Fig.5 Effect of improper use of a checklist (left:50% improper use,
right:0% improper use)

Fig.6 Comparison between three-
times-a day-inspection and
once-a day-inspection

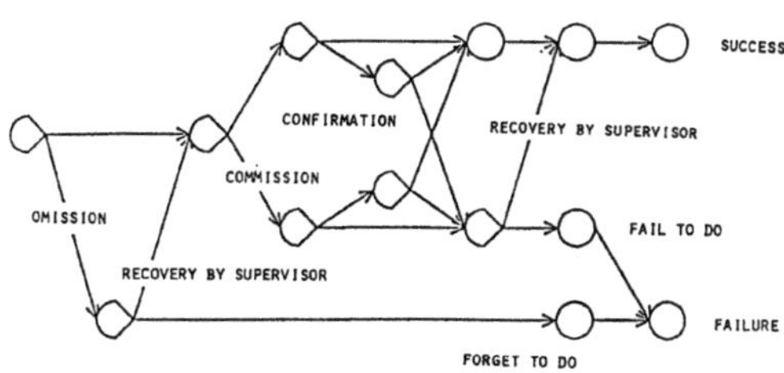

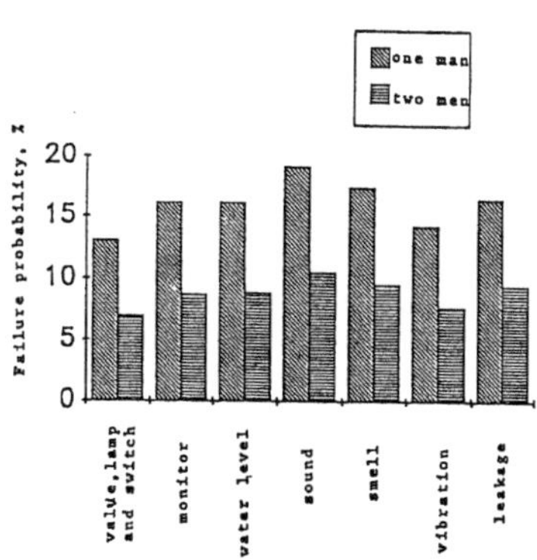

Fig.7 Supervisor-model
(two-inspectors model)

Fig.8 Comparison of one-man and
two-men inspection

(1) remote control robot
    The robot sends the information to a man. The decision is made by
    a man. The reliability of the inspection contains human factor
    problems.
(2) self-recognition robot
    The robot recognizes the deviant condition by itself. The inspec-
    tion will be free from the human factor problems.

The latter type of robot is considered in the present study.

Generally, the reliability of a robot is high than that of a man.
It has no dependence to the previous inspection. Recovery of the error,
however, seems to be very difficult for the robot. Once the robot fails
to detect the deviant condition, it will fail again next time if the
circumstances do not change. This will be the weak point of the inspection by the robot.

Failure probability of the inspection by a man, a robot and their
combination is calculated under the following assumptions.

(1) The robot and the man have identical visual capability of recogniz-
    ing deviant condition.
(2) The robot cannot recover 80 % of the errors of the previous in-
    spection. The rest, 20 % can be recovered.
(3) The inspection frequency is three times a day, every 8 hours.
(4) The error probabilisties of the robot for recognizing deviant con-
    dition are the parameter for the analysis. They are 1 %, 0.1 %
    and 0.01 %.
(5) The dependence of a man on the previous inspection by the robot
    changes with the error probability of the robot, zero-dependence
    (ZD) for 1 % error, low-dependence (LD) for 0.1 % and medium de-
    pendence (MD) for 0.01 %.
(6) The dependence of a man on the previous inspection by a man is as-
    sumed to be the LD (low depenence). The robot does not have the
    dependence.

The calculated failure probabilities of the inspection are shown in
Fig.10 and Fig.11 for 1 % and 0.01 % error probability of the robot
respectively. The following combinations of a robot and a man are chosen
as the cases of the analysis.

case 1 : man only, Every inspection (three times a day) is done by a
         man.
case 2 : man-man-robot, The inspection is carried out three times a day
         sequentially by a man, a man and a robot.
case 3 : man-robot, Alternate inspection by a man and a robot.
case 4 : man-robot-robot, The inspection is carried out three times a
         day sequentially by a man, a robot, a robot.
case 5 : robot only, Every inspection is done by a robot.

The result shows that the combination of a man and a robot gives
the lowest failure probability after a few times of inspections even for
the case of 0.01 % error probability of a robot. The inspection of a man
only shows high probability at first, but it gradually decreases due to
the recovery of a man. The inspection by a robot only shows low probability at first, but it remains high after several times of inspections. it is attributed to the small recovery rate of a robot. In order
to increase the recovery factor of the robot, two types of completely
unsimilar(independent) robots should be used for the inspection.

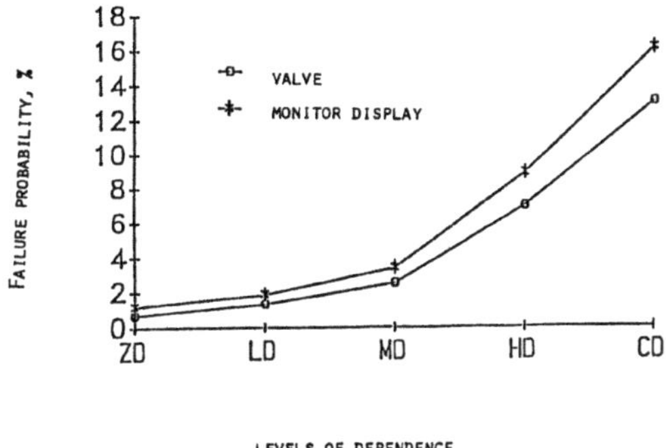

Fig.9 Change of the failure probability of the inspection with the level
of dependence of the supervisor on the inspector

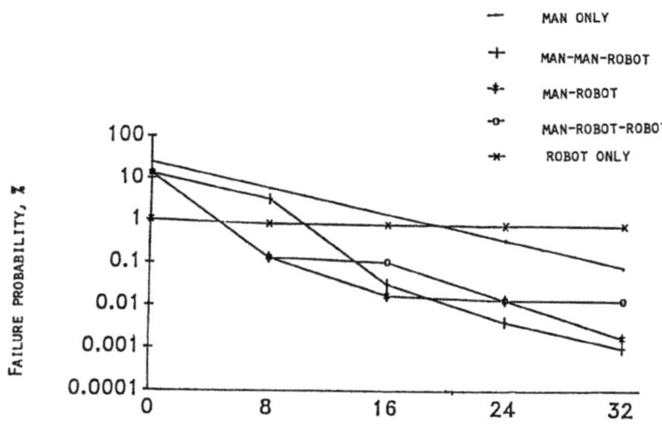

Fig.10 Failure probabilities of the inspection by a man only, a robot
only and their combination (1% error probability of a robot)

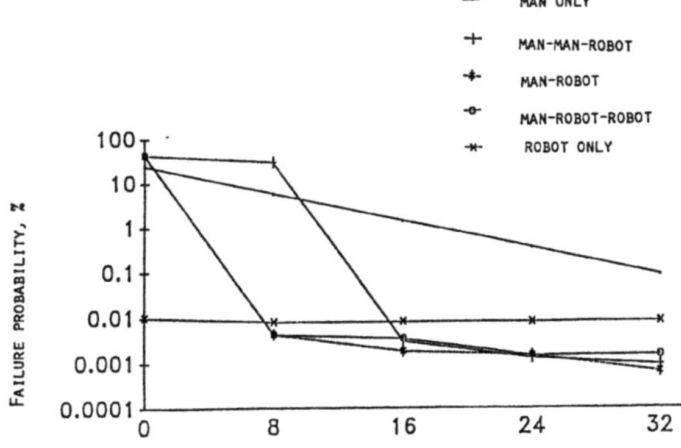

Fig.11 Failure probabilities of the inspection by a man only, a robot
only and their combination (0.01% error probability of a robot)

The present result in Fig.10 and Fig.11 shows the failure probability when the deviant condition is left not to be detected. Generally the occurence of the deviant condition has very low probability. For example if 2000 items are inspected, probably, 2 or 3 are the deviant. The error probability of the robot of recognizing normal condition should be very low compared with that of recognizing deviant condition. If both error probabilities are identical, for example 1 %, the robot recognizes 20 normal cases to be abnormal and 0.2 or 0.3 abnormal cases to be normal among the 2000 items inspected. That robot cannot be used for the inspection.

At present the robots which are being developed for the inspection cannot inspect all the items of a nuclear power plants. The robots, however, can inspect the apparatus under high radiation. The role of men and robots may be complementary to each other in the near future. The present analysis has been carried out under several assumptions, but it gives us directions of future development of the robot for the inspection of a nuclear power plant.

SUMMARY

Probabilistic network models of tour inspectors are proposed for the reliability analysis of the inspection of a nuclear power plant. Probability of omission and commission errors are the basic input data of the model. Recovery is expressed by the nodes showing confirmation of the inspections. Performance shaping factors are considered by properly selecting the input data. Dependence is also taken into account by using conditional human error probability.

Reliability of two-men inspection is higher than that of the one-man inspection, but the difference decreases with the degree of dependence. The analysis also shows that the presence of some stress increases the reliability and that the strict use of the check sheets makes the inspection more reliable.

The reliability of the inspection by a robot, a man and their combination is analysed. Alternate inspection by a robot and a man is more reliable than the inspection by a robot or by a man alone. The increase in reliability is attributed to the combination of the recovery of a man and no dependence of a robot.

ACKNOWLEDGMENTS

The authors express their thanks to Mr.K.Moriya and Mr.R.Kubota for the valuable discussions and also thanks to Miss M.Ataka for typing the manuscript.

REFERENCES

Swain A.D. and Guttman H.G. " Handbook of human reliability analysis with emphasis on nuclear power plant applications " NUREG/CR-1278 (1980).

AUTOMATIC SCHEDULING OF MAINTENANCE WORK IN NUCLEAR POWER PLANTS

Takayasu Kasahara, Yasuo Nishizawa,
Kanji Kato, and Takasi Kiguchi

Energy Research Laboratory, Hitachi, Ltd.
1168 Moriyama-cho, Hitachi-shi 316, Japan

ABSTRACT

An automatic scheduling method for maintenance work in nuclear power plants has been developed using an AI technique. The purpose of this method is to help plant operators by adjusting the time schedule of various kinds of maintenance work so that incorrect ordering or timing of plant manipulations does not cause undersirable results, such as a plant trip. The functions of the method were tested by off-line simulations. The results showed that the method can produce a satisfactory schedule of plant component manipulations without interference between the tasks and plant conditions.

INTRODUCTION

In the maintenance of nuclear power plants, work scheduling is an important task of the operators.[1)-3)] Time schedules of various takes must be carefully adjusted so that incorrect ordering or timing of plant component manipulations does not cause such problems as a plant trip. However, examining the effect of the work on plant conditions is sometimes complicated because of the number of documents which must be refered to. Computer application to this scheduling is then desirable.

Nowadays various computer-aided plant maintenance support systems are being developed actively.[4)] However, precedent tasks are object to almost every maintenance scheduling systems, but these systems cannot judge interferences among taks and plant conditions. To support operators in maintenance scheduling, checking for interference among tasks and plant operations is important function.

Considering this requirement, the basic automatic scheduling method for maintenance work has been developed using an AI (artificial intelligence) technique. Its functions were tested by off-line simulations.

SYSTEM CONFIGURATION

Figure 1 describes the configuration of the automatic scheduling system for maintenance work. This system makes a plant maintenance schedule by consideration of interference between tasks and plant conditions. Here, inteference is defined as an undesirable effect, from plant component manipulations included in a task, on another task or plant conditions.

As shown in Fig. 1, the scheduling procedure consists of four steps.

(1) Initial scheduling, considering personnel limitations.
(2) Checking for interference between tasks and plant conditions.
(3) Searching for a method to remove interference.
(4) Readjusting the time schedule, and adding new operations to remove
    interference.
Steps (2) to (4) are iterated until no interference can be found.

This method checks for interferences by simulating change of plant state in performing plant maintenance operations, not using decision table about interferences among tasks. So this method can check for interferences among tasks which are not expected.

As shown in Fig. 1, the data used are: ① plant process data; ② pairs of the task names and planned times of execution which are input manually by the operator; ③ knowledge data about work contents; ④ plant structure; ⑤ workers information; and ⑥ an operation plan retrieved from the memory. And output information is work schedule, procedure of each task, and etc.

CHECKING FOR INTERFERENCES

A checking method for interference between tasks are as follows. This method uses a knowledge base which consists of knowledge about connectivity (which means the way how each components are connected each other) and transfer functions of each component. Transfer function is simplified model of component. It represents how plant process data propagate in component. The scheduling system makes a connection graph for the object component, and checks for interference between tasks by using the graph. The process is shown in Fig. 2.

At first, a task conditions list is taken from the work schedule data. Task condition is operational condition which must be satisfied when operations or tasks are to be performed. Next, the connection graph of the components which are concerned with the task conditions is made. The connection graph represents connectivity for the object component. In step ③ , using the connection graph, an inference is made on how the fluctuation of the state value (plant process data) is transferred to the object component.

Lastly, in step ④ the simulation results are compared with task conditions. If every task condition is satisfied, the system concludes that there are no interferences between tasks.

Figure 3 describes the process of interference judgement. In the left part of this figure, maintenance work for filter replacement in the control rod drive system (task A) and calibration of flow rate monitor (task B) is shown. Changes in the operation conditions are memorized in the work schedule data. At first, the task condition list is made for all tasks which are scheduled to be done simultaneously. The task condition of task A is "fluctuation of flow rate monitor is 0". In order to judge whether this task condition is satisfied or not, the connection graph in Fig. 3 is used. Valve operation included in task B, causes a fluctuation of the flow. To calculate how this influence is propagated, transfer functions of the components in the connection graph are used. Examples of transfer functions are shown in Table 1.

After checking influences from all components in the connection graph, the task condition is compared with the simulation result. If all the task conditions are satisfied, the system concludes that there are no interferences between tasks.

METHOD FOR FINDING ISOLATION PROCEDURE

The isolation procedure separates systems by valve operation or electrical jumping operation. A method to find an isolating procedure for each system using the standard isolation pattern is shown in Fig. 4, which includes patterns for the valve and piping system.

In order to reduce the search area of the connection graph for the object component, pattern matching is done by the procedure shown in Fig.

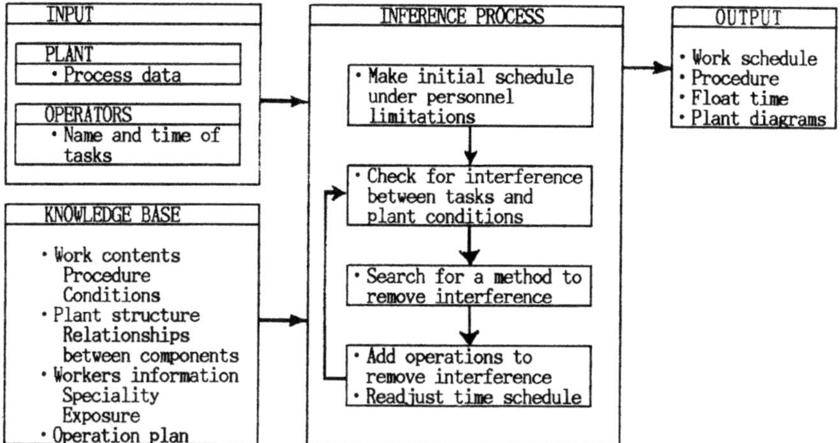

Fig. 1 Configuration of Automatic Scheduling System of Maintenance Work

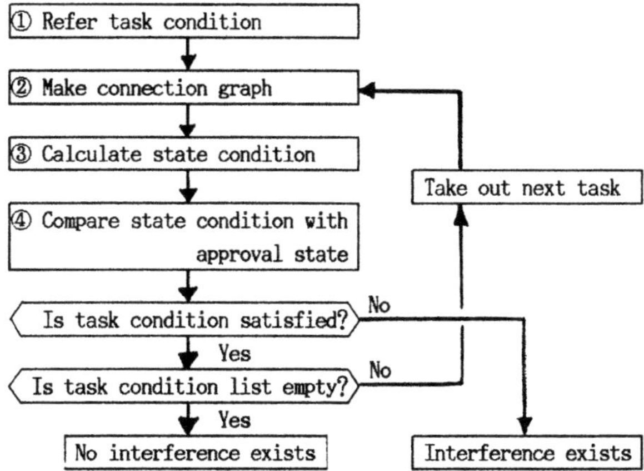

Fig. 2 Checking for Interference between Tasks

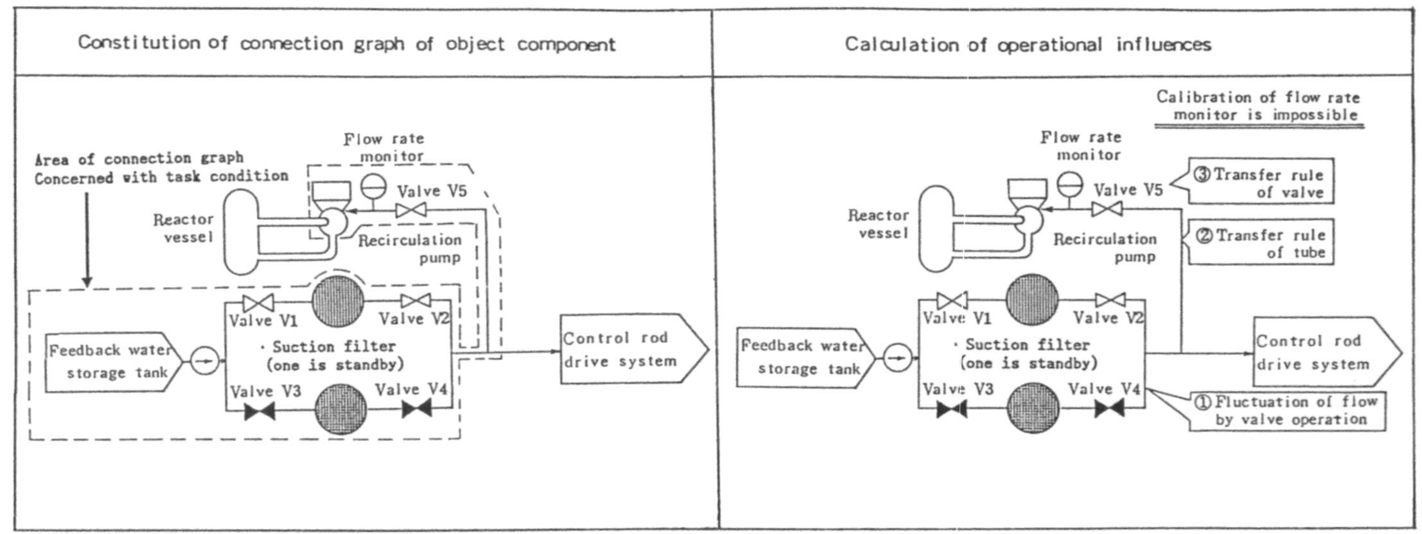

Fig. 3 Example of Interference among Tasks

5. This process includes pattern matching, enlarging the search area on
the connection graph, and simplification of the graph. Figure 6 shows
examples of transformation rules. These rules simplify the connection
graph for the object component by synthesizing components which are
connected parallel or directly.
    The results of interference judgements and a method for avoiding
interferences between tasks are recorded in network of tasks. This
network includes information about precedence and interference
relationships, and is used in adjusting the maintenance schedule.

ADJUSTING MAINTENANCE SCHEDULE

    A branch and bound method was used to make the optimum work
schedule, based on the results of interference judgements between tasks.
The branch and bound method searchs for the optimized task schedule which
satisfies the restrictions which are described in network of tasks.
Branching is done if there are interferences among tasks which are
scheduled to be done simultaneously. After adjusting the schedule,
checking for interferences is done. If no interferences are found, the
schedule is optimu. If interferences are found which were not described
in network of tasks yet, adjusting schedule is done after recording
interference relations. Iterations are done until optimum schedule is
composed. There are two merits of this procedure. Only interferences
appearing in the interation must be checked, not all interference
relations. The number of times needed to check interferences is reduced
by using results of the previous interference checks.
    In the branching process the lower bound is used as a bound evalua-
tion function. For example when objective function is total working
time, the lower bound is calculated by assuming there no restrictions in
the remaining schedule.

EXAMPLE OF WORK SCHEDULING

    The functions of the basic automatic scheduling method for
maintenance work were tested by off-line simulations. Test case is
chosen from maintenance tasks in nuclear power plants. Figure 7 shows
the results of a case in which a surveillance test of an average power
range monitor (APRM), (task A), and tasks for valve repairs in a turbine
ground steam system (task B) and boiler system (task C), were scheduled.
    Each of the tasks was divided into subtasks as shown in Fig. 7. In
the initial schedule, these tasks were planned for simultaneous execution. However, in checking for interference, it was found that carrying
out subtasks B-1 and C-1 at the same time would bring about reactor
scram. Therefore, tasks to remove the interference (I-1 and I-2) were
searched for and added to the adjusted schedule shown on the right of the
figure. Subtask I-1 is isolation procedure and I-2 is restoration
procedure. By subtask I-1, turbine grand system and boiler system are
isolated each other, and subtask B-1 and subtask C-1 can be performed
simultaneously.

CONCLUSION

    The basic automatic scheduling method for maintenance work has been
developed using an AI technique. The functions of the method were tested
by off-line simulations and it was concluded that schedules of plant
component manipulations could be produced without interference between
the tasks and plant conditions. In the future, the automatic scheduling
system for maintenance work will be realized in actual plants, by adding
maintenance knowledge data and by developing high speed knowledge data
processing techniques.

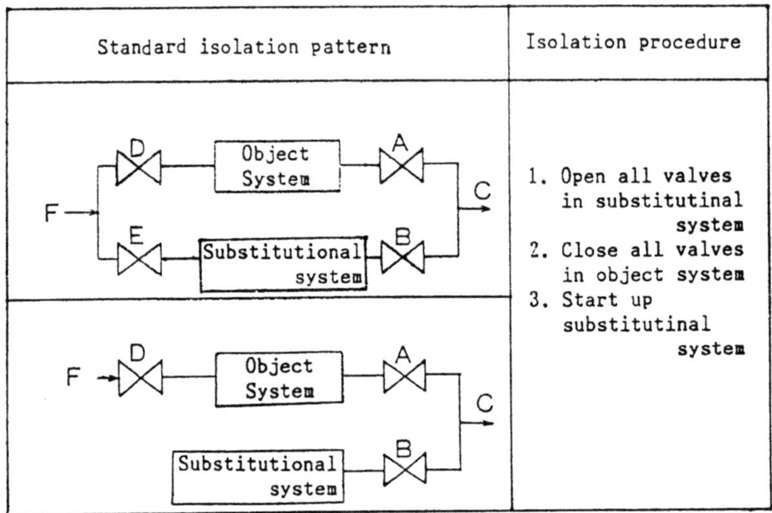

Fig. 4 Standard Isolation Pattern and Isolation Procedure

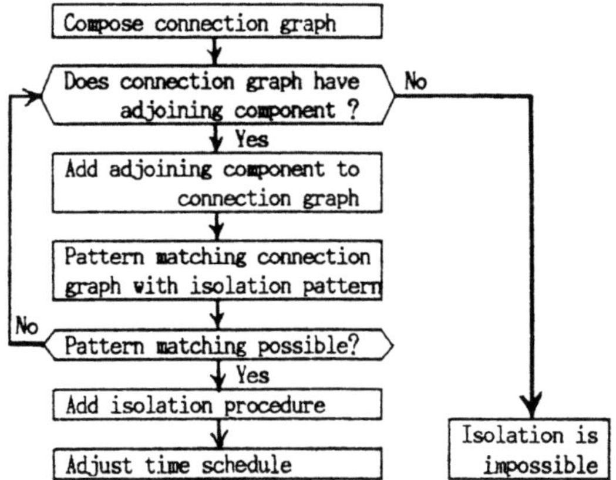

Fig. 5 Finding Isolation Procedure

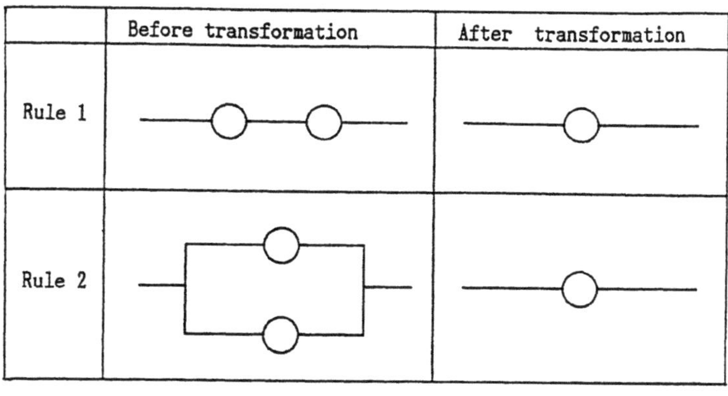

◯ : Component   ──────── : Connectivity

Fig. 6 Transformation Rules of Connection Graph

Initial Schedule   Schedule Considering Interference

Task A | A-1 | A-2 | A-3 | A-4 | Task A | A-1 | A-2 | A-3 | A-4 |

Task B | B-1 | B-2 | B-3 | Task B |I-1| B-1 | B-2 | B-3 |I-2|

Task C | C-1 | C-2 | C-3 | Task C | C-1 | C-2 | C-3 |

                         TIME   TIME

Task A : APRM Surveilance Test   I-1 :Isolation Procedure
Task B : Valve Repair   4 Valve Open & Close
            (Turbine Grand System)   Operations
Task C : Valve Repair   I-2 :Isolation Restoration
            (Boiler System)   Operation
A-i : Subtask of Task A
B-i : Subtask of Task B
C-i : Subtask of Task C
Subtask B-1 & subtask C-1 have interference

Fig. 7 Scheduling Example

Table 1 Example of State Transfer Rules

COMPONENT	State Transfer Rule
VALVE	Output State Value = Coeff. • Input State Value Coeff. : function of valve openness, valve specifications
TUBE	Output State Value = Coeff. • Input State Value Coeff. : function of tube specifications
FILTER	Output State Value = Coeff. • Input State Value Coeff. : function of filter specifications
TANK	Output State Value = Coeff. • Input State Value Coeff. : function of tank specifications

Coeff. : Coefficient

REFERENCES

1) McKenzie, A.R., and MacLeod, J.; Work planning and maintenance management at the Point Lepreau Nuclear Generation Station ; IAEA Nuclear Power Plant Outage Exp., pp. 113-123 (1984).

2) Kozusko, A.M. ; Computerized plant maintenance management; IEEE Transactions on Nuclear Science 33 (1), pp. 962-965 (1986).

3) Leiniger, E., and Carp, P. ; Davis Besse Maintenance Management System; ASME 84-NE-16, pp. 1-5 (1984).

4) G. Butterworth, and T.M. Anderson ; Managed maintenance, the next step in power plant maintenance ; IAEA Nuclear Power Plant Outage Exp., pp. 171-178 (1985).

THE INSTRUMENTATION CALIBRATION REDUCTION PROGRAM (ICRP) AT

NORTHEAST UTILITIES

Richard Wyckoff and Paul Blanch*

Wytek Corporation, 50 West Street, Portland, Maine

*Northeast Utilities Service Co., 107 Selden Street
Berlin, Connecticut

ABSTRACT

The desire to substitute some program for the "Calibration" of the
Core Exit Thermocouple (CET's) at Millstone 3 (MP3) prompted Northeast
Utilities (NU) to fund a project to study the feasibility of determining
the state of CET calibration without having to have direct access to the
CET's. The project, carried out by WYTEK and Northeast Utilities, was
commenced in October, 1986 and just recently completed. Although the
CET's were the prime focus, other safety related sensors were investigated
as well.

This paper describes presumptions and methods employed in the first
phase, the feasibility study. Additionally, it describes the cost/benefit
analysis which can be used by any utility to determine ICRP payback.

INTRODUCTION

The CET's are listed in MP3 Technical Specifications as needing to be
calibrated every refueling interval (18 months).

The definition of CHANNEL CALIBRATION "...the adjustment, as necessary, of the channel output such that it responds with the necessary range
and accuracy to KNOWN VALUES (emphasis added) of the parameter which the
channel monitors...", as spelled out in MP3 Technical Specifications.

The probability is quite good that the average of 50 CET's, under
known reactor zero power conditions, closely reflects the KNOWN VALUE of
core exit temperature. Therefore, it was felt that if a means could be
found to monitor each CET channel output over a selected time period then
a deviation comparison could be made between individual CET's and the
group average. This, then, was the basis upon which the ICRP was
initiated.

Once begun, the research showed significant stability of most (46
total) of the CET measurements. The other 4 CET's were found to be failed
and were therefore not in the average calculation.

The approach was then expanded to include some other measured parameters of ultimate interest to NU and MP3 personnel alike. What follows
is a description of the NU ICRP and potential benefits to users.

ICRP SYSTEM DESCRIPTION

The need to acquire data for eventual comparison was filled through the use of the Offsite Facilities Information System (OFIS), a program running on the NU corporate IBM mainframe which acquires MP3 plant data and displays it for emergency response personnel use. The OFIS database contained all of the parameters needed for the ICRP feasibility study. Unfortunately, the OFIS program didn't allow for individual data point extraction other than to a CRT screen or a dedicated printer. The solution was provided through the use of an IBM PC-AT equipped with communication hardware and software so as to emulate an IBM 3279 S3G colorgraphics terminal. A matching communication software package was installed in the IBM mainframe which then allowed NU to develop a file extraction program through which the PC-AT user could specify data time/date and parameters for eventual downloading. This being done, the OFIS itself was used to help identify selected plant conditions (e.g. plant heatup) so that the necessary data could be specified and downloaded into the PC-AT hard disk.

Once the data files were resident in the PC-AT, powerful software tools such as LOTUS 123 (Lotus Development Corp.) and STATGRAPHICS (Statistical Graphics Corporation) were employed to average, compare for deviation, store and plot the results. One such plot is shown in Figure 1 and is described further in this paper.

ICRP Hardware Description

This description does not include the IBM mainframe computer or any of its peripherals except to say that the OFIS program relies on the existence of the MP3 process computer data link to the NU corporate IBM mainframes. Previous papers, describing OFIS, have been presented and published at various ANS and EPRI Conferences.

* 3 Mbyte extended memory

* 2.5 Mbyte expanded memory

* EGA graphic card (132 column LOTUS worksheet compatible)

* NEC Multisync monitor

* (2) serial and (2) parallel ports

* IRMA communication board

* FORTE Graph board for IBM 3279 S3G emulation

* 650 Kbytes low memory

* 80287-8 math coprocessor

* 30 Mbyte hard disk, 1.2 Mbyte and 360 Kbyte floppy drives

The PC-AT peripherals required to support the ICRP are:

* Hewlett Packard 7475 plotter

* EPSON FX-185 dot matrix printer

MP3 HEATUPS 04/09/87 AND 06/09/87
CET DRIFT COMPARISON (FILTERED)
DEVIATION DIFFERENCE
AVERAGE CET TEMPERATURE
CET3
CET4
CET6
CET7
CET11

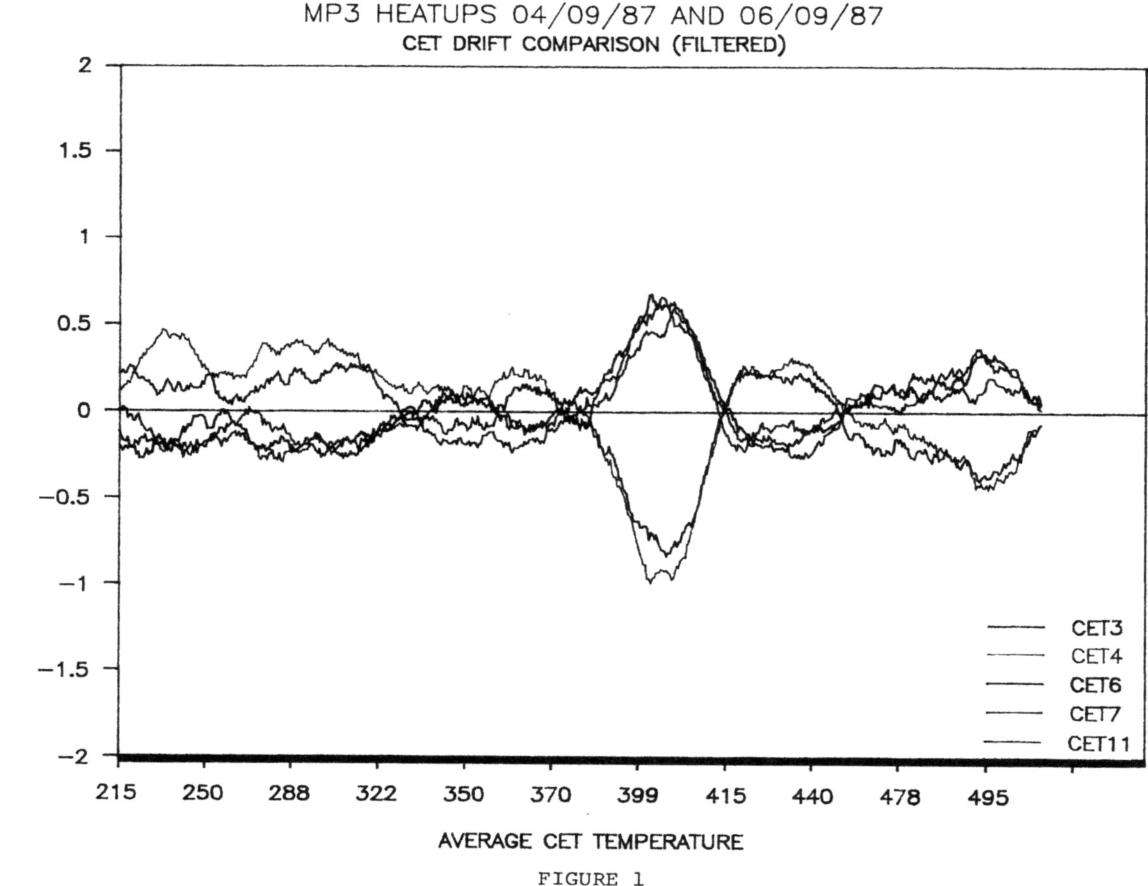

FIGURE 1

ICS Software Description

Again, the software resident in the IBM mainframe will not be
addressed except to indicate that custom-built application programs exist
to allow for data selection and download to the PC-AT. The following
software is commercially available and formed the nucleus of the ICRP
software "tools" for our data analysis efforts.

* LOTUS 123 (Lotus Development Corporation) Version 2.01

* STATGRAPHICS (Statistical Graphics Corporation) Version 2.1

* IBM PC DOS (International Business Machines Corp.) Version 3.2

* NOTE-IT (Turner Hall Publishing) Version 2.0

* Various utility programs and drivers common to most PC's

PARAMETER EVALUATED AND METHODS EMPLOYED

This section of the paper deals with the actual processes involved
in the NU ICRP approach. Where problems were encountered they will be
described as well as the solutions to them. It is organized by parameter
of discussion, method(s) used, problems encountered and final results.

Core Exit Thermocouples

These signals, as well as all to be subsequently discussed, are made
available to the OFIS with 30 second time intervals between samples. In
the CET's case the method employed was to acquire data that time-bounded
plant cooldowns and heatups. The idea being to see as much "span" of
temperature as possible. The actual data acquisition system span for the
CET's is 200 to 2300 degree F. Our data was restricted to a more reasonable 200 to 600 deg. range. The CET's were analyzed using ANOVA techniques
and any values greater than 2 sigma variance were flagged and removed
from averaging. The remaining CET's (in most cases the same 46) were
imported into a LOTUS template which was designed to compute averages of
time synchronous data sets, calculate individual deviations from the
average, and store them in another PRN file for plotting purposes. Plots
were then made of each individual CET's deviation over the span of temperature.

Problems encountered included loss of data during data link failures,
ICC cabinet communication problems with the process computer, incorrect
plant conditions (initially) such as power operation where the exit temperature deviate due to enthalpy differences between fuel bundles. By
screening the datasets we were able to eliminate all data transmission
loss problems. The obvious solution to the enthalpy difference problem
was to sample only when the reactor was shutdown and decay heat was
minimal in its effects (this occurs within a few hours after power
operations is ceased).

The results are quite heartening. Over a period of two months (as
shown in Figure 1) the difference between the two sets of individual
deviations from the averages rarely exceeds one half of one degree F.
This kind of data convinces NU that the CET's are stable and that no need
for calibration exists for those CET's examined. Figure 2 is included to
show a "bad" CET and how its deviation looked when examined.

MILLSTONE III HEATUP 04/07/87
CET#14 DEVATION FROM CET AVERAGE
DEVIATION (DEGF)
16
14
12
10
8
6
4
2
0
-2
-4
-6
-8
-10
-12
-14
-16
486 464 439 411 383 360 337 312 286 257 224 205 188
TEMPERATURE

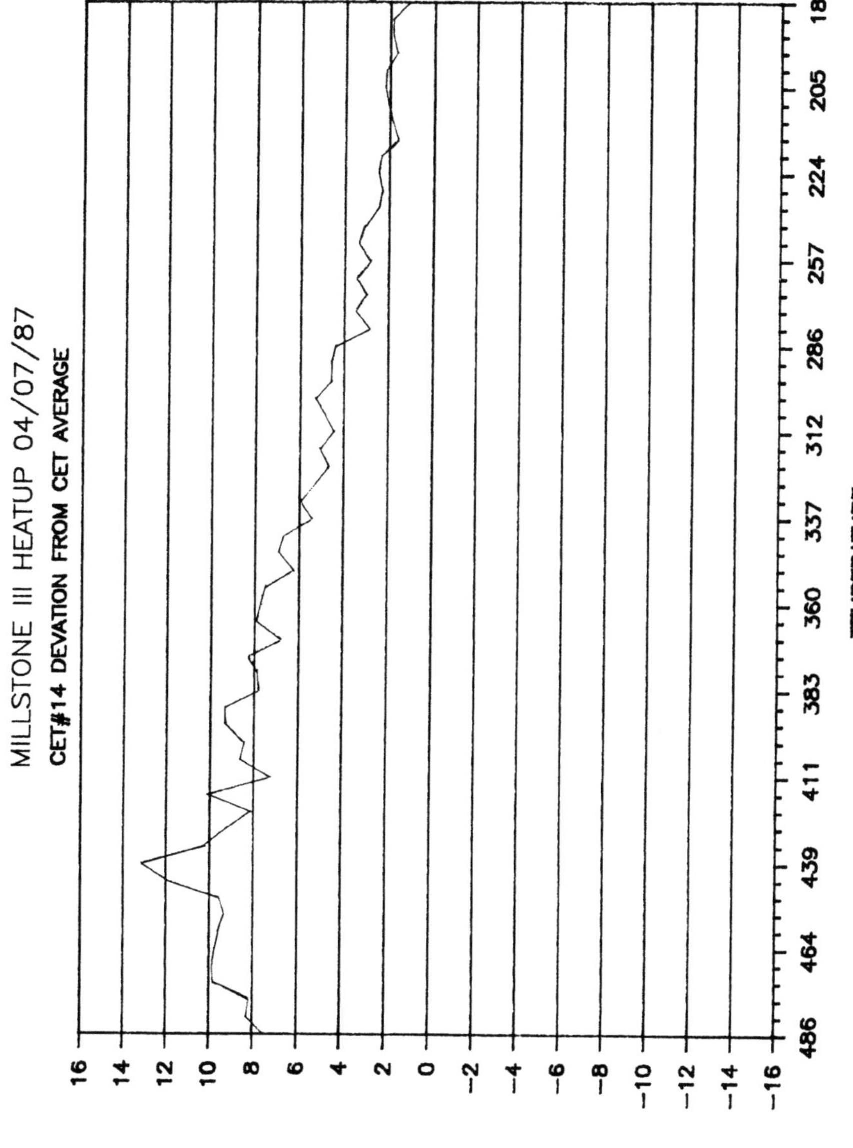

FIGURE 2

Hot Leg Temperature Channels (RTD's)

The Reactor Coolant system (RCS) T-HOT channels employ platinum RTD's, in thermowells, as sensors. Typically, very few plants actually remove their RTD's every refueling for calibration as the procedure for so doing can and has damaged the RTD's as well as creates a significant radiation exposure problem for the technicians doing the work. Most "calibrations" actually performed by I/C people involve a lead balance and then the insertion of a precision decade box to simulate the RTD itself, in order to calibrate the resistance to current (or voltage) converters. The presumption of the RTD inherent stability is implicit within this method. Furthermore, a number of plants employ some kind of a time response check (generally based on the Loop Current Step Response technique) on their installed RTD's. This, again, presumes stability of the measuring element but it does check the thermal transfer coefficients between the sensor and the fluid. The NU ICRP method was two-fold; first a time coincident comparison of T-HOT vs CET average was performed for deviation analysis and secondly the T-HOT signals were averaged and individual deviations examined. This was performed during plant heatups and cooldowns.

Problems encountered were similar to those experienced in all cases of OFIS data transmission losses. Additionally, the T-HOT signals showed more significant deviations (as a percent of instrument span) when compared to the T-HOT average than they did when compared to CET average. The deviations also showed a sensitivity to steaming rate. It is felt that some stratification of the RCS fluid is occurring and that this is the reason for the change in deviations. Therefore, we excluded data sets from calculation that were time coincident with steam demand (such as occur when the secondary side is being warmed up during plant heatup).

The results were quite good, with deviations not much greater than 3 or 4 degrees F and, more importantly, with event to event repeatability. This method does eliminate some data points within the "calibration" but it is felt that the probability of a sensor or conditioner discontinuity extent within a range (perhaps 10 to 15 degrees F) and not also occurring elsewhere within the instrument span is extremely low.

Cold Leg Temperature Channels (RTD's)

The description and information previously given for the Hot Leg RTD's applies equally as well to the Cold Leg RTD's. The data and results were similar with the exception that the stratification effects seem to be less problematic thus allowing full data span comparisons.

Reactor Coolant System Flow Channels (Differential Pressure)

The RCS flow channels are normally near to full span output during all plant conditions when the single speed coolant pumps are running. Here the approach we employed involved deviation comparisons at the normal signal range and a snapshot comparison of deviations during flow coastdown after the respective pumps were shut down for plant cooldown and refueling preparatory operations. Similar snapshots of data were taken for pump startup events.

The only problems encountered were that our "span" datasets (pump S/D and S/U conditions) were, of necessity, short due to the 30 second data time interval inherent in the OFIS database. However, we still had enough discrete points to track all the channel outputs.

Again, the results were reassuring in terms of data repeatability over time and we had the opportunity to track a failing transmitter which gave us some real time information for future use in an automated ICRP system.

Reactor Coolant System Pressure Channels (Absolute Pressure)

MP3 has mix of wide and narrow range pressure channels, both of which we were able to check using deviation comparison. The method was applied over plant heatup and cooldown events.

We had little problems with comparison techniques and were able to cross compare wide range data with narrow range, where there was overlap between the two ranges.

The results were deviations in the order of 5 to 7 psi and repeatable over events.

Containment Pressure Channels (Absolute Pressure)

As there are 4 narrow range pressure channels and only two wide range channels we averaged the narrow range signals and used that average for all comparisons.

No major problems were encountered except that no significant span information was available due to the fact that the MP3 containment is slightly subatmospheric in pressure. Thus, there is only an approximate 5 psi change in pressure when the containment is vented to atmosphere for refueling operations.

The limited span aside, the tracking and deviations were very good, which allowed us to resolve to better than 0.1 percent of span. This is sufficient resolution for trend detection should long term drifting occur.

Steam Generator Pressure (Absolute Pressure)

The method used for the SG pressure channels was the straightforward signal average and comparison technique.

The pressure channels were somewhat "noisy" and subject to deviation errors due to steam flow. The resolution was to digitally filter the data using a moving average technique in the LOTUS worksheet. This reduced the noise, greatly. Care was taken to not acquire data during steaming conditions which was easy to recognize from the steam flow signals.

The deviations were in the order of 3 to 5 psi ($\pm$.25%) and very repeatable over events such that it would be easy to resolve very small drift in signals.

Steam Generator Level (Differential Pressure)

Both wide range and narrow range channels were analyzed for deviations from the narrow range averages, similar to the method used for RCS Pressures.

The major problem encountered revolved around a peculiar "noise" anomaly extent mostly at power conditions and caused by improper condensate pot design. We restricted the data to ranges during heatup and cooldown as well as certain level transients known to any PWR control room operator during secondary side startup maneuvers.

NU is in the process of correcting the condensate pot problem. The
other data that we obtained, however, was enough to certify that the level
channels would be monitorable using the deviation method. Most deviations
were within 1 to 2 percent of signal span.

NRC Licensing Perspective

Presently, the NRC requires the calibration of certain "Safety
Related" instruments at an interval of every 18 months. This interval was
initially selected as it was longer than normal refueling cycles and
historically, instruments did not change in their calibration during this
interval.

A review of equipment histories at two power plants has indicated that
the calibration conducted during the 18 month interval are not required in
more than 90% of the cases. At the present time there is no data which
can demonstrate whether an instrument is calibrated or not. The ICRP
provides documented data which is equivalent to a 1000 point calibration.

The ICRP has been informally reviewed with the NRC and no major
problems have been identified.

Potential Program Benefits

The potential benefits of a successful program are obvious and a few
of these are as follows:

* Reduced exposure

* Reduced manpower

* Quantification of drift

* Setpoint realization

* Actual calibration of CET's and RTD's

* Reduced potential for technician errors (Reactor Trips)

* Misaligned valves

* EEQ degradation

* "Qualified life" extension

* Reduced calibration induced failures

* Identification of sticking and noisy instruments

* Improved plant availability

* Reduced critical path time

CONCLUSION

The deviation of an individual process signal from the average value
of a sufficient number of physically redundant based signals is a reliable
and sensitive indicator to instrument loop calibration. Where the redun-

dancy is less than 3 for a given set of process channels it may be possible to create a mathematical model form diverse parameter which could yield a suitable analog to the process parameter of concern, but that is beyond the scope of the NU ICRP Feasibility Study that this paper describes.

The claim is not that the average value of, say, three redundant signals is statistically defensible as the equivalent "KNOWN VALUE" referenced in the definition of "CALIBRATION" provided in the Technical Specifications. The basis for using this average is supportable when viewed in the context of obtaining such data, immediately after the channels had been calibrated with National Bureau of Standards traceable tertiary standards. The average, then, can be viewed as a transfer standard. As long as there exists no common mode mechanism to invalidate this assumption it is perfectly fair to employ this technique to track the deviations over time and over matching parameter conditions. If the change in the deviations is less than the inaccuracy of the entire monitored channel it is safe to state that the channel need not be recalibrated, regardless of the elapsed time since the last NBS traceable calibration.

The case of the Core Exit T/C's is different that there are enough CET's, all monitoring the same temperature (during certain plant conditions), that one can state that the Root Sum Square inaccuracy method yields the fact that the measurement uncertainty of the average of the CET's (approximately 0.7 percent of instrument span) is nearly a factor of ten less than the measurement uncertainty of any single CET channel (approximately 5 percent).

Other methods have been investigated to achieve the results we have described, such as the employment of temporary, high accuracy, measurement devices from time to time in order to track calibration changes. While some are, perhaps, inherently more accurate due to type and technique none to our knowledge are as easy to use or document as a computer based data acquisition and comparison system which can run unattended and flag potential problems before they result in Licensee Event Reports or Reactor Trips.

A KNOWLEDGE-BASED ASSISTANT FOR VALVE MAINTENANCE PLANNING

Michael J. Winter, R.A. Danofsky, B.I. Spinrad,
and Kenneth Howard

Iowa State University, Ames, IA
Iowa Electric Light and Power Co., Cedar Rapids, IA

ABSTRACT

A knowledge-based program is being developed to assist
engineers in maintenance planning for safety related, motor-
operated valves at a boiling water reactor. The purpose of
this project is to develop the general framework for a
prototype system that will demonstrate the capabilities for
diagnosing valve symptoms and prescribing corrective
maintenance, completing a portion of the Corrective
Maintenance Action Request (CMAR) form which must be prepared
for each job, and managing an interactive valve data base.
Minimizing user input and providing output in a form that is
familiar to the maintenance planning engineer are important
goals for the program. This paper describes the present
features of the valve maintenance advisory system which is
currently being tested.

INTRODUCTION

Valves and valve maintenance play a very important role
in the operation and availability of a nuclear power plant.
Verna (1), for example, points out that valve related
problems cost U.S. utilities $100 million per year in loss of
plant availability and up to 30% of the industry's annual
maintenance budget. A knowledge-based system that would aid
maintenance planning could contribute to improved plant
availability.

Valve maintenance planning is a very time consuming task
because of the large number of valves in a plant and the high
level of quality assurance required. The steps required in
the process typically include diagnosis of the problem,
prescription of maintenance, the determination of a number of
factors which affect the maintenance task, and the
identification of required post-maintenance tests of the
valve. The maintenance engineer must consult various sources

to gather information necessary for the analysis and complete
required forms. Factors which affect maintenance planning
include requirements for cleanliness control, tag out,
radiation work permits, and lifted motor lead evaluation.
Operational testing following maintenance is required by the
ASME Boiler and Pressure Vessel Code (2) which states,

> "When a valve or its control system has been
> replaced or repaired or has undergone maintenance
> that could affect its performance and prior to the
> time it is returned to service, it shall be tested
> to demonstrate that the performance parameters
> which could be affected by the replacement, repair
> or maintenance are within acceptable limits."

We can see that these various aspects of valve
maintenance planning serve to indicate that a knowledge-based
advisory system, that could reason about valve
characteristics and access a valve data base, would be a
useful aid for the maintenance engineer.

This paper describes a cooperative effort between
faculty members at Iowa State University and engineers at The
Duane Arnold Energy Center (a 545 MWe BWR operated by Iowa
Electric Light and Power Co.) to explore the development of a
valve maintenance planning assistant. To restrict the size
of the required domain of knowledge, the system is presently
limited to safety related motor-operated valves
(approximately 117 valves).

DEVELOPMENT OF SYSTEM

The initial effort has been directed to identifying the
general framework and characteristics for an advisory system
to diagnose valve symptoms,prescribe maintenance, complete a
portion of the Corrective Maintenance Action Request (CMAR)
form, and maintain an interactive valve data base. It was
recognized early in the development that it was desirable to
keep requested user input to a minimum and to provide a
printed report from a session that is similiar in form to the
CMAR that the maintenance engineer is familiar with.

A prototype system has been programed using the expert
system tool INSIGHT2+ (3). INSIGHT2+ is a production rule
system using a backward chaining inference strategy. The
syntax of the INSIGHT2+ Production Rule Language (PRL) is
simple and easy to learn resulting in rapid program
development. The PRL source code is compiled to produce a
rapidly executable knowledge base. Confidence factors can be
implemented, allowing for the specification of uncertainties
in facts and conclusions. One of the features of this tool
that was found to be very useful for this project is the
ability to call external programs while executing a knowledge
base. For this application, a Turbo Pascal program is called
to retrieve valve data from an external data base. Another
Turbo Pascal program handles the naming of report files that
are generated during a session. Data base editing is also
performed from within the knowledge base by a call to an
external program. A Line Of Reasoning Report (LORR) for a
session can also be obtained.

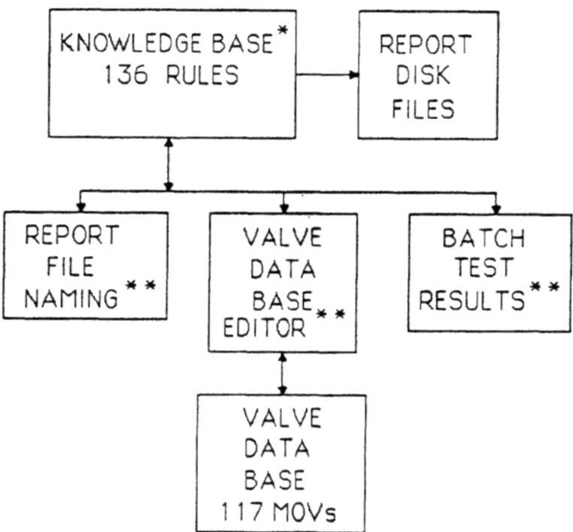

Fig. 1.   Block diagram of system

```
            Valve Maintenance Planning Assistant

Select your required task:

     Determine maintenance planning information for a valve

     Diagnose problems and prescribe maintenance for a valve

===> Diagnose problems, prescribe maintenance, and determine
     maintenance planning information for a valve

     Edit the valve data base

     Run the batch testing knowledge base

     Leave knowledge base
```

Fig. 2.   Selection of task

Knowledge acquisition is considered to be of the more difficult steps in the development of knowledge-based systems. For this phase of the project, we used tape recorded lectures given by a valve maintenance expert and held discussions with maintenance engineers. Several times during the development effort we gave demonstrations of the prototype system. Feedback received from these interactions was incorporated into the system.

The original purpose of the program was to determine post-maintenance testing requirements. The list of desired functions has now grown to include the determination of nearly all the items in one section of the CMAR form. These items include cleanliness control, tag out requirements, primary containment requirements, radiation work permit requirements, lifted motor lead evaluation, etc. In early versions of the program these items were evaluated assuming that the maintenance action was known. We have now added rules which provide a limited amount of diagnostic capabilities and then evaluate maintenance actions and generate the CMAR form.

A block diagram of the system is shown in Fig. 1. The knowledge base consists of compiled modules chained (an INSIGHT2+ feature equivalent to linking) together to form a single knowledge base which provides the capability to diagnose valve problems, prescribe maintenance, determine maintenance planning information, edit the valve data base, and run the batch testing cases. The batch testing capability allows any number of test case inputs to be processed by the main knowledge base. The results are stored in a data file, and can be used to study the effect of changing rules in the knowledge base, or to validate the knowledge base using real experts as a bench mark. Disk files which contain reports related to maintenance planning are written from the knowledge base. These along with a partially completed CMAR form can be printed. External programs written in Turbo Pascal provide naming of files, data base editing, and comparing of batch test results. The valve data base contains valve information including the valve ID number, piping and instrumentation drawing (P&ID) coordinates, size and type, class and category, and applicable operational tests.

USE OF THE MAINTENANCE PLANNING ASSISTANT

An example session with the maintenance planning assistant is shown in Figs. 2-6. The first screen (Fig. 2) presents the task to be performed. In this example "Diagnose problems,prescribe maintenance, and determine maintenance planning information for a valve" is selected. The user would then be asked to enter the valve ID number (not shown). Since the system is to diagnose valve problems, the next screen (Fig. 3) asks for input of the valve system exhibiting the symptom. Two systems are shown, the valve body and the valve operator, where either one or both may be selected. The use of the function key F4 to indicate done implies that the fact involved is multiattributed so that it can have more than one value. The user selects "valve body" and the third

```
                    Valve Diagnosis Knowledge Base

    Select all the valve systems which contain the
    parts that are exhibiting problem symptoms


  ===> valve body

        valve operator




                    After making your selections press F4 for DONE
```

Fig. 3.   Selection of valve system


```
                    Valve Diagnosis Knowledge Base

     Select all the problem symptoms evident in
     the valve body


        Excessive handwheel effort

  ===> Leakage between valve body and bonnet area

        Leakage through stuffing box and around stem

        There is no more packing adjustment left

        Valve stem is binding when operated

        Leakage between valve disc and seat area

                    After making your selections press F4 for DONE
```

Fig. 4.   Selection of symptoms

```
                    Valve Diagnosis Knowledge Base


    - The gasket between the valve body and bonnet has been
      diagnosed as needing replacement from the symptom of
      leakage through the gasket area.  The maintenance
      action prescription is to replace the body to bonnet
      gasket (valve disassembled).
```

Fig. 5. Valve diagnosis report

```
          ***  Valve Maintenance Planning Results  ***

           Valve ID: MO 2003    P&ID: M-120  F-4

       QL:          pri cont: N    NPRDS:         MIF:
   tagout: Y    req for S/U:         CCP:      proc#1:
    J&LL:        heavy load: N      ASME: Y    proc#2:
     RWP:          fire prot:         EQ:      proc#3:


   ---------------- post-maintenance requirement ----------------

           BTC stroke close test is required
           BTO stroke open test is required
           PIT position indication test is required
           AT1 leakage test is not applicable
           AT5 leakage test is required
           CTCC check valve stroke test is not applicable
           System pressure test is required
```

Fig. 6. Maintenance planning summary

screen (Fig. 4) requests all the symptoms evident in the
valve body. This also is a multiattributed fact so any
number of symptoms may be selected. In this case a single
symptom, "Leakage between valve body and bonnet area" is
selected.

Based on input information the report shown in Fig. 5 is
produced giving the diagnosis and recommended maintenance.
The partially completed CMAR form (Fig. 6) is also generated
for the recommended maintenance. The upper portion of the
CMAR form indicates that tagout of the valve is required, the
valve is not part of primary containment, a heavy lifting
load is not involved, and the valve is in the ASME code list.
The post-maintenance requirements involve testing of the
valve closing and opening stroke time (BTC,BTO), position
indication test (PIT), pressure isolation test (AT5), and a
system pressure test. The valve ID number and P&ID
coordinates are also shown on the form. Other information on
the form either does not apply to this valve or is not
currently in the knowledge base.

It can be seen from this example that the required user
input is minimal. The present framework for the knowledge
base can be expanded without adding significantly to the
complexity of the user input. The report generated from the
maintenance diagnosis can be easily expanded to provide any
desired detail of maintenance instructions.

CONCLUDING REMARKS

A knowledge-based program which is to serve as an aid to
the maintenance engineer has been described. The present
program provides a framework so that the refinement and
addition of new knowledge can be easily provided. The
example session illustrates that minimal user input is
required and output of the CMAR form provides information
that is familiar to the maintenance engineer. The next step
in the development of the planning assistant will be to
incorporate feedback we receive from the maintenance
engineers. We want to expand the system to have more
diagnostic capabilities and to provide more information on
the CMAR form.

REFERENCES

1.   B. J. Verna, A series of articles on motor-operated
     valves appearing in Nuclear News: Vol. 25 No.
     15 p53 (1982), Vol. 26 No. 2 p56 (1983), Vol.
     26 No. 5 p46 (1983), Vol. 26 No. 8 p70 (1983),
     Vol. 26 No. 10 p68 (1983), Vol. 29 No. 3 p59
     (1986), Vol. 29. No. 7 p48 (1986), Vol. 29 No.
     9 p49 (1986).

2.   ASME Boiler and Pressure Vessel Code, Sec XI,
     Notebook A, Subsection IWV-3200 Valve
     Replacement, Repair, and Maintenance.

3.   INSIGHT2+ Copyright by Level Five Research Inc.,
     Indialantic, FL 32903.

STEAM GENERATOR INSPECTION PLANNING EXPERT SYSTEM

Peter Rzasa

Nuclear Services Technology Development Group

Combustion Engineering, Inc.
1000 Prospect Hill Road
Windsor, Connecticut 06095

ABSTRACT

Applying Artificial Intelligence technology to steam generator
non-destructive examination (NDE) can help identify high risk locations in
steam generators and can aid in preparing technical specification compliant
eddy current test (ECT) programs. A Steam Generator Inspection Planning
Expert System has been developed which can assist NDE or utility personnel
in planning ECT programs. This system represents and processes its
information using an object oriented declarative knowledge base, heuristic
rules, and symbolic information processing, three artificial intelligence
based techniques incorporated in the design. The output of the system is an
automated generation of ECT programs. Used in an outage inspection, this
system significantly reduced planning time.

STEAM GENERATOR INSPECTION PLANNING EXPERT SYSTEM

INTRODUCTION

The Steam Generator Inspection Planning Expert System is a
computer-based system which automates the process of selecting steam
generator tubes that are to be eddy current tested (ECT) as part of
technical specification surveillance or expansion programs. The Steam
Generator Inspection Planning Expert System is a tool for both outage
inspection personnel and those utility personnel responsible for timely,
accurate, and technical specification compliant steam generator inspection
programs. This system has been used in actual outage inspections and has
enabled outage planners to generate initial and expansion inspection
programs in minutes. This system was developed by Combustion Engineering's
Nuclear Services to automate their efforts in non-destructive examination
and to codify the expertise of a few key individuals.

One main component of a nuclear power plant is the steam generator. A
responsibility of the utility is making sure by periodic inspections and
repair that the integrity of the steam generator is intact to allow the
power plant to continue operating. Steam generator tubing constitutes a
substantial portion of the primary system pressure retaining boundary. In
order to function as an effective barrier, this tubing must be free from
cracks, perforations and general deterioration. The eddy current
examination is designed to evaluate the integrity of the tube pressure
boundary.

To avoid unplanned downtime, nuclear power plant shutdowns are done on a outage schedule, roughly once a year for inspection, maintenance and repair. Eddy current examinations are usually performed during the outages. The planned outage is usually scheduled to take place within a critical window of time, typically four to six weeks long. Downtime beyond the planned window can cost a utility company a quarter of a million dollars each day because of lost power generation. To avoid exceeding the scheduled outage, the inspection and repair process must be efficient. Equally important the inspection must be accurate and not overlook important data. The Steam Generator Inspection Planning Expert System is a tool to help minimize eddy current testing time by providing speed and accuracy in the planning stage of the process.

Deciding which steam generator tubes to test prior to or during an outage is a complex human decision process. The unique expertise required to plan these inspections includes understanding the current operating conditions, recommending test plans, suggesting appropriate inspection equipment, taking measurements with various types of instruments (ultrasonic signals, eddy current probes, radiography, vision inspection), interpreting the data and identifying the problems. There are a few experts who are able to effectively prepare steam generator inspection plans and provide the appropriate recommendations. The approach to the problem does not entail a closed-form solution. Recognizing this constraint, a novel approach to automating one task in the planning phase of the ECT process was taken in the development of the inspection planning tool.

The Steam Generator Inspection Planning Expert System is based on a combination of an emerging computer technology called Artificial Intelligence (AI) and a repository of human expertise gained from many years of nuclear power plant inspection services. AI technology is a class of computer software techniques designed to solve problems which require human reasoning and expertise which are typically not solvable by mathematical or algorithmic approaches. Applying AI technology to steam generator NDE has helped in identifying high risk locations in the steam generator and has aided in preparing ECT programs.

The Steam Generator Inspection Planning Expert System can be beneficial in a number of ways. The system produces accuracy that was not possible before in the interpretation of thousands and thousands of data points. The system can cut down on mistakes and inconsistencies. It can allow the expert to use the computer as a tool for developing the plan and provides documented records of the inspection. Automating steam generator inspection planning can be beneficial in a number of additional ways:

o   It gives the utility a consistent approach to steam generator inspection.

o   It provides a means for validating the tube selection process.

o   It can serve as a tool for preparing inspection bid specs by providing a means for quickly scoping the inspection.

o   It can be used as a training aid for utility and outage personnel.

o   Most importantly, it automates the inspection planning process and reduces plan preparation time, freeing up those individuals whose expertise are typically in high demand.

The Steam Generator Inspection Planning Expert System was developed on a Symbolics LISP Machine and is written entirely in LISP.

SYSTEM DESCRIPTION

The Steam Generator Inspection Planning Expert System requires, as does
the human planner, a variety of information; such as steam generator design,
results of previous inspections, operating history, repair history,
technical specifications and regulatory guidelines. Much of this
information currently resides in conventional databases at utility sites.
Taking this into consideration, the system was designed to interface with
conventional databases. This feature is attributed to the system's
methodology for representing and processing information through object
oriented data representation and symbolic data processing, two artificial
intelligence based techniques incorporated in its design.   .

Initially the planning system reads plant specific historical and
design information from data base files and creates an object oriented
representation of the tube map and associates the historical data with each
respective tube object. Additional plant specific information may be
entered manually. The system then determines which flaws to look for and in
which regions of the steam generator to look for the flaw by consulting a
knowledge base of rules reflecting the planning expertise. The rules are
divided into 3 sets:

1.   Rules about what action should be taken when a particular flaw is
     positively identified in a region.

2.   Rules about what action should be taken when a particular flaw in
     a region is suspected via past eddy current test results.

3.   Rules that determine which flaws are probable based on other
     sources of information such as plant chemistry, leak rates, loose
     parts, vendor designs, etc.

Here are some examples of a rule from each category:

<u>Confirmed Rule</u>: If a defect has been confirmed in a region, then
assume that any indications found in that region are indications of that
defect type.

<u>Indicated Rule</u>: If there is an indication in the steam blanket region
of a C-E generator, then suspect IGA/IGSCC.

<u>Probable Rule</u>: If the chemistry is (or was) phosphate, then suspect
wastage at all supports and in the sludge region.

Figure 1 is an example of an output from this process. Defects are
classified on a per region basis and additionally categorized by which set
of rules identified them. This information provides a valuable template for
identifying the actual sets of tubes to inspect. The system uses the
information from this table to dynamically build a menu of problem regions
specific to the steam generator. An example of a menu of problem regions is
shown in Figure 2.

Once the planner knows what defects to look for and in what regions to
look in, the process of creating the test pattern can be very tedious and
may involve many hours. The system can aid in automating the generation of
these test sets by providing a flexible set of menu items and mouse
sensitive tools. An example of how a planner could use these tools is
explained below.

Initially the planner will consult the technical specifications online
spreadsheet shown in Figure 3 to determine the size of the test set. This

tool maintains an internal calculation of the target size of the test set
based on the number of steam generators in the unit, the number of steam
generators to be inspected and the result of previous test cycles within a
given ECT project. The formula used incorporates the technical
specifications for eddy current testing.

DEFECT TYPE	CONFIRMED *Region of Steam Generator*	INDICATED *Region of Steam Generator*	PROBABLE *Region of Steam Generator*
THINNING			STEAM-BLANKET SLUDGE
PITTING			SLUDGE
PS-SCC			ALL-EXPANDED
IGA			TUBE-SHEET-CREVICE SLUDGE

Figure 1. Defects Location Table

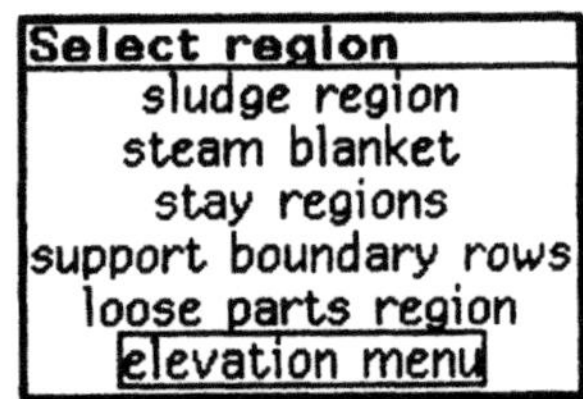

Figure 2. Problem Region Menu

1st SAMPLE INSPECTION			2nd SAMPLE INSPECTION		3rd SAMPLE INSPECTION	
Sample Size	Result	Action	Result	Action	Result	Action
SG's in unit: 3	C-1	*NONE*				
SG's inspected:1	C-2	*inspect 2S tubes*	C-1	*NONE*		
Sample%:9.8		*2S%: 18.8*	C-2	*inspect 4S tubes*	C-1	*NONE*
ntubes:514		*ntubes: 1028*		*4S%:36.8*	C-2	*NONE*
				ntubes: 2056	C-3	*do 1st C-3 action*
			C-3	*perform 1st sample C-3 action*		
	C-3	*inspect all tubes in this SG*	all other SG's C-1	*NONE*		
		2s tubes in each other SG	some SG's C-2 no additional C-3	*perform 2nd sample C-2 action*		
			additional SG C-3	*inspect all tubes in each SG*		

Figure 3. Technical Specification Spreadsheet

Once the target statistics are calculated, the planner exits to the main interface of the system, a mouse-sensitive map of the steam generator (See Figure 4). Each tube displayed on the map is an object within the object oriented environment and is mouse selectable. The object representation provides a window to all the available information on the tube, e.g. member regions, deposit depths, thruwall indications, denting information, repair data, and inspection status. (See Figure 5).

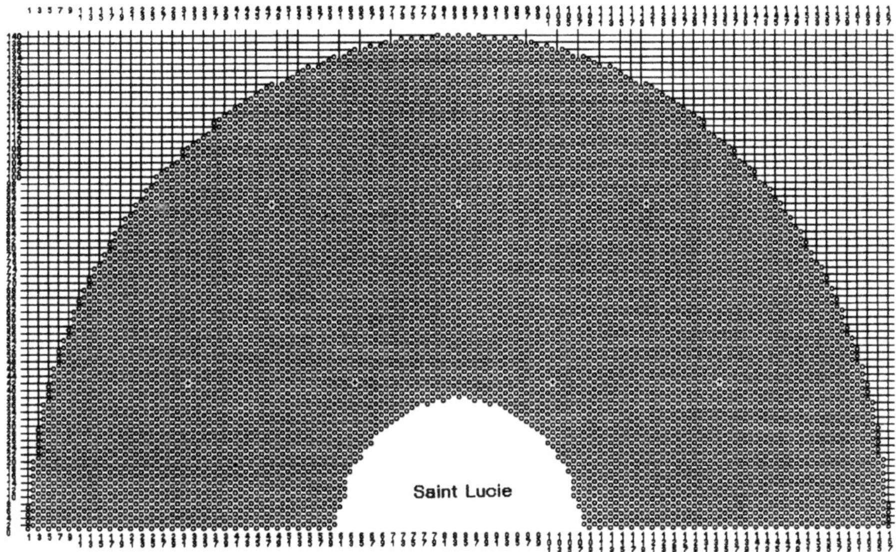

Figure 4. Tube Sheet Map

```
TUBE92-51

regions: EGG-CRATE#1
         EGG-CRATE#2
         EGG-CRATE#3
         EGG-CRATE#4
         EGG-CRATE#5
         EGG-CRATE#6
         DRILLED-SUPPORT-#7
line:    92
row:     51

hot deposit:  NIL
cold deposit: 11.1
          indication records
elv   side   loc   indc   ampl   extent
NIL    H     2.2    38    1.04   F/L

status:   INDICATIONS




tube window
```

Figure 5. Online Data for Each Tube

With this map the planner now has access to a number of planning aids
to develop the inspection program. These aids include: Add and delete
functions for single tubes, predefined problem regions, user-defined
regions, and historical subsets; screen printouts of programmed inspection
patterns; on-line and hardcopy reporting facilities; and interfaces to
existing historical data and inspection database packages. (See Figure 6).

An example of how a planner would utilize these functions follows. A
planner may want to develop an initial test program which contains: all
tubes adjacent to tie rods (stays), 50% of the #7 partial support structure,
and 20% of a user defined region. Establishing this criteria is currently
part of the planner's task. Once these objectives are established the user
would begin by selecting the stay region menu option. The mouse-sensitive
barometer icon shown in Figure 7 will then appear allowing the planner to
select the appropriate percent of the stay region (in this case 100%) to add
to the test program. The planner would then select the elevation menu to
program the #7 partial support. This is accomplished by mousing on the
graphical representation of the #7 support shown in Figure 8.

The planner would then use the barometer to select 50% of this region.
A random number generator is used to determine the tubes in this pattern.
Finally the planner would define his own region by outlining its border with
the mouse and selecting 20% with the barometer. A sample 1000 tube program
generated with this system is shown in Figure 9. By providing a disk
storage and retrieval facility, the system allows planners to mix subsets of
programs using exclusive and inclusive set management. This gives the
planner the capability to build on existing programs, or completely exclude
past test sets.

It has been described through example how we have computerized both the
process of identifying steam generator problem regions and the process of
creating ECT programs. We now will describe how the system was used in an
application.

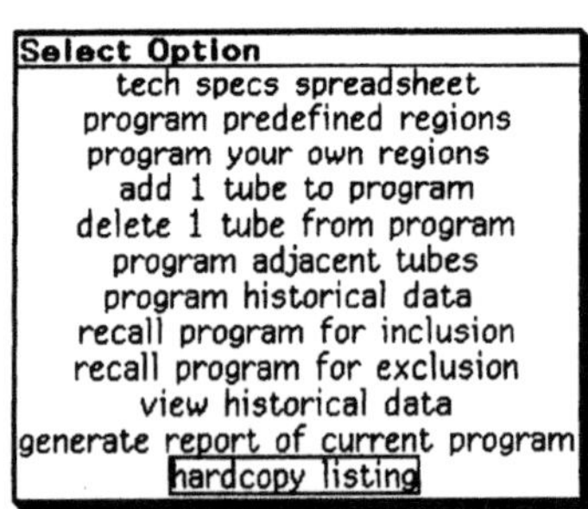

Figure 6: Menu of Planning Aid Tools

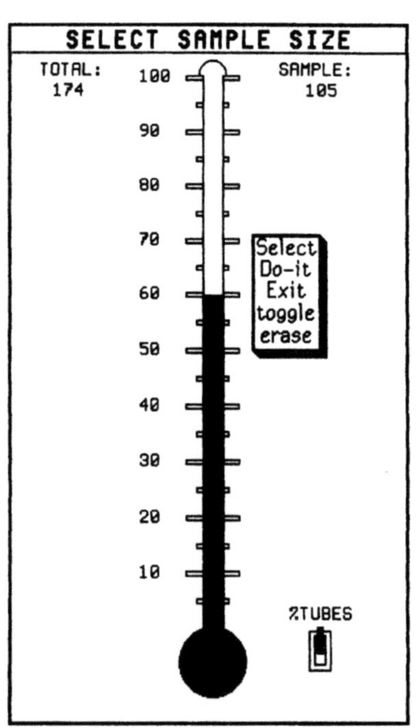

Figure 7: Barometer Icon

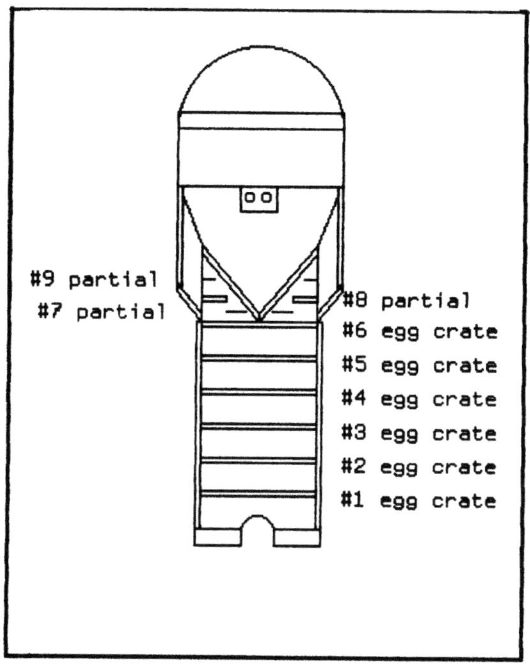

Figure 8. Steam Generator Elevation Menu

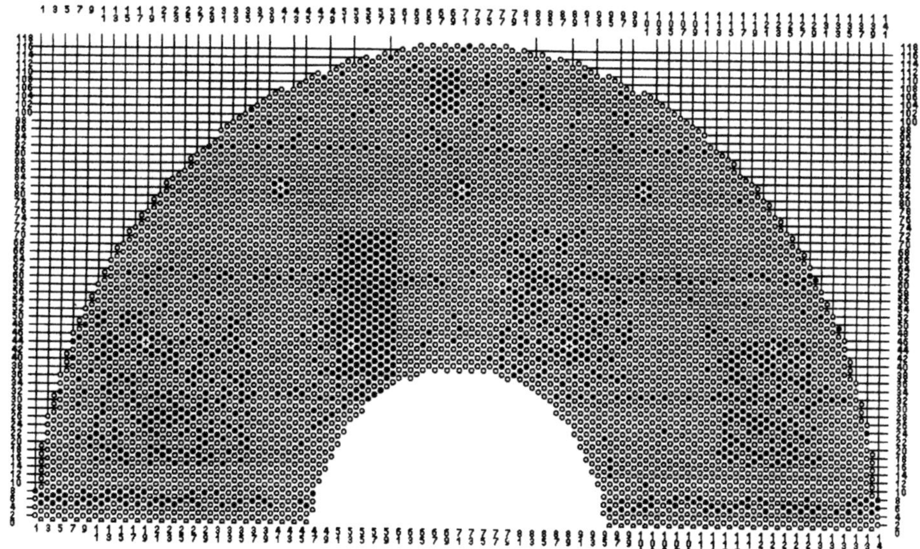

Figure 9. Sample ECT Program

EXPERIENCE

As part of the development effort, C-E made the Steam Generator Inspection Planning Expert System ready for field use during an actual plant outage at Maine Yankee Atomic Power (MYAP). This required loading the site specific design and historical data. The intent was to field test the concept and prototype to obtain field directed design ideas and feedback and to expose the technology to field and utility personnel. It turned out that the system performed so well that the inspection programs generated were actually utilized as part of the MYAP outage (not as a parallel side effort, but as part of the outage mainstream).

The system was used at the C-E main facility to generate the original 250 tube inspection program (prior to chemical cleaning) and the first expansion (to 550 tubes) program (after chemical cleaning). The system, utilizing a portable Symbolics Machine, was subsequently installed at the Maine Yankee CE Outage Services data center and used in the field to expand the inspection to a 1000 tube program. The system is currently undergoing configuration for use at a number of other C-E client utilities.

SUMMARY AND CONCLUSIONS

A Steam Generator Inspection Planning Expert System has been described. This system assists planners in identifying problem regions of the steam generator and in selecting tubes for eddy current testing. This system has been implemented using artificial intelligence based methods for representing and processing information. The system has been developed with a symbolic programming language (LISP) and incorporates object oriented programming techniques and rule based programming. The system was field tested at Maine Yankee Atomic Power and enabled field service personnel to generate timely technical specification complaint initial and expansion ECT programs throughout the inspection.

Generating the inspection plan is the first step in an eddy current inspection. Large amounts of data, the time critical nature of the task and the need for highly accurate inspection plans are but a few reasons for automating the inspection planning process. The system developed to date focuses on steam generator inspection planning. The inspection planning concept can be applied to a number of other nuclear power plant applications such as steam line weld inspections, pressure vessel inspections and bond inspections. In addition, it is important to note that planning is just one phase of an inspection program. Expert Systems technology can also be useful in the actual inspection process, data analysis, diagnostic process, corrective action programs, equipment recommendations, and scheduling.

ACKNOWLEDGEMENTS

I would like to thank Charlie Ashman, Carl Neuschaefer, and Jack Lareau for providing the expertise and management support required in the development of this project.

CHAPTER 3

ALARM AND SIGNAL VALIDATION

A REAL TIME KNOWLEDGE-BASED ALARM SYSTEM EXTRA

J. Ancelin, J.P. Gaussot, and P. Legaud

Electricité de France
Direction des Etudes et Recherches
6, quai Watier, 78401 Chatou Cedex

ABSTRACT

EXTRA is an experimental expert system for industrial process control. The main objectives are the diagnosis and operation aids. From a methodological point of view, EXTRA is based on a deep knowledge of the plant operation and on qualitative simulation principles. The application concerns all the electric power and the Chemical and Volume Control System of a P.W.R. nuclear plant.

The tests conducted on a full-scope simulator representative of the real plant yielded excellent results and taught us a number of lessons. The main lesson concerns the efficiency and flexibility provided by the combination of a knowledge-based system and of an advanced mini-computer.

A new version of EXTRA, with an extended application domain, is being developed and will be installed at one BUGEY unit (a 900 MW P.W.R. plant) in June of 1988.

INTRODUCTION

One of the most critical moments of the operation of nuclear power plants is when one is faced with accident conditions involving lots of alarms working simultaneously. Indeed, whereas an isolated alarm attracts the operator's attention on a local problem, the presence of numerous alarms makes the diagnosis of a situation and its follow-up more difficult.

Therefore, when expert systems have been developed ([1], [2], [3], [4], the Service de la Production Thermique and the Direction des Etudes et Recherches decided to explore their potentialities for alarm processing.

By the end of 1985, the merits of knowledge-bases systems in this field had been demonstrated on a research prototype [5]. To determine whether the system could be extended to the entire facility, a fairly large field prototype was built to perform tests on a simulator used for operator's training.

OBJECTIVES AND METHODOLOGY

EXTRA is aimed at two objectives :

- To identify the plant conditions on line and to check the consistency
  of the collected data [6].

- To make a diagnosis and, possibly, on the basis of this diagnosis, to
  select one operating document (procedure, alarm sheet) among those
  available.

From a methodological point of view, EXTRA is based on a deep knowledge
of the plant operation and of the qualitative simulation principles.

DESIGN

We have chosen two expert system tools for building EXTRA.

The first is used during the design phase. It is the S2.BOOJUM [7]
language (inference engine based on predicate calculus using forward
chaining), it automatically generates rules bases from the topological
and functional description of the facility. These rules bases will be
used during on line processing.

There are written in LRC [8] (inference engine based on proposition
calculus using forward chaining) and perform the on-line processing,
strictly speaking.

This kind of compilation has the merit of combining the rapid execution
time of the LRC language, for the processing operation, and the power of
the S2.BOOJUM language for designing very large knowledge bases.

APPLICATION TO A NUCLEAR POWER PLANT

The objective of this application was to test the above-defined
processing method under representative conditions and to determine
whether the system can be extended to a whole plant.

It was therefore decided to develop a field prototype covering a fairly
significant part of a nuclear power plant (approximately 15 %) and to run
it on a minicomputer with a view to linking it with a full scope
simulator of the Bugey Electricité de France training center.

<u>Application field</u>

The EXTRA application concerns all the electric power supplies and a
complex thermohydraulic system, with its auxiliary systems, in a 900 MW
PWR nuclear unit.

Quantitatively, it means :

  - 250 alarms,
  - 100 mechanical components (valves, pumps, heat exchangers),
  - 600 electric components (busbars, diesel generators, circuit
    breakers...),
  - 900 data acquired on line,
  - 1200 potential failures.

<u>Knowledge-Base</u>

This knowledge base incorporates three subsets that will be briefly described. This description will be completed by a simple example.

a) <u>Topological and functional description of the facility</u>

These principles are also formulated as S2.BOOJUM facts and correspond to general operating and diagnosis principles. By combining with rules these principles to the facility description, a specific rules base for the studied plant is automatically generated. Nowadays, there are 40 such principles.

By applying these principles to the description of the plant, some 5000 LRC rules can be automatically generated. These rules are simple. On the average, they made up of two premises and two conclusions.

c) <u>Specific rules base of the facility</u>

This rules base includes, first, rules automatically generated, as described above, and, second, directly handwritten rules (about 5 %). The handwritten rules actually describe very specific operating conditions.

d) <u>Example</u>

We will illustrate the different points mentioned above with a straight forward example. Let us take the following electric diagram.

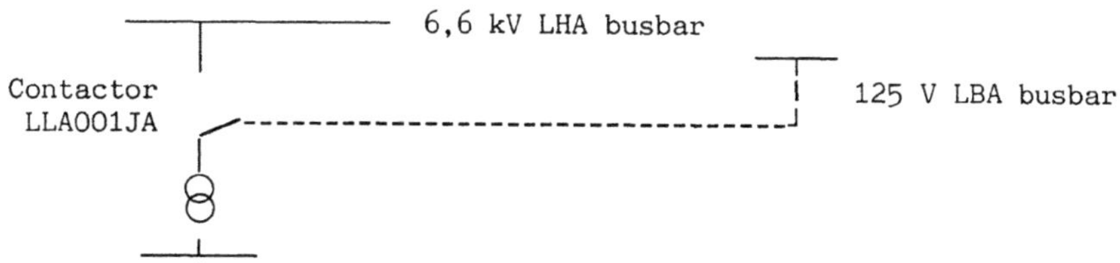

* Description

LHA	NATURE	BUSBAR
LBA	NATURE	BUSBAR
LLA	NATURE	BUSBAR
LLA001JA	NATURE	CONTACTOR
LHA	SUPPLIES	LLA001JA
LLA001JA	SUPPLIES	LLA
LBA	CONTROLS	LLA001JA
LLA001XU	CONCERNS	LLA
LLA001XU	MONITORS	POWER LOSS
LLA001XU	NATURE	ON-OFF INFORMATION

* Generic principle

IF

     T is a busbar
and   T is energized
and   T with N supply lines
and   N - 1 supply lines are loss (open breaker or loss of power supply
     to upstream busbar)

THEN the Nth supply line is operating (closed breaker and upstream busbar energized).

* Specific rule

When the previously defined principle is applied to our figure, the following rule is obtained :

```
RULE    :  LHA - LLA
IF         LLA POWER LOSS  = NO
THEN       LLA001JA STATUS = CLOSED
           LHA POWER LOSS  = NO
```

IMPLEMENTATION ON A REAL TIME MINI-COMPUTER

Organization

We will briefly describe the implementation of this application on a computer.

As regards the hardware, the application has been implemented on a BULL SPS 7 machine in the following configuration : monoprocessor MOTOROLA 68010 running under UNIX. The GOULD - SPS 7 communication is achieved by an RS 232 link with a 9600 baud rate. (The GOULD computer is the simulator).

Operating characteristics

a) Memory

The inference engine and the communication modules require 48 K of memory. Fifty K are necessary for the sole presentation module.

The rules and facts take up less than 128 K.

b) Computing time

The average cycle time is 12 sec. (about 5 sec. for the inference). This varies slightly with the number of times the rule base is consulted but is always below 15 sec.

LESSONS DRAWN FROM THE TESTS

The tests conducted during some 100 hours on a full-scope simulator representative of the real plant yielded excellent results and taught us a number of lesssons.

The first lesson is of a general character. It concerns the efficiency and flexibility provided by the combination of a knowledge-based system and of an advanced mini-computer. Thus, the cost of this prototype is assessed at 18 engineers x months.

The other lessons are more specifically related to the EXTRA application itself. The system has the following advantages :

- The capability of the system to detect inconsistencies concerning failures of the instrumentation.
- The modular design of the system.
- The system adaptability even in conditions resulting in information losses.

CONCLUSION

The tests of EXTRA on a simulator demonstrated that the knowledge-based
system approach is promissing when applied to the diagnosis and
processing of alarms in industrial processes.

Now the evaluation an development of EXTRA is continuing in real
operating conditions.

A new version of EXTRA, with an extended application domain
(quantitatively it means about 5000 components and 3000 data acquired on
line) will be installed at one BUGEY unit (a 900 MW PWR plant) in June
of 1988.

REFERENCES

[1] GONDRAN M.
    Introduction aux Systèmes Experts.
    Eyrolles - Paris - 1985.

[2] WATERMAN D.A.
    A Guide to Expert System.
    Addisson - Wesley Publishing Company - 1986.

[3] NELSON W.R.
    REACTOR : an expert system for diagnosis and treatement of nuclear
    reactor accidents.
    Proceedings AAAI - 82 p 296 - 301 - 1982.

[4] R.L. ENNIS ET AL.
    A continuous real-time expert system for computer operations
    IBM JOURNAL RESEARCH DEVELOPMENT - n° 1 - January 1986.

[5] ANCELIN J. - LEGAUD P.
    Un système expert pour le traitement des alarmes d'un réacteur
    nucléaire.
    Sixième Journées Internationales - Les systèmes experts et leurs
    applications - Avignon - 1986.

[6] OSBORNE R.L. - GONZALEZ A.J. - BELLOWS J.C. - CHESS J.D.
    On line diagnosis of instrumentation through artificial intelligence -
    1985.
    Westinghouse Electric Corporation - Power Generation Operations
    Division - Orlando, FL. 32817.

[7] DORMOY J.L.
    Notice du langage et guide d'utilisation de S2.BOOJUM.
    Note EDF-DER HI/5551/02 - Septembre 1986.

[8] HERY J.F. - LALEUF J.C.
    Notice d'utilisation et analyse des logiciels LRC.
    Note EDF-DER 6 HT 14/23/85 - HI/5040-02 - 1985.

A NUCLEAR REACTOR ALARM DISPLAY SYSTEM UTILIZING AI TECHNIQUES FOR

ALARM FILTERING*

Dan Corsberg and Larry Johnson

Idaho National Engineering Laboratory
Idaho Falls, Idaho 83415

A control room upgrade is currently being performed at the Advanced
Test Reactor (ATR) at the Idaho National Engineering Laboratory (INEL).
As part of the control room upgrade, an improved alarm monitoring and
display system is being installed. This system consists of completely
redundant termination boards, sequential events recorders (SERs), and
fiber optic serial alarm transmission from the SERs to the window panels
and to "host" computers. Residing on these host computers will be
knowledge-based alarm filtering systems that will pass filtered and
prioritized information on to several CRT displays. Further processing
will be performed at the display workstations to provide the most
effective display to the operators.

The knowledge-based Alarm Filtering System (AFS) is an integral part
of the new monitoring and display system. AFS aids operators by focusing
attention on the most significant information relative to the current
plant state.

INTRODUCTION

The ATR is located at the INEL near Idaho Falls, Idaho. EG&G Idaho,
Inc. operates ATR for the United States Department of Energy. ATR is a
test reactor providing controlled neutron environments for a wide variety
of materials testing, fuels testing, and engineering studies. The dynamic
nature of this work requires an on-going process of adding and deleting
alarms. The facility itself is twenty years old and has a life expectancy
that extends beyond the year 2000.

The objective of this upgrade is to provide a near state-of-the-art
control room display system for the remainder of ATR's projected life.
Because of ATR's operating characteristics, several requirements must be
satisfied by the display system:

- The system, from terminal boards to annunciator panels, must be able
  to adapt to changing alarm display requirements and configurations.

- ATR operations personnel often need to reconstruct sequences of
  events and alarms. The new monitoring system must be able to time
  tag information to a resolution of one millisecond.

*Work supported by the Department of Energy under DOE Contract
No. DE-AC07-76ID01570.

- Since ATR is a testing reactor, there are many scheduled shutdowns per year (ten to twenty) to allow experiment changeout and changes in fuel placement. During these shutdowns, many alarms are activated that are meaningless when the reactor is in shutdown. This "clutter" of alarms reduces the effectiveness of the alarm display. The new display system must allow for mode switching so that alarms can be disabled if they are not appropriate for the current plant mode.

- The display of alarms on the CRTs must be better and more effective than the display provided by the annunciator panels. Operators did not want alarms displayed on the CRTs unless a significant improvement in display capabilities could be demonstrated.

This paper describes the hardware and software used to support and meet the requirements listed above. The hardware architecture will be described first, followed by a discussion of AFS, and of the alarm display on the CRTs.

The alarm monitoring and display system being installed at ATR is a significant improvement over the previously existing display system. It represents the integration of several state-of-the-art technologies into a large, cooperating, display-oriented environment. The flexibility of the architecture and of the equipment being used will allow the system to evolve over its projected lifetime.

SYSTEM ARCHITECTURE

The hardware configuration was designed to minimize the impact on existing plant wiring, to expedite installation, and to be flexible and powerful enough to be useful over the remainder of ATR's projected life.

Fig. 1 is a simplified block diagram of the system. The vendor will supply input termination boards with physical dimensions and electrical terminations emulating the input termination section of the existing backlighted window panels. The new input terminals will be labeled identically to existing input terminals so that no relabeling of plant wiring will be required. Wiring and conduit will be removed from the existing panels and reinstalled in the same position in the vendor supplied input termination boards. These measures were taken to minimize impact on existing wiring and to allow for incremental installation.

Vendor supplied connectors and cross-linked polyethylene cabling will bring the field inputs to the Sequential Events Recorders (SERs) cabinet. This cabinet will house parallel redundant SERs. These SERs record contact closures and openings with a 1 millisecond resolution. Two independent unidirectional fiber-optic serial communications ports from each SER will transmit alarm information to the control room window panels. Fig. 1 shows only the termination boards and cabling for one of seven alarm panels in the control room. The time span from one field contact operation to window panel reception varies from 10 to 522. milli-seconds, with an average time of 266 milliseconds. If multiple alarms occur simultaneously, there is an extra 3 milliseconds per point in each port required for transmission.

Annunciator response switches (acknowledge, test, reset) will communicate with the window panels via existing wiring to the input termination boards and new connections to the window panels. These response switches also communicate with the host computers described below.

A fiber-optic port from each SER will provide communication with

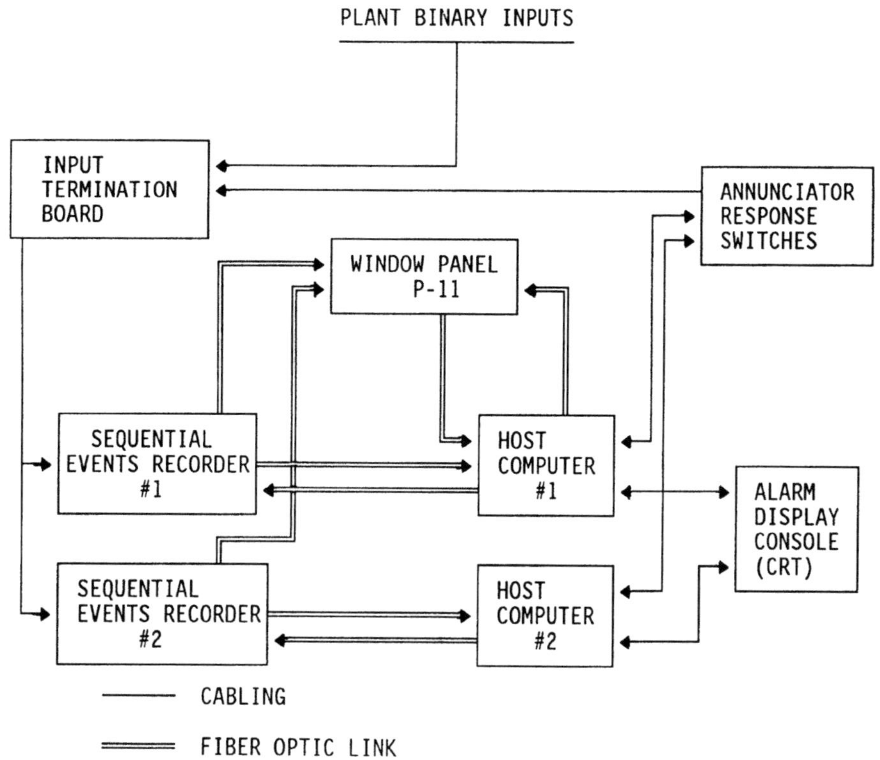

Fig. 1. System Architecture.

redundant host computers. These computers control system configuration, mode switching, self testing, and the interface to the console display system. Configuration and interrogation of the SERs will be accomplished via the bi-directional fiber optic link between the SERs and host computers. The window panels are also configured from the host computers. The ability to reconfigure the entire alarm display system contributes to the system's flexibility and usefulness over the life of the facility. The host computers will perform mode switching (enabling and disabling of alarms to match plant operating conditions) by sending the required information to the SERs.

The SERs and window panels continuously self-test for proper response. If an SER detects problems in itself, a message is transmitted to its host computer to initiate corrective action. The SER testing includes a full end-to-end check of a simulated input contact operation. If an individual window panel detects problems with itself or in either of its communication links, both the associated SER and host computer are notified. The window panel testing includes testing of the alarm light bulbs in each of the redundant channels.

The host computers act as an interface between the alarm system and the console display system. The console display system consists of several workstations connected together through an ethernet system. Each workstation has as much computing power as the host computer. Some of the high resolution color displays will be equipped with touch screens to improve the human-computer interface. Besides displaying alarms as described later in this paper, the consoles are used to provide operational information about power, core configuration, and plant parameters.

ALARM FILTERING AND PRIORITIZATION

A major problem at nuclear reactors is the on-rush of alarms that
occur following a major transient[1,2,3]. While operators can process
and assimilate this information, undue stress is placed on the operator
in doing so. This stress can lead to mistakes. Pertinent information
must be presented in a manner that is clear and concise so that it does
not compound the stress. The approach used to process raw information
and determine what is pertinent is described in this section. The presentation of that information is discussed in the next section.

The goal of alarm filtering is to identify the most important
information and focus operator attention on it. In the past, various
methods have been used to accomplish this; functional grouping of annunciators; static prioritization; and first-out logic. Each of these have
been useful in some circumstances, but none have completely solved the
problem. AFS does not solve all the problem either, but it is a positive
step in the right direction.

The basic idea is to determine how important a piece of information
is relative to the current state of the plant. At ATR, we are dealing
exclusively with alarms. Thus, AFS assigns a level of importance to an
alarm's activation based on the set of currently activated alarms. As
the set changes (other alarms are activated or deactivated), that level
of importance is changed as needed. The level of importance (or prioritization) is dynamic, changing as the plant state changes. This
closely models the way operators process alarms that are activated or
deactivated.

There are just a few types of relationships that need be defined and
used in processing these alarms. These relationships are functional in
nature because they are based on the way the plant functions rather than
on a set of physical principles. Each type of relationship has a set of
rules associated with it concerning how the alarms should behave and what
levels of importance should be assigned. The four types of relationships
are level and direct precursors, required actions, and blocking conditions. These are discussed in greater detail in references 4 and 5. For
now, we will briefly describe the level precursor relationship as an
example. This relationship is usually found when there are two setpoints
on the same parameter. If alarm A is a level precursor to alarm B, then
certain behaviors are expected. Alarm A should always be activated
before B. If both are activated, then B should deactivate prior to A.
Alarm A's occurrence does not imply B's just as B's clearing does not
imply A's.

We found that we could separate these generic rules from the specific alarms. Thus, the rules stated in the previous paragraph could be
applied to any two alarms that had a level precursor relationship. This
separation of generic knowledge from specific information about the alarms
themselves contributes greatly to the flexibility of the approach. Defining a new alarm is simply a matter of gathering the information about what
types of relationships it might have with other alarms in the system. The
rules about behavior and prioritization remain unchanged.

The implementation of AFS uses two types of representation for the
two types of knowledge used. Rules are used to represent the generic
knowledge about how alarms should behave based on their relationships. A
technique called object-oriented programming is used to represent the specific knowledge about alarms. Object-oriented programming supports the
integration of both types of knowledge into a cooperating software system

that emulates human processing. Each object representing an alarm
acts as a relatively independent entity, invoking and stopping processing
as needed for the specific alarm represented. If an alarm is activated,
only those alarms that might be concerned about the activation are notified. They, in turn, can invoke further processing to readjust their own
priority. When an alarm is deactivated, only those alarms that might
need to readjust their priority are notified. Unnecessary processing and
long searches of pre-enumerated sequences are avoided by this approach.
There is no need to define those possible sequences ahead of time because
priorities can be adjusted as the information becomes available.

Currently, AFS is being installed at ATR and at a chemical processing plant. The generic nature of the functional relationships and
their associated rules allows AFS to be applied to a wide variety of
processes. Other applications include telecommunications and aerospace.
In addition, the Department of Energy has applied for a patent on the
approach used in AFS.

In the introduction to this paper, we mentioned four requirements
that the display system had to meet in order to be a viable interface for
the expected life of ATR. As ATR evolves, so too will the alarm configuration as well as the knowledge about relationships that alarms have.
AFS is flexible enough to handle these modifications without changing the
rules that its processing is based on. While mode switching is handled
in the sequential events recorders, it could also be handled in AFS on a
software level. Finally, AFS provides the underlying processing that
makes the CRT alarm display better than the annunciator panels. AFS
provides the added information needed to focus attention on the most
important alarms.

CRT-BASED ALARM DISPLAY

This section describes how alarm messages are displayed on the CRTs.
The display of alarms on the CRT's supplement annunciator window panels.
Of the five CRTs in the control room, one will be dedicated to alarm display. The color red will be used to emphasize higher priority items while
yellow will show lower priority alarms. The CRT screen dedicated to alarm
display will have the basic layout shown in Fig. 2.

One of the major problems with the display of ATR's alarms is the
fact that there are over 400 alarm tiles on the various control room
panels. It would not be possible to put all of these tiles on a single
screen in a useful manner. We had to avoid paging screens of alarms
because of problems in getting to the most important alarms in the
quickest manner. A dual approach has been taken to solving the problem.

The main panel display area is the focal point of the interface. Its
format will be in one of two types, depending on what the operator wants
to see. The default display will be a "dynamic" alarm panel, showing only
the activated alarm tiles. This dynamic display will expand, shrink, and
change as alarms are activated, deactivated, or priorities are changed.
Higher priority alarms will be grouped at the top and displayed in red.
Lower priority alarms will be displayed in yellow below the other tiles.
If more tiles are activated than can be displayed, lower priority alarms
will be removed until tiles are deactivated and removed from the display.
The advantage of this approach is that no deactivated tiles are shown.
Only activated alarm tiles are displayed and even those are grouped and
maintained such that the most important alarms are always on the screen.
The disadvantage of using a dynamic display is that operators lose the
spatial relationships they use when dealing with window panels.

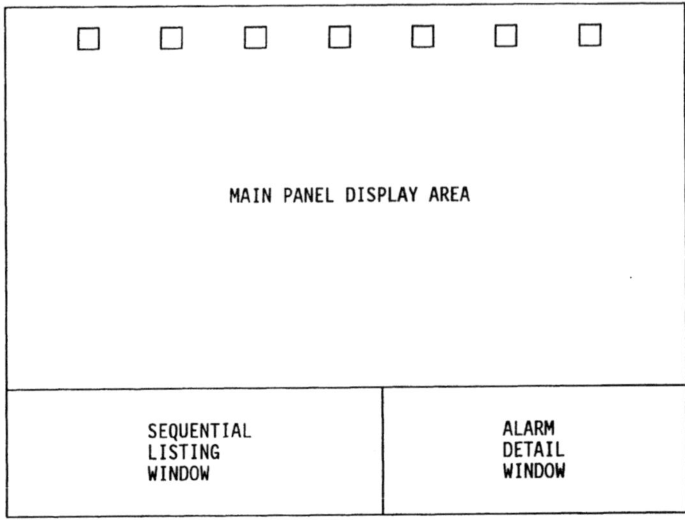

Fig. 2. ATR Alarm Console Display Layout.

In order to address the problem of spatial relationships, the main
panel display area can also be used to display a mimic of a specific window panel. The selection of a panel is made by the operator using
miniature panels that range across the top of the screen. Each miniature
panel corresponds in relative size, shape, and position to the window
panels in the control room. When an operator wants to look at a panel
(rather than the dynamic display), the operator touches the corresponding
miniature panel and it is expanded into a full size mimic. When the
default display is desired, the mimic is shrunk back up into its
miniature panel.

The use of miniature panels works for selecting a mimic of a window
panel to expand and display. We have added another feature to show operators what priority AFS assigns to the activated alarms on each panel.
Tiles within the miniature panels will not be engraved, but will be
illuminated with the appropriate color coding as each alarm is activated.
Thus, the operator can quickly scan the miniature panels, see what panel
has the most important alarms, and expand it.

There are two other areas on the display screen, the detail window
and the sequential listing window. The detail window will be used to display setpoints and readings for an alarm selected by operators. Selection
will be made by touching the desired alarm tile. The sequential listing
window simply lists the last few alarm messages that have come in. As
more arrive, the oldest are scrolled off the window. Operators can examine sequences in further detail by calling a utility program to look at
stored data. This utility resides on one of the other display
workstations.

The design of this display is meant to be simple and to the point,
providing the most useful information to the operator with minimal stress
and effort. As with the previously described portions of the display
system, this part is also designed to support the evolution of ATR. The
contents of the displays can be changed very easily. The alarm tiles are
changed automatically when modifications are made to the configuration of
the window panels. The use of colors to emphasize information helps to
make the CRT display more effective than the window panels. Finally, the
touch screen provides an interactive capability that is easy and natural.

IMPLEMENTATION AND INSTALLATION

One of the requirements placed on this project was that reactor downtime be minimized. The system architecture selected will allow panel by
panel replacement of the existing window panels during normal shutdown
periods. During this replacement process, there will be no alarm filtering and limited capabilities for mode switching and sequential event
display. The computer based portion of the display system will also be
installed incrementally with minimal impact on ATR's operational schedule.

A complete replica of the display system is being installed in a
simulator for ATR. Software development will be performed on the simulator's display system. The simulator will be used for testing, evaluation, and verification of the software prior to placement on the actual
ATR control room display system. An AI programming environment is being
used in the implementation of AFS for ATR's computer system.
In addition, several databases are used to initially configure AFS, the
displays, and the communications system. Other display programs are being
developed for single and multiple parameter displays, system mimics, and
core power level displays. Thus, AFS can be seen as one small process in
a much larger environment of display and data analysis programs.

SUMMARY

This paper has described an innovative architecture of hardware and
software for displaying alarm information. The system has been designed
to meet the four goals of flexibility, mode switching, sequential events
recording, and improving the operator interface to the alarm system. The
control room display system is designed with the future in mind, allowing
for the expansion of capabilities as technology advances.

The use of a knowledge-based approach to alarm filtering is
significant in that it is one of only a few real-time applications of
AI. AFS is a first step towards the development of more powerful
diagnostic aids for operators. Just as AFS now filters and prioritizes
alarms for humans, in the future it can perform the same function for
other diagnostic processes that can afford to take more time and to
reason on a deeper level about the problems that the process may be
having.

REFERENCES

1.   P. Visuri, A Minicomputer-Based Multivariate Alarm System and
     Results From Operation in a Nuclear Power Plant, IEEE Conference in
     Decision and Control, Las Vegas, Nevada, December 12-14, 1984.

2.   W. W. Banks and Boone M. P., "Nuclear Control Room Annunciators:
     Problems and Recommendations," NUREG/CR-2247, EGG-2103,
     September 1981.

3.   A. Krigman, Alarms, Operators, and Other Nuisances: Cope . . . or
     Court Catastrophe, InTech, December, 1985, pp. 33-40.

4.   D. R. Corsberg and D. Wilkie, An Object-Oriented Alarm Filtering
     System, Power Plant Dynamics, Control & Testing Symposium,
     Knoxville, TN, April 14-16, 1986.

5.   D. R. Corsberg, Alarm Filtering: Practical Control Room Upgrade
     Using Expert System Concepts, InTech, April, 1987, pp. 39-42.

AN EXPERT SYSTEM FOR ALARM DIAGNOSIS AND FILTERING

J. T. Robinson and P. J. Otaduy

Instrumentation and Controls Division
Oak Ridge National Laboratory[*]
Post Office Box X
Oak Ridge, Tennessee 37831-6008, U.S.A

ABSTRACT

This paper describes an alarm processing system being developed at the Oak Ridge National Laboratory for implementation in the High Flux Isotope Reactor. The purpose of the system is to perform two related functions, alarm diagnosis and filtering, to aid nuclear plant operators in responding to the large number of alarms typically activated during a major plant transient. The alarm diagnostician determines the root cause of a pattern or sequence of alarms and generates an explanation of the events leading to the alarm sequence. The alarm filter identifies and deemphasize those alarms which are either irrelevant to the current plant condition or contribute no significant new information.

INTRODUCTION

The human factors problems associated with current alarm systems are well known. In a typical plant, the operator has over 2000 alarms and displays available to him, of which 500 or more may be activated during the first 5 s of a transient.[1] The alarm-processing system being developed at ORNL for the High Flux Isotope Reactor (HFIR) will aid the operator in interpreting and responding to an influx of alarms by performing two related functions, alarm filtering and diagnosis. The alarm filtering scheme is designed to inhibit or deemphasize alarms which are either irrelevant or add no significant new information to an existing alarm pattern. The diagnostic portion of the system attempts to identify a root cause and a probable sequence of events leading to a pattern of alarms.

The main components of the system are the Knowledge Base, Network Compiler, and Inference Engine. The key features of each of these components are described below.

---

[*]Operated by Martin Marietta Energy Systems, Inc., under contract DE-AC05-84OR21400 with the U.S. Department of Energy.

KNOWLEDGE BASE

The underlying knowledge base of the system is in the form of a cause-consequence network linking alarms with conditions in the plant. A simplified portion of the network for the HFIR is illustrated in Figure 1. Nodes in the network are represented as frames. Two types of frames are used, one representing alarms and the other representing plant conditions. These are illustrated in Figure 2. The cause-consequence network is developed by creating instances of these frames, which involves supplying (at a minimum) the name and possible causes or possible effects of the node, then operating on these instantiations with the network compiler described below. After compilation, the instantiations become nodes in the network.

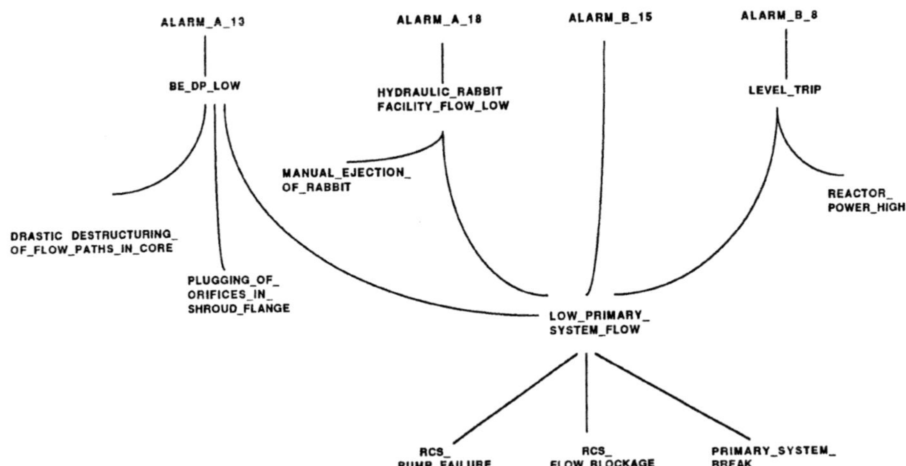

Figure 1.   Simplified cause-consequence network.

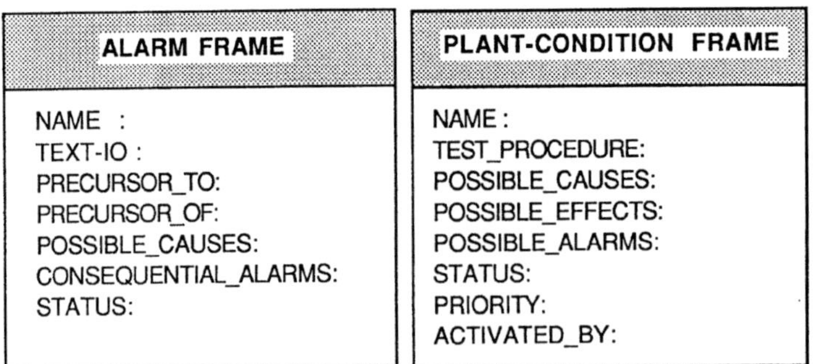

Figure 2.   Knowledge base frames for alarms and plant conditions.

After creating the nodes the knowledge base is run through a
knowledge network compiler. The compiler performs the following
functions:

1. links nodes into a cause-consequence network,

2. checks for consistency and completeness of the network,

3. determines the alarms that each plant-condition node is capable
   of activating (directly or indirectly), and

4. prioritizes branch nodes in the network.

Nodes are linked into a cause-consequence network by filling vacant
possible_effects slots from information contained in the possible_causes
slots and vice-versa. The consistency and completeness of the network is
checked, and the user is notified of (a) nodes that are referenced but
have not been created and (b) nodes that are isolated from the network.

The final step in the compilation procedure is to prioritize the
branch nodes. Branch nodes are nodes that have multiple possible
effects, for example low-primary-system flow in figure 1. These nodes
are placed in a special queue and ordered according to the number of
alarms each are capable of activating. This queue is used by the alarm
diagnostician described in the next section.

Figure 3 illustrates the values of the frame "low primary system
flow" before and after compilation. Note that the only information
supplied by the developer were the name, possible_causes, and
test_procedure values. The remaining slots were filled automatically by
the compiler.

INFERENCE ENGINE

The first step in each cycle of the inference engine is a status
check of all alarms. Active alarms are then filtered and displayed. Next
a test is performed to determine if all active alarms are explained by a
previously diagnosed root cause. If this test fails, the diagnostician
is invoked. The filter and diagnostician are described below.

Alarm Filter

The purpose of the alarm filter is to emphasize significant alarms
and deemphasize those alarms which are either irrelevant or contribute no
significant new information to the already existing alarm pattern. This
is a key concept of the alarm handling schemes discussed by Corsberg and
Wilkie[1] and Visuri et al.[2] At present, three methods are used to
accomplish this. The first is to simply inhibit alarms which are not
applicable to the current operating state, for example the low
temperature alarm during start-up.

A second filtering scheme is based on the concept of precursors
presented in reference 1. The alarm frame (see Figure 2) contains the
slots "precursor_to" and "precursor_of," which implies the following:

If A is a precursor to B, Then A should _always_ occur before B.

For example, the high-power alarm is a precursor to the very high power
alarm. If alarms A and B occur together, then alarm A is deemphasized.

```
              PLANT-CONDITION  FRAME

NAME                      : LOW_PRIMARY_SYSTEM_FLOW
TEST_PROCEDURE            : READ_CAMAC(A29)
POSSIBLE_CAUSES           : (RCS_PUMP_FAILURE
                              RCS_FLOW_BLOCKAGE
                              PRIMARY_SYSTEM_BREAK)
POSSIBLE_EFFECTS          : NIL
POSSIBLE_ALARMS           : NIL
STATUS                    : NIL
PRIORITY                  : 0
ACTIVATED_BY              : NIL
```

Figure 3a.   Before compilation.

```
                 PLANT-CONDITION  FRAME

NAME                   : LOW_PRIMARY_SYSTEM_FLOW
TEST_PROCEDURE         : READ_CAMAC(A29)
POSSIBLE_CAUSES        : (RCS_PUMP_FAILURE
                           RCS_FLOW_BLOCKAGE
                           PRIMARY_SYSTEM_BREAK)

POSSIBLE_EFFECTS       : (LEVEL_TRIP
                           LOW_HYDRAULIC_RABBIT_FACILITY_FLOW
                           LOW_BE_PRESSURE_DROP
                           LOW_REACTOR_PRESSURE_DROP
                           ALARM_B-15)

POSSIBLE_ALARMS        : (ALARM_A-13
                           ALARM_A-18
                           ALARM_B-15
                           ALARM_B-8
                           ALARM_1E-22)

STATUS                 : NIL
PRIORITY               : 5
ACTIVATED_BY           : NIL
```

Figure 3b.   After compilation.

The third method for deemphasizing alarms is based on separating
primary cause from consequential alarms. Primary cause alarms are
understood to be those alarms more closely related to the initiating
event. For example, in the case of a primary coolant pump trip a number
of alarms should occur including "low primary flow," "reactor pressure
drop low," and "strainer pressure drop low." The last two conditions may
be caused by low primary flow and are thus deemphasized relative to the
low-flow alarm. The relationships among primary and consequential alarms
are automatically generated from the cause-consequence network during the
network compilation stage and stored in the associated slot of each node.

<u>Diagnostician</u>

The purposes of the alarm diagnostician are as follows:

1. Determine the root cause of a pattern or sequence of alarms.

2. Generate an explanation of the events leading to the alarm
    sequence.

Two levels of reasoning are implemented. The top-level reasoner
uses knowledge about the structure of the cause-consequence network to
narrow the search space. This is accomplished by matching the alarm
pattern against patterns associated with the branch nodes within the
network. (The branch nodes and their associated alarm patterns are
automatically identified by the network compiler during the knowledge
base development stage). Branch nodes with the largest list of possible
alarms are tested first. This pattern matching typically narrows the
search space to a small region of the network.

After narrowing the search space based on the alarm pattern, a
second-level inference mechanism is invoked to determine the root
cause(s) of the event. This mechanism is in essence a modified depth-
first searching scheme. The first step in this scheme is to execute the
procedure (referenced in the test procedure slot) for verifying the truth
of a suspected node. This procedure might involve, for example,
requesting the current reading of a sensor from a data acquisition
program. If the procedure verifies the truth of a suspected node, its
possible causes are tagged as suspect and the cycle is repeated.

The search is considered complete when all activated alarms have
been explained in terms of root causes. A message is then given to the
operator of the form "Alarms ... were caused by ...." The operator may
then ask for an explanation in either printed or graphical form.

It is important to note that the alarm pattern is used only to
direct the search for a probable cause, not to draw conclusions. To
verify a suspected cause the second-level inference engine must be
invoked. This prevents the program from drawing inaccurate conclusions
based on an incomplete alarm pattern.

EXAMPLE PROBLEM

Figure 4 illustrates the conclusions of the filter and diagnostician
following a simulated transient instigated by a cold slug in the
secondary system. This event led to a cold slug in the primary system
and within 15 s resulted in the activation of 6 alarms related to a
decrease in primary system pressure and a power excursion. An event
graph constructed by the diagnostician is illustrated in Figure 5. The
alarm filter, using information in the possible_consequential_alarm slots

supplied by the network compiler, determined that the "core inlet temperature low" alarm was a primary cause alarm and that the other five were consequential. This is because a decrease in primary temperature can lead to a simultaneous decrease in pressure and an increase in power, but the reverse is not true. The diagnostician, using the alarm pattern, determined that the condition cold_slug_in_primary_system was a probable cause of the transient. This was confirmed by an analog reading of the primary temperature.

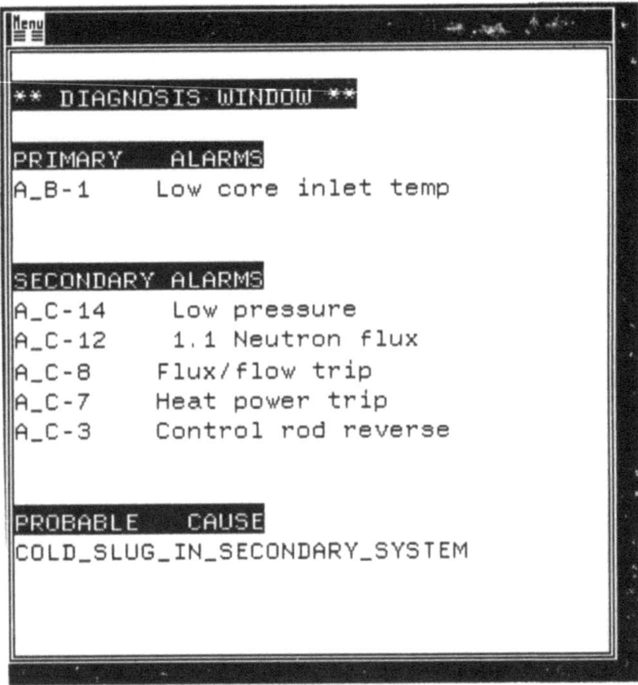

Figure 4. Sample diagnosis window.

SUMMARY

The alarm system described in this paper integrates a frame-based knowledge representation scheme with a specialized inference engine developed and optimized for alarm filtering and diagnosis. A key feature of this system is the use of a network compiler, which eases knowledge base development and debugging and optimizes the knowledge base for the inference engine.

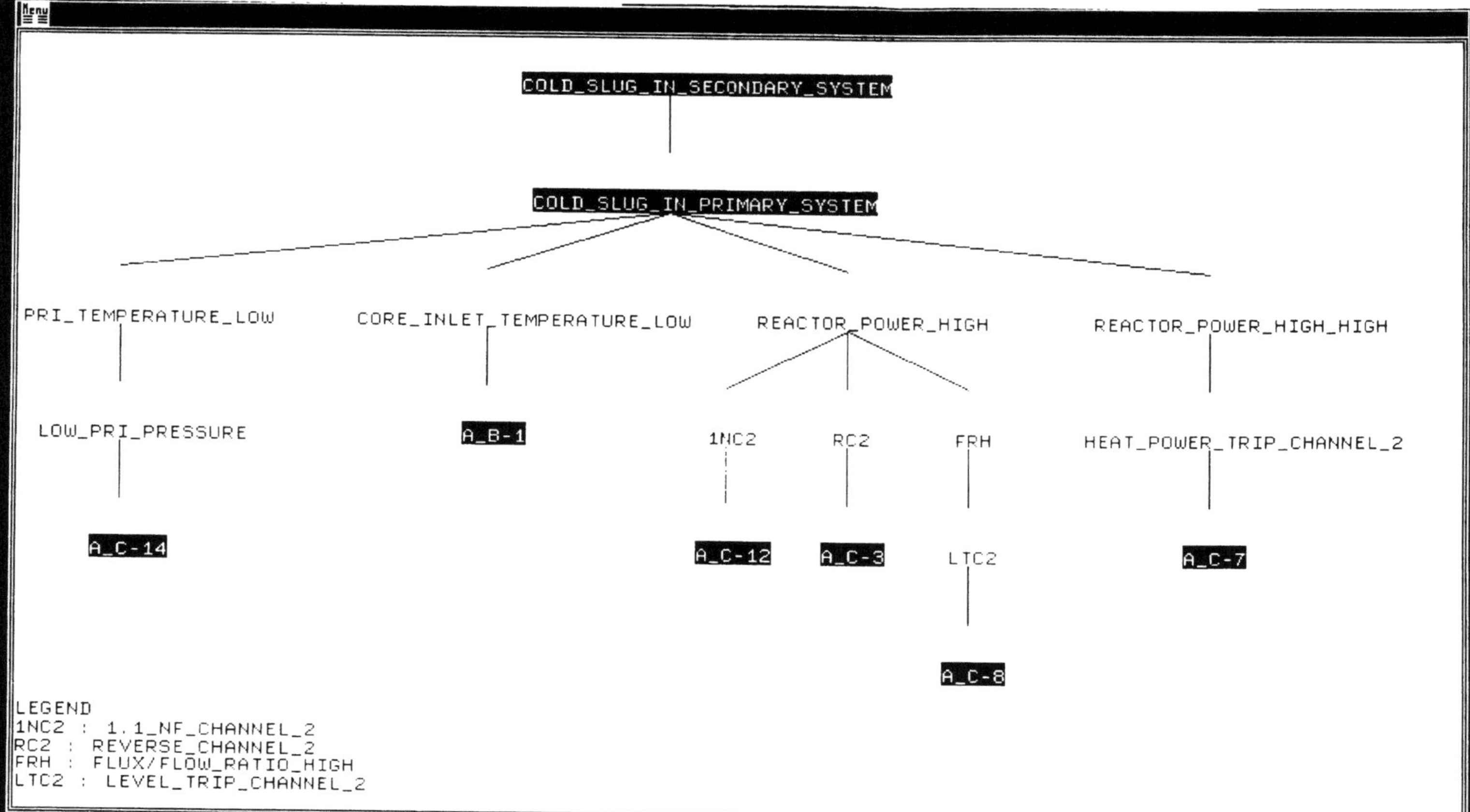

Figure 5. Diagnostic event graph.

REFERENCES

1. D. R. Corsberg and Dean Wilkie, "An Object-Oriented Alarm-Filtering
    System," Proceedings of the Sixth Power Plant Dynamics, Control, &
    Testing Symposium, Knoxville, Tennessee, April 14-16, 1986.

2. Visuri, P. J., Thomassen, B. B., and Owre, F. A., _Handling of alarms
    with logic (HALO) and other operator support systems_, Halden Work
    Report HWR-24, Halden, Norway, May 1981.

APPLYING AI TECHNIQUES TO IMPROVE ALARM DISPLAY EFFECTIVENESS[1]

J. M. Gross[2], S. A. Birrer[2], and D. R. Corsberg[3]

Idaho National Engineering Laboratory

SUMMARY

The Alarm Filtering System (AFS) addresses the problem of information overload in a control room during abnormal operations. Since operators can miss vital information during these periods, systems which emphasize important messages are beneficial. AFS uses the artificial intelligence (AI) technique of object-oriented programming to filter and dynamically prioritize alarm messages. When an alarm's status changes, AFS determines the relative importance of that change according to the current process state. AFS bases that relative importance on relationships the newly changed alarm has with other activated alarms. Evaluations of alarm importance take place without regard to the activation sequence of alarm signals. The United States Department of Energy has applied for a patent on the approach used in this software. The approach was originally developed by EG&G Idaho for a nuclear reactor control room.

AFS will soon be installed in the control room of the Flourinel and Storage (FAST) facility at the Idaho National Engineering Laboratory. The facility is part of the nuclear fuel reprocessing plant which is operated by Westinghouse Idaho Nuclear Company (WINCO). FAST began operations in 1986 and contains a complex, dynamic, chemical process. Operators have a sophisticated display system, the Data Processing System Enhancement (DPSE), to aid in monitoring that process. The DPSE is a powerful, computer based system that collects, archives, and displays most of the needed process information. It also guides operators in following specific sequences of process operations and performs process calculations.

While there are panel mounted alarm displays in the control room, the DPSE has its own alarm display. Each of its five CRTs contains a window which is dedicated to alarm display. Experience gained during initial operation showed that, for several reasons, operators were not making full use of that window.

---

1   Work supported by the U.S. Department of Energy Idaho Operations Office
     under contract DE-AC07-84-ID12435

2   Westinghouse Idaho Nuclear Co.   P.O. Box 4000   Idaho Falls, Id 83403

3   EG&G Idaho, Inc.   P.O. Box 1625   Idaho Falls, Id 83415

This was the basic problem addressed by AFS. Specifically, we wanted
to increase the usefulness and effectiveness of that window. Our objective
was to display and emphasize the most important alarms while de-emphasizing
and possibly not displaying those of less importance. Finally, we wanted to
simplify the alarm display on the DPSE while providing more useful information to operators.

INTRODUCTION

The purpose of FAST is to dissolve nuclear fuel elements and produce a
liquid from which uranium can be recovered by a solvent extraction process.
The dissolution process is performed in batches in three parallel systems.
During the process, operators must monitor parameters which include: temperatures and liquid levels in various vessels; temperatures and pressures in
components of the off-gas system; position and operation of numerous valves;
and pressures controlled by the building ventilation system. Optimum values
for some of these parameters depend on the stage of the process.

The DPSE displays alarm messages when these parameters are beyond prescribed limits. The alarm window is a limited resource, fixed in size at 36
characters wide by 16 lines. Since most messages are two lines each, they
may be visible only for a short time before being scrolled off the window.

Another function of the DPSE is to grant permission for valve operation
when such operation would not violate procedures. Operators indirectly open
and close valves by remote handswitches. The DPSE is able to sense the
position of the handswitches. And the valves themselves are instrumented so
that their status may be known by the DPSE. Valve operation was a major
source of messages to the alarm window since separate messages were generated for granting of permission, handswitch operation, and valve operation.

Other messages were sent to the window when alarms were acknowledged
and when actual alarm conditions returned to normal. Routing these informational messages to the alarm window also contributed to scrolling off of
important information.

The off-gas system includes a condenser, scrubber, mist eliminator,
heater, and high efficiency particulate air (HEPA) filters. The HEPA
filters remove particulates, the scrubber removes chemical vapors, and the
other components are required to keep the filters dry. A single problem in
the off-gas system may cause several alarm messages to be generated. Prioritizing these messages can help the operator determine the root problem.

In addition to the alarm window of the DPSE display, other alarm indications are available to the operator. These include lighted annunciator
panels, an audible signal, and a printed alarm listing from the DPSE. Since
these would still be available after AFS is installed, the alarm window can
be made more effective by preventing some messages from being displayed and
possibly changing the order of others.

The most significant problem of the DPSE alarm display system was the
large number of messages being sent to the window, the majority of which
were informational rather than actual alarms. Often, status signals such as
valve position indicators were the sources of messages to the alarm window.
We could have eliminated these status messages entirely, but there are situations where a valve's position would be an alarm condition. Since valves
are operated indirectly by remote handswitches, an alarm condition would
exist if a valve failed to open or close after a handswitch operation. AFS
is able to detect this condition and generate an appropriate message.

136

Another problem was the activation of alarms when actual alarm conditions did not exist. Examples from this category include alarms that do not apply to the current operating state and alarms that are expected due to current conditions. These nuisance alarms reduced the effectiveness of the entire alarm system. Operators tended to ignore or, at the very least, discount the significance of the nuisance alarm's activation. This is a natural reaction to repeated false alarms.

The objectives of AFS are to emphasize information that is immediately important; de-emphasize information that is of less importance to operators; and reduce the number of messages to the alarm window. Color coding of alarm messages achieves emphasis. Instead of displaying a message stating that an alarm condition has cleared, the alarm message is removed from the window. Combining messages for permission, handswitch, and valve operation into a single status message which is not sent to the alarm window further reduces message traffic to the alarm window.

We also decided that when the alarm window becomes full, the least important message is scrolled off rather than simply the oldest. As space becomes available in the alarm window due to cleared alarm conditions, messages which have been scrolled off are scrolled back onto the window. In this case the message selected to be returned to the window is the most important of the scrolled off messages rather than simply the most recent. Although this may result in messages being displayed out of chronological order, FAST operators felt that the most important alarms should always be displayed in the alarm window. Since the sequence of alarms is available from a printer in the control room, they did not consider this to be a problem.

Conventional programming techniques could provide solutions to these problems. But they would not be easy to implement and maintenance would be even more difficult. Object-oriented programming can be used to simplify both the handling of dynamic interaction of alarms and maintenance of the system.

OBJECT-ORIENTED PROGRAMMING

Conventional programming treats procedures and data separately. In object-oriented programming data and operations are combined into entities called objects. An object is a collection of information and a description of how that information is operated upon. Manipulation of information and possibly some occurrence of action are the result when objects receive messages. Different object types may receive identical messages but respond differently.

The points to be made are: the object sending a message tells another object to do something; the message does not specify how to do it; and the object which receives a message carries out the action in a way specific to the type of object it is. This allows data abstraction which means that solutions are in terms more closely related to the problem. As a result, underlying implementations may be changed without changing calling programs or procedures. Individual instances of the same type of object are part of the same class and may share common procedures. This inheritance of behavior and attributes eases the definition of new instances of the same type and simplifies global changes. It is easy to define a subclass for a group of objects which are only slightly different from an already defined class. Individual instances in the class "Alarm", have variables which include identifier, short name, and other alarms which are related. The subclass called "Analog" contains these and variables for value and units.

Individual objects in AFS represent each possible alarm condition. For
example, the object for low flow of cooling water could send a message to
the object for high temperature. Based on a type of relationship, AFS could
determine that the high temperature alarm is of secondary importance. If we
were interested in process control in addition to alarm filtering, the low
flow object might send a message to an object representing a valve. The
valve object would respond to the message by opening the correct valve in
the plant.

FOUR TYPES OF ALARM RELATIONSHIPS

Four different relationships between alarms have been defined for AFS.
They are level precursor, direct precursor, required action, and blocking
condition. These are used to determine the relative importance of activated
alarms. We decided to have only primary and secondary levels of alarm priority at FAST. Their messages are color coded red and yellow respectively.
In addition, some conditions are reduced in priority to a notification and
their messages are not displayed in the alarm window.

The simplest relationship is level precursor. If X is a level precursor of Y, then X should occur before Y. However, the occurrence of X does
not imply that Y will follow. If alarms are activated for both X and Y,
then X is a secondary level alarm and Y is primary. The level precursor
relationship usually occurs when there are two alarm setpoints on a single
parameter. For example, a low alarm is a level precursor to a low-low
alarm. A high alarm would have this relationship to a high-high alarm. If
no other relationships apply, a low alarm is displayed as primary until the
corresponding low-low alarm is activated. When this happens, the low alarm
is changed to secondary and the low-low alarm is of primary importance.

The direct precursor is basically a causal relationship. It is similar
to the level precursor but has an important difference. If X is a direct
precursor of Y, then not only should X occur before Y, but the occurrence of
X implies that Y will follow after some time delay. If alarms are activated
for both X and Y, then X is primary and Y is secondary. Furthermore, there
may be other direct precursors to Y in addition to X. Either low flow of
cooling water or high temperature of cooling water may be considered to be
the cause of a high temperature in a vessel. If so, both are direct precursors to the high vessel temperature alarm. If only a precursor or the high
vessel temperature alarm is activated, it is of primary importance. But if
both are activated, the direct precursor (cause) is primary and the high
vessel temperature alarm (effect) is secondary.

The required action relationship differs in that alarm messages are not
filtered or prioritized. It is used to notify operators that an important
expected event has not happened within a prescribed time limit. If Y is a
required action of X, then if X occurs, Y had better follow. In this type
of relationship, the initiating event may be considered to be a direct precursor of the required action. In some situations, the fact that an action
is missing may be more important than any of the activated alarms. A valve
operation after its corresponding handswitch operation is an example of this
type of relationship. If the valve fails to open or close after a few seconds, AFS generates an alarm message stating that the valve was expected to
operate.

The blocking condition relationship prevents the display of alarm messages which are not true alarms in the current state. If X is a blocking
condition of Y, then Y is not displayed when X is true. X is usually a
state or condition rather than another alarm. A pump which is not operating

may be a blocking condition for low flow rate. Another example of this type
occurs in the building ventilation system. Different pressures which are
maintained in various plant areas are disturbed when doors are opened. A
timer is started when the alarm is activated. The timer is a blocking condition for the alarm. After the timer expires, the blocking condition no
longer exists and the alarm message is displayed. If the door only remains
open for several seconds, correct pressures are restored before the timer
expires and the alarm message is never displayed. It is more cost effective
to add this type of delay to an existing system than to add hardware delays.

An individual alarm may be related in various ways to several other
alarms. In general, when an alarm is emphasized by one relationship and
de-emphasized by another, the de-emphasis takes precedence. Additionally, a
few alarms have been defined to be always primary and others as always
secondary.

AFS contains two types of knowledge. The first is generic knowledge
about prioritizing alarms based on the four types of relationships which
have been defined. It is encoded into the Lisp methods which handle each of
the possible relationships. The second type of knowledge is specific to the
FAST facility and is encoded into the objects which represent the alarms.
Because these two types of knowledge are kept separate, AFS can be used for
other processes at our facility. All that needs to be done is to define the
appropriate relationships between alarms so that a new set of objects can be
constructed.

DEVELOPMENT AND IMPLEMENTATION

Implementing a knowledge based approach on an existing, conventional
computing system is possible. However, special software supporting the
object-oriented technique used in AFS must be available. The first phase of
the development of AFS consisted of defining relationships between alarms
and building a prototype. Relationships were determined by interviewing
senior operators and process engineers.

We used a stand alone workstation to develop the prototype in order to
avoid interfering with normal operation of the facility. Four months of
effort over six calendar months were required to complete the prototype.
Development of the AFS prototype was carried out on a Xerox 1186 AI workstation. The Interlisp-D and Loops environment proved to be excellent for
prototype development. It was not difficult to simulate the relevant DPSE
processes on the development workstation. The AFS prototype successfully
demonstrated the advantages of the object-oriented approach to improving the
effectiveness of the alarm display. A study of actual alarm message data
showed that in an eight hour period, the number of messages to the alarm
window would have been reduced from three hundred to only fifteen. With the
support of WINCO management, we are proceeding with the implementation of
AFS on the DPSE.

The DPSE consists of a network of six Hewlett-Packard 9000 computers.
Transporting AFS to the DPSE is not expected to be too difficult since the
Hewlett-Packard AI environment has many similarities to the Xerox environment. The conversion of AFS to the DPSE is expected to require two
months of effort. This will include defining some additional relationships
and developing an interface to existing DPSE software.

Operators and process engineers will verify that the defined relationships for the alarms and signals are correct. After the relationships are
reviewed and approved, they will be entered into AFS. System response will

then be evaluated to make sure that it corresponds to the approved relationships prior to actual use of AFS during facility operation.

FUTURE ENHANCEMENTS

Deciding which message to scroll is an area which may be enhanced.
Although we have currently defined least important as the oldest low priority message and most important as the newest high priority message, more
complex definitions can be added later. A primary or secondary alarm with
fewer relationships to other activated alarms may be more important than
other alarms of the same priority. Another possibility is an alarm which is
secondary because it is a level precursor, is less important than other secondary alarms. As additional plant operating experience is gained, engineers may be able to assign relative importance factors (perhaps on a scale
of one to three) to alarms. These factors would be used for prioritizing
alarms of the same level.

Other enhancements could involve correlations of conditions and alarms.
For example, alarms for high and low liquid level in a vessel could be correlated to open transfer valves.

Engineers may determine that additional alarm relationships should be
added to the system as experience is gained in operating the FAST process.
Implementation will be easy since no programming will be required; only
editing of object descriptions is necessary to add relationships between
alarms. Addition of new alarm signals to the DPSE will only require creation of new objects in AFS. This is also a simple task. Additional programming would only be required in the event that new types of relationships
are to be added to AFS.

DEVELOPMENT OF KNOWLEDGEBASE SYSTEM FOR ASSISTING

SIGNAL VALIDATION SCHEME DESIGN

M. Kitamura, T. Baba, T. Washio and K. Sugiyama

Department of Nuclear Engineering
Tohoku University, Aoba, Sendai, 980 Japan

INTRODUCTION

The purpose of the present study is to develop a knowledgebase system
to be used as a tool for designing signal validation schemes. The outputs
from the signal validation scheme can be used as;

(1) auxiliary signals for detecting sensor failures,
(2) inputs to advanced instrumentation such as disturbance analysis and
diagnosis system or safety parameter display system, and
(3) inputs to digital control systems.

Conventional signal validation techniques such as comparison of
redundant sensors, limit checking, and calibration tests have been
employed in nuclear power plants[1]. However, these techniques have
serious drawbacks, e. g. needs for extra sensors, vulnerability to
common mode failures, limited applicability to continuous monitoring, etc.
To alleviate these difficulties, a new signal validation technique has
been developed by using the methods called analytic redundancy[2,3] and
parity space[4,5]. Although the new technique has been proved feasible as
far as preliminary tests are concerned, further developments should be
made in order to enhance its practical applicability.

First, the procedure of designing the signal validation scheme should
be systematized. The schemes of redundant signal estimation have been
designed based on designer's empirical knowledge of the plant and of
mathematical or computational techniques. For a better acceptance of the
technique, establishment of a more systematic and unified guideline to
scheme design is imperative. Second, the technique should be expanded to
cover wider class of redundant information, since diversity in measurement
principles is one of the key issues to attain high reliability in signal
estimation. It is also necessary to enhance the applicability of the
technique to the plant under transient conditions.   Though a nuclear
power plant is mostly operated under steady state, it can experience
significant transients as well. The function of signal validation should
not be interrupted even during such operations.

A new procedure of signal validation scheme design to satisfy these
needs is proposed in this paper. The signal validation schemes are
synthesized in a customized, rather than prefixed, manner. In other

words, the schemes are synthesized in response to the request of plant operator. The reason is because the objective signal to be validated and available information are quite diverse , depending on situation, operator, plant design, etc.   It is the authors' belief that higher acceptance by operating clue can be obtained when the signal validation system is provided as the flexible, on-demand assistant than as a black box to be believed.   Development of a knowledgebase system was perceived to be a natural approach to satisfy these requirements.   The system suggests the user how to combine the software modules stored in other process computer.   The experienced designer's knowledge is utilized efficiently in this approach.

It should be noted that the present signal validation scheme has been developed mainly for the usage of (1), detection of sensor failures, which is regarded to be the most important issue to upgrade overall man-machine performance of nuclear power plant instrumentation.   The knowledge obtained through this stage can be effectively utilized in subsequent developments for the usage of (2) and (3).   The actual development procedure was broken down into three tasks given below:

Task-1:   Collect and classify possible means to extract redundant information.

Task-2:   Develop a knowledgebase to store the heuristic knowledge acquired at task-1 in a structured manner to allow efficient search of possible schemes.

Task-3:   Develop an expert system for supporting the plant operator to design and install a customized validation scheme.

ESTIMATION PROCEDURES

In conducting the task-1, available procedures of redundant signal estimation are classified into three categories as;

(a)   conservation-law-based,
(b)   functional-relationship-based, and
(c)   derivative-relationship-based redundancy.

The category-(a) covers estimation procedures by utilizing physical principle (e.g. mass and energy conservation) models.   Simple model-based predictions as well as the system theoretic prediction by Kalman filtering or Luenberger observer[3,4] are classified to this category irrespective of algorithmic complexity.   Estimation of feedwater flow rate by measuring water level and steam flow rate of steam generator is a typical example of category-(a) redundancy.

The category-(b) covers the redundant information derived from intrinsic nature of plant components.   The feedwater flow rate is also estimated indirectly by monitoring the behavior (e. g. rotating speed and pump head) of feedwater pump. Opening of a flow control valve is estimated by measuring pressure difference across the valve.   These are examples of the category-(b) redundancy.

The category-(c) implies the estimation of a plant variable by measuring its derivative, indirect effect on other variables[6].   For example, the coolant pump speed and the coolant temperature are estimated by monitoring specific resonance frequencies observed in power spectral

density function of primary coolant pressure signals. The category-(c)
redundant information, mostly contained within the noise (i.e.
fluctuating) components of plant variables is found to be highly helpful
in synthesizing a dependable signal validation scheme, since it provides
us with additional diversity in estimation principles.

KNOWLEDGEBASE DESIGN

    Applicability of a knowledgebase system is influenced not only by the
amount and quality of the stored knowledge but also by the actual design
of the system. Several design issues of practical importance are
described in this section.

Organization

    The descriptions of empirical knowledge collected in task-1 are
organized in task-2 to have a multi-level structure. The first (i. e.
bottom) level corresponds to the direct and concrete descriptions of the
method for obtaining a specified signal. The higher level contains more
general or abstracted descriptions of the lower-level knowledge. The
knowledge of feedwater flow rate estimation cited in II, for instance, can
be generalized as "an estimate of fluid flow rate at downstream of a
driving pump is estimated by monitoring rotating speed and exhaust head of
the pump". The abstracted knowledge allows us to seek out unawared but
possible sources of redundant information in a more exhaustive manner.

    Note that each of the estimation procedures is evaluated to assess
its applicability to nonstationary state of the plant. Maximum
permissible rate of change in process variables was adopted as one of the
measure of applicability. The minimum time duration needed to collect a
sufficient amount of data is another measure. These supplementary data
are also attatched to the stored knowledge of estimation procedures.

Representation

    Representation of the knowledge is one of the most important issues
in designing knowledgebase systems. The widely-used production-rule type
representation is usable in principle. In order to attain a higher
efficiency in knowledge data handling, however, certain systematization or
structurization of the knowledge representation is crucial. The multilevel knowledge described in the preceding paragraph was implemented by
taking the advantage of <u>frame-representation.</u>

    The well-known "inheritance" function of the frame representation is
highly useful in avoiding unnecessary increase in the amount of knowledge
to be stored. The search of interested knowledge data as a "chunk" of
information is performed quite efficiently by adopting this structured
knowledge representation. A more concrete idea of the frame
representation can be obtained in terms of the typical example given in
Fig. 1.

    Note that production-rule type knowledge representation is also
adopted to describe guidelines, or meta-knowledge, for selecting
appropriate frame to be examined. This modification is introduced to
improve the data-handling performance of the knowledgebase system.
However,the main body of the knowledge is solely described by the frame
representation.

```
(a) coolant temp. ──┬──hot leg temp. ──┬── hot  leg  temp.  in  loop1
                    │                   ├─ hot  leg  temp.  in  loop2
                    │                   ├─ hot  leg  temp.  in  loop3
                    │                   ├─ "hltest1"
                    │                   ├─ "hltest2"
                    │                   └─ "hltest3"
                    │
                    ├─ cold  leg  temp.
                    └─ core  exit  temp.


(b)   ------------------------------------------------------------------
      Frame:coolant temp.
            measuring object                value:water
            sensor output                   value:voltage
            sensor type                     value:T.C.
                                                  R.T.D.
      ------------------------------------------------------------------
      Frame:hot leg temp.
            upper frame                     value:coolant temp.
            hardware redundancy             value:yes
            number of loops                 value:3
            sensor at upstream              value:core exit temp.
            sensor at downstream            value:cold leg temp.
            availability of sig.            value:yes
            common mode failure             value:possible
            output range                    value:unknown

            measuring object                value:water
            sensor output                   value:voltage
            sensor type                     value:T.C.
                                                  R.T.D.
      ------------------------------------------------------------------
```

Fig.1 Examples of frame representation
(a)Tree structure
(b)Inheritance

Data Handling

An essential function of the present knowledgebase system is to
provide the analyst with the knowledge about the available methods and
procedure of signal validation for a specific set of signals.   In other
words, the generic task[7] in this system is to find out the data which
fulfill the specified conditions imposed by the analyst.   In terms of the
frame representation, the task is defined as the search of a specific
"slot" of an appropriate frame.

Data handling is carried out by applying the conventional inference
procedures, i.e. forward reasoning and default reasoning, complementary.
The forward reasoning is applied to the production rules to find out the
appropriate frames.   The default reasoning is realized by the inheritance
capability of the frame representation.   When a slot value of a frame is
not predefined, the slot is regarded to have the value identical to that
of the corresponding slot of the higher level frame.   These two inference
procedures are quite helpful in upgrading the efficiency of data handling
in this knowledgebase system.

TABLE 1 Examples of redundant information sources obtained
        through three categories;
        (a):conservation law,
        (b):functional relationship,
        (c):derivative relationship.

Estimation object	Input information	Estimation category	
Reactor power	Rod position, Cold leg temperature, Primary pressure, Boron concentration	Reactor kinetics	(a)
Hot leg temperature	Reactor power, Coolant flow, Cold leg temperature	Energy conservation	(a)
Coolant flow	Rotating speed, Exhaust head	Pump characteristic	(b)
Coolant verocity	Incore neutron flux, Core exit temperature	Time delay	(c)
Pressurizer level	Primary pressure	Spectrum	(c)

<u>Implementation</u>

Following the development of knowledgebase, the expert system to
assist signal validation scheme design was developed in task-3 using
PROLOG-based shell called TELL(Iwasaki Tech. Co.) on a 16-bit personal
computer PC-9801 (NEC Co.).   The shell allowed us easy implementation
and modification of the multi-level knowledge in an interactive manner.

APPLICATION

A prototype expert system was developed based on our knowledge about
the redundancy categories-(a),-(b), and -(c) for a pressurized water
reactor.   Some examples of the redundant information sources are listed
in Table-1.   A partial record of user-machine interaction is given in Fig.
2.   The user is requested to answer a set of inquiries presented in the
window-(a).   The system reports the list of possible methods for
redundant estimation, together with the information about availability of
hardware redundancy.   The list, including brief description of estimation
methods, is presented in the window-(b).   The name of each candidate for
estimation corresponds to the module name in the process computer in which
the signal validation scheme is to be realized.   Though the default
knowledge is presented at first, the user can continue conversation with
the system to find out more desirable sets of methods.   Through numerical
experiments, it was judged that the performance of the present system is
satisfactory as a prototype.

The performance of the signal estimation modules was evaluated through
simulation tests.   The output of the estimation modules were compared
with that of a detailed simulation code.   An example of the comparison is
shown in Fig. 3.   In spite of algorithmic simplicity of the estimation
method, the magnitude of discrepancies is small enough to allow synthesis
of a signal validation scheme.   Some of the modules were evaluated to be

(a)

```
Cold leg temp. available ?
  yes
  no
  unknown

Coolant flow available ?
  yes
  no
  unknown

Excore n. flux available ?
  yes
  no
  unknown

Core exit temp. available ?
  yes
  no
  unknown
```

(b)

```
Estimation candidate : hardware redundancy
Cold leg temp. is unavailable
Coolant flow is available
Excore n. flux is available
Rate of change(%/min.) is less than 1%

Coolant flow is available
Core exit temp. is available
Rate of change(%/min.) is less than 1%
Estimation candidate : hltest2
  dhlt/dt=const*(cet-hlt)*cf
```

Fig.2 Example of user-machine interaction. The shaded
parts correspond to response selected by the
user.
(a)Question window
(b)Result window

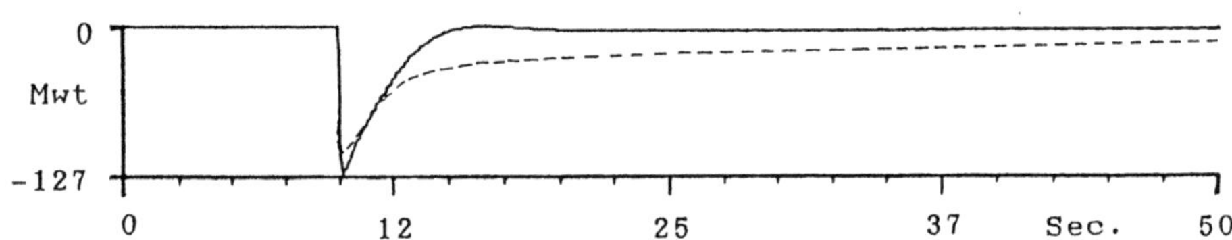

Fig.3 Response of reactor power(deviation from
steady state).
Dashed line : simulation code
Solid line : estimation module

satisfactory even for transient conditions. Potentiality of the proposed
methodology was well exemplified by these observations.

CONCLUDING REMARKS

The knowledgebase system proposed in this study was found to be
promising as a tool for assisting the user (i. e. operator) who wishes to
organize a specific-purpose signal validation scheme. Through
discussions with the knowledgebase system, the scheme is iteratively
synthesized as a transparent, rather than black, box realized on the other
computer. This approach can lead to higher flexibility, modifiability,
user-friendliness and operator acceptance of the validation scheme. The
capability of the present knowledgebase system can be upgraded further by
expanding memory capacity of the system.

REFERENCES

(1) J. A. Thie, Surveillance of Instruments by Noise Analysis, _Nucl._
Tech. 22:738 (1981).
(2) R. N. Clark and B. Campbell, Instrument Fault Detection in a
Pressurized Water Reactor Pressurizer, _Nucl. Tech._ 56: 23 (1982).
(3) J. L. Tylee, A Generalized Likelihood Ratio Approach to Detecting and
Identifying Failures in Pressurizer Instrumentation, _Nucl. Tech._ 56:484
(1982).
(4) C. H. Meijer, et al., "On-Line Power Plant Signal Validation Technique
Utilizing Parity-Space Representation and Analytic Redundancy", EPRI, Palo
Alto, NP-2110, (1981).
(5) O. L. Deutsch, et al.,"Validation and Integration of Critical PWR
Signals for Safety Parameter Display Systems", EPRI, Palo Alto, NP-4566,
(1986).
(6) M. M. R. Williams, (ed.), _Progr. in Nucl. Energ.,_ 9:(1984).
(7) B. Chandrasekaran, Generic Tasks in Knowledge-Based Reasoning: High
Level Building Blocks for Expert System Design, _IEEE EXPERT_, Fall:23
(1986)

AN EXPERT SYSTEM FOR SENSOR DATA VALIDATION AND MALFUNCTION DETECTION

Siavash Hashemi[1], Brian K. Hajek, and Don W. Miller

Nuclear Engineering Program
The Ohio State University
Columbus, Ohio 43210

Introduction

Nuclear power plant operation and monitoring in general is a complex task which requires a large number of sensors, alarms and displays. At any instant in time, the operator is required to make a judgment about the state of the plant and to react accordingly. During abnormal situations, operators are further burdened with time constraints. The possibility of an undetected faulty instrumentation line, adds to the complexity of operators' reasoning tasks.

Failure of human operators to cope with the conceptual complexity of abnormal situations often leads to more serious malfunctions and further damages to plant (TMI-2 as an example). During these abnormalities, operators rely on the information provided by the plant sensors and associated alarms. Their usefulness however , is quickly diminished by their large number and the extremely difficult task of interpreting and comprehending the information provided by them. The need for an aid to assist the operator in interpreting the available data and diagnosis of problems is obvious.

Recent work at The Ohio State University Laboratory of Artificial Intelligence Research (LAIR) and the nuclear engineering program has concentrated on the problem of diagnostic expert systems performance and their applicability to the nuclear power plant domain. We have also been concerned about the diagnostic expert systems performance when using potentially invalid sensor data. Because of this research, we have developed an expert system that can perform diagnostic problem solving despite the existence of some conflicting data in the domain. This work has resulted in enhancement of a programming tool, CSRL[1, 2], that allows domain experts to create a diagnostic system that will be to some degree, tolerant of bad data while performing diagnosis. This expert system is described here.

Diagnosis in CSRL

The developed expert system is capable of diagnosing the coolant system malfunctions of a simplified General Electric Boiling Water Reactor-6 (BWR/6) with Mark III containment. This model is developed by using Final Safety Analysis Report (FSAR) and other related documents of the Perry Nuclear Power Plant, near Cleveland, Ohio.

---

1. Currently at IntelliCorp, MountainView, CA

Problem solving in this domain has a generic character and typically has the task of classifying a given case into a diagnostic statement. In classification, the form of the required knowledge is of the form of a hierarchy of malfunction hypotheses[3]. In this hierarchy of malfunctions, the top levels are more general diagnostic classes, and the bottom levels correspond to more specific cases (a type/subtype arrangement of the malfunctions). The set of the tip nodes in this hierarchy will then be thought of as the diagnostic conclusions relevant to the system.

The coolant system malfunctions are arranged in a class/subclass hierarchical malfunction tree (Fig. 1). Each node on this tree is called a malfunction specialist and contains domain specific knowledge (or a knowledge base). Each specialist also contains information about how to utilize its knowledge base which is called the control knowledge.

Utilizing the above two sources of knowledge, diagnosis is performed on this malfunction hierarchy. The problem solving strategy of this classification system is a top-down examination of the hierarchy, termed "Establish-Refine"[4].

Establish refine can be thought of as the examination of the "appropriateness" of a particular malfunction of the hierarchy. If that malfunction is found appropriate (i.e. Established), then its submalfunctions are also asked to establish (i.e. Refine). If a specialist is not established, then none of its subnodes need be examined and that part f the tree can bee pruned. Thus, the arrangement of the specialists on this hierarchical tree is of importance.

The evaluation of malfunction appropriateness is based on knowledge embedded in each malfunction node of the hierarchy. In essence, each node is capable of performing a simple kind of problem solving that evaluates whether the malfunction at this node exists, given the currently available data. This is accomplished by the knowledge groups of each node (Fig 2). Each knowledge group contains domain expertise in the form of, "what pattern of data must exist for this malfunction to be labeled appropriate? Furthermore, if the indicated pattern found, how must the appropriateness of the malfunction be modified?"

Appropriateness measure of each node is obtained by utilization of the confidence factors. These confidence factors range from -3 to +3 with significances ranging from "DEFINITELY NOT" to "DEFINITE", respectively. Confidence values are represented in the "VALUE" column of the Figure 2.

Should the confidence value of a node be determined as +2 or +3, then the subnodes of that particular specialist are also examined. If the node's confidence factor is determined to be between -1 to +1, further analysis of its subnodes are only possible if further data be available, or no other conclusive diagnostic conclusion is obtained. If the confidence value of a node is -2 or -3, the subnodes of that particular specialist are ignored until later steps into the diagnostic reasoning.

Up to this point, it has been assumed that the input data are a set of valid and reliable data. These input data are assumed to have gone through the standard data validation routines (Fig. 3) and are in general, reliable. However, due to common cause failures, the input data may be faulty, and thus, some of the data may require further analysis. The past studies have also shown that sensor data validation and diagnosis are integral components to one another and one can not be accomplished without the other, in an efficient manner. Thus, sensor data validation is added to this expert system and is described next.

Sensor Data Validation

Sensor data validation can be thought of as a two stage function. Each stage is capable of identifying a certain class of faulty sensors. The two stages are as follows:

1. This stage of data validation which is mainly dependent on comparison of redundant sensor signals (such as "like sensor"comparisons, fail safe assumptions, auctioneering, and so on) and is aimed at identification of gross failures of the instrumentation channels (such as shorts, open circuits, connector or detector failures). These techniques are based on various kinds of redundancy in sensor hardware and are discussed elsewhere[5].

150

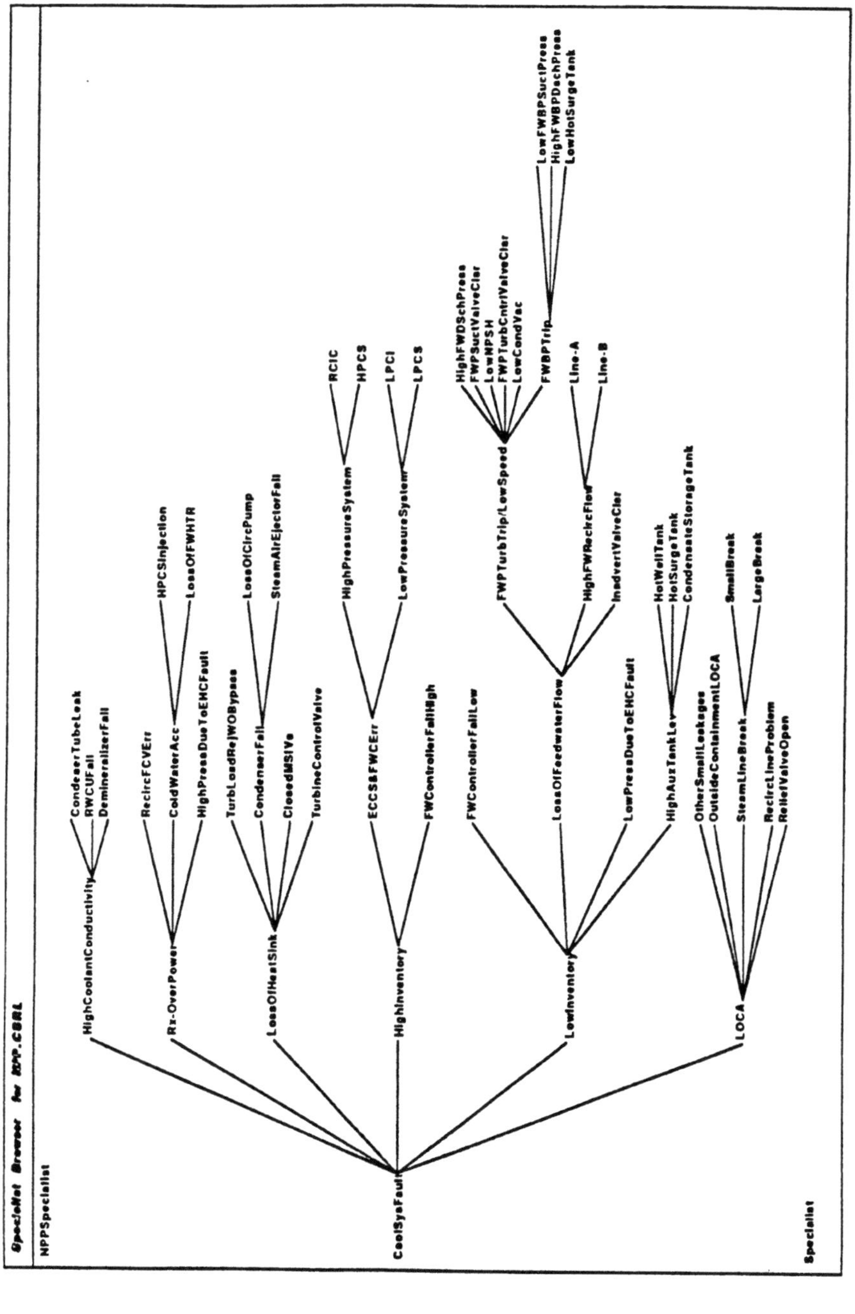

FIG 1. HIERARCHY OF MALFUNCTION SPECIALISTS

SRVopen KG of ReliefValveOpen

Expressions of table
1- (AskYNU? "Are there any SRVs open")
2- (AskHLN? "What is the suppression pool temperature")
3- (AskTrend? "What is the suppression pool temperature trend")
4- (AskHLN? "What is the suppression pool water level")

1	2	3	4	VALUE	SENSOR
T	(or H HH)	(or I II)	(or H HH)	3	---
(or F U)	(or H HH)	(or I II)	(or H HH)	3*	1
(or F U)	N	(or I II)	?	1*	1 4
?	?	?	?	-3	---

FIG 2. AN EXAMPLE OF A KNOWLEDGE GROUP IN MODIFIED CSRL

2. The second level of validation is beyond hardware redundancy and is accomplished through expectations derived during malfunction diagnosis. "That is, in the process of exploring the space of possible malfunctions, initial data and intermediate conclusions set up expectations of the characteristics of the final answer"[6]. This method of validation uses analytical redundancies and local versus system loop condition and is aimed at more subtle failures (such as instrumentation calibration drifts), and failures that are not detectable through the first stage of validation (such as common cause failures). This technique is described below.

During the design stages of this system, a great deal of attention was paid to ensure the system's ability of reaching a diagnostic conclusion based on partially complete data sets as well as potentially incorrect input parameters. To reach this goal, all relevant and useful sources of information for each malfunction hypothesis (specialist) were identified. Next, the relationship between them are determined and proper combinations of this information are utilized in establishing or rejecting the malfunction hypotheses. As an example, in order to detect a large Loss Of Coolant Accident (LOCA), the containment and the reactor pressure vessel parameters as well as the turbine-generator, suppression pool, main steam line and feedwater parameters are considered. In this fashion, having only a partial list of valid parameters, a fairly accurate diagnosis is still possible.

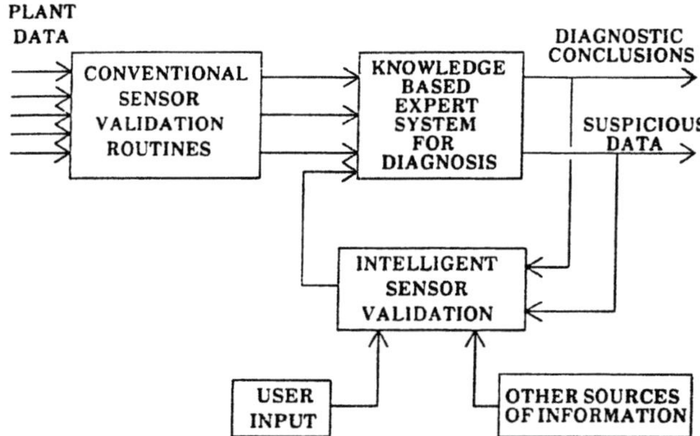

FIG 3. SENSOR VALIDATION ROUTINE BLOCK DIAGRAM

One of the important factors to be noted here is that some combinations of the input data values can be marked as not capable of occurring or highly unlikely, due to violation of "normal expectations" of malfunctions[6, 7]. That is, in the process of matching the malfunction patterns to data, certain combinations of data do not make sense in the given domain. As an example, turbine/generator electrical output must be zero (or decreasing rapidly), if the Main Steam Isolation Valves (MSIVs) are closed.

Thus, the process of faulty sensor identification involves three steps. These steps are:

1. The first step is to set some expectations by using local knowledge, or the context of other nodes.

2. The second step is to use these expectations to flag particular data values as questionable. And,

3. The third step is the process of validating (accepting) or rejecting the questionable sensor values based on the relational redundancies available in other parts of the plant.

The expectations derived during diagnosis are utilized as an extra source of redundancy, in addition to various conventional sensor hardware redundancies. These expectations are embodied in the specialists' knowledge groups and are formed and encoded a priori. The knowledge groups were then modified to check the malfunction pattern fit to the input data, and flag the parameters that lie outside the expectations as suspicious. The suspicious data are then subjected to further analysis. This concept is demonstrated in the last column of Figure 2.

The knowledge group rows that do not meet the expected symptoms are identified as containing suspicious data by addition of a "*" in their confidence values, as can be seen from Figure 2. The * alarms the computer (and the user) that this row may contain invalid data and thus, the confidence factor of this knowledge group may change as a result of sensor data validation.

Should a knowledge group row be identified as containing suspicious data, one of the following two cases must be valid:

- The set of expectations are not valid at that particular instant of time (eg. IF LARGE LOCA THE RPV WATER LEVEL=LOW is no longer valid if we are on the recovery path of a large LOCA).

- The set of symptoms are not valid at this particular instant of time which leads to the conclusion "THE XYZ (GROUP OF) SENSOR(S) HAVE FAILED".

In order to determine which one of the above situations is true, further analysis on the suspicious sensors is necessary. This analysis utilizes the following resources:

- Logical conclusions drawn from diagnosis,

- Logical conclusions drawn from available data,

- Analytical calculations,

- Causal relationships between unlike parameters

These extra sources of information about the suspicious sensors are used to either validate or reject suspicious data and are arranged in the order of preference for each individual sensor. In this fashion, the routines with minimum cost, highest reliability, or minimum effort can be executed first, depending of each individual sensor. Note that this methodology is relying on unlike sensor data comparison and thus, common cause failures or instrumentation channel drifts are no longer problems.

One important factor to avoid in this routine is that validation of one sensor should not rely on information supplied by another sensor which is also in the

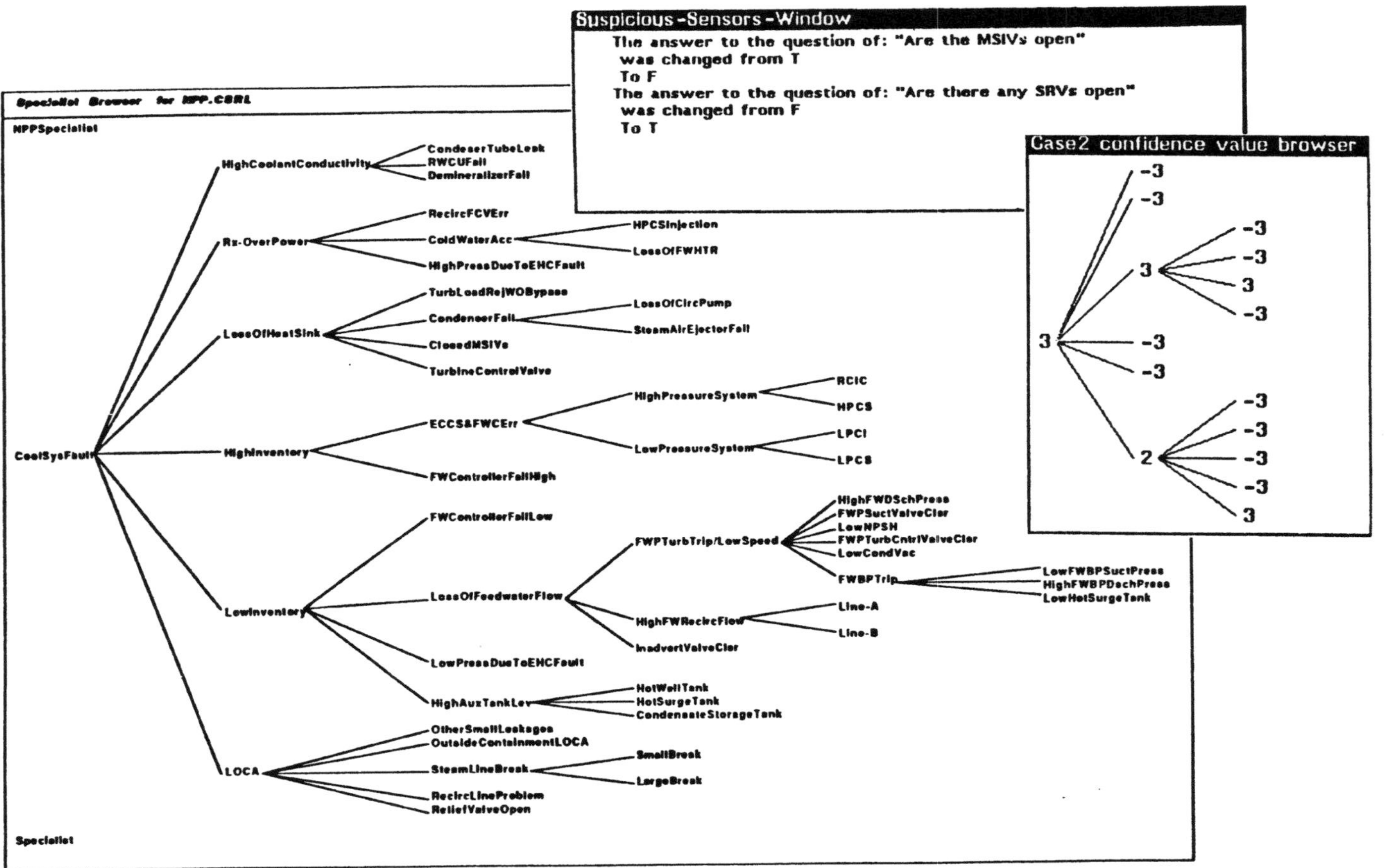

FIG 4. INADVERTENT CLOSURE OF MSIVS AFTER VALIDATION

suspicious sensors list. That is, one has to rely on the "potentially valid" data as redundancy, as opposed to "potentially faulty".

Once the faulty sensors are identified, there are two routes that can be taken:

1. Replace that faulty data with the new (and validated) value and resume diagnosis after informing the user, or

2. Ignore that faulty data (mark it as Unknown) and resume diagnosis after informing the user.

These two methods must lead to the same diagnostic conclusion (because a real plant can only have one state at a time). However, the first method is adapted for this project, due to ease of implementation.

An Illustrating Example

An inadvertent closure of the MSIVs (Closed MSIVs malfunction node of Fig. 1) is considered as an example. If the MSIVs close, the Safety Relief Valves (SRVs) should cycle, in order to relieve and maintain pressure in the RPV. For this scenario, it is assumed that one of the SRVs fails to reclose after opening.

TO make this scenario more complicated for malfunction diagnosis, it is also assumed that the direct indications of SRV and MSIV positions have failed and read CLOSED and OPEN, respectively. That is, the control board lights do not change status from their pre-event indications.

During the first run of the diagnostic system, the LOCA node has fired with a confidence factor of 0* to show more evidence is required to either establish or reject the LOCA possibility. Further up in the tree, the ClosedMSIVs and TurbineControlValve nodes fire with confidence factors of 0* and 3*, respectively. This means that there is a mild indication of MSIV closure and therefore, a strong evidence for the wrong turbine control valve position.

During the sensor signal validation step, the list of suspicious data is inspected and validated. Backup information for the SRV position is obtained from the local suppression pool temperature profile and SRV-downcommer pressure (Perry specific). The backup information for the MSIV position was obtained from matching the steam flow rate, RPV pressure and turbine electrical output.

After validation of the suspicious data, the new values are replaced in the database and new diagnosis indicated the correct state of the plant [Fig 4]. The new confidence factors indicate that no more suspicious data exists.

Conclusions

This expert system is capable of performing diagnosis and sensor data validation, in an efficient manner. This expert system is tolerant to faulty input parameters to a certain extent, and provides an additional sensor validation routines to the nuclear power plant personnel.

Acknowledgments

The authors wish to acknowledge Professor B. Chandrasekaran for his insights and assistance in developing these ideas. We would like to also acknowledge Mr. John Stasenko for his attempts to expand the knowledge base of this expert system. This work was supported in parts by NSF, DOE, and internal grants.

References

1. T. Bylander, S. Mittal, B. Chandrasekaran, "CSRL: A Language for Expert Systems for Diagnosis", Proc. of IJCAI 83, Los Altos, CA, PP 218-21.

2. B. Chandrasekaran, "Decomposition of Domain Expert Knowledge into Knowledge Sources: the MDX Approach", Proc 4th National CSCSI/SCEIO, Sakatoon, Canada, May, 1982.

3. B.K. Hajek, S. Hashemi, D.D. Sharma, B. Chandrasekaran, D.W. Miller, Artificial Intelligence Enhancement to Safety Parameter Display Systems", Proc. 6th Power Plant Dynamics, Control and Testing Symposium, Knoxville, Tenn, April, 1986.

4. B. Chandrasekaran, "Towards Taxonomy of Problem Solving Types", AI Magazine, Vol 4, No. 1, Winter/Spring 1983. J.J. Deyst, R.M. Kanazawa, J.P. Pasquenza, "Sensor Validation: A Method to Enhance the Quality of Man/Machine Interface in Nuclear Power Stations", IEEE Transactions on NS, Vol NS-28, No. 1, Feb. 1981.

5. B. Chandrasekaran, W.F. Punch, "Data Validation During Diagnosis, a Step Beyond Traditional Sensor Validation", Proc. AAAI, Seattle, WA, 1987.

6. B. Chandrasekaran, W.F. Punch, "Hierarchical Classification: Its Usefulness for Diagnosis and Sensor Validation", Proc 2nd AIAA/NASA/USAF Symposium on Automation, Robotics, and Advanced Computing, Feb 1987.

SMART SENSOR APPLICATION TO NUCLEAR PLANT THERMOCOUPLE CHANNELS

N. S. Cannon, J. D. Martin, B. A. Patterson,
J. L. Stringer, and B. D. Zimmerman

Westinghouse Hanford Company
P. O. Box 1970
Richland, WA 99352

H. L. Heinisch

Pacific Northwest Lab
P. O. Box 999
Richland, WA 99352

C. P. Cannon

Cannon Technology
320 N. Johnson
Kennewick, WA 99336

ABSTRACT

The application of smart sensor technology to thermocouple (TC)
channels at the Fast Flux Test Facility (FFTF) has been accomplished. The
objective of this work was to provide a first step toward the development
of nuclear plant sensor systems that not only produce a transducer output
signal, but also perform diagnostics on the sensor channel to provide
signal validation. When TC impairment was detected, the system provided
an analysis of the severity of the damage, its location, and probable
cause.

Two systems were constructed: a laboratory system for development
purposes, and a follow-on system for installation at FFTF. Since its
installation, the FFTF system continues to provide TC trending data for
lifetime predictions, and a test facility for new concepts. The laboratory system has been adapted to include a new sensor type, an FFTF low-
level neutron flux monitor.

INTRODUCTION

Operation of nuclear power generating facilities has and will
continue to be enhanced by the application of computer technology. A
relatively new frontier for the operation of complex plant equipment in a
variety of industries is the development of computer-based smart sensor
technology. This work details the application of smart sensor technology
to thermocouple (TC) channels at the Fast Flux Test Facility (FFTF), a
sodium-cooled test reactor located in Richland, Washington (operated for
the Department of Energy by Westinghouse Hanford Company).

In 1985, Westinghouse Hanford Company (WHC) started work to develop
smart sensor instrumentation; the concept was a sensor system that would
not only provide the transducer output (such as temperature, pressure,
etc.) but would also perform diagnostics on the sensor channel (transducer, cabling, connectors, etc.) to validate the transducer output.

Additionally, sensor channel aging would be monitored, with recommendations for component replacement made as failure is approached. In the
case of nuclear plant applications, these systems promise both significant
economic gain by reducing unscheduled downtime, and increased public and
industry confidence in plant safety control during off-normal events. The
strategy was to develop a smart system for FFTF, using that reactor as a
test bed for sensor demonstration to other industries, e.g., Light Water
Reactors (LWRs).

The initial effort concentrated on the development of a Smart
Thermocouple System (STS). The STS work was performed in two steps.
First, a laboratory demonstration system was developed as a "proof of
principle," and to gather data on FFTF-certified type K TCs. These data
were formed into rules for a modest expert software system. Secondly, the
laboratory STS was transposed into an online system at FFTF.

LABORATORY SYSTEM

The laboratory portion of the project included 1) evaluating the
baseline characteristics of the TCs, 2) determining critical values of
parameters that reflect the impairment or failure of a TC, 3) generating
rules for an expert system to diagnose TC status, and 4) demonstrating
system performance on simulated failure modes.

Initially, the parameters investigated were the TC loop resistance
(LR) and insulation resistance (IR); later, capacitance and dissipation
factor were added. An illustration of an ungrounded junction TC is given
in Fig. 1; LR measurements were made by measuring the resistance between
points A and B. IR measurements were made by measuring the resistance
between points A and C. Capacitance and dissipation factor measurements
were made first between points A and C, then between B and C.

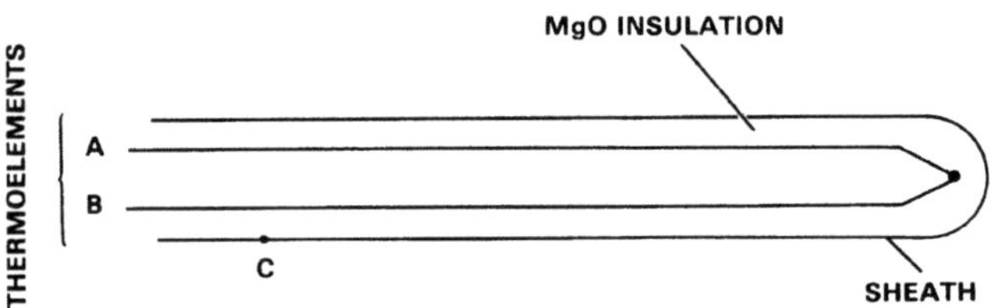

Fig. 1. Schematic of FFTF type K ungrounded TC.

A flow chart of the STS hardware is given in Fig. 2. This system
allowed four TCs to be monitored; the computer-controlled relays, which
switched the appropriate TC leads into various configurations, allowed the
individual (TC amplifier, LR, IR, or capacitance) instrumentation to
obtain data on that particular TC.

Characterization of Thermocouples

The TCs used for testing were off-the-shelf, 1/16-inch diameter by
six-feet long, stainless-steel sheathed, type K, ungrounded TCs with
compacted MgO insulation. The FFTF TCs are of the same type, but they
have been qualified for use in the reactor, and are typically 33 to 36

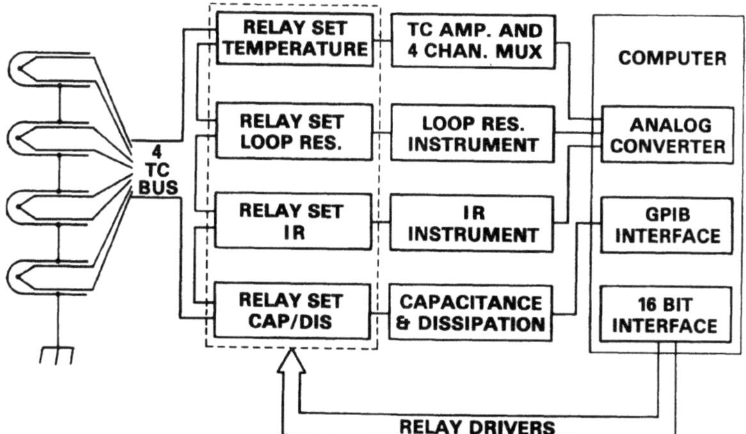

Fig. 2.   Smart TC system flow chart.

feet long.   Three FFTF spare parts TCs were tested in the laboratory and compared to the off-the-shelf TCs.

Distributed loop resistance, insulation resistance from wire to sheath, and thermocouple emf were measured in each test of a TC.   The TCs were first characterized by heating the last six inches, including the measurement junction, in a furnace and making measurements as a function of temperature.   The temperatures ranged from room temperature to 1200°C. The temperature of the furnace was measured independently with a type R TC.   Additional tests were run with the measurement junctions held at 400°C, while the middle two feet of the TCs were heated to temperatures up to 1150°C.

As an example, Figs. 3 and 4 show LR, IR, and the error in the temperature reading for an off-the-shelf TC with its measurement junction at 400°C and the central third of its length at higher temperatures.

LR per unit length as a function of temperature was deduced from the data taken during heating of the central sections of the TCs.   The LR per unit length increased by roughly 60% from room temperature to 1200°C.

The IR as a function of temperature during heating of the central section was constant up to approximately 500°C.   The resistance then dropped rapidly, decreasing exponentially with temperature from 800°C to 1200°C.   Both LR and IR recovered to near their original values upon return to room temperature.

Diagnostic Rules

Expert system diagnostic rules were developed from laboratory data on TC failure characteristics using an Expert-Ease ® expert system construction package.   Expert-Ease produced two logic trees, one designed to give diagnostic message information, and one based on the red-yellow-green traffic-signal-style code.   A "green light" was indicated for a TC channel

---

® Expert-Ease is a registered trademark of Intelligent Terminals Limited, Glasgow, Scotland.

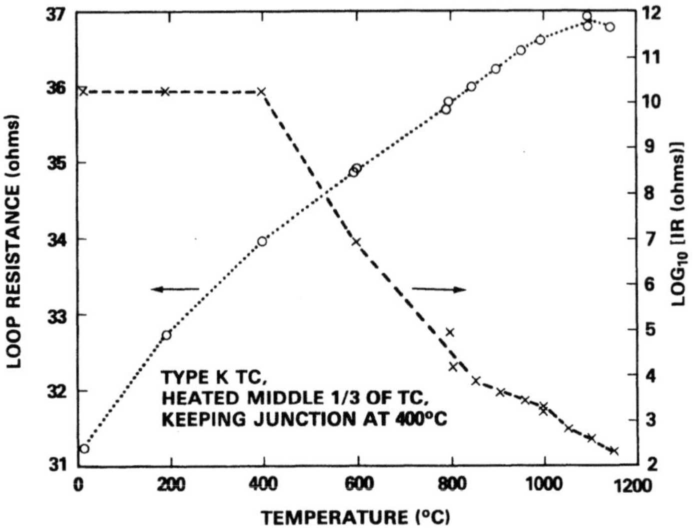

Fig. 3. LR and IR dependence on temperature for type K TC.

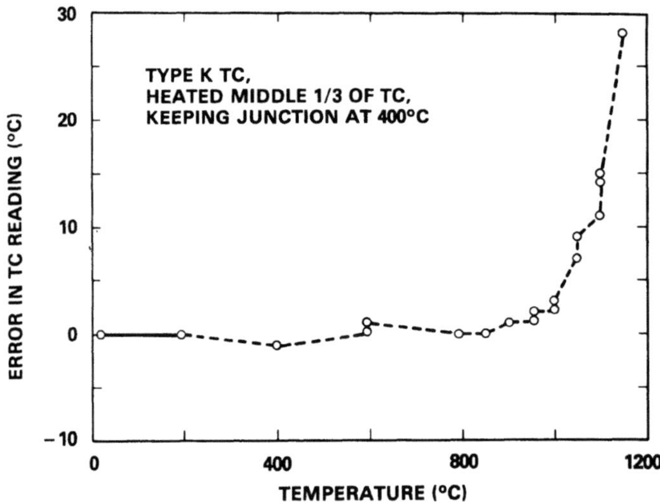

Fig. 4. Temperature error versus virtual junction temperature.

with parameter values within the range determined as acceptable; a "yellow light" was indicated for a TC that was impaired but giving a valid temperature reading; and a "red light" was indicated for a failed TC. The resulting fault tree is given in Table 1. The software was constructed so that the operator was continuously given a light indication for each of the four TC channels, and a diagnostic message would be displayed when a special key was struck.

Laboratory Demonstration System

The laboratory demonstration system combined the hardware shown in Fig. 2, data acquisition software, and the modest expert system described in Table 1. The laboratory demonstration system was used to demonstrate the STS concept on several simulated TC failures. Because inservice conditions (especially aging) could not be easily duplicated in the laboratory, the demonstration failure modes were laboratory simulations of generic failure modes.

Table 1. Laboratory STS Fault Tree

	Loop Resistance	Insulation Resistance	Temperature	Light	Message
Highest Rank	$(1-x) \leq R_L \leq (1+x)$	$R_I > C$	NA	Green	Good; all values within limits.
	$R_L \leq B$	$A > R_I$	+	Red	Low IR; virtual junction.
	$R_L \leq B$	$A > R_I$	–	Red	Low IR; shunt formed.
	$(1-x) > R_L$	$R_I > C$	+, 0	Red	Low IR shorting (in high temperature region).
	$R_L > B$	-----	NA	Red	Conducting path open.
	$(1-x) > R_L$	$R_I > C$	–	Red	Low IR; shorting (in low temperature region).
	$B \geq R_L > (1+x)$	$R_I > C$	NA	Yellow	High LR; conducting path degrading.
	$R_L \leq B$	$C \geq R_I \geq A$	+, 0	Yellow	Low IR; virtual junction forming.
Lowest Rank	$R_L \leq B$	$C \geq R_I \geq A$	–	Yellow	Low IR; shunt forming (low temperature region).

R_L = Measured loop resistance/calculated value at temperature, x = 0.2 (values can be changed by operator).
R_I = Measured insulation resistance, A = 10^3, B = 10^2, C = 10^5 (values can be changed by operator).
Temperature = Difference between the current temperature and the oldest temperature reading available; + indicates temperature is increasing, etc.

The "nearly open" failure mode was achieved by using oxidized connector
contacts that resulted in a circuit having resistance much higher than
normal, while still carrying a valid signal. One of the most common failure
modes of TCs in many applications is the failure of connectors, which the
STS has the capability of spotting before total failure of the instrument
occurs.

The "virtual junction" failure mode is a type of shunt that causes the
TC temperature reading to be high. This failure mode occurs when some part
of the TC other than the measurement junction is at a temperature high
enough to reduce the resistance of the insulation substantially, creating an
additional conducting path in the high-temperature region. This type of
failure might occur during off-normal or accident conditions, or it may be
due to poor routing of a TC through a normally hot region.

The third failure mode simulated was a simple short in a part of the TC
at a lower temperature than the measurement junction. This short caused an
erroneously low-temperature reading. A short could be the result of
manufacturing errors, mishandling during installation, or mechanical forces
during an accident.

Laboratory System Upgrade

The laboratory demonstration system was transposed into a second
system for installation at FFTF, as will be discussed. However, efforts
to improve the laboratory STS capability continued after the FFTF system
was built and installed. Capacitance measurement was added to the
laboratory demonstration system; as well as providing additional diagnostic information to determine that a fault condition exists, capacitance
allowed the determination of a fault location, particularly in the case of
a completely open TC. Fig. 5 illustrates a TC with an open lead. Capacitance is measured first from the +TC lead to sheath, then from the
-TC lead to sheath. A simple derivation gives Equation (1):

$$b/L = (C_b - C_a)/C_{TC} \tag{1}$$

where b/L provides the distance to the fault, C_b is the capacitance
measured on the long lead, C_a is the capacitance measured on the short
lead, and C_{TC} is the capacitance before the break occurred.

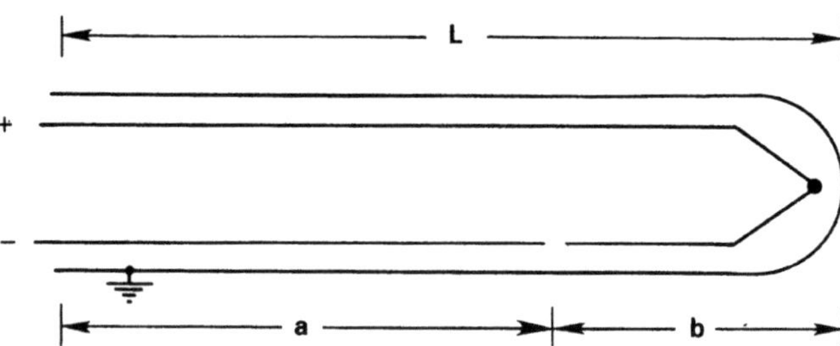

Fig. 5. Sketch of TC with an open lead at a distance b from junction.

A mockup of an FFTF TC channel that had recently failed (TE22008B)
was constructed in the laboratory, and is diagrammed in Fig. 6. The last
227 feet of the channel were included in the mockup, along with two
connectors and the TC thermal well. The BTB 556 connector was 72 feet
from the thermal well; the other connector was for the thermal well

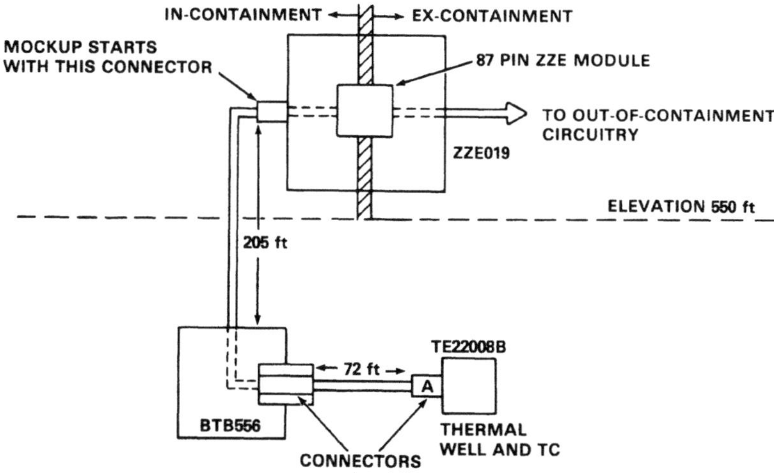

Fig. 6. Sketch of mockup of FFTF channel TE22008B. The mockup didn't
include the out-of-containment portion of the channel.

attachment. The actual failed thermowell connector (A) from FFTF was
included in the mockup.

To test the ability of the STS to use capacitance to locate open
faults in the mockup, combinations of pins were individually disconnected
at the BTB 556 connector, and later at the thermal well connector. The
results are given in Table 2 for comparison. The worst error observed in
prediction of an open location was about 3%. The error increased when the
open lead was near the thermal well; also, the ability to predict which
lead was open decreased at the thermal well as $C_b \sim C_a$.

Table 2. Capacitance Fault Location

Open Lead	Connector Site	Smart Predicted Location (Feet From Thermal Well)	Actual Location (Feet)
Red	BTB 556	68.9	72
Yellow	BTB 556	72.4	72
Both	BTB 556	77.7	72
Red	Thermal Well	7.0	0
Yellow	Thermal Well	7.0	0
Both	Thermal Well	8.4	0

Had this system been available at FFTF when Channel TE22008B failed,
about a man-month of effort to find the fault would have been saved. The
connector (A) at the thermal well apparently was damaged by a periscope,
but did not fail until weeks later (when normal operation mechanical
vibrations caused it to work loose).

FFTF SYSTEM

The FFTF system as adapted from the laboratory system was installed at
the reactor and placed into operation in February 1986; it has been online
since then, with occasional shutdowns to change configuration. The system
currently monitors nine TCs, six Temperature Removable Instrument Assembly
(TRIA) TCs from two Instrument Trees, and three TCs from two fast Prox-
imity Instrumented Open Test Assemblies (PIOTA). The FFTF STS is still
considered to be in a development stage and is thus transparent to plant
systems. The FFTF STS's modifications from the laboratory STS dealt
primarily with adjustment to the actual FFTF TC channels; capacitance
measurement has not yet been installed on the FFTF STS. The capability to
archive data was added to provide trending information that will be useful
for "teaching" the STS normal FFTF TC operating parameters and predicting
TC life by modifying expert software as experience is gained.

Diagnostic Rules

Table 3 gives the presently used decision logic tree and limit values
for the FFTF system. The values were chosen to allow a large channel
parameter variation until some experience is gained from the trending
data. As experience is gained, the decision logic can be improved to
include parameter temperature dependence, as was done with the laboratory
STS.

Table 3. FFTF STS Fault Tree

	Loop Resistance (ohms)	Insulation Resistance (ohms)	Light	Message
Highest Rank	$Rx >= 25K$	NA	Red	TCXX bad; conducting path open.
	$Rx <= 100$	NA	Red	TCXX bad; conducting path shorting.
	$100 < Rx < 500$	$Ri <= 1K$	Red	TCXX bad; low IR.
	$500 <= Rx <= 25K$	$Ri > 10K$	Yellow	TCXX impaired; low IR.
	$100 < Rx < 500$	$1K < Ri <= 10K$	Yellow	TCXX impaired; low IR.
Lowest Rank	$100 < Rx < 500$	$Ri > 10K$	Green	TCXX good; all values within limits.

$<= 100$ Ohms: Low fault value for loop resistance (Rx).
$>= 500$ Ohms: Low impaired value for loop resistance (Rx).
$> 25K$ Ohms: High fault value for loop resistance (Rx).
$<= 1K$ Ohms: Low fault value for insulation resistance (Ri).
$< 10K$ Ohms: Low impaired value for insulation resistance (Ri).

<u>Discussion</u>

The FFTF STS is online and providing diagnostic information on the TC
channels being monitored. By way of example, Table 4 gives a summary of
the data obtained on two dates (2/18/86 and 3/13/86) on the eight (seven
core) TCs then being monitored. The numbers under the message column in
Table 4 are in the order given in Table 3, with 1 = Highest Rank. Note
that four of the seven core TCs are indicated as failed, a result of low
IR. Engineering drawings show all seven of these TCs to be ungrounded;
thus the STS has brought attention to this problem.

It should also be noted that the temperature readings for these four
TCs appear consistent with the other three TCs rated as good; the readings
might indeed be valid if the TCs were grounded at only one point. The STS
doesn't currently have the ability to distinguish multiple grounds from a
single ground, from a virtual junction. This could be accomplished with a
greater refinement of LR and IR trending data, and additional parameter
measurement capability [such as capacitance and time domain reflectometry
(TDR)].

Table 4. Smart System Data Summary

TE No.	Loop Resistance (ohms)	Insulation Resistance (ohms)	Temperature (°C)	Status Color	Message No.
		2/18/86			
TE10412A	347	24.3 M	448	Green	(1)
TE10412B	329	< 10^3	430	Red	(4)
TE10418A	335	< 10^3	450	Red	(4)
TE10418B	338	< 10^3	451	Red	(4)
TE10422	347	< 10^3	449	Red	(4)
TE10428	314	89.3 M	455	Green	(1)
TE10431	354	> 100 M	455	Green	(1)
N/A	200	1.09 M	96	Green	(1)
		3/13/86			
TE10412A	334	> 100 M	355	Green	(1)
TE10412B	318	< 10^3	348	Red	(4)
TE10418A	324	< 10^3	349	Red	(4)
TE10418B	327	< 10^3	351	Red	(4)
TE10422	337	< 10^3	353	Red	(4)
TE10428	303	> 100 M	346	Green	(1)
TE10431	342	> 100 M	346	Green	(1)
N/A	199	1.10 M	93	Green	(1)

CONCLUSIONS

A Smart Thermocouple System has been demonstrated in the laboratory
and installed at FFTF as a preliminary step in applying Smart Sensor Technology to nuclear plants. As the laboratory system demonstrates new
techniques with TCs and other types of sensors, these techniques can be

transferred to the FFTF system for evaluation under actual reactor
operating conditions. The laboratory STS has been prepared to branch to
other sensor types; a mockup of an FFTF low-level neutron-flux monitor has
been made.

Addition of new parameter measurements such as capacitance and TDR
have the potential to improve diagnostic sophistication so that virtually
all TC failure modes can be detected, and in many cases located. These
additions can be made in steps, folding in trending knowledge to the
expert software to provide an increasingly powerful diagnostic capability.

Also, the incorporation of additional sensor types would provide an
opportunity for synergistic analysis based on parameters that may be
interrelated. For example, a smart system-type analysis of related
sensors such as TCs, control rod position indicators, neutron detectors,
and coolant flow/pressure meters for a reactor might well identify off-
normal response of a defective sensor channel with respect to the other
sensors. This information, in conjunction with the regular analysis of
the individual sensor parameters, would strengthen the diagnostic process.

AN INTEGRATED APPROACH FOR SIGNAL VALIDATION IN NUCLEAR POWER PLANTS

B. R. Upadhyaya, T. W. Kerlin, O. Glöckler, Z. Frei, L. Qualls
and V. Morgenstern

Department of Nuclear Engineering, The University of Tennessee
Knoxville, Tennessee 37996-2300

ABSTRACT

A signal validation system, based on several parallel signal processing modules, is being developed at the University of Tennessee. The
major modules perform (1) general consistency checking (GCC) of a set of
redundant measurements, (2) multivariate data-driven modeling of dynamic
signal components for maloperation detection, (3) process empirical
modeling for prediction and redundancy generation, (4) jump, pulse, noise
detection, and (5) an expert system for qualitative signal validation. A
central database stores information related to sensors, diagnostics rules,
past system performance, subsystem models, etc. We are primarily concerned
with signal validation during steady-state operation and slow degradations.
In general, the different modules will perform signal validation during all
operating conditions. The techniques have been successfully tested using
PWR steam generator simulation, and efforts are currently underway in
applying the techniques to Millstone-III operational data. These methods
could be implemented in advanced reactors, including advanced liquid metal
reactors.

1. INTRODUCTION

In large systems such as nuclear power plants, chemical plants and
others, process signals are used in control systems, protection systems and
plant monitoring systems[1,2]. Implementation of signal validation techniques is useful in minimizing plant downtime, to increase reliability of
operator decision and to help schedule plant maintenance. A program was
initiated in October 1986 at the University of Tennessee to develop a
comprehensive signal validation system for implementation in current and
future power plants[3]. Consideration must be given for incorporating such
systems during the plant design phase so that necessary data highways and
data acquisition systems may be planned.

The signal validation system employs an architecture which implements
diverse signal processing modules (see Fig. 1). The measurement degradation information supplied by each of the modules is evaluated by a logical
program and uses the LISP language environment. This architecture has the
flexibility of adding new signal processing channels and deleting those
that are found to be ineffective. The system emphasizes validation during

steady-state, or quasi steady-state operation. For methods that require
models, plant models will be "learned" from operational data (training set)
and will be updated periodically. The availability of large memory and
large storage in current computer workstations makes it possible to use a
large database of plant operating conditions, and to characterize them by
empirical patterns, such as empirical prediction models. In a broader
sense, these have implications in predictive maintenance and signal
trending. We must also point out that these patterns may be developed
either in the time-domain or in the frequency-domain.

The boxes defined by acronyms in Fig. 1 are the parallel signal validation modules. The signal conditioning for each of the modules will be
different. The system and sensor database, signal history, and outputs of
one or more of the modules are stored in a large database. This is
accessible to all the modules and the system executive. There is no direct
communication among the modules. The system executive provides proper
sequencing of the operation of individual modules, and if necessary resche-
dules their activity. The signal validation information from all the modules is processed in the system executive and a decision about the signal
status is made. This operation of the system executive and its interface
with the various modules is being developed using the LISP language
environment.

The description of the generalized consistency checking module and an
example of its application are given in Sect. 2. Section 3 contains a
description of the data-driven modeling and steady-state prediction
modeling. An example of predicting process variables related to a PWR
steam generator subsystem is described. Section 4 contains an outline of
the jump, pulse, noise detection module and the expert system module.
Summary and concluding remarks are given in Sect. 5.

2. GENERALIZED CONSISTENCY CHECKING AND SEQUENTIAL ERROR TESTING MODULE (GCC)

2.1 Outline of GCC

This module performs consistency checking of a set of redundant
measurements (direct and/or analytical) of a specified process variable.
Based on the comparison of a measurement with the others, an index is
assigned which is related to its inconsistency with respect to other redundant measurements. The principle of the procedure is shown in Fig. 2, and
provides a sample-by-sample evaluation of the sensor inconsistencies.
Additional signal redundancy is generated using unlike signals and either
an empirical or a physical model. Two measurements m_i and m_j are said to
be inconsistent with each other if

$$| m_i - m_j | > n_i + n_j \tag{1}$$

where n_i and n_j are the specified tolerances of the two measurements.
Based on this, an inconsistency index I_i is calculated[4] for each measurement. A procedure was developed to compare the inconsistency indices and
discard those measurements for which the values of I_i exceed limits. A
weighted average $\hat{m}(k)$, using the remaining signals is calculated at every
sampling instant.

To improve the reliability of the consistency checking algorithm, we
integrated this with the sequential probability ratio test (SPRT) of
Wald[5]. A cumulative error function, which is the ratio between the joint

density of signal residuals for the degraded and the normal sensor operational modes, is calculated sequentially. Defining this residual as

$$S(k) = m_i(k) - \hat{m}(k) \tag{2}$$

the likelihood ratio is given by

$$LR_N = \frac{p[S(1),S(2),\ldots,S(N)/H_1]}{p[S(1),S(2),\ldots,S(N)/H_o]} \tag{3}$$

According to Wald's test, accept hypothesis H_1 if $LR_N > B$, and accept hypothesis H_o if $LR_N < A$. The bounds A and B are defined as

$$A = \ln(\frac{\beta}{1-\alpha}), \quad B = \ln(\frac{1-\beta}{\alpha}) \tag{4}$$

where α and β are false alarm and missed-alarm probabilities, respectively. For bias error detection, the SPRT was modified by Chien and Adam[6]. The one-sided test applied to the bias degradation, computes a boundary B* for a specified meantime, T, between false alarms. For large T, the bound B* is given by [6]

$$B^* = \ln[\frac{\theta_1^2}{2\sigma^2} T] \tag{5}$$

where θ_1 is the average degraded mean, and σ^2 is the variance of $S(k)$. The modified SPRT minimizes the delay in detecting a failed measurement.

2.2 Results of Implementation of GCC-SPRT

This algorithm was tested using experimental data from a hot-air loop where outputs of thermocouples measuring air temperature were monitored[7]. The algorithm was tested for the detection of bias, noise, and pulse-type sensor errors. A detailed analysis of the algorithm has been made using five redundant signals with various combinations of errors. The five signals are shown in Fig. 3. The cumulative log-likelihood ratio for signal number 4 is shown in Fig. 4. Figure 5 shows the inconsistency index for signal 4 as a function of sample numbers. The samples where the signal is excluded from the set are shown by cross markers. The status of all the signals, weighted average of the process variable, and the type of failure are displayed to the operator.

3. EMPIRICAL MODELING TECHNIQUES

Two forms of empirical modeling methods have been developed and implemented for signal validation. The multivariate data-driven modeling (MDM) is used to establish the cause-and-effect relationship among a set of signals from a subsystem by multivariate autoregressive (MAR) modeling of signal dynamic components. The MAR model may be used for detecting sensor maloperation or process diagnostics. The second approach, called process empirical modeling (PEM), is primarily used for characterizing steady-state or quasi steady-state system behavior. A general nonlinear polynomial fit of a process variable of interest is obtained in terms of a set of "input" variables. This model is then used to predict the system output, for use as analytical redundancy, or for signal trending.

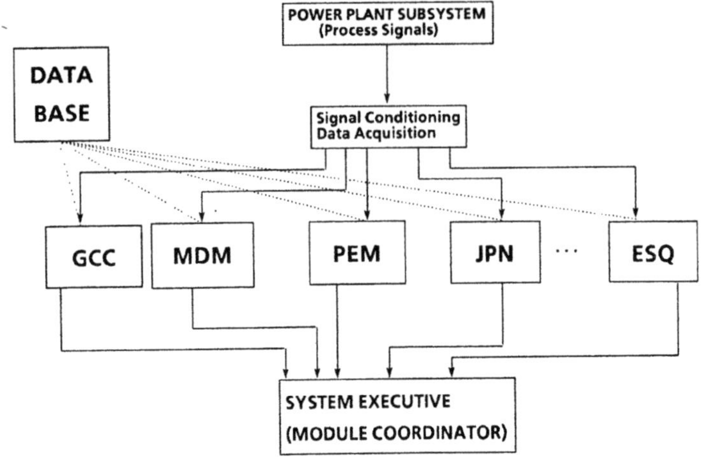

Fig. 1. Signal validation system architecture.

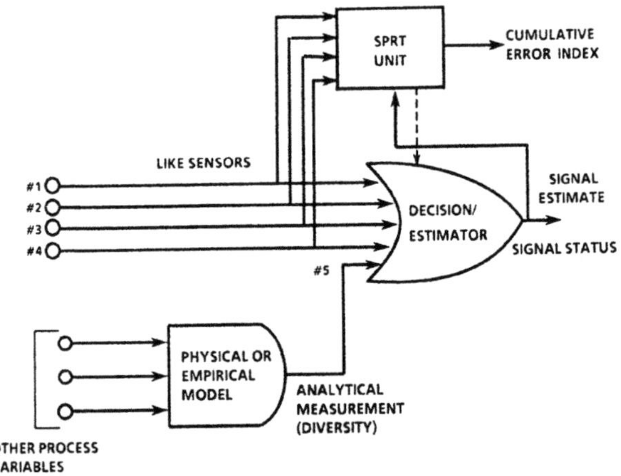

Fig. 2. Redundancy management module combining GCC and SPRT units.

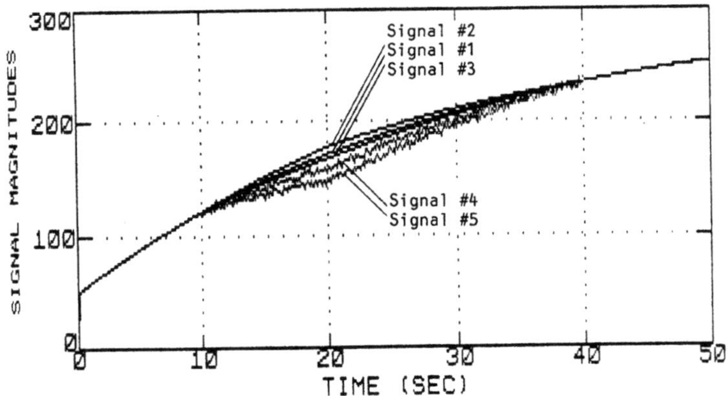

Fig. 3. Five redundant temperature signals (deg F) with bias, noise, or both errors in the various signals.

3.1. <u>Multivariate Modeling of Process Dynamics</u>

The fluctuating or dynamic components of a set of signals are often modeled by the MAR model in the form

$$\underline{X}(t) = \sum_{i=1}^{n} A_i X(t - i\Delta t) + \underline{V}(t) \tag{6}$$

where $\underline{X}(t) = (x_1(t), x_2(t)\ldots, x_m(t))$ represents the vector of measured signals, $\underline{V}(t)$ is a vector of driving noise with covariance Σ and $\{A_i\}$ are (mxm) parameter matrices. The MAR model for a subsystem can be established on-line using computationally efficient algorithms[8]. Consideration must be given to the proper combination of data sampling rate, frequency range, and the choice of model order.

The time domain model is transformed to the frequency domain and the following signatures are calculated.

(a) Auto- and cross power spectral density functions, and inherent noise spectra.
(b) Ordinary and partial coherence functions and phase angle relationships. The coherence function determines the commonality between two signals.
(c) Ordinary and partial noise source signal contribution ratio functions. These functions provide the measure of contribution from one signal to another signal.

A systematic interpretation of these quantities provides diagnostic information about sensor maloperation and process anomaly.

3.2 <u>Process Empirical Modeling (PEM)</u>

Independent prediction of critical signals is required in consistency checking of sensor outputs, for isolating common-mode failure, and for signal trending. Empirical models are updated periodically using plant data. If the purpose is to develop steady-state models, then a certain minimum set of measurements is necessary. General nonlinear models are being developed to characterize the subsystem behavior. Based on an earlier work[9], we have developed a systematic procedure for fitting nonlinear polynomial prediction models. Let y be a process variable to be represented by the inputs $\underline{X} = (x_1, x_2, \ldots, x_m)$. We want to fit a polynomial function in $\underline{X}$, based on N sets of steady-state measurements of the variables $\underline{X}$ and y. In a physical system the components of $\underline{X}$, in general, are not independent of each other. Let

$$y = \sum_{i=1}^{n} c_i \phi_i(\underline{X}) + c_o \tag{7}$$

where n is the number of terms in the polynomial fit, and c_i is the coefficient corresponding to the nonlinear term ϕ_i. For a specified maximum order of the polynomial, the problem is to fit the polynomial that minimizes the error between the measured and predicted y. The fitting algorithm is a successive approximation method. Let M be the total number of cross-product terms for the given number of inputs, and for the specified maximum polynomial order. After choosing the best one-term fit, the remaining (M-1) cross-products and y are projected on to the plane orthogonal to the first term. From this projection again a best polynomial term is selected. The procedure is continued until the error between successive projected

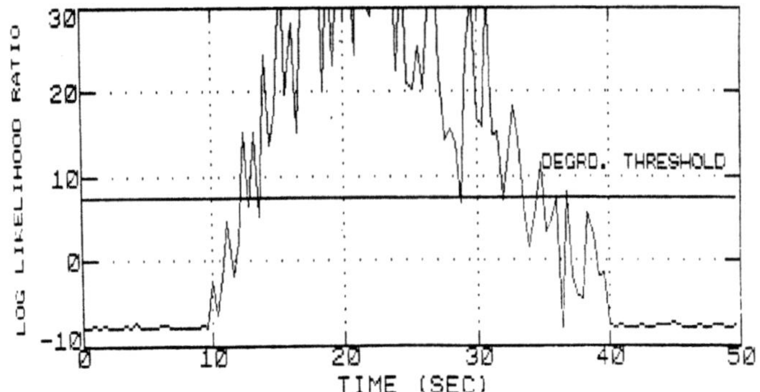

Fig. 5. Inconsistency index for Signal No. 4.

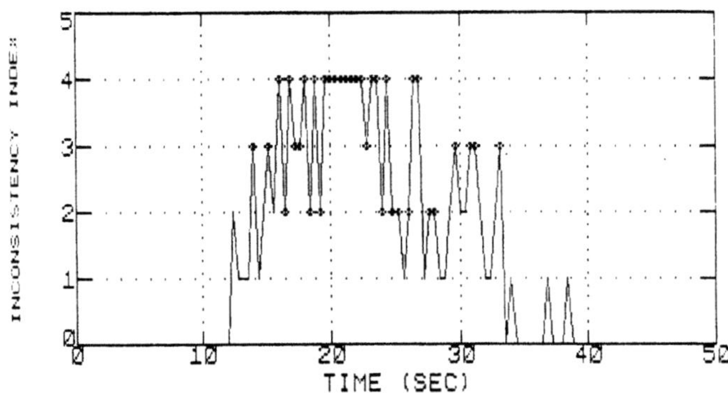

Fig. 4. Log likelihood ratio for bias error (SPRT) in Signal No. 4.

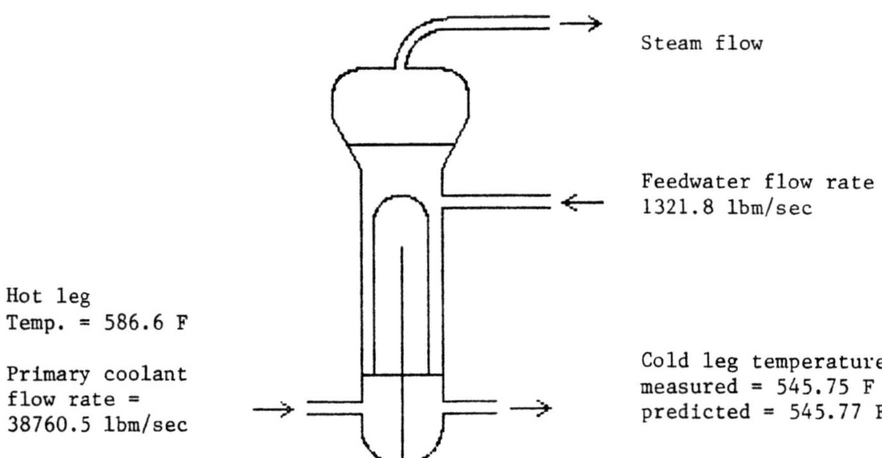

Fig. 6. U-tube steam generator display showing cold leg temperature prediction and other signals used in the nonlinear model fit.

values of y and $\phi_1(\underline{X})$ is less than a desired limit. The method has the
form of a layered or tree-like structure, and does not require all possible
nonlinear terms.

Remarks

(1) The order in which the nonlinear terms are selected does not influence
     the final modeling error.
(2) It was claimed earlier[9] that the algorithm provides the optimal n-
     term model in the first n iterations. Our study showed that this is
     incorrect and it is necessary to have more than n iterations to
     obtain an optimal n-term fit.
(3) It is necessary to scale some of the input variables to the algorithm
     in order to avoid numerical instabilities. This is due to the large
     differences in the magnitudes of process variables.
(4) Error propagation due to the fitting procedure and measurment uncer-
     tainties are incorporated in the algorithm.

3.3 Prediction of a U-Tube Steam Generator Response

The algorithm has been tested using a known nonlinear input-output
relationship, and data from the Combustion Engineering Nuclear Transient
Simulation (CENTS) code. Using data from the primary and steam generator
subsystems we generated nonlinear prediction models of the hotleg temperature, coldleg temperature, and the steam mass flow rate. As an example
of the process empirical modeling, we used primary flow rate, lbm/sec (x_1),
hot leg temperature,F (x_2), and feedwater flow rate, lbm/sec (x_3) to predict the coldleg temperature, F (y). The following fit is obtained:

$$y = 0.141x_1 - 0.2707x_2 - 1.15*10^{-5}x_2x_3 - 3.61*10^{-11}x_1{}^3 + 4.31*10^{-7}x_2{}^3 - 2734$$

Figure 6 shows an example of the display generated by the empirical model
to predict the cold leg temperature. Both the measured and predicted
values are shown in the figure. The input variables are also shown in the
figure.

4. JUMP, PUSLE, NOISE (JPN) DETECTION AND EXPERT SYSTEM MODULES

4.1. JPN Module

The purpose of the JPN module is to detect changes in a signal behavior. The following anomalies will be detected by this module:

(a) jumps in signals,
(b) pulses or spikes in signals,
(c) changes in signal broadband noise,
(d) presence of single frequency noise.

The database includes previous failure information, sensor characteristics,
allowable signal-to-noise ratio, process and sensor time constants, etc.

Some of the variables included for signal anomaly detection are (a)
mean-square or RMS value, (b) signal threshold, (c) signal rate of change,
(d) frequency of occurrence of anomaly, (e) time duration of level change,
(f) power spectrum to detect isolated frequencies, (g) amplitude probability density to determine skewness in the signal behavior, and (h) common
behavior in multiple channels. When a signal anomaly is detected by this
module, further diagnostic analysis is necessary, and the system executive

will activate one or more of the signal processing modules.

4.2. Expert System Module for Qualitative Signal Validation (ESQ)

The primary objective of this module is to evaluate the process signals in a qualitative manner to determine the status of sensor and/or component operations. One of the expert system modules consists of a rule-based inference engine which matches all available normal system operations with the current conditions. The subsystem conditions stored as strings are used to perform this "pattern matching" by forward chaining. A database of a PWR steam generator operating conditions, and a sensor database consisting of failure information, sensor characteristics and a history of past events is being developed. This will also help in deciding which sensors are more likely to fail in a given situation. A knowledge base and diagnostic rules are also being compiled from interview and in-plant experience at the Millstone-2 PWR.

5. SUMMARY AND CONCLUDING REMARKS

A comprehensive signal validation system that integrates various parallel signal processing channels is described in this paper. The use of diverse techniques for detecting measurement degradation will increase the reliability of operator decisions. The approach permits on-line implementation with minimum interference to plant operation. The different signal processing modules have varying signal requirements. In future advanced reactors, such as the advanced liquid metal reactors, consideration must be given to data acquisition requirements and to the use of computer workstations during the plant design phase.

REFERENCES

1. O. L. Deutsch et al., "Development and Testing of a Real-Time Measurement Validation Program for Sodium Flowrate in the EBR-II," Charles Stark Draper Lab., CSDL-R 1952 (Oct. 1982).
2. B. J. Benedict et al., "Validation and Integration of Critical PWR Signals for Safety Parameter Display Systems," EPRI, Report NP-4566 (May 1986).
3. B. R. Upadhyaya et al., "Signal Validation in Nuclear Power Plants," Annual Progress Report prepared for the U.S. Department of Energy, Univ. of Tennessee, Rept. No. DOE/NE/37959-1 (May 1987).
4. A. Ray, M. Desai and J. Deyst, "Fault Detection and Isolation in a Nuclear Reactor," J. of Energy, Vol. 7, pp. 79-85 (Jan.-Feb. 1983).
5. A. Wald, Sequential Analysis, John Wiley, New York (1947).
6. T. T. Chien and M. B. Adams, "A Sequential Failure Detection Technique and Its Applications," IEEE Tran. Aut. Control, Vol. AC-21, pp. 750-757 (1976).
7. B. R. Upadhyaya et al., "An Integrated Approach for Sensor Failure Detection in Dynamic Systems," Research Report, Univ. of Tennessee, Rept. No. NE-MCEC-BRU-87-1 (March 1987).
8. O. Glockler and B. R. Upadhyaya, "Reactor Noise Diagnostics Based on Multivariate Autoregressive Modeling," Fourth Conf. on Utility Experience in Reactor Noise Analysis, Rocky Hill, Connecticut, DOE/NE/37959-5 (May 1987).
9. A. Desrochers and S. Mohsen, "On Determining the Structure of a Nonlinear System," Intl. J. Control, Vol. 40, pp. 923-938 (1984).

CHAPTER 4

EMERGENCY RESPONSE

POTENTIAL APPLICATIONS OF ADVANCED INFORMATION

TECHNOLOGY IN EMERGENCY MANAGEMENT

Håkan Andersson and Conny Holmström

Studsvik Energiteknik AB
S-611 82 Nyköping, Sweden

BACKGROUND

Within nuclear-,offshore- and petrochemical industries there is
always a potential risk for severe incidents and accidents.On such
occasions the staff in charge have to face decisions that:

- must be taken relatively fast
- involve complex judgements
- involve making trade-offs between partly incompatible demands
- must be based on technical data that is difficult to get access to
- involve many decions-makers and experts

It is a commonly shared belief that timely and correct decisions in
these situations could either prevent an incident to develop into a a severe
accident or to mitigate the negative consequences of an accident. It is also
a commonly shared belief that in those cases where poor decisions have been
taken it has been because of insufficient access to information and expert
knowledge when the decisions were taken. These are the background
experiences for the joint Nordic research program on the use of advanced
information technology in emergency preparedness organisations which is
scheduled to run from 1985 through 1989.

Important initial research tasks in the program have been to identify
and specify the needs for advanced information technology applications in
emergency preparedness organisations. Sofar a couple of studies aiming for a
needs assessment of advanced information technologies in nuclear power
emergency preparedness organisations in Sweden and Finland have been
completed. The conclusions from these studies will be presented in this
paper.

NEEDS ASSESSMENT

In the needs assessment studies emergency scenarios were analysed
from a technical as well as an organizational point of view.
Particularly the flow and content of information between different
Nuclear Emergency Preparedness Organisations were studied.See figure 1.

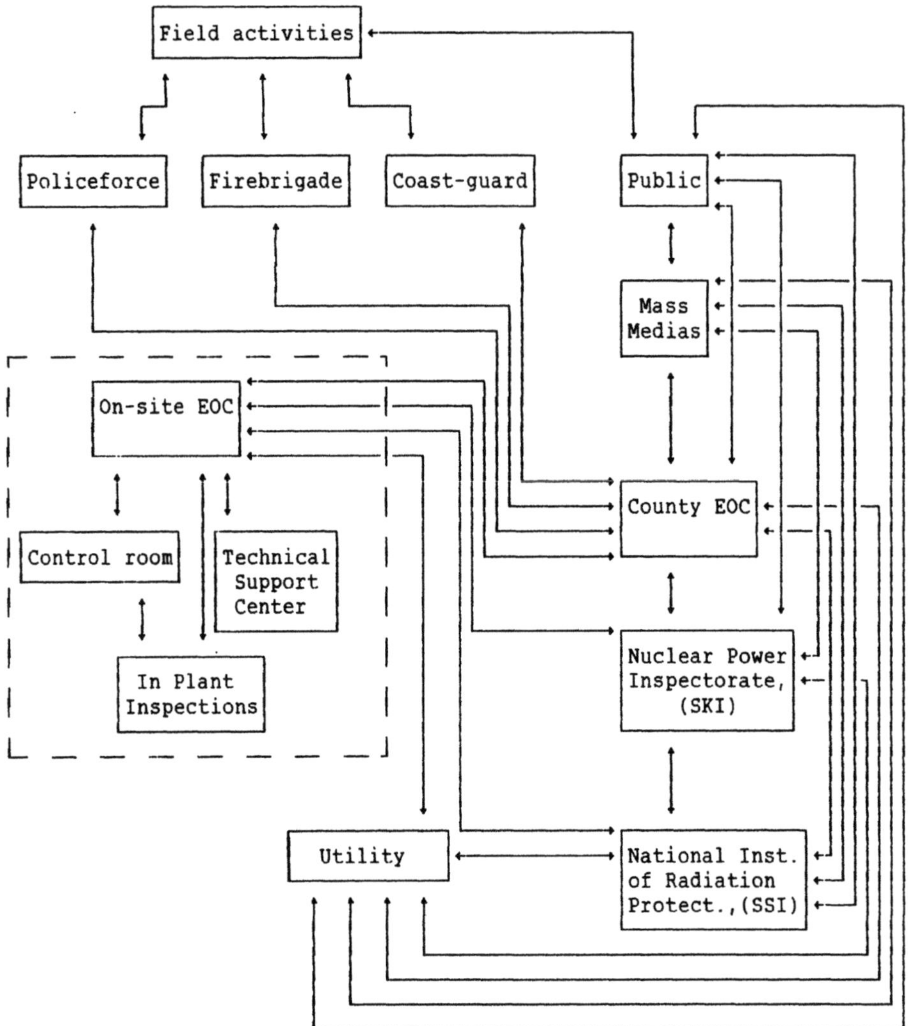

Fig. 1. The flow of information between on-site and off-site
        participants in Swedish Nuclear Power Emergency
        Preparedness Organisations

The information flow was studied during a number of full-scale
emergency management exercises involving all participants in figure 1. For
the most important Emergency Operations Centers (EOPs), that is, the County
EOC, the On-site EOC, the Nuclear Power Inspectorate EOC, and the National
Institute of radiation protection EOC, the following information was
collected:

- tasks and subtasks
- information received
- information generated /decisions/proposals/judgements
- necessery situation overviews
- expert knowledge required
- rules for data collection and distribution
- information sent

- problems attributable to inadequate information management tools,
  unsufficient procedures and lack of expert knowledge

A short summary of the results for the Nuclear Power Inspectorate will
be given below. More detailed information can be found in (1,2,3). As can be
seen in figure 1, the Nuclear Power Inspectorate receives and sends
information to the On-site EOC, the County EOC, the National Institute of
radiation protection EOC, the mass media and the public.

The main tasks are:

- to rapidly collect information about the course of events of the
  accident and the development of main technical parameters
- to evaluate the technical implications of the event to be
  able to give advice and participate in decisions regarding the safe
  operation of the plant
- to evaluate the consequences of counteractions taken and/or
  proposed at the plant
- assess the risk for a radiation release and/or the need for a
  controlled venting within the nearest future
- in case of a radiation release, assess how extensive it could be
  and in what way it could be reduced
- participate in the approval of deviations from the technical
  safety regulations and other improvisations in the use of the
  safety systems
- to consider the consequences of the accident at other units on the
  actual nuclear installation and at other installations

Data received:

- detailed information about the plant status
- continously updated information about main technical parameters
- topical data about the safety systems and the critical safety
  functions
- information about counteractions taken at the plant
- radiation status
- meteorological data

Data generated:

- the current plant status
- plant predictions and risk analysis about the consequences of
  specific technical situations

Data sent:

- requirements concerning more detailed informatin about plant
  status, plant predictions, risk and consequence analysis
- advice regarding safest possible way of plant operation
- judgements and evalutations regarding the course of events and
  other technical plant aspects
- information to the mass media
- telephone service for the public

Desired functions:
Continuously updated information about the:
- detailed plant status
- topical technical plant predictions and evaluation
- radiation status
- meteorological situation
- topical status of information being distributed
- current status of the communication systems

Status and global overview:
- plant status
- predictions and evaluations regarding the technical plant
  situation
- start of potential releases or already occured releases
- radiation levels
- isotopes involved
- meteorological data

Expert knowledge:

- technical plant aspects
- technical plant predictions and evaluations
- reactor risk and consequence analysis
- plant operation
- emergency operation procedures

THE INTERDEPENDENCE OF INFORMATION MANAGEMENT TASKS IN THE EMERGENCY
PREPAREDNESS ORGANISATIONS

A very important aspect for the functioning of a such a complex
organisation as the one shown in figure 1 is how the dependent the
participants are on information and advice from others to be able to solve
their assigned tasks. The studies have shown a number of various
information management tasks that have to be coordinated in case of a
Nuclear Power Plant (NPP) accident.

In a situation when measures for public protection have to be
considered the specific decisions regarding protective actions have to be
taken by the County EOC due to their general responsibility for the safety
of the public in case of nuclear accidents. The expertise on health risks
and radiation, reside , however in another organisation,the National
Institute of Radiation Protection. The latter organisation have a
consultative/ supportative functions versus the County EOC. Field
measurements of dangerous releases are carried by the fire brigade under the
control of their ordinarie management which take orders from the County EOC.
The results are reported to the County EOC and from there passed on to the
National Institute for Radiation Protection. Assessments of the plant status
from a radiological and technical point of view are carried out by experts
at the plant and reported primarilty to the county EOC and to the Nuclear
Power Inspectorate. Weather conditions at the site are indicated by
meter-readings in the central control room at the plant. This information is
collected by the staff at the On-site EOC and reported primarily to the
county EOC and where meteorological experts are present but also to the
Nuclear Power Inspectorate. Experts at the plant also supplies the county
EOC with predictive calculations about the movements of the radioactive
plumes in case of a radiological release. The Nuclear Power Inspectorate
needs information about the status and ongoing activities at the plant to be
able to fill their consultative/supportative function versus the county EOC
and also to meet the requests for information from the mass media and the
public. Similar types of information thus have to be continously updated by
the situation departments at the various EOCs in the emergency preparedness
organisations. Some of the information types may only be available at one
EOC but be needed by analysis groups or decisionmakers at other EOCs. The
result of an analysis or a decision may in turn have to be reported back to
the the first EOC and/or other participants in the total organisation.
For several of the EOCs shown in figure 1 dependencies exists for
many of the most important types of information.

A good example of such a chain of dependencies can be found for the
task of predicting the radiological impact on people of radioactive outlets.
In such a case people at the plant are the only ones to have access to
data about the content and magnitude of the outlet and the local
meteorological conditions. That information is then reported to the County
EOC in charge of the protection of the public and to the Nuclear Power
Inspectorate. Before decisions about protective action are taken at the
County EOC the staff in charge need to consult radiological expertise at the
National Institute for Radiation Protection which therefore need access to
the information about the radiological situation that the County EOC
receives.

A similar but even more complex situation exists in those situations
when a controlled venting can be anticipated at the plant. In such
situations facts and judgements about the technical situation made by the
staff on the site and the staff at the Nuclear Power Inspectorate have to
be coordinated with judgements regarding biological consequences of a
controlled venting made by experts at the National Institute for Radiation
Protection. The practical possibilities to make people take adequate
protective actions also have to be considered. For the latter aspects the
County EOC will have to rely on information from the local police, the
coast-guard and experts on public transportation and sheltering.

For all situations involving a nuclear accident the EOCs involved also
will have to serve the public and the media with general information about
the accident and about ongoing activities and statements made by the
utilities and the authorities involved. The performance of most of the tasks
in a nuclear accident from identifying, diagnosing and evaluating a
technical transient at the plant to those of assessing and evaluating the
environmental consequences and to propose adequate protective actions rely
critically on the access to correct and adequate information as well as
subject matter expertise.

In the organisations that are activated during a nuclear power accident
a large number of people are engaged. They have different professions,
knowledge, location, organisational affiliation and varying degrees of
familiarity with their assigned tasks and they have to exchange information
about facts and judgements in a fast, precise and reliable way, in order to
make the common tasks successful. For these reasons omissions, confusions
and misunderstandings frequently have been found to cause many of the
problems observed during the emergency drills (1) and therefore are likely
to be found in real emergencies as well.

PROBLEMS OBSERVED WITHIN THE EMERGENCY PREPAREDNESS ORGANISATIONS

A frequent problem in the NKA studies has been delays in the flow of
information between the centres. An immediate consequence if delays is less
time for decisions and counteractions. Other consequences have been that the
order of occurance of important events have been mixed up and in that way
a false picture of the situation have been created. In such cases
decision makers frequently have been found to discuss the "situation" based
on different facts without being aware about this. Confusing discussions and
further delays have been the most common results.

Often information is missing and have to be requested with further
delays as a consequence. This is often the case for the experts in the
analysis groups in the centres. During those phases of the event when the
important decisions have to be made they often don't have access to the

facts about the situation they need to use their expertise and consequently
they can't contribute very much.

Another common problem is to understand information that is
communicated. The technical and radiological terminology is often
misunderstood which leads to very different perceptions and judgements about
the seriousness of a given situation.

POTENTIAL BENEFITS FROM ADVANCED INFORMATION APPLICATIONS IN AN
EMERGENCY PREPAREDNESS ORGANISATION

Since many of the problems that have been observed during emergency
drills can be traced back to information problems it is important to see in
what ways information technology could improve the situation.

A Common Computerised Database

The strong interdependence between the emergency preparedness
organisations regarding their access and use of information suggests that a
common computerised database for the main participants in the emergency
preparedness organisation could have many advantages. Such a database would
make information that is fed into the system immediately available to all
EOCs at the same time and some information could also be automatically
entered into the database by computers. In this way the reliability and the
efficiency of the situation reports could be increased. All participants
involved would continously get access to information as soon as it is
updated and people who now are busy collecting and exchanging routine data
on telephone lines could be relieved for more important tasks.

A common database would also increase the possibilities to detect and
correct errors that enters the system. That way problems in the decision
process caused by mistakes in the information management activities could
possibly be reduced. Situation departments have turned out to be key
departments in the information management activities and those that probably
would benefit most from a common database in exchanging, storing and
presenting information (1).

For these reasons the Nordic project on emergency management is
presently specifying a pilot system for the situation departments at the
On-site EOC, the Nuclear Power Inspectorate, the County EOC and the
National Institute for radiation protection (3, 4). The common information
types that have to be kept updated at those departments have been identified
in the needs assessment studies (2). The pilot system should be built and
tested in 1988 and 1989.

Special Analytical and Predictive Functions

The needs assessment studies also have identified subtasks in the
accident and emergency management activities which could be supported by
special analytical and predictive functions in some of the EOCs.

Examples of such subtasks are:

- detection and monitoring of emergency action levels
- monitoring and evaluation of the status of safety systems
- prediction of the impact on people of various radioactive
  outlets
- predictions of time to containment breaks and or controlled
- outlets (pressure relief)

In the Nordic program work on special functions have been restricted to expert system functions for the evaluation of the status of safety systems. Presently the only systems at use in the EOCs are dose forecasting systems which predicts body doses for a sample of accident types that lead to radioactive outlets.

<u>Training</u>

The needs for expert knowledge in the emergency preparedness organisations are the normal case met by gathering experts into expert and/or analysis groups in the EOCs. In these groups of experts, however, the problem solving and decision making process have turned out to be a problem in itself (1).

The lack of every day experience of the kind of problem solving and decision making situations that are faced in Nuclear accidents often causes the experts to make things worse by premature decisions and recommendations. This happens mainly because many of the experts neither are familiar with the tasks nor with each others areas of expertise. Some times the expertise is in fact also missing. In this way the knowledge that de facto exists in the group is not used properly, hypothesis are mistaken for facts, suggestions mistaken for decisions, and sometimes charlatans are mistaken for experts. These phenomena are all well known and documented from research on decision making in groups.

Modern information technology could also here be used to improve the situation. The technology could in this case be used to design cost-effective training tools. Key staff personell could then be trained on specific problems and the need to engage large parts of an emergency preparedness organisation just in order to train limited groups of people could be reduced. The quality and the frequency of drills could then be increased without a corresponding increase in the costs.

Many of the specific advisory functions mentioned in this paper and other papers at this conference could most likely with only little modifications by used as training devices, since an advisory system and a training device often can be looked upon as different sides of the same coin. This is particularly true if one consider the obvious demand for training that goes with the implementation of new advisory systems in an emergency preparedness organisation.

REFERENCES

(1) JOHANSSON, R., ANDERSSON, H. & HOLMSTRÖM, C. A DESCRIPTIVE ANALYSIS OF THE MANAGEMENT OF NUCLEAR POWER PLANT EMERGENCIES. Studsvik Technical Notes, NI-86/7.

(2) Holmström, C. IDENTIFICATION AND ANALYSIS OF INFORMATION NEEDS WITHIN NUCLEAR POWER PLANTS' PREPAREDNESS ORGANIZATIONS BY THE USE OF EMERGENCY SCENARIOS.Studsvik Technical Notes, NI-87/10.

(3) Berg,O., Larsen,J.S., Westgaard,T., Holmström,C., Andersson,H., Jacobsson,S., Heinonen,R. & Kukko,T. A PRELIMINARY SPECIFICATION FOR A MULTIFUNCTIONAL DECISION SUPPORT SYSTEM IN EMERGENCY MANAGEMENT USING EXPERT SYSTEM TECHNIQUES. OECD Halden Reactor Project, HWR-205.

(4) Heinonen,R., Jakobsson,S.,& Kukko,T. DESCRIPTION OF A MULTIFUNCTIONAL DECISION SUPPORT SYSTEM FOR ON-SITE AND OFF-SITE EMERGENCY MANAGEMENT. Technical Research centre of Finland, Electric Engineering Laboratory and Imatran Voima, 1987.

OPA : OPERATOR ADVISOR, AN EMERGENCY RESPONSE GUIDANCE AND
MONITORING SYSTEM

Lucas Mampaey and Patrick Moeyaert
Tractebel, Brussels, Belgium

Frédéric Bastenaire and Fred Casier
A.I. Systems, Brussels, Belgium

Ngo Duc Chi
Trasys, Brussels, Belgium

INTRODUCTION

Since the TMI II accident much investigation has been done into
human factors affecting emergency response. This has resulted in
improvements in accident management, like the use of symptom based
Emergency Operation Procedures and the installation of Safety Parameter
Display panels.

Even though current training programs meet high quality standards,
the use of written procedures still puts the operator under considerable
strain in post-accident situations : he has to find the correct procedure
entry, read each procedure step, read instruments and compare them to set
values, respond correctly to situation changes and complications, keep in
mind cautions and remain aware of what is effectively happening.

The advent of Artificial Intelligence techniques makes it now
possible to build expert systems that make better use of the - already
existing - expert knowledge, than just putting it into written
procedures.

THE OPERATOR ADVISOR APPROACH

In order to tackle the problem of operator aid during post-accident
situations, Emergency Response Guidelines and related background
documents have been subjected to a thorough knowledge engineering
exercise. This has resulted in an appropriate knowledge representation
scheme and an efficient inference mechanism, that permit the Operator
Advisor expert system to act as a truly dynamic task scheduler rather
than like a conventional computerized procedure display.

The OPA development is partly funded by the Belgian Institute for
Research in Industry and Agriculture (IRSIA/IWONL).

This is illustrated in fig. 1, where the following interface features can be recognized :

- A graphical view of the applicable procedure, and an assessment of the state of execution through the use of color codes.
- A compact, action-oriented "what-to-do-urgently" punch list.
- A history of past events that are relevant to the applied procedure.
- Detailed information and explanation upon request.

After proper validation testing on a full scope simulator, it is intended to connect OPA to the plant process computer ; the system will operate in a data driven way without any keyboard dialogue required, to supply the following functionalities to the Operator :
- Status assessment : accident diagnosis and status tree monitoring.
- Automatic sequence monitoring
- Dynamic response management :
  . Context dependent advice on actions to be taken.
  . Event driven triggering of concerns and cautions.
  . Continuous check of context validity (rediagnosis).
  . Evaluation of operator actions.
  . Display of background information upon request.

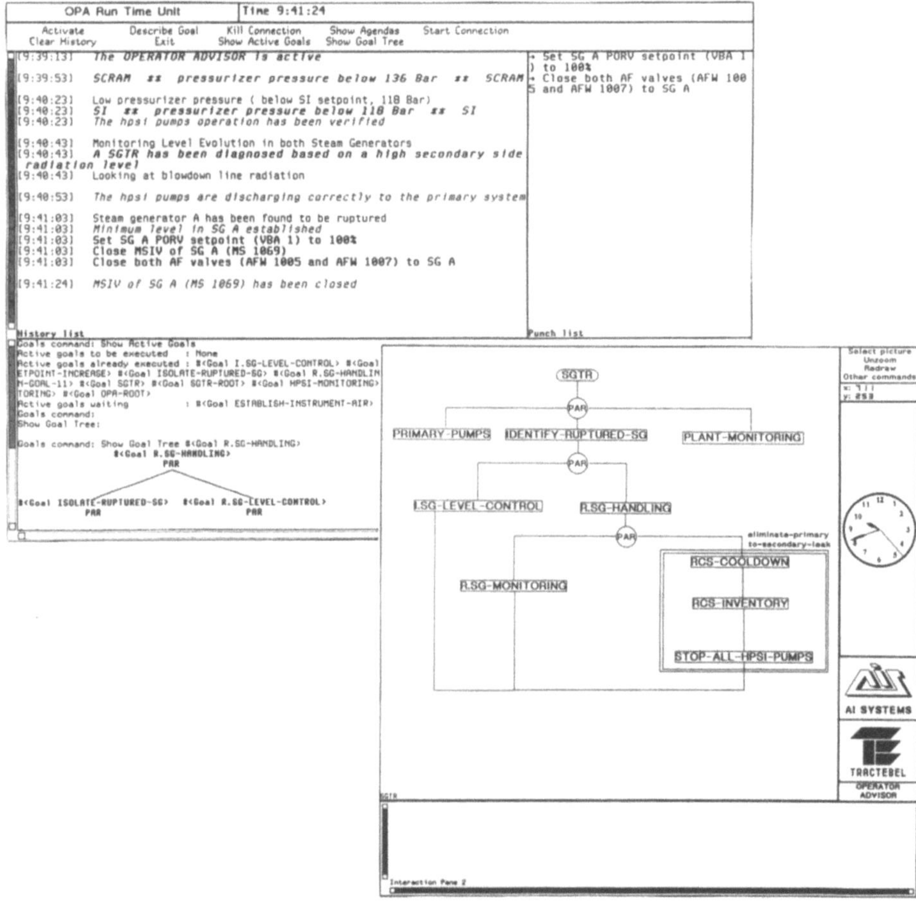

FIGURE 1

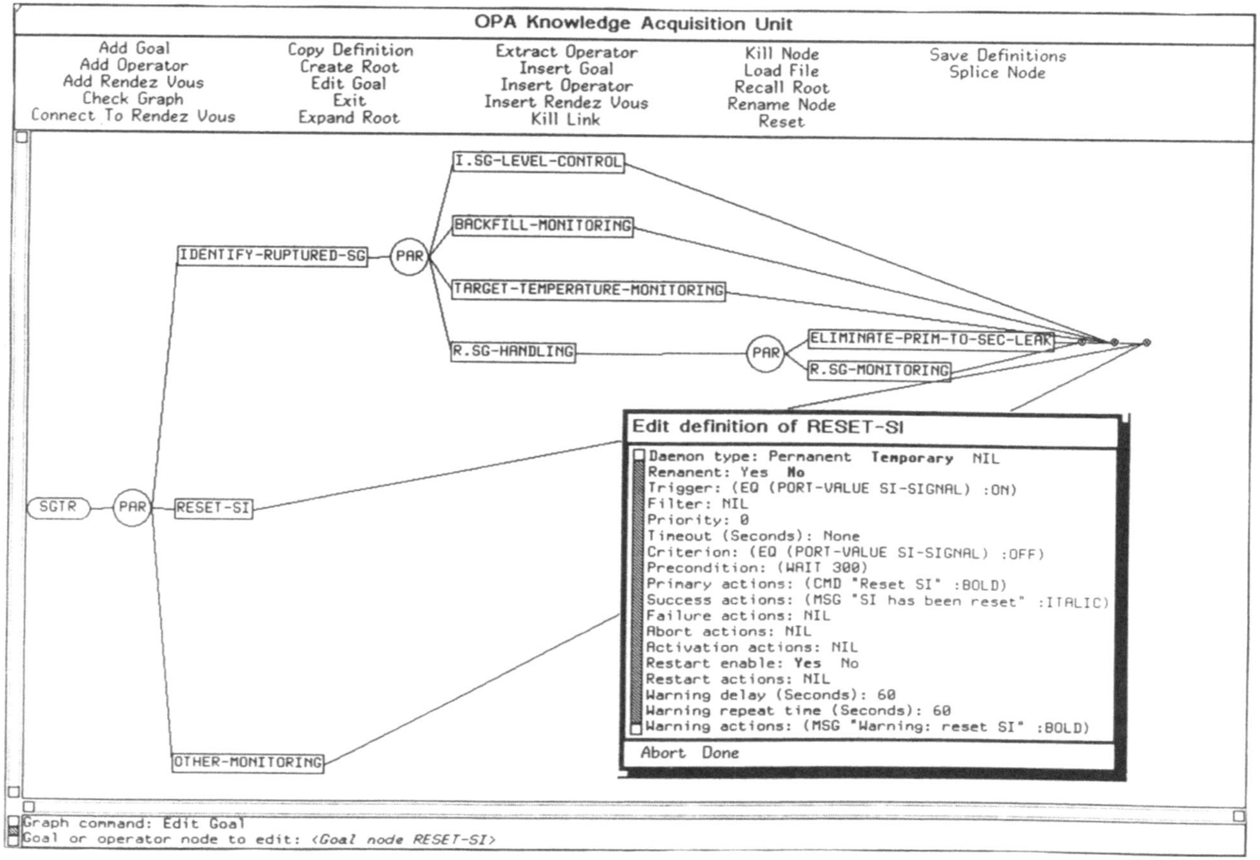

OPA Knowledge Acquisition Unit

Add Goal
Add Operator
Add Rendez Vous
Check Graph
Connect To Rendez Vous

Copy Definition
Create Root
Edit Goal
Exit
Expand Root

Extract Operator
Insert Goal
Insert Operator
Insert Rendez Vous
Kill Link

Kill Node
Load File
Recall Root
Rename Node
Reset

Save Definitions
Splice Node

I.SG-LEVEL-CONTROL
BACKFILL-MONITORING
IDENTIFY-RUPTURED-SG
PAR
TARGET-TEMPERATURE-MONITORING
R.SG-HANDLING
PAR
ELIMINATE-PRIM-TO-SEC-LEAK
R.SG-MONITORING

SGTR
PAR
RESET-SI
OTHER-MONITORING

Edit definition of RESET-SI
Daemon type: Permanent Temporary NIL
Remanent: Yes No
Trigger: (EQ (PORT-VALUE SI-SIGNAL) :ON)
Filter: NIL
Priority: 0
Timeout (Seconds): None
Criterion: (EQ (PORT-VALUE SI-SIGNAL) :OFF)
Precondition: (WAIT 300)
Primary actions: (CMD "Reset SI" :BOLD)
Success actions: (MSG "SI has been reset" :ITALIC)
Failure actions: NIL
Abort actions: NIL
Activation actions: NIL
Restart enable: Yes No
Restart actions: NIL
Warning delay (Seconds): 60
Warning repeat time (Seconds): 60
Warning actions: (MSG "Warning: reset SI" :BOLD)
Abort Done

Graph command: Edit Goal
Goal or operator node to edit: <Goal node RESET-SI>

FIGURE 2

To enter all the knowledge related to this into the system, OPA has
a knowledge acquisition interface (fig. 2) that provides the Procedure
Designer with an effective tool to describe accident treatment in a
natural way.

OPERATOR ADVISOR DEVELOPMENT

For demonstration purposes, a PC/AT containing an interactive real-
time simulation of a steam generator tube rupture accident (*) has been
coupled with OPA. A reduced part of the SGTR ERG procedure and related
background knowledge has been introduced in OPA's knowledge base. During
a 20 minutes demonstration, it is possible to "play" the operator's part,
and to test the monitoring and advice giving capabilities of OPA.

The next step in development is to connect OPA to an interactive
version of RELAP (**). This will allow to cover the complete spectrum of
accidents ; it will allow validation of the procedures, and quasi real-
time testing of the expert system itself.

Finally, for validation testing with operators and for optimizing
the system's man-machine interface, it is intended to connect OPA to one
of the Belgian NPP training simulators.

___

(*) Simulation developed by J. Defloor of Doel 1-2 Nuclear Power Station

(**) RELAP5/mod2 is a best estimate Thermal-Hydraulics Simulation Code
     developed by EG&G, Idaho Falls, for the USNRC. An interactive
     version has been developed in-house at Tractebel.

REACTOR EMERGENCY ACTION LEVEL MONITOR

Robert A. Touchton

Technology Applications, Inc.
Jacksonville, FL 32216

ABSTRACT

The Reactor Emergency Action Level Monitor (REALM) Expert System is designed to provide assistance in the identification of a nuclear power plant emergency situation and the determination of its severity. REALM has been developed to operate in a real-time processing environment. REALM embodies a hybrid architecture utilizing both rule-based reasoning and object-oriented programming techniques borrowed from the Artificial Intelligence discipline of Computer Sciences.

The rulebase consists of event-based rules and symptom-based rules. The symptom-based rules go beyond the current EAL structure to address the more problematic scenarios and entail a more symbolic representation of the plant information. The results to date have been encouraging that expert system technology can provide improved emergency decision-making capability in nuclear power plants. In this context, the REALM prototype can analyze plant conditions at the root level, evaluate those conditions and, based on an overall perspective, identify an emergency situation and classify its severity in relationship to emergency action levels.

INTRODUCTION

In late 1983, EPRI began a program to encourage and stimulate the development of Artificial Intelligence applications for the nuclear industry. As a part of that effort, EPRI contracted Technology Applications, Inc. (TAI) to develop an artificially intelligent Emergency Classification System Prototype. This paper presents a description of and a status report on this recently completed prototype.

EMERGENCY CLASSIFICATION DOMAIN OVERVIEW

Because there exists the possibility of an accident at a nuclear facility which could cause adverse health effects in

the areas surrounding the plant, a great deal of planning and
preparation takes place to reduce or eliminate public health
consequences of a reactor accident, should one occur. One of
these planning elements involves the classification of
emergencies into one of four standardized classes by use of
Emergency Action Levels (EALs). The four classes of
emergencies are:

o   Notification of Unusual Event
o   Alert
o   Site Area Emergency
o   General Emergency

     EALs represent the observables, thresholds, and logic for
mapping plant and site conditions into the appropriate
emergency class. A great deal of information is available in
the power plant (both sensors and manual observables) to help
the staff identify and classify an emergency situation.
However, this wealth of data must be reduced in an expeditious
manner during a time when the staff is already busy responding
to an off-normal condition. This would seem to be a problem
well-suited to computerization, and yet, at most plants,
manual look-up tables are used in the emergency classification
process. Attempts to computerize this process are hindered by
the fact that the data contains significant uncertainties.
There may be missing, conflicting, or ambiguous sensor
readings or there may be false alarms. Therefore, judgement is
required for proper interpretation of the information. The use
of conservative assumptions within a computer program (a
technique often used to solve this problem in many
conventional safety-related programs) does not offer a
solution because over-reacting is just as undesirable as
under-reacting.

     The use of an expert system approach can overcome these
problems by incorporating domain knowledge (heuristics) into
the computer program such that it can begin to make judgements
regarding the data. Such a solution not only computerizes the
data acquisition and processing, but also can appropriately
compensate for the uncertainties contained in the data.

REALM OVERVIEW

     The Reactor Emergency Action Level Monitor (REALM) Expert
System is being designed to provide expert assistance in the
identification of a nuclear power plant emergency situation
and the determination of its severity. REALM is has been
developed to operate in a real-time processing environment.
REALM embodies a hybrid architecture utilizing both rule-based
reasoning and object-oriented programming techniques borrowed
from the Artificial Intelligence discipline of Computer
Sciences. (For background on the underlying computer sciences
concepts relevant to expert systems, reasoning or inferencing
with rules and object-oriented programming, refer to
References 1 and 2, or equivalent.) The rulebase consists of
two general classes: event-based rules, which codify the logic
embodied in the current EAL tables, and symptom-based rules,
which draw inferences even when no specific problem events can
be identified. The symptom-based rules go beyond the current
EAL structure to address the more problematic scenarios and

entail a more symbolic representation of the plant
information.

REALM has been designed to ultimately interface with the
plant in some manner in order to collect sensor data. The most
likely interface is with the safety parameter display system
(SPDS) computer, since it is already charged with the
collection of safety and emergency-related data. However,
during the prototype development stage, REALM utilizes
simulated sensor data which is stored on disk and read in as
if it were actual instrument readings. REALM must also
interface with its user(s). Approximately 10% of the necessary
information must be manually entered into the system. However,
the manually derived inputs relevant to any given emergency
will be small and, thus, the user will be prompted only for
those relevant inputs.

REALM has four operating modes of which two are on-line
(actual and trial) and two are off-line (scenario development
and training). In actual mode, REALM interprets the input data
to identify and classify emergencies in real-time. Thus, REALM
displays to the user its current finding on the appropriate
emergency class and the reasoning which led to that finding.
The trial mode allows the user to make assertions in order to
assess their impact (e.g. see if deenergizing a certain piece
of equipment would cause a higher emergency class to be
invoked).

The scenario development mode serves as a step-wise
training and exercise file builder. This mode operates off-
line and in non-real-time, thus, allowing a user from the
emergency planning or training staff to assert a set of plant
conditions, see the impact on the EALs, look ahead to find
events which would cause the next higher level of emergency,
and finally store on disk that set of conditions if desired.
The training mode uses a training file (developed using the
scenario development mode) to simulate the plant's behavior
during an emergency in order to train operators in emergency
classification and users of REALM in its operation. The
training mode operates in real-time (as represented on the
simulated accident scenario) but is off-line and includes the
scoring and archiving of the trainee's response. Within each
mode, the user may request either a vulnerability analysis or
the system's rationale for the current plant classification.
The vulnerability analysis, knowing the current knowledge
context, reviews all of the rules relevant to the next higher
emergency class and determines what events would cause that
higher class to be declared. Conceptually, this involves
special processing to identify which rules are partially
satisfied and to list the missing antecedents. The processor
then translates the list and displays it in the vulnerability
window. The rationale processor collects all of the readings,
conditions, threats, and any other facts which led to the
current findings, processes the list into a near-natural
language format, and displays the information in the rationale
window.

The REALM Prototype was developed on a Xerox 1108-121
Lisp Machine computer within IntelliCorp's Knowledge
Engineering Environment (KEE), with extensions and special
features developed directly in the Interlisp-D language. (For

background on the KEE and Interlisp-D programming environments, refer to (3) and (4).) The power of KEE's hybrid
architecture for knowledge representation (frames,
ActiveValues/demons, or rules) along with the powerful
man/machine interfaces (graphics, windows, KEE editor, and
mouse), were utilized to produce a successful emergency
management decision aid prototype.

<u>REALM Structure</u>

REALM has been structured to model an emergency
classification process which might be used by the emergency
director and his technical support group during an actual
emergency. Thus, REALM incorporates a team of experts and is
designed to handle a well-behaved situation quickly and
accurately using a minimum set of reasoning and resources and,
at the same time, is prepared to handle a situation with
missing or conflicting data and still arrive at the best
possible conclusion using all available resources. The system
will, upon demand, provide the user with an explanation of the
reasoning used to obtain that conclusion. The system, thus,
consists of distinct but interactive elements: interface,
objects, "a team of experts," a series of message boards, and
rules (as shown in Figure 1, REALM Expert System Prototype-
Schematic Diagram).

<u>Interface</u>. The REALM interface consists of the system
displays and windows for communicating with users. Figure 2
shows an example display. It also provides the user with a
control menu which changes dynamically to alleviate
unnecessary choices based on current conditions. Through this
menu, the user can select scenarios, run REALM, change mode,
or change the current operation of REALM. The interface is
also the home of several message boards used to store data,
such as changed units, instruments out of range, failed
instruments, etc. The message boards are used by REALM.FP, the
top level Lisp program shown in Figure 1, to form an agenda
for polling the REALM experts. REALM.FP will not call an
expert unless a message has been placed in one of the message
boards from which it can determine that pertinent information
has been added or changed since its last execution.

<u>Objects</u>. REALM contains knowledge bases (KBs) of
instruments, systems and subsystems, components, observables,
threats and resources, all of which are represented by
"objects" within the knowledge base. These objects are logical
storage locations for keeping important physical information
about the object and the current state of the plant. Each
attribute of the object is contained in a "slot" which itself
can be further defined by "facets." (For example, the
PRESSURIZER.PRESSURE object has a CURRENT.TREND slot which, in
turn, has a SLOPE facet.)

While data are loading into the specific object slots, a
series of automatic functions (referred to as "ActiveValues"
or "AVs", which are Lisp programs that execute as they are
needed, based on triggers strategically tied to the objects)
perform one or all of the following operations on each of the
slot values: unit conversion; out of normal range
determination; trend analysis using least-squares slope to
determine "decreasing," "increasing," "rapidly increasing," or

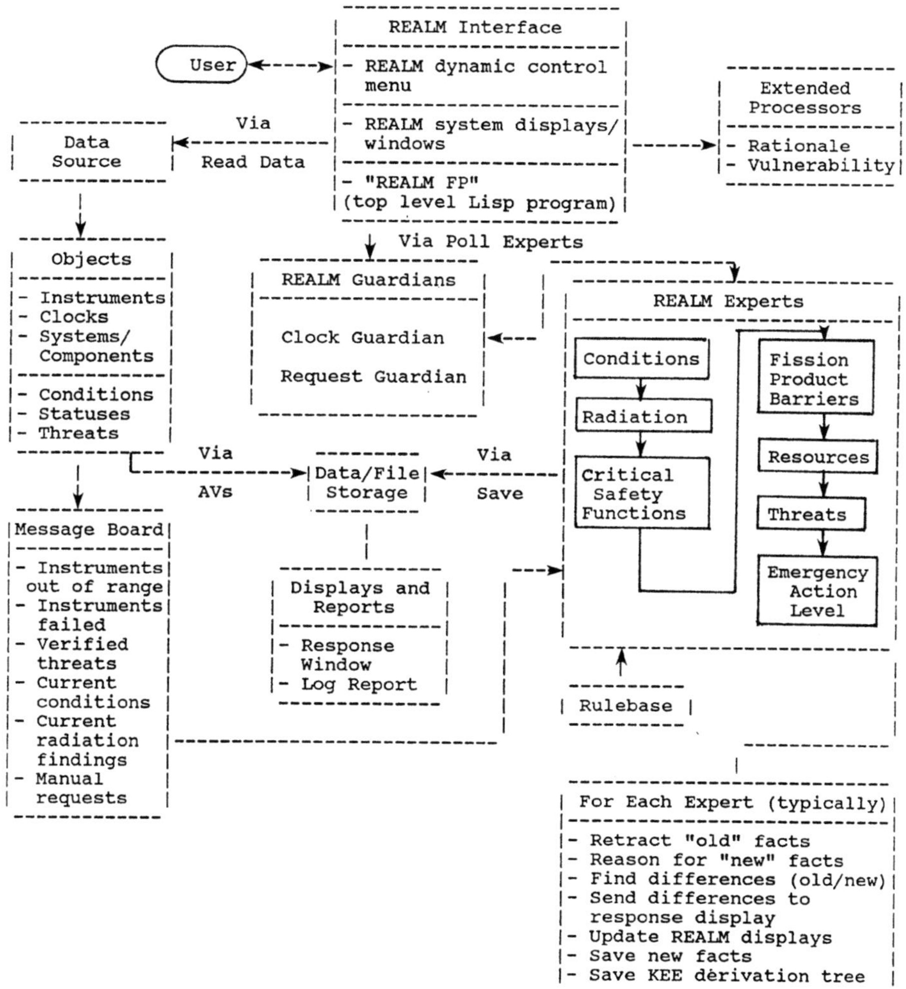

Fig. 1. REALM Expert System Prototype-Schematic Diagram

193

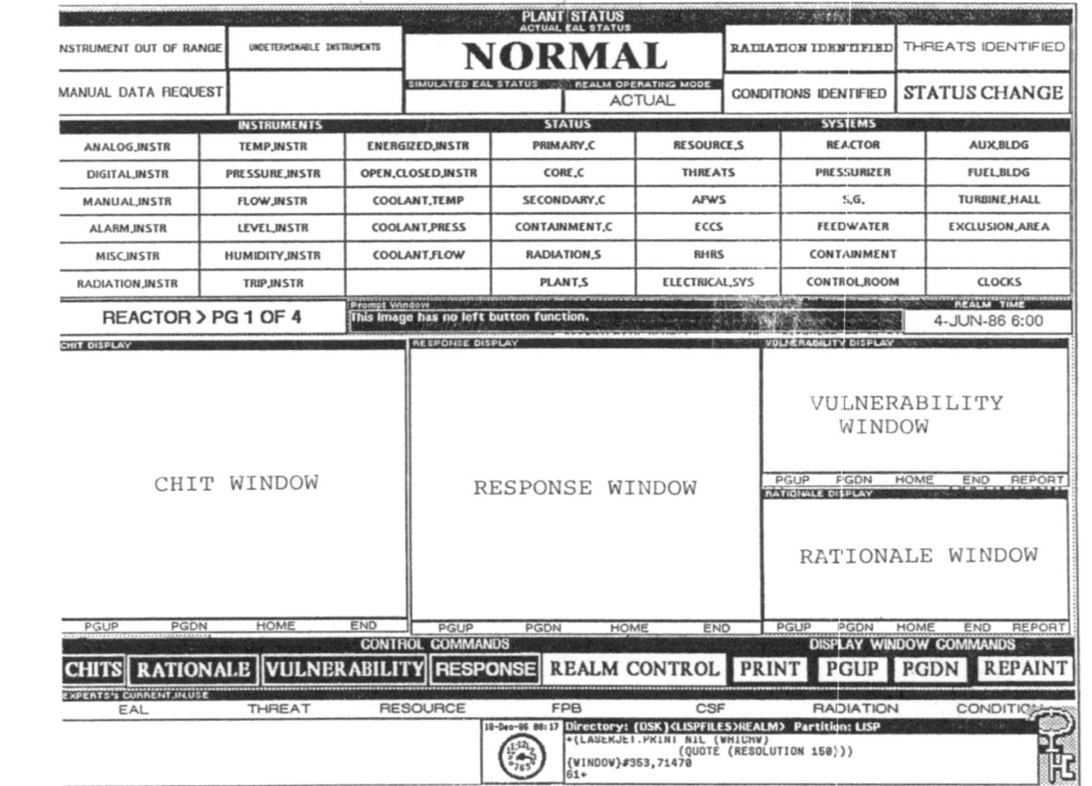

Fig. 2. REALM Interface: Display Window showing location of the four primary text windows.

"rapidly decreasing"; validity tracking; and alternate means
determination for substituting a value for an invalid input.
In this way, each object (particularly instruments) is
responsible for monitoring itself and informing REALM.FP, via
the message board, of any problems.

   <u>Experts</u>. REALM contains several objects realized as Lisp
functions which serve as "experts" on specific aspects of the
emergency classification domain. The conditions expert, the
radiation expert, the CSF (critical safety functions) expert,
the FPB (fission product barriers) expert, the threats expert,
and the resources expert each have fairly deep knowledge about
their respective areas. They all derive information pertinent
to emergency classification to be utilized by the EAL expert
in actually determining the proper emergency classification.
The EAL expert has a fairly shallow knowledge of the overall
plant, relying upon the other experts to supply results upon
which to make a classification recommendation. This diversity
of information allows REALM to classify an emergency condition
even if no specific threat or event can be identified.

   Each expert has detailed knowledge about itself, the
instruments that are important to it, and the rule classes it
affects. With this information the expert can then efficiently
process the proper set of rules (referred to as a "dynamic
agenda").

   <u>Message Boards</u>. REALM message boards provide a facility
for message passing and information sharing among experts.
Information typically found in a REALM message board includes:
(1) instruments out of range, invalid, or rapidly changing,
(2) current threats and conditions, and (3) current findings.

   <u>Rules</u>. Rules are condition-action pairs of the form:

IF condition1 and condition2
THEN action1

   They are processed by an "inference engine" which
basically performs the actions if the conditions are true.
REALM contains such rules divided into separate classes for
use by each expert. Each class is further divided into
contextual subclasses, so that only those subclasses of rules
relevant to a given situation are processed.

REALM Control Strategy

   In emulating the human process for classifying
emergencies, the REALM control strategy places REALM.FP in the
emergency director role to query and dialogue with the team of
experts to properly classify the emergency. Thus, REALM.FP
will assess the observables and, if a problem is found, will
poll his team of experts. REALM.FP operates in a cyclic
fashion, reading new data and polling the experts. In more
detail, the classification process proceeds in the following
manner.

   REALM.FP reads in a new set of data containing any
associated signal validation information available from the
host computer (simulated on a disk file for this prototype).
As data are received, the appropriate ActiveValues within each

object are triggered, enabling the object to monitor its own
state. The instrument out of range function will notify
REALM.FP, via the instrument message board, if its value
indicates a problem and that further investigation is
required. Likewise, the instrument validity function will
notify REALM.FP (via the indeterminable instrument message
board) if it judges itself to be unusable by or unavailable to
the experts. A message in either of these two message boards
will trigger REALM.FP to start evaluating and classifying the
current plant status. On the other hand, an instrument that
has recovered from being invalid or out of range will be
removed from these message boards and will also trigger the
classification process. This process proceeds as follows:

o   REALM.FP first activates the clock guardian and
        request guardian to determine if time dependent data
        has expired or if manual data should be requested.
        Then it queries the conditions expert, radiation
        expert, CSF expert, FPB expert, resources expert,
        and threats expert to "report in." Lastly, the EAL
        expert is called to determine the EAL status.

o   The conditions expert forward chains on its relevant
        rule classes to determine any intermediate
        conclusions which may be of use to the other
        experts.

o   The radiation expert forward chains on its relevant
        rule classes to determine if any radiation has been
        released and, if so, how much and where it is
        located.

o   The CSF expert forward chains on its relevant rule
        classes to determine the status of the critical
        safety functions.

o   The FPB expert forward chains on its relevant rule
        classes to determine the status of the fission
        product barriers.

o   The resources expert forward chains on its relevant
        rule classes to assess the status of plant resources
        needed to counter any problems the plant may
        encounter.

o   The threats expert forward chains on its relevant
        rule classes to determine if any threats to the
        plant exist (e.g. LOCA).

o   Finally, the EAL expert backward chains on its rule
        classes using information gathered by the prior
        experts to determine the appropriate.

REALM PROTOTYPE DEVELOPMENT RESULTS

    Development of the REALM Expert System Prototype has been
completed using rapid prototyping software development
techniques. The objectives of the REALM Prototype development
process were:

o   to determine the feasibility of REALM functions and
      features,

o   to evaluate the usefulness and importance of REALM
      functions and features,

o   to identify real-time/on-line applications
      development issues early,

o   to refine applications development techniques,

o   to clarify/refine/define functional requirements,
      and

o   to demonstrate REALM concepts to EPRI and utility
      personnel.

This initial effort has met these objectives and the
results to date have been encouraging that expert system
technology can provide improved emergency decision-making
capability in nuclear power plants. In this context, the REALM
Prototype can analyze plant conditions at the root level,
evaluate those conditions and, based on an overall
perspective, identify an emergency situation and classify its
severity in relationship to emergency action levels.

In its present stage of development, REALM can serve as a
compact, desktop training aid to sharpen the analysis and
decision-making skills of emergency response personnel. REALM
can also be used to aid emergency planners in the creation of
challenging, yet realistic emergency exercise scenarios.

The ultimate goal of the REALM project is to provide an
on-line system for aiding decision-making during actual
emergencies. Although much work remains to refine the system,
to validate the decision-making logic, and to optimize
performance, the feasibility of the REALM system has been
established. This finding was further confirmed by the
feedback from a field test of REALM during an annual site
emergency drill (June 6, 1986) at the Indian Point 2
generating station, operated by Consolidated Edison of New
York.

REFERENCES

1.   Frederick Hayes-Roth, Donald A. Waterman, and Douglas B.
      Lenat. _Building Expert Systems_. Reading, Massachusetts:
      Addison-Wesley Publishing Company, Inc., 1983.

2.   Patrick Henry Winston. _Artificial Intelligence Second
      Edition_. Reading, Massachusetts: Addison-Wesley
      Publishing Company, Inc., 1984.

3.   _Interlisp-D Reference Manual_. Xerox, October, 1985.

4.   _The Knowledge Engineering Environment_. IntelliCorp, 1984.

DEVELOPMENT OF AN ACCIDENT MANAGEMENT EXPERT SYSTEM

FOR CONTAINMENT ASSESSMENT[a]

W.R. Nelson, D.E. Sebo, and L.N. Haney

Idaho National Engineering Laboratory
EG&G Idaho Inc.
P.O. Box 1625
Idaho Falls, ID 83415 USA

ABSTRACT

The United States Nuclear Regulatory Commission (USNRC) is sponsoring a program at the Idaho National Engineering Laboratory (INEL) to develop an accident management expert system. The intended users of the system are the personnel of the NRC Operations Center in Washington, D.C. The expert system will be used to help NRC personnel monitor and evaluate the status and management of the containment during a severe reactor accident. The knowledge base will include severe accident knowledge regarding the maintenance of the critical safety functions, especially containment integrity, during an accident. This paper summarizes the concepts that have been developed for the accident management expert system, and the plans that have been developed for its implementation.

INTRODUCTION

One of the most useful features of expert system technology is the ability to gather large amounts of expert knowledge and make it usable in complex decision environments. NRC research programs have generated substantial knowledge regarding the behavior of reactor systems during severe accidents. Unfortunately, a large fraction of this knowledge is not currently accessible by the people who are responsible for the management of accidents when they actually occur. For example, the personnel of the NRC Operations Center in Washington, D.C. are responsible for assessing the seriousness of accidents in progress and communicating their assessment and recommendations to the appropriate decision makers. Expert system technology has the potential to provide the Operations Center with access to the knowledge available from NRC research programs in a form that would be usable during actual accidents. The NRC is supporting the development of concepts for an accident management expert system for the Operations Center. This expert system will concern itself with the assessment of containment performance during severe accidents.

a. Work supported by the U.S. Nuclear Regulatory Commission, Office of Nuclear Regulatory Research under DOE Contract No. DE-AC07-76ID01570.

The application of containment assessment was chosen for the following reasons:

1. The control of containment function during a severe accident has a critical influence on accident consequences, including offsite protective measures.

2. Decisions regarding containment function are highly complex, and involve highly specialized expertise in the areas of physical phenomena and containment performance. These decisions will be made in the face of considerable uncertainties in both the knowledge of severe accident phenomena, and uncertainties regarding the current plant status. However, a reasonable knowledge base does exist for addressing questions about containment function even though these uncertainties exist.

3. Decisions regarding the control of containment function will likely involve personnel in addition to or other than the control room operating crew. For example, in the United States, the onsite technical support center (TSC) would be one of the primary locations where these decisions would be made. Also, the containment experts at the NRC Operations Center would be responsible for assessing threats to the containment function during the course of the event. In addition, most accidents in which containment control becomes vital, occur over a substantial period of time so that these facilities would be activated.

Development of the accident management expert system will occur in three phases over a period of three years. The first phase, currently in progress, is the development of a detailed specification for the expert system. The second phase will consist of the implementation of the expert system in prototype form. The third phase will involve test and evaluation of the expert system during actual drills at the NRC Operations Center.

FUNCTIONAL ANALYSIS

Interviews were conducted with NRC Operations Center personnel to determine the functions and tasks performed during containment assessment. Interviews were conducted individually to obtain different perspectives of the process. These perspectives were combined to determine the overall functions performed by the Operations Center decision makers. The three major functions are performed:

1. Identify plant status. The current status of the plant is defined to determine the nature and severity of the accident. Two approaches can be used in the process: event-oriented analysis and function-oriented analysis. Event-oriented analysis is used to determine (if possible) what event sequence is in progress by assessing the status of plant parameters and equipment. Function-oriented analysis is used to determine the status of the critical safety functions, i.e., core heat removal, reactor coolant system (RCS) inventory control, etc. This analysis permits the severity of the accident to be assessed even if the specific event cannot be identified.

2. Identify containment threats. Current and potential challenges to containment integrity are identified using either function- or event-oriented analysis. This includes determination of the

estimated time and magnitude of potential releases as well as
possible containment failure modes. If possible, critical
actions that would reduce the likelihood of loss of containment
integrity are identified.

3. Identify offsite threats. The nature and magnitude of offsite
    threats are also evaluated. This requires the assessment of
    meteorological conditions and population distributions to deter-
    mine the impact of the threat. Potential protective measures
    such as sheltering and evacuation are identified and the provi-
    sions of the facility emergency plan are reviewed to determine
    what actions, if any, are appropriate to minimize the offsite
    consequences.

Initial development of the accident management expert system will
focus on a subset of the first two functions described above. That is,
the initial prototype will be designed to determine what components or
systems should be operated to maintain the critical safety functions and
thereby minimize the consequences of the event.

PRELIMINARY SYSTEM DESIGN

The approach segments accident management into three generic tasks:
1.) determination of the severity of the incident and the challenges to
the critical safety functions, 2.) determination of the nature of challenges to containment integrity, and 3.) determination of the equipment
that should be operated to maintain the critical safety functions. The
knowledge required to perform these tasks is partitioned into different
sources. The following paragraphs give a short description of each
knowledge source in the model.

Critical safety function knowledge source. To assess threats to the
critical safety functions, it is necessary to functionally define the
safety functions and determine how the plant's resources are used to maintain them. A logic structure that has been developed to accomplish this
is the critical safety function response tree. Response trees illustrate
how the plant's resources are organized into different success paths that
can be used to maintain each safety function. The response trees can be
evaluated to determine the effects of plant status on the availability of
success paths when a safety function is challenged. The response trees
will be used to determine the availability or unavailability of each
success path, determine what success paths are preferable for restoring
the safety function, and identify critical components or systems whose
failure or recovery would have a significant impact on the availability
of success paths.

Heuristic knowledge source. This portion of the knowledge base will
include heuristics and experiential knowledge used by the containment
experts of the Operations Center. This may include short "back of the
envelope calculations" such as estimation of core damage based on containment radiation readings or estimation of the containment pressure
increase resulting from a hydrogen burn.

Offsite actions knowledge source. This knowledge source will use
the estimates of the challenges to containment integrity and the current
situation to provide estimates of the relative value of different recommendations for offsite actions. This module will be developed after the
other modules have been developed and tested.

MAN-MACHINE INTERFACE

Established human factors guidelines will be considered in the development of the expert system specification. These guidelines provide extensive information for all aspects of the user-computer interface including screen layout; information and data formatting, input devices, command structure and type; and ergonomic, visual, and auditory considerations. Applicable guideline documentation and human factors input will be provided throughout the development of the expert system specification.

Containment assessment requires the use of system parameter values, equipment status, and operator actions. The ability to provide the expert system with this information in a quick, concise, and error free format will be crucial to the success of the system. This requires that the expert system ask the user for only a minimum of information. Otherwise, the system may detract from the ability of the Operations Center personnel to carry out their tasks by impeding the flow of information with requests for unnecessary or irrelevant data.

Expert system output must be both valid and relevant. Conclusions that are true but nonessential may hinder decision making. The capability for explanation of conclusions is also critical for user confidence and acceptance of the expert system. The design of the output format must provide the user with a concise summary of the current situation with the capability for more detailed explanations of expert system conclusions when desired.

SYSTEM EVALUATION

Once the specification for the accident management expert system has been completed, the system will be implemented in prototype form. The prototype system will then be tested using sample problems from previous Operations Center drills to determine the correctness of the conclusions reached. Identified deficiencies will be corrected.

SYSTEM IMPLEMENTATION

The effectiveness of the prototype expert system will then be evaluated. The expert system will be tested during actual drills of the NRC Operations Center. Performance data will be collected and analyzed to assess the degree to which the expert system has fulfilled its design objectives and to assess its overall effect on the decision making performance of Operations Center personnel.

CONCLUSIONS

An accident management expert system for containment assessment is being developed for use at the NRC Operations Center. The expert system will be used to help NRC personnel monitor and evaluate the status of the containment during a severe accident, using knowledge regarding the maintenance of critical safety functions. Development of the functional and performance specification for the expert system is underway. The expert system will be implemented and then evaluated during drills at the Operations Center.

AN EXPERT SYSTEM FOR IMPROVING NUCLEAR EMERGENCY RESPONSE[†]

A. Salame-Alfie,[1] G.C. Goldbogen,[2] R.M. Ryan,[1]
W.A. Wallace[3] and M.L. Yeater[1]

1. Department of Nuclear Engineering & Engineering Physics
2. Center for Manufacturing Productivity & Technology
   Transfer
3. Decision Sciences & Engineering Support
   Rensselaer Polytechnic Institute
   Troy, New York   12180-3590

ABSTRACT

The accidents at TMI-2 and Chernobyl have produced initiatives aimed
at improving nuclear plant emergency response capabilities. Among them
are the development of emergency response facilities with capabilities for
the acquisition, processing, and diagnosis of data which are needed to
help coordinate plant operations, engineering support and management under
emergency conditions [1].

An effort in this direction prompted the development of an expert
system. EP (EMERGENCY PLANNER) is a prototype expert system that is
intended to help coordinate the overall management during emergency
conditions.

The EP system was built using the GEN-X [2] expert system shell.
GEN-X has a variety of knowledge representation mechanisms including
AND/OR trees, Decision trees, and IF/THEN tables, and runs on an IBM PC-XT
or AT computer or compatible. Among the main features, EP is portable,
modular, user friendly, can interact with external programs and interro-
gate data bases. The knowledge base is made of New York State (NYS)
Procedures for Emergency Classification, NYS Radiological Emergency
Preparedness Plan (REPP) [3], and knowledge from experts of the NYS
Radiological Emergency Preparedness Group and the Office of Radiological
Health and Chemistry of the New York Power Authority (NYPA).

In its initial configuration, for an onsite user EP can determine the
emergency level classification of the event by asking questions about
plant parameters (based on the Emergency Action Levels or EALs). If the
user is offsite, EP provides a list of the necessary actions required by
the REPP for individuals in each organization participating in the
emergency response.

[†] Based on the Doctoral Research of A. Salame-Alfie.

Currently, EP can access an external file with plant data. Future
work includes reading data directly from the plant computer or Safety
Parameter Display System (SPDS). The system is being validated by
implementing various drill scenarios and comparing the results obtained to
those predicted.

We believe that the use of expert systems by plant operators, when
making decisions of potential safety risk, can act as a check and a source
of valuable information, and help increase the overall safety of the
plant.

BACKGROUND

Declaring, planning and conducting an emergency response requires
reliable prediction and correct assessment of plant parameters,
meteorological data and radiological conditions. To correctly and timely
respond to emergencies, this information must be delivered with
considerable lead time, thus enabling complex decisions to be made in
relatively short times.

The main phases in the emergency preparedness process are prevention/
mitigation, response and recovery [3]. Each one of these plays an
important role, and decisions are required at each phase.

In the prevention/mitigation phase the objective is to eliminate or
reduce the effects of radiological emergencies by means of logistical and
technical assistance, as well as by off-site monitoring.

The response phase deals with the reaction to an emergency which
encompasses the Federal, State, Local and Private sectors as well as the
coordination among them. It includes activities such as direction and
control, communication, public notification, accident assessment,
protective response action, public information, evacuation, etc. Typical
step-by-step, or simultaneous response proceedings during a radiological
emergency are depicted in Fig. 1.

In the recovery phase, an effort is made to restore the quality of
life to the community.

The operators of nuclear facilities have primary responsibility for
accident assessment. This responsibility includes taking prompt action to
evaluate any potential risk to public health and safety, both onsite and
offsite, and making timely recommendations to state and local governments
concerning protective measures. Since those protective measures could be
needed within a short time - a half hour or perhaps less - after determination that a hazard exists, prompt accident assessment at the source is
important.

The Nuclear Regulatory Commission (NRC) has established four classes
of emergencies in ascending order of seriousness; they are:

- notification of unusual event
- alert
- site area emergency
- general emergency

Exceeding specific plant instrument readings would indicate that a

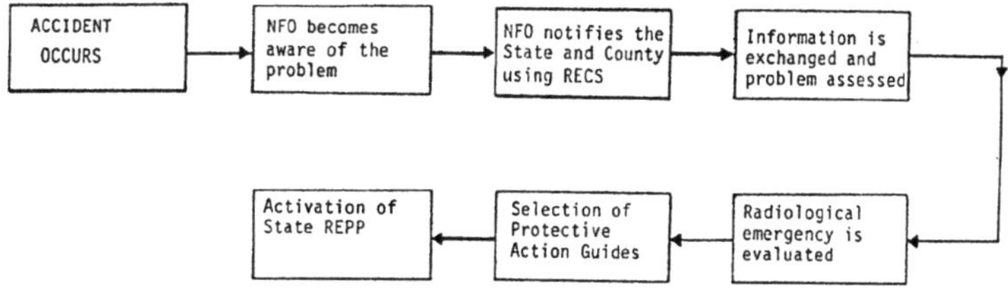

Fig. 1 Typical step-by-step, or simultaneous response proceedings during a radiological emergency

given initiating condition has been met and that the appropriate class of emergency must be declared. The plant-specific instrument readings are called Emergency Action Levels (EALs). Their purpose is to provide a clear basis for the rapid identification of a possible problem and for the notification of offsite authorities that an emergency exists.

Since the TMI accident, the nuclear industry has undertaken a major initiative to improve nuclear plant emergency response capabilities. Among the major improvements is a shift from the single failure criterion into the multiple failure concept, and further the shift to event-oriented emergency procedures. This new approach has led to the development of the Safety Parameter Display System (SPDS) concept [1], whose main function is to provide a continuous indication of the plant status relative to safety functions, in one centralized location. Continued research in that direction showed the importance of human-factors in the SPDS design, which has resulted in the system-oriented procedures and the use of artificial intelligence systems for function maintenance [7].

Additional initiatives aimed at improving the response capabilities resulted in the development of Emergency Response Facilities (ERF's). The design of those facilities includes computerized capabilities for acquisition, processing and diagnostics of data needed to help coordinate plant operations, engineering support and management under emergency conditions.

Further studies [8] show that a need exists for a greater emphasis on improvements aimed at reducing "cognitive errors" rather than "skill-based performance", because the former are more likely to be the dominating factor in causing or aggravating accidents than the misoperation of equipment.

It has also been suggested [5] that if a computerized information system is provided in the control room to assist during high level plant monitoring and decision making, that is, allowing them to proceed in parallel with control board actions, then a better crew performance could be achieved, especially under high stress and time constraints. This in turn eases the burden of classifying emergencies (one of the tasks in emergency preparedness), and consequently facilitate every day operation through the decrease of uncertainty in decision making.

A step further in this direction is the use of expert systems that will help predict when an emergency is likely to occur, and that once an emergency situation is discovered, will speed up the recovery.

The above need, coupled with the availability of new tools for decision making, prompted the development of EP (EMERGENCY PLANNER).

METHODOLOGY

It has been pointed out that there is a need for computerized decision-support systems in the control room, because of the volume of information that the operator and supervisor have to deal with on a daily basis, and the fact that it increases tremendously when there is an emergency. Since the introduction of the SPDS in the control room, the operator can look at the key parameters with different levels of detail. The development of systems that couple the SPDS design with the human factors has resulted, among others, in systems such as the CFMS [4], DASS [5,9] and RSAS [6]. A natural step forward will be a system that can look at the plant parameters (either from the plant computer, SPDS, or direct interrogation of the operator), ask a few "judgmental" questions to the operator and determine if an emergency exists.

The EP system was designed as an attempt to help in the coordination of overall management during emergency conditions. This prototype was intended to aid in the response phase, the part of the emergency preparedness that involves the reaction and coordination of the various agencies. It was found that this phase could be represented (to a certain extent) with an expert system.

The knowledge base was taken from the Emergency Level Classification for the Indian Point Nuclear Power Plant the procedures from the State of New York Radiological Emergency Preparedness Plan [3], and expert's knowledge from the New York State Radiological Emergency Preparedness Group (REPG). Additional information will be obtained from interviews with personnel from the Indian Point Power Plant.

The EP system addresses issues that have arisen with the improvements in the emergency preparedness. They are:

- Aid in problem detection by determining the condition at the plant based on selected parameters.

- Help in the response planning and control evaluation by presenting the personnel of different agencies (for example, Department of Transportation or Civil Defense Administrator in the counties) with a checklist of actions that have to be taken during different levels of emergency.

- Help in the prompt recognition of plant disturbances.

- Use of a symptom-oriented approach. This is helpful because it supports procedures and training.

- Use of a modular top-down approach. This feature makes it possible to have an overview of the plant condition.

Several expert system shells were tried, and it was concluded that the GEN-X shell was the best suited for this type of project. Among other reasons, because it has the ability to combine three types of modules:

AND/OR trees, Decision trees and IF/THEN tables. These modules share a
global fact base and information is transferred among them.

The EP knowledge base consists of various parts, and each can be
represented with a different scheme, for example, the EALs can be "easily"
converted into AND/OR trees, the procedures that have to be followed by
the Emergency Coordinators can be represented with IF/THEN tables and the
overall control of the program is done with the use of Decision trees.

During the development of EP it was found that AND/OR trees were a
more natural way of translating the EALs, than the use of rules. It was
also found that the rule representation was adequate only when there were
mostly AND conditions, because all the facts under an AND gate are
represented with one rule, but each fact under an OR gate implies a
separate rule. It was also easier for the developer to visualize the
logic by looking at the AND/OR trees, especially if one is familiar with
the fault and/or event tree methodology. An extensive discussion of this
topic can be found in [10]. An example showing these differences is given
in Fig. 2.

The EP system, using the GEN-X expert system shell, consists of 65
modules, of which 37 are AND/OR trees. The overall logic is depicted in
Fig. 3. It can run in two modes: interactive or semi-automated. In the
interactive mode, questions are asked to the user at each step of the
inference process, whereas in the semi-automated, a data base or file is
interrogated and the user is only required to confirm or supply
"judgmental" answers.

Functionally, EP will ask the user about the location. The available
choices are offsite and onsite. For the offsite case there are two
choices, namely County Emergency Operations Center (EOC) or State EOC.
For each one of those, there is a checklist depicting the actions that the
Director of Emergency Preparedness is expected to follow. The whole
purpose of this list is to be a mind jogger. There is a list for each of
the emergency levels, and the selection of the appropriate list is done
using IF/THEN tables.

For the onsite case, there are two possible locations, namely, the
Control Room/Technical Support Center (TSC), and the Emergency Operations
Facility (EOF). The locations were chosen based on the amount and type of
information available, and since the control room and the TSC have access
to the same information they are considered as one location.

Once provided with the location, EP asks the user for the event type.
The events that result from the EALs have been grouped into six major
groups. They are:

- Thermohydraulic
- Radiological
- Power Failure
- Safety Related
- Natural Hazard, and
- Other.

This last type of event is further divided into

- Fuel Handling Accident
- Fire
- Security Threat, and
- Other Limiting Conditions.

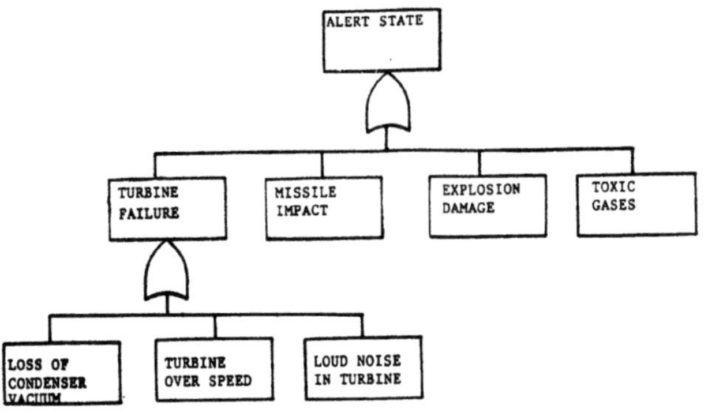

RULES

1.   IF
     LOSS OF CONDENSER VACUUM IS TRUE
     THEN
     TURBINE FAILURE IS TRUE

2.   IF
     TURBINE OVERSPEED IS TRUE
     THEN
     TURBINE FAILURE IS TRUE

3.   IF
     LOUD NOISE IN TURBINE IS TRUE
     THEN
     TURBINE FAILURE IS TRUE

4.   IF
     TURBINE FAILURE IS TRUE
     THEN
     ALERT STATE IS TRUE

5.   IF
     MISSILE IMPACT IS TRUE
     THEN
     ALERT STATE IS TRUE

6.   IF
     EXPLOSION DAMAGE IS TRUE
     THEN
     ALERT STATE IS TRUE

7.   IF
     TOXIC GASES IS TRUE
     THEN
     ALERT STATE IS TRUE

Fig. 2   Comparison of AND/OR Tree and Rule Representation

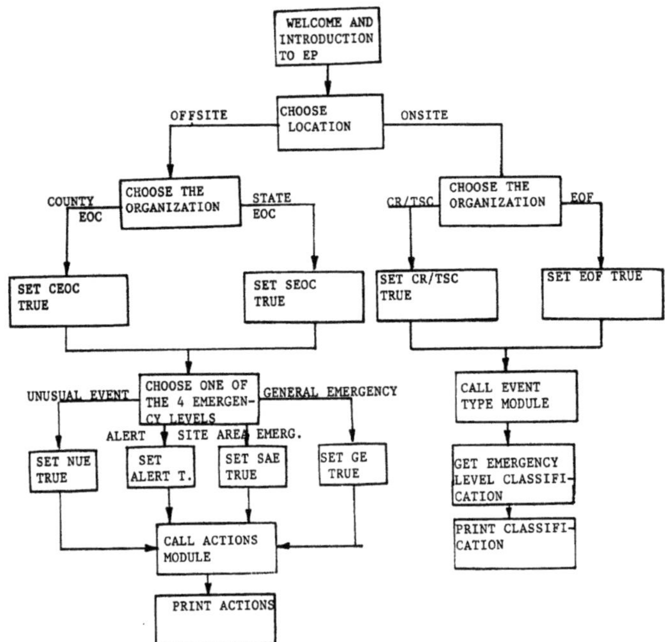

Fig. 3   Emergency Planner (EP) Overall Logic

All of the above event types exist for the four emergency classes and
the user is first presented with questions leading to determine a general
emergency. If that fails, then a site area emergency is tried, then an
alert and finally a declaration of an unusual event. Questioning in this
order streamlines the process, and also the experts agree that it is most
appropriate to start from the worst case scenario. Since the facts are
shared by means of a global fact base, once a question is asked and a fact
is set, then it is not asked again. By looking carefully at the EALs one
can see that some events are present at, say, both the site area emergency
and the general emergency. The general emergency, however, has additional
conditions as shown in Fig. 4. Here is one case where the hierarchical
top down approach is used by means of specialization of nesting modules.

For the onsite location, then, the end result is a classification of
the emergency. Future work includes adding some of the expected responses
and presenting the user with a checklist.

RESULTS

The EP system is expected to be a useful tool before and during
emergency situations. Among other applications, it can be used for
training and for playing "what-if" games. The above can be accomplished
by taking advantage of its modular design as well as its ability to run
forward and/or backward chaining. It is designed using a hierarchical top

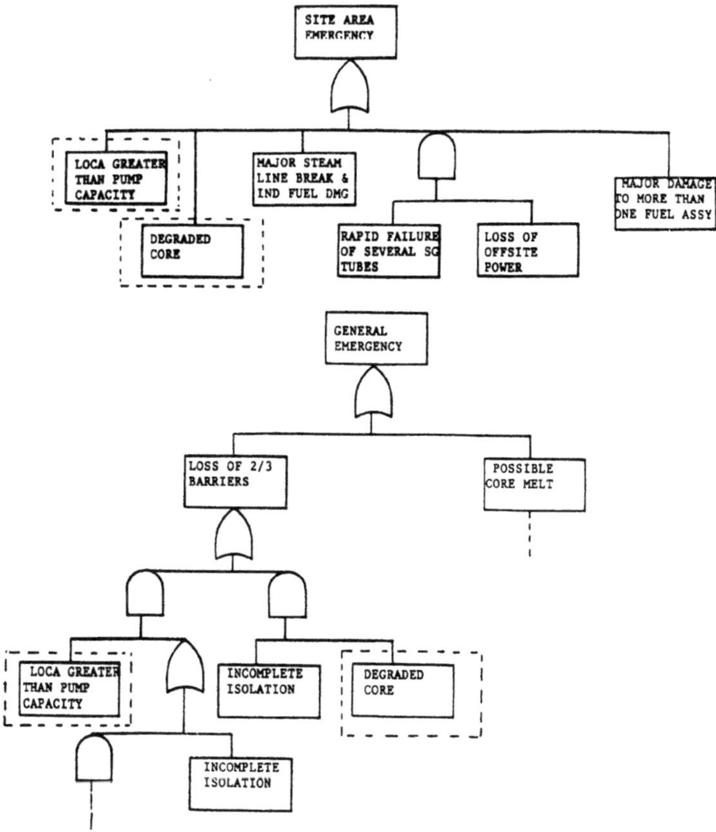

Fig. 4 Example of Module Sharing

down approach; i.e., going from general to specific questions until an answer is obtained or can be inferred. It has two running modes: interactive, asking questions at every step, and semi-automated, interrogating a fact/data base and asking the user only a few questions.

This prototype system is being validated by implementing various plant scenarios from past drills, and comparing the answers with those obtained in the field.

Future work includes expanding the prototype into a complete system, to be used in the control room or any of the offsite agencies, as a checklist or confirmation device, especially when stress is high and time is limited.

ACKNOWLEDGEMENTS

The authors wish to thank the following agencies: The New York State Radiological Emergency Preparedness Group, especially Mr. J. Baranski; The New York Power Authority, in particular Mr. J. Kelly and Mr. B. Methe; and' The General Electric Company, especially Mr. P. Dietz.

REFERENCES

1. Workshop on Emergency Response Facilities Prepared by Nuclear Software Services, Inc., Report # NSAC/57, Feb. 1983.

2. GEN-X Version 1.0 User's Manual, General Electric.

3. New York State Radiological Emergency Preparedness Plan Prepared for the Disaster Preparedness Commission of the State of New York by the Radiological Emergency Preparedness Group, July 1984.

4. Critical Function Monitoring System Algorithm Development, D.L. Harmon, IEEE Transactions on Nuclear Science, Vol. NS-31, No. 1, Feb. 1984.

5. DASS: A Decision Aid Integrating the Safety Parameter Display System and Emergency Functional Recovery Procedures, EPRI NP-3595, Aug. 1984.

6. Reactor Safety Assessment System, D.E. Sebo, M.A. Bray, M.A. King, Second Expert Systems Ingovernment Conference, McLain, VA, EGG-M-11886, Preprint.

7. Design of an Artificial Intelligence System for Safety Function Maintenance, D.D. Sharma, D.W. Miller, B. Chandrasebaran, ANS Winter Meeting, San Francisco, CA, Nov. 10-14, 1985, Vol. 50, p. 294.

8. Computerized Operator Decisions Aids, A.B. Long, Nuclear Safety, Vol. 25, No. 4, July-Aug., 1984.

9. Summary and Evaluation of Scoping and Feasibility Studies for Disturbance Analysis and Surveillance Systems (DASS), EPRI NP-1684, Dec. 1980.

10. Salame-Alfie, A., "An Application of Expert Systems Methodologies to the Analysis of the Safety Systems in a Nuclear Power Plant", Rensselaer Polytechnic Institute, Ph.D. Thesis, (December 1987).

THE KNOWLEDGE-BASED OFF-SITE EMERGENCY RESPONSE SYSTEM

FOR A NUCLEAR POWER PLANT

Li-Wei Ho, Wei-Whua Loa and Chiu-Lung Wang

Institute of Nuclear Energy Research
P.O. Box 3-18 Lungtan 32500 Taiwan ROC

ABSTRACT

A knowledge-based expert system for a nuclear power plant off-site emergency
response system is described. The system incorporates the knowledge about
the nuclear power plant behaviors, site environment and site geographic
factors, etc. The system is developed using Chinshan nuclear power station
of Taipower Company, Taiwan, ROC as a representative model. The objectives
of developing this system are to provide an automated intelligent system
with functions of accident simulation, prediction and with learning
capabilities to supplement the actions of the emergency planners and
accident managers in order to protect the plant personnel and the
surrounding population, and prevent or mitigate property damages resulting
from the plant accident. The system is capable of providing local and
national authorities with rapid retrieval data about the site
characteristics and accident progression. The system can also provide the
framework for allocation of available resources and can handle the
uncertainties in data and models. The prototype of the system is being built
at INER using the off-the-shelf microcomputers. The Chinese user interface
is also being developed. The system is also being implemented on an AI
machine for a larger knowledge base and for the speed requirement.

I. INTRODUCTION

Artificial Intelligence (AI) has come into nuclear business area for
many years, application of this technology to the nuclear power plant off-
site emergency response system was studied and a prototype expert system has
been developed. The major components of the system include a knowledge
acquisition interface, an inference engine, a knowledge base, and an user
friendly interface.

The expert system shell, ES1[1], is the expert system building tool. The
language used for the tool is Common Lisp[2]. The knowledge base includes the
Chinshan nuclear power station[3] plant behaviors, site environment and
geographic factors, etc. Since the expert system uses the build-in knowledge
to solve the specific problems, so that the knowledge acquisition is a very
important task in building the system[4]. The process of extracting knowledge
from an expert and transferring it to a program is termed knowledge
acquisition. It involves problem definition, implementation, and refinement,
as well as representing facts and relations acquired from an expert. In this

paper, a very efficient method called Operational Evaluating Modeling (OpEM)[5] was used to perform this task.

A very important aspect of the user interface of the system is using native language, Chinese, as the user language. A high resolution bit-mapped raster display for color graphics and a monochrome display for texts are used. The software used for the user interface includes a user-friendly, interactive menu driven software, which manages the display of various activities on the screen in Chinese and graphics, and directs input from the key board to the appropriate program.

II. SCENARIO DESCRIPTION

Since The 1979 accident at the Three Mile Island (TMI) nuclear power plant in Pennsylvania, emergency planning has become a major research task to the nuclear utility industry. Prior to the TMI accident, all commercial nuclear power reactors in ROC had an emergency response plan[6] which was based on the National Nuclear Emergency Response Plan (NNERP)[7]. It had been redrafted in 1980 by Atomic Energy Council, ROC (ROC-AEC) and approved by the Executive Yuan in 1981. The representative model used for this study is Chinshan nuclear power station of Taipower Company (TPC).

Chinshan nuclear power station is the first nuclear power station in Taiwan. The station utilizes two identical boiling water reactors, each with its associated Nuclear Steam Supply System furnished by General Electric Company. The rated power for each reactor is 1775 MWt, with a gross electric output of 635 MWe. The units were designed with a minimum of shared facilities, utilizing a combination structure reactor building and radwaste facility. Each unit utilizes a Westinghouse turbine-generator. The units were stared up in December 1978 and July 1979, respectively. The Chinshan site is located at Chien Fua Tsun, about 28 km northeast of Taipei, the capital of ROC.

III. THE DESCRIPTION OF THE EXPERT SYSTEM

This section describes the structure and the inference mechanism of the system, some examples are used to give a brief idea about the system. The system includes several software components: a knowledge acquisition mechanism, an inference engine, a knowledge base and an user interface.

III.1 The Knowledge Acquisition Mechanism

The knowledge acquisition process is one of the most difficult phases of expert system building, acquiring knowledge from an expert is a gradual process that can stretch over months or years. The authors tried to introduce the method of the Operational Evaluation Modeling (OpEM) which was developed by Dr. John R. Clymer in 1970 as the knowledge acquisition mechanism for this system. The method was primarily used as a management tool which can assist a manager, whose job is to improve the mission effectiveness of an organization or system, to make decisions about where to focus his scarce resource in order to better perform his job. OpEM is based on the concept of a parallel process, which is defined as the collection of all possible sequence of states and events representing a system or organization. The states represent periods of time where either: (1) mission functions are being performed or (2) the process is waiting for a specified logical condition to be satisfied before it can continue. The events mark points in time where a change of state occurs. A directed graph model is used to provide an easy to understand description of model operation for use by non-programmers or non-knowledge engineers.

Figure 1 illustrates an OpEM directed graph model with the parallel process for a part of the emergency notification procedures. The circles represent periods of time when a task is being performed on a step at a subsystem or a step is waiting in a queue for a resource to become available. The events (E1, E2, E3, ...) are instants in time when a change in state occurs. There are several possible sequences of system states and events representing the procedure of the notification of the emergency, depending on whether the wait for a message or the wait for an order. The parallel process is the set of all sequences of system states and events which represents system operation. The system state is comprised of the discrete state of each process in parallel plus the value of each state variable defined. Events are changes of system states, and they are represented by directed line segments in the directed graph model. This graphical language provides a means to represent sequential processes and to describe the types of interactions which can occur as these parallel processes are executed simultaneously. The language is intended to allow the process level description of system operation to be specified. A knowledge engineer can understand the system operation, learn how to communicate performance improvement ideas to non-modelers and learn how a simulation program which emulates directed graph model description of system operation, operates, etc.

III.2 Inference Engine

The inference engine used for the system is an expert system developing tool, ES1, using the programming language of Common Lisp, a new dialect of Lisp. The tool has been developed at INER for over a year. It has the capabilities of forward chaining and backward chaining which are used to decide which portions of the knowledge base are applied to a given problem and what additional data must be supplied by the user. The inference engine can also provide explanation of the system's behavior. Through this engine, the knowledge engineer or the system user can interact with the domain knowledge base which stored in the mass storage device. Forward chaining is often described in terms of bottom-up or event-driven reasoning. In forward chaining the system does not start with any particular goals defined. That is no initial subgroup of production rules are required to establish a starting point. Instead the system begins with a set of evidence and proceeds to invoke production rules in forward direction until no further rules can be invoked. Backward chaining is described as top-down or goal-driven reasoning. In backward chaining the system starts with a set of initial goals. then, a limited set of production rules are invoked in reverse order.

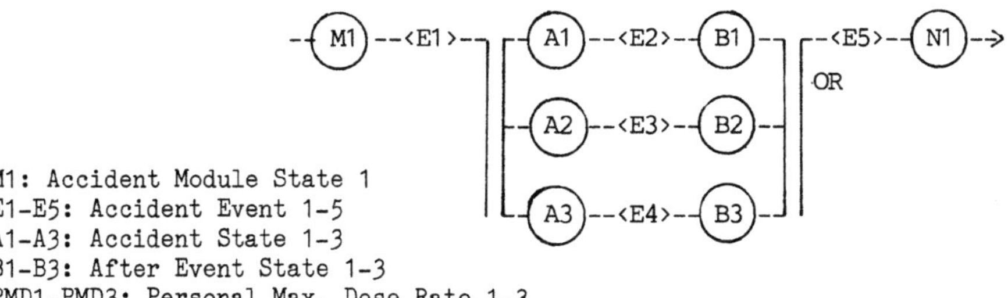

M1: Accident Module State 1
E1-E5: Accident Event 1-5
A1-A3: Accident State 1-3
B1-B3: After Event State 1-3
PMD1-PMD3: Personal Max. Dose Rate 1-3
N1: End State 1

Figure 1. The OpEM Directed Graph Model for the Emergency Notification Procedures

The knowledge base of the system includes the production rules, the equation rules and the data base. The knowledge which stored in equation rule base and data base are called shallow knowledge, inference engine uses these knowledge bases to infer the answer first. If the information of the shallow knowledge is not enough to get the goal, it is necessary to augment shallow knowledge with deep knowledge. The deep knowledge could help to solve the deeper problems and some conflict problems. It is designed as a set of production rules. The production rule is an " IF .....THEN " structured statement with a condition followed by an action or decision, the inference is made or the action is taken if the condition holds true. The knowledge, equations and data integrated in the knowledge base include the transportation model and data, population data, topograph data, hydrology data, atmospheric transport models, accident sequence models, dose/response models, agricultural production and water use data bases, location of emergency response centers and routes for obtaining emergency response aid and source term and radiation release models, etc.

Figure 2 is the knowledge base structure of the system, nine modules of subsystems were integrated in the prototype. The subsystems can be activated by the main menu according to user's requests. Each of the subsystem also has deeper systems, or deeper knowledge, these parts can be activated by requests or by the inference procedures. Figure 3 shows some equation rules, figure 4 is some examples of production rules and figure 5 is an example of look up table of data base. About 250 production rules, 50 equation rules and 30 sets of data base had been built in the prototype of the knowledge base. It is expected that thousands of rules and hundreds sets of data base is needed for the delivery system in the future.

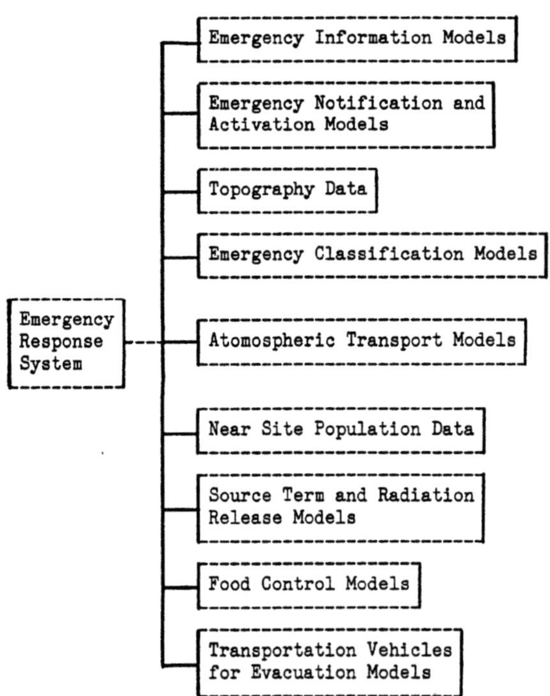

Figure 2. Knowledge Base Structure of the Prototype System

```
                            Equation rules

(PUT I TRANSLATION '(I - 131 RADIOACTIVITY CONCENTRATION))
(PUT I UNIT '(PCI / L))
(PUT RI TRANSLATION '(RADIUS OF EPZ - INGESTION))
(PUT RI UNIT '(KM))
(PUT MWBER TRANSLATION '(MAXIMUM WHOLE BODY EXPOSURE RATE))
(PUT MWBER UNIT '(MREM))
```

Figure 3. Example of Equation Rules


```
                          Production Rules

; (PR11 (($AND (= AZC CC4)) ((IS AZ AZ4))))
"(If the AFFECTED ZONE CODE (AZC) is equal to 4.0 then the  AFFECTED
  ZONE (AZ) is SS TO EE (AZ4) .)"
; (PR12 (($AND (= AZC CC5)) ((IS AZ AZ5))))
"(If  the AFFECTED ZONE CODE (AZC) is equal to 5.0 then the AFFECTED
  ZONE (AZ) is SE TO EN (AZ5) .)"
; (PR13 (($AND (= AZC CC6)) ((IS AZ AZ6))))
"(If the AFFECTED ZONE CODE (AZC) is equal to 6.0 then the  AFFECTED
  ZONE (AZ) is EE TO NW (AZ6) .)"
; (PR15 (($AND (= AZC CC8)) ((IS AZ AZ8))))
"(If  the AFFECTED ZONE CODE (AZC) is equal to 8.0 then the AFFECTED
  ZONE (AZ) is NN TO WW (AZ8) .)"
```

Figure 4. Example of Production Rules


```
                            Data Base

Primary key :
    (SETQQ *P-K-CONTEXT* (WIND DIRECTION / CODE))
    (SETQQ *P-K-VARIABLE* WDC)
Libruary of attribute structure :
    (SETQQ *LIB-F-NAME* ((DWZC N (DOWNWIND ZONE CODE))))
    (PUT WDC LIB-F-NAME '((DWZC N (DOWNWIND ZONE CODE))))
Libruary of attribute number :
    (SETQQ *F-NAME-NUM* 1)
Libruary of context variable :
    (SETQQ *LIB-V-CONTEXT* (SE SS WS WW NW NN EN EE))
Database :
    (PUT WDC SE '(SOUTH - EAST))
    (PUT SE DWZC "8.0")
    (PUT WDC SS '(SOUTH))
    (PUT SS DWZC "7.0")
    (PUT WDC WS '(WEST - SOUTH))
    (PUT WS DWZC "6.0")
    (DBO (RDS))
```

Figure 5. Example of a Set of Data Base

In addition to use English as a nature language, a very important aspect of the user interface is using native language, Chinese, as one more option for the system[8]. The Chinese sentences and clauses are bound as patterns for the different outcomes by inferences. A high resolution bit-mapped raster display for color graphics and a monochrome display for texts are used. Figure 6 is an example of Chinese, English and graph merged display, users can catch the main idea from the display easily. The graphic displays are linked with the knowledge base, the goals of the consultations can activate the appropriate graphic displays. The software for the user interface includes a user-friendly, interactive menu driven software, which manages the display of various activities on the screen in Chinese and English, and directs input from the key board to the appropriate program. Many on-line help facilities guide the user in using the system and provide quick reference to specific information.

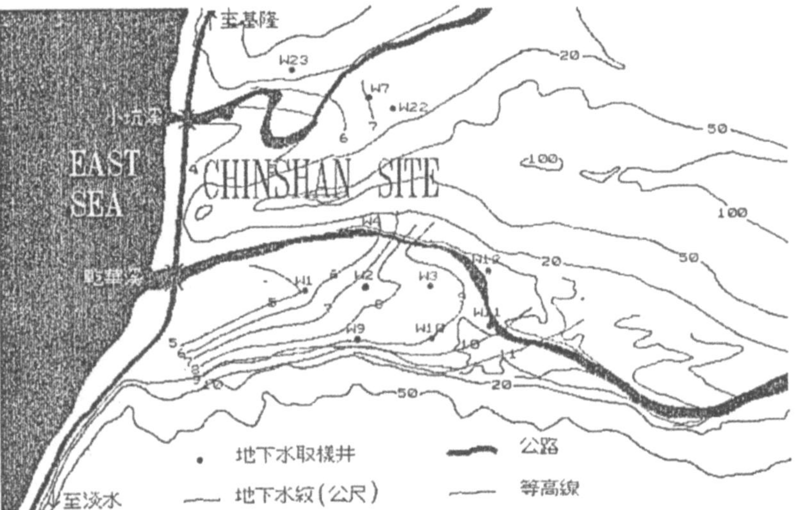

Figure 6. An example of Chinese, English and Graph Merged Display

III.5 <u>The Hardware of the System</u>

The prototype of the system is being built on IBM PC XT/AT compatible microcomputers. The system is also being implemented on an AI machine[9] for a larger knowledge base and for the speed requirement. The AI machine (also called Lisp Machine) has a better developing environment than the environment of PCs. For example, AI machine has the utilities of the Lisp Listener, the ZMACS editor, the Window system and the Lisp Debugger, etc. The on-line graphics system is implemented by a direct Multibus-to-Multibus channel between the AI machine and the graphics system[10], it is a faster connection than the standard Ethernet.

IV. CONCLUSIONS

The results of the work indicate that the expert system technique and the OpEM method are applicable tools to the building of nuclear power plant off-site emergency response system. Although the system performance is limited by the number of rules due to the capacity of microcomputers, but an AI machine can demonstrate the real power of the system. If using AI machine as a developing tool, to speed up the R&D period, then down load the developed software to a microcomputer with the extended memory boards and plug-in processors, e.g., 386-based board, that could be an alternative to the delivery system of AI machine.

In addition to the off-site emergency response system, the on-site emergency response system is also an important issue for the nuclear utilities. The techniques used for the system and the developing environment of the system can be applied to the on-site emergency response system. The same techniques can also help the non-nuclear industries to make more effective decisions to deal with the complexities of the operational problems.

V. ACKNOWLEDGEMENTS

The authors wish to thank the staffs of Artificial Intelligence Lab., INER, ROC for providing of the AI facilities and the financial support. The helpful informations from ROC-AEC and Taipower Company are also gratefully acknowledged.

VI. REFERENCES

1. M. C. Han and T. C. Lyie, " ES1, The Expert System Building Tool ", Information Science Division, Institute of Nuclear Energy Research, the Republic of China (March 1987).

2. G. L. Stelle Jr., " Common Lisp ", Carnegie-Mellon University, Tartan Laboratories Incorporated (1984).

3. " Final Safety Analysis Report, Chinshan Nuclear Power Station Unit 1 & 2 ", Taiwan Power Company (March 1986).

4. F. Hayaes-Roth et al., " Knowledge Acquisition, Knowledge Programming, and Knowledge Refinement ", Rand Paper, P-6241, Rand Corp., Santa Monica, Calf. (1978).

5. J. R. Clymer, " Operational Evaluation Modeling ", California State University at fullerton (May 1982).

6. " Chinshan Nuclear Power Station Emergency Plan ", Taipower Nuclear Emergency Planning Executive Committee (February 1986).

7. " National Nuclear Emergency Response Plan ", Atomic Energy Council, the Republic of China (1981).

8. C. L. Lyie, " Chinese Language on Lisp Machine ", Engineering Test and Measurement Quarterly, Vol.24, pp. 46-56 (January 1987).

9. M. Cohen and R. Ingria, " Introduction to the Lambda ", Lisp Machine Incorporated (1987).

10. P. Cann, " Vista Library User Manual ", Lisp Machine Incorporated (1987).

REACTOR SAFETY ASSESSMENT SYSTEM[a]

D.E. Sebo, M.A. Bray, and M.A. King

Idaho National Engineering Laboratory
Idaho Falls, ID 83415 USA

ABSTRACT

The Reactor Safety Assessment System (RSAS) is an expert system
under development for the United States Nuclear Regulatory Commission
(USNRC). RSAS is designed for use at the USNRC Operations Center in the
event of a serious incident at a licensed nuclear power plant. RSAS is a
situation assessment expert system which uses plant parametric data to
generate conclusions for use by the NRC Reactor Safety Team. RSAS uses
multiple rule bases and plant specific setpoint files to be applicable to
all licensed nuclear power plants in the United States. RSAS currently
covers several generic reactor categories and multiple plants within
each category.

INTRODUCTION

The Reactor Safety Assessment System (RSAS) is an expert system[1]
under development for the United States Nuclear Regulatory Commission
(USNRC).[2] RSAS will be used at the USNRC Operations Center during
serious incidents at licensed nuclear power plants and during emergency
response drills. RSAS is a heuristic classification expert system which
converts plant parametric data into plant status information useful to
USNRC emergency response personnel. The key technical challenge of RSAS
was to produce useful conclusions and cover all licensed nuclear power
plants without writing separate expert systems for each nuclear power
plant.

The purpose of the Operations Center and the function of the Reactor Safety Team, the intended users of RSAS, are covered in this paper.
A system description includes the rule bases, plant files, and user
interface. Rule bases have been developed for all pressurized water
reactors (PWRs) and an initial boiling water reactor (BWR) rule base has
been developed. The system currently includes plant setpoint files for
approximately 35 nuclear power plants. RSAS is currently installed on a

---

a. Work supported by the U.S. Nuclear Regulatory Commission, Office of
Analysis and Evaluation of Operational Data under DOE Contract
No. DE-AC07-76ID01570.

Xerox 1186 Artificial Intelligence Workstation at the Operations Center
for user testing. Continuing work and future plans for system
development complete the paper.

NRC OPERATIONS CENTER - REACTOR SAFETY TEAM

The NRC Operations Center, located in Bethesda, Maryland serves as
the operational communications center for the NRC. The center houses the
communications equipment, computer, and reference material needed by NRC
emergency response personnel. The Operations Center is the focus for NRC
response to an incident until an NRC team arrives at the licensee site.
A key NRC function is to provide an independent technical assessment of
existing and projected plant status, offsite actions, and licensee
actions. The NRC maintains a 24 hour watch at the Operations Center to
accept communications from the licensees. If a licensee reported incident is sufficiently serious that a potential threat to the public
exists, the NRC Duty Officer will take the actions necessary to activate
the Operations Center.

The Operations Center consists of three primary teams; the Executive Team, the Reactor Safety Team, and the Protective Measures Team.
The Reactor Safety Team (RST) is responsible for assessing the current
status and projecting the future status of the core and containment at
the affected site. Of particular interest are those aspects of plant
status which could lead to offsite radioactive release. Team members
are chosen to provide an appropriate mix of disciplines including reactor and containment systems, risk assessment, and familiarity with the
affected plant. The team analyzes parametric data from the affected
plant, formulates questions to be sent to the licensee via a dedicated
phone line, and generates assessments of current and possible future
plant status

The Reactor Safety Team is headed by a Director and Deputy Director. The Director periodically briefs the Executive Team on RSAS conclusions. The Deputy Director controls the functioning of the RST. The
Deputy Director directs analysis tasks to team members, filters questions
for transmission to the affected plant, and focuses team conclusions.

The phone talker is located in a different room than the RST and
speaks with a designated licensee phone talker. The licensee phone
talker is in the plant control room or Technical Support Center. All
plant status data and information currently is received from the licensee phone talker. Parametric data consists of approximately 20-25
parameters updated every 15 minutes. The current communications
arrangement is slow and can introduce errors. The transcription of the
data into the Operations Center Information Management System is another
possible source of errors. The current system will be upgraded to automated parameter acquisition using the Emergency Response Data System
(ERDS). ERDS data varys from plant to plant but generally correspond to
Safety Parameter Display System data.

RSAS DESCRIPTION

RSAS is a situation assessment expert system for use during a
serious incident at a commercial nuclear power plant. The basic constraint in RSAS development was the requirement that RSAS be applicable
to any licensed nuclear power plant. As a result, RSAS is a heuristic
classification, rule based expert system.[3] The system models the
cognitive processes of a reactor expert rather than the physical

processes at the affected plant. The applicability of RSAS to all plants
precluded, in general, the use of plant structural information such as
response trees,[4] fault trees, or causal models.[5] RSAS is not
designed as a diagnostic system to identify specific transient initiators
or faulted plant equipment. The goal is to aid the RST in monitoring
the course of the transient.

The initial stage of RSAS development was the categorization of
licensed nuclear power plants into generic categories. This decomposition was based on vendor type, containment type, etc. This was
necessitated to avoid the production of individual rule bases for each
individual plant within the United States. This categorization was
feasible due to both the limited data available to the expert system and
the goal of the expert system, situation assessment, not diagnosis. The
simplified basic system architecture is shown in Fig. 1.

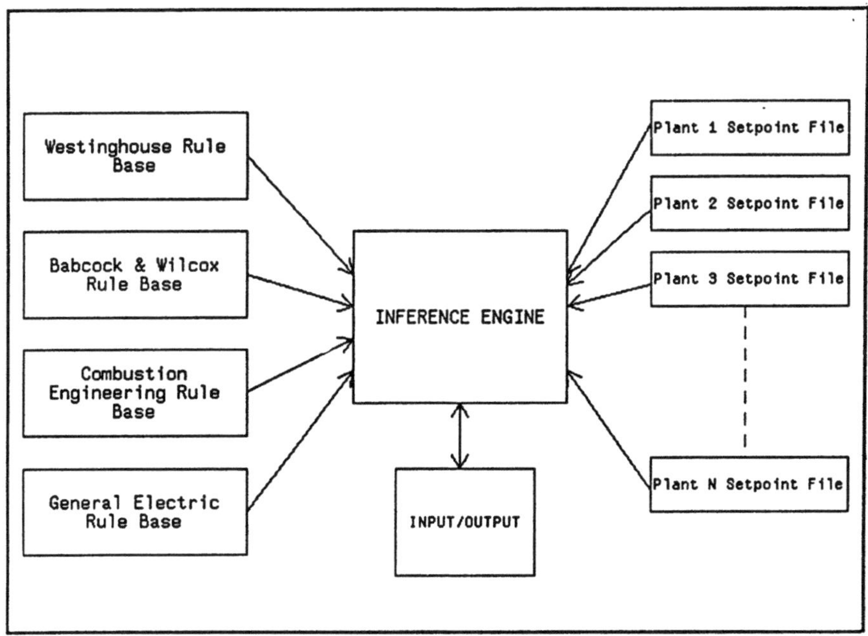

Fig. 1. RSAS System Architecture.

The initial system display screen is a menu of all commercial
nuclear power plants in the United States. The user chooses a plant and
the appropriate rule base for that plant is loaded into working memory.
A plant specific setpoint file is also loaded. The plant specific
setpoint file contains all plant specific data used by RSAS. This file
contains data such as instrument ranges, setpoint information such as
PORV setpoints, etc. Plant specifics were segregated from plant
type generic rules to minimize the changes to extend the system to
include additional plants. In the case that plant specific rules are
required, these rules will also be stored in the plant specific files.

The inferencing procedure is a forward chaining, data driven procedure. That is, all appropriate conclusions are produced for each data
set. Confidence factors are not generated with each conclusion. A high

confidence factor is implicit in the presence of a rule in the rule
base. The goal of RSAS system output is to produce only factual output
or conclusions with significant supporting evidence.

RSAS RULE STRUCTURE

The basic knowledge representation used by RSAS is the data structure shown in Fig. 2. Each data structure or rule is a "chunk" of knowledge and contains all of the knowledge required to determine how it
should be applied in each situation. The design of the rule structure
was intended to model the manner in which an expert interprets the parametric data. This resulted in each rule being more complex and to a great
extent independent of each other. This permits the use of a simple
inferencing procedure and avoids the use of long inference chains. The
shallow inferencing chain has two major benefits. It makes it possible
for an expert to understand and review the rule base without extensive
understanding of system details. It also simplifies the task of modification of existing rules and addition of additional rules. Alterations
do not "cascade" through a long inference chain.

IF	Activating conditions
THEN	Action to be performed if IF is true
ELSE	Alternative actions if IF is not true
UNLESS	Specific condition when rule does not hold
SUPERSEDE	Other rules that supersede the existing rule
OFF	Conditions to deactivate rule
SUPPORT	Supporting evidence for the existing rule
ID	Rule identifier
CLASS	Output classification
OUTPUT	Output message generation

Fig. 2. RSAS Rule Structure.

Consideration in rule development included the high probability of
missing or incomplete data and the masking of symptoms due to multiple
failures or operator actions. This required multiple rules based on
various combinations of parameters, where possible, to avoid not correctly identifying a particular plant condition. Also positive parametric evidence, the OFF condition, was required to withdraw a conclusion.
Thus, missing data or the effect of operator actions would not incor-
rectly imply recovery from upset conditions.

The rules within each rule base fall into three basic categories.
The first category are diagnostic rules, that is, rules which generate

significant inferences concerning the major concerns of the Reactor
Safety Team such as indications of core damage, loss of coolant accident
(LOCA), or lack of heat transfer. Diagnostic rules require significant
data. In most cases these rules require data from multiple data sets.
This approach was taken based on feedback from RST members. The concern
was expressed that the generation of a "shopping list," i.e., the listing of a large number of possible hypothesis could mislead the RST
rather than ensure that no possible hypotheses are overlooked.

The other rule categories are factual in nature and serve as data
filters to highlight significant parameter values and relationships.
There are data filter rules to interpret parameter values relative to
plant setpoints, e.g., pressure exceeds a relief valve setpoint. These
are straightforward, but significant, conclusions that are easy for the
RST to overlook due to unfamiliarity with the particular plant. This
will eliminate the need for RST members to perform bookkeeping tasks
such as looking up plant setpoints. The third rule category highlights
significant relationships between plant parameters which may aid in the
identification of a transient and its possible consequences. An example
of this type of rule is that cold leg temperature is no longer coupled
to steam generator pressure.

RSAS SYSTEM OPERATION

RSAS is currently installed on an artificial intelligence workstation, the Xerox 1186. This system was chosen because of the user
interface features and software developmental environment available with
the 1186. RSAS has a mouse and menu driven user interface. This is
particularly appropriate for a system such as RSAS which will be used
infrequently. Also the windowing and graphics capability allow flexibility in system design. A typical RSAS screen is shown in Fig. 3.

All user operations, with the exception of entering and correcting
data which requires keyboard input, are initiated from menus using mouse
operation. Left mouse buttoning on the menu choices will activate the
associated menu function. Right mouse buttoning will produce a help message indicating the function of that menu item. Interface considerations
are as important to overall system success as the technical competence
of the system.

The following is a summary of the major interface features associated with RSAS operation.

- On line parameter monitoring: The user can enter bounds on
  the magnitude and rate of change of parameter values. A
  message is generated if these bounds are exceeded.

- What if: The user can alter parameter values and rerun RSAS
  to determine changes in system operation. RSAS is then reset
  to the original state. This can be used as a sensitivity tool
  as well.

- Correct past data: Incorrect parameter values are possible,
  especially if parameter values are obtained via a phone
  talker. This option enables parameter values to be corrected
  for any previous data set.

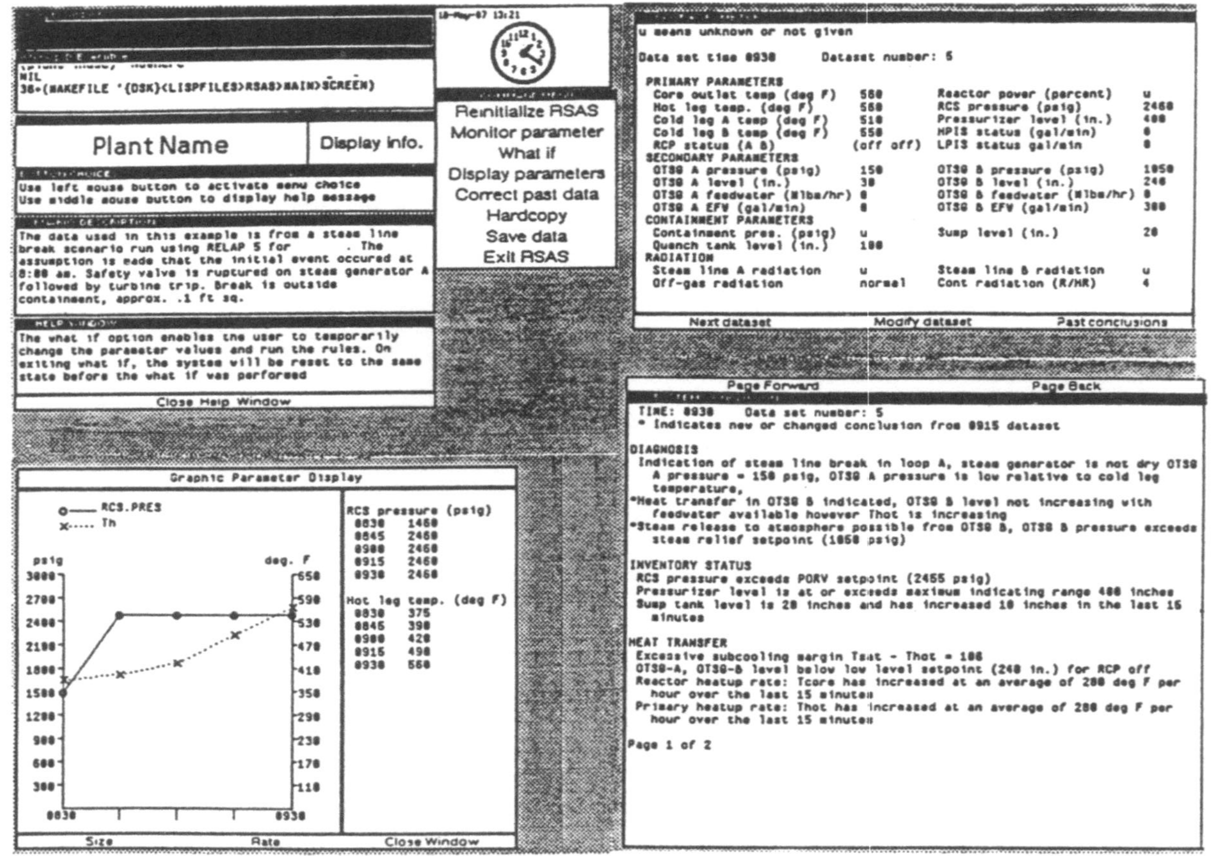

Fig. 3. RSAS Screen Image.

- Hardcopy: A hardcopy is available for all screen images.

- Parameter display: Parameter values can be displayed either
  in tabular form or graphically, rates of change or magnitude.

- Save data: The user can save all data input to the system for
  review and evaluation at a later date.

- Display info: A short description of the plant, trip
  setpoints will be displayed on the screen.

Eventually RSAS will obtain all data directly from the Operations
Center Information Management System. However, as the future Operations Center computer environment has not been determined, RSAS currently requires manual data input in a tabular format using mouse
selection. There is an input checking routine to detect gross errors in
input parameter values. Extensive error checking and parameter
validation is difficult due to the limited amount of data available.

VALIDATION AND TESTING

Validation of the technical competence of RSAS is a major concern
in system development. Areas of concern include: completeness of the
rule base, consistency of the rule base, and accuracy of the rule base.
The need to explicitly structure the knowledge became clear early in
rule development. A knowledge structure tree was developed using the
top level concerns of the RST; core damage, loss of coolant, loss of
heat transfer, etc. An example branch of the knowledge structure tree
is shown in Fig. 4.

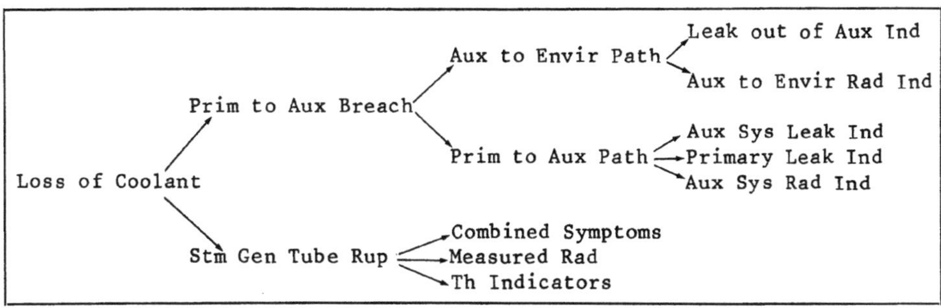

Fig. 4. RSAS Knowledge Structure Branch.

This structure is functional and is not plant or plant type specific. It
facilitates knowledge acquisition by providing a framework for interaction
with domain experts and helps ensure completeness and consistency in the
rule bases for each generic plant category.

A second type of error is inconsistent conclusions. This was
addressed originally by using only rules with an implicit high confidence
level and/or requiring significant evidence. However, testing was
required to demonstrate consistency. A random parameter generator was
developed and RSAS exercised to detect inconsistent conclusions and
unanticipated rule interactions.

Validation of system technical competence is an ongoing process.
This validation is based on performance evaluation using computer generated plant scenarios. A major difficulty is that accurate reactor simulations are expensive, do not contain all plant parameters used by RSAS,
and are available for only a small number of nuclear power plants. Consequently, validation is also being undertaken by rule reviews using
experts not involved in system development. System validation also
requires extensive testing by reactor experts to manually input parameter
input and verify system output.

FUTURE WORK

RSAS development is continuing with the development of rule bases
for the remaining generic reactor categories. Also plant specific setpoint files will be developed for each plant. User testing and modification will continue based on user needs. The availability of more and
better parameter data using automated data acquisition will permit the
extension of the existing rule bases. The data filtering capabilities of
RSAS will become more useful as the amount of data increases.

The availability of the RSAS workstation and associated plant setpoint files can lead to further enhancements to support the Operations
Center staff. An intelligent workstation can be developed which will aid
experts in nuclear plant behavior without usurping their decision making
functions. Currently, prediction is a very heuristic task because the
input required for deterministic calculations is not available. This is
an appropriate area for expert system techniques and RSAS extension.

REFERENCES

1. F. Hayes-Roth, "The Knowledge-Based Expert System: A Tutorial
    Computer," September, 1984.

2. M. A. Bray, D. E. Sebo, and B. W. Dixon, "Reactor Safety Assess-
    ment System - A Situation Assessment System for USNRC Emergency
    Response," IEEE Proceedings of the Expert Systems in Government
    Symposium McLean, Virginia, October 24-25, 1985.

3. W. J. Clancy, "Heuristic Classification," STAN-CS-1066 Department
    of Computer Science, Stanford University, Stanford, CA.

4. W. R. Nelson, "Response Trees for Nuclear Reactor Operations,"
    National Technical Information Service, Springfield, VA,
    NUREG/CR-3631, 1984.

5. W. E. Underwood and R. E. Swanson, "Causal Reasoning and Explana-
    tion of Nuclear Power Plant Operation," ANS Topical Meeting on
    Computer Applications for Nuclear Power Plant Operation and
    Control, September 8-12, Pasco, WA, 1985.

6. C. H. Meijer, "A Critical Function Expert System for Nuclear Power
    Plants," The Enlarged Halden Programme Group Meeting on Fuel

Performance Experiments and Analysis and Computerized Man-Machine
Communication, Loen, Norway, May 23-28, 1985.

NOTICE

A PRODUCTION-RULE ANALYSIS SYSTEM FOR NUCLEAR PLANT MONITORING AND EMERGENCY RESPONSE APPLICATIONS

Magdi Ragheb, Lefteri Tsoukalas,
Timothy McDonough and Michael Parker

University of Illinois at Urbana-Champaign
103 S. Goodwin Ave., Urbana, Illinois 61801

Illinois Department of Nuclear Safety
Division of Engineering, 1035 Outer Park Drive
Springfield, Illinois 60801

ABSTRACT

A Production-Rule Analysis System for Nuclear Power Plant Monitoring
is presented. The signals generated by the Zion-1 Plant are considered
for Emergency Response applications. The integrity of the Plant
Radiation, the Reactor Coolant, the Fuel Clad, and the Containment
Systems, is monitored. Representation of the system is in the form of a
goal-tree generating a Knowledge-Base searched by an Inference Engine
functioning in the forward-chaining mode. The Goal-tree is built from
Fault-Trees based on plant operational information. The system is
implemented on a VAX-8500 and is programmed in OPS-5.

INTRODUCTION

The described system uses plant parameter data continuously updated,
with a time lag, from the power plant sites in the state of Illinois.
It uses the production-rule paradigm to construct Model-Based systems.
The system uses OPS-5[1] as a programming tool, in the data-driven or the
forward-chaining mode. Earlier work along this line used an inference
engine on a microcomputer in the backward-chaining modes.[2] Realizing
that this approach is very promising for the intended application, but
was limited in speed and capability, the present effort was undertaken
to explore the implementation of the system of a minicomputer, and
assess its capability in dealing with a combination of numerical and
symbolic plant data.[3]

OBJECTIVES

The objectives of plant monitoring and emergency preparedness are to
detect abnormal situations in Nuclear Power plants that have a potential

of adversely affecting surrounding communities, such as might be caused
by the inadvertent release of radiation. In this unlikely event, it
would be necessary to assess the accident's potential for causing a
release of radiation, determine the consequences of the radiation
release, and if necessary coordinate the evacuation of any endangered
population.

In this work, the goal is to automate a part of the monitoring process, alleviating humans from the repetitive task of analysing large
amounts of data that are continuously updated on a twenty-four hours
basis. The human effort can thus be directed towards meaningful analysis, faster decision-making in emergency situations, and expanding the
amount of information that could be analysed. One approach to monitoring is the use of setpoint alarms for critical plant parameters, such
as fuel temperatures from sensors and area radiation levels from monitors, as shown in Fig. 1.

Considering the Zion-1 plant, Fig. 2 shows a goal tree describing
the possible fault states of four of its systems:

1. Plant Radiation System
2. Reactor Coolant System
3. Fuel Clad System
4. Containment System

The OR gate inputs to M1, M2, ... , M8 and Mp, describe subsystem
states whose failures affect the above-mentioned systems. For instance,
the Reactor Coolant System Integrity is jeopardized if the reactor core
is not solid(M7), OR the reactor heatup rate exceeds the technical
specifications(M3). These subsystems states can be further described in
more detail based on plant data and the corresponding Fault Trees as
shown in Fig. 3 in terms of signals originating directly from plant sensors' information. Figure 1 displays a subset of 41 sensors descriptions with their associated setpoints and units. Some signals have
numerical values, and some others possess symbolic values such as the
HI/NORMAL reading for the YD2516 sensor. The present analysis considers
those 41 plant sensors from up to 1100 sensor data points per plant continuously updated every two minutes.

PROGRAM STRUCTURE

The goal-tree of Figs. 2 and 3 are translated into a Production-Rule
Analysis system using the Rule-Based paradigm from the field of
Knowledge Engineering. The main body of the generated system consists
of several "literalize" statements and production-rules. The literali-
zations are used to declare classes. Classes are the information structures that OPS-5 uses in making "working memory elements".

Production-rules begin with a "p" followed by the name of the production, a premise for the rule on the left hand side, an implication
symbol, and a consequent of the rule on the right hand side. The premise has a number of patterns called condition elements and corresponds
to the "if" part of an "if ... then" rule. The consequent part has a
number of actions, which are executed when the premise patterns are
satisfied. The consequent corresponds to the "then" part of a rule.
Typical productions pertaining to the plant radiation system and to the
containment system pressure, respectively, are here shown:

SENSOR	DESCRIPTION	UNITS	SETPOINT
P1000	Containment 1 pressure	psig	> 1
P1001	Containment 2 pressure	psig	> 1
P1002	Containment 3 pressure	psig	> 1
P1003	Containment 4 pressure	psig	> 1
R0011	Radiation – containment air particle	mREM/hr	> 10
R0012	Radiation – containment gas	mREM/hr	> 10
R0019	Radiation – steam generator blowdown drain	dkcounts/min	> 10
R0053	Radiation – containment purge exhaust stack effluent radiation rate	counts/min	> 5600
R0054	Radiation – containment purge exhaust stack effluent iodine radiation rate	pCi/cc	>.0001
R0055	Radiation – containment purge exhaust stack effluent air particulate radiation rate	counts/min	> 5600
U0081	Reactor heat–up/cool–down rate	°F/hr	> 100
U0090/R	Rate of change of sensor U0090, incore thermocouple (instantaneous value of hottest)	°F/min	> 30
U0932	Incore thermocouple temperature average of 10 highest sensors	°F	> 1000
U0932/R	Rate of change of sensor U0932, incore thermocouple average of 10 highest sensors	°F/min	> 20
U1170/R	Rate of change of sensor U1170, incore thermocouple time average value	°F/min	> 10
U9009	Radiation – containment hi–range areas (maximum of)	REM/hr	> 10
X0003/R	Rate of change of sensor X0003, reactor coolant system average temperature	°F/hr	> 100
X0005	reactor vessel level	%	< 94
X0005/R	Rate of change of sensor X0005, reactor vessel level	%/min	< −1
X0011	Radiation – containment	mR/hr	> 10
X0013	Radiation – steam generator blowdown	cpm	> 10
Y9021	Radiation – containment hi–range area 1	kREM/hr	> 10
Y9022	Radiation – containment hi–range area 2	kREM/hr	> 10
Y9033	Radiation – main steam line monitor RIM-PR58 calc	mR/hr	> 2.5
Y9034	Radiation – main steam line monitor RIM-PR59 calc	mR/hr	> 2.5
Y9035	Radiation – main steam line monitor RIM-PR60 calc	mR/hr	> 2.5
Y9036	Radiation – main steam line monitor RIM-PR61 calc	mR/hr	> 2.5
Y9072/R	Rate of change of sensor Y9072, reactor vessel water level – wide range	%	< −1
Y9073/R	Rate of change of sensor Y9073, reactor vessel water level – narrow range	%	< −1
Y9074/R	Rate of change of sensor Y9074, reactor vessel water level – wide range	%	< −1
Y9075/R	Rate of change of sensor Y9075, reactor vessel water level – narrow range	%	< −1
Y9118	Radiation – U1 containment noble gas hi range	uCi/cc	> 10
Y9119	Radiation – U1 containment noble gas medium range	uCi/cc	> 10
Y9120	Radiation – U1 containment noble gas low range	uCi/cc	> 10
Y9121	Radiation – U1 containment beta particulate	uCi/cc	> 10
Y9122	Radiation – U1 containment iodine	uCi/cc	> 10
YD2515	Radiation – containment purge exhaust stack effluent gas radiation rate	HI/NORMAL	HI
YD2516	Radiation – containment purge exhaust stack effluent iodine radiation rate	HI/NORMAL	HI
YD2517	Radiation – containment purge exhaust stack effluent air radiation rate	HI/NORMAL	HI
YD9030	Containment hi pressure cause safety injection & reactor trip	TRIP/NOTTRIP	TRIP
YD9031	Containment hi hi pressure cause phase B isolation (RCP cooling water isolation)	TRIP/NOTTRIP	TRIP

Fig. 1: Plant Sensors Descriptions and associated setpoints

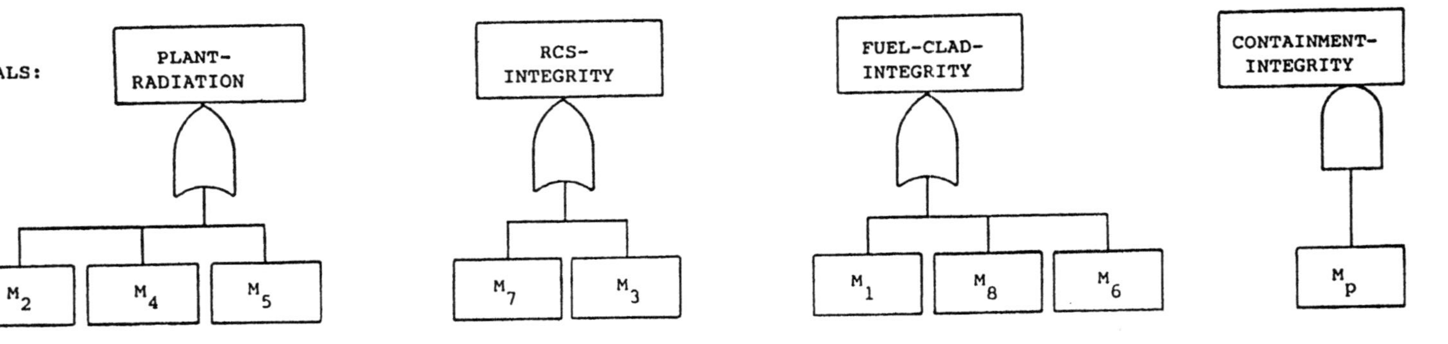
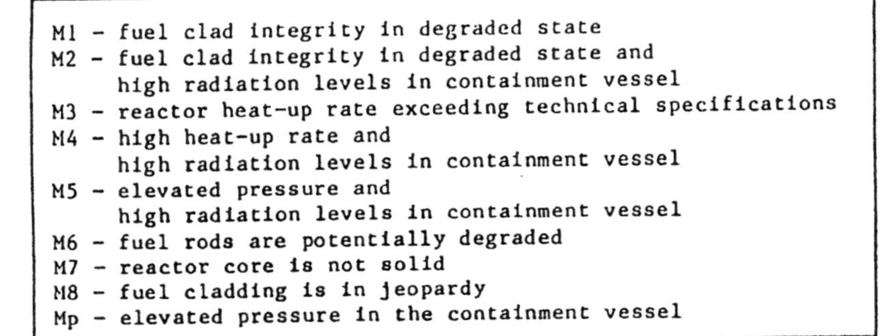

Fig. 2: Goal-Tree for the plant systems under monitoring

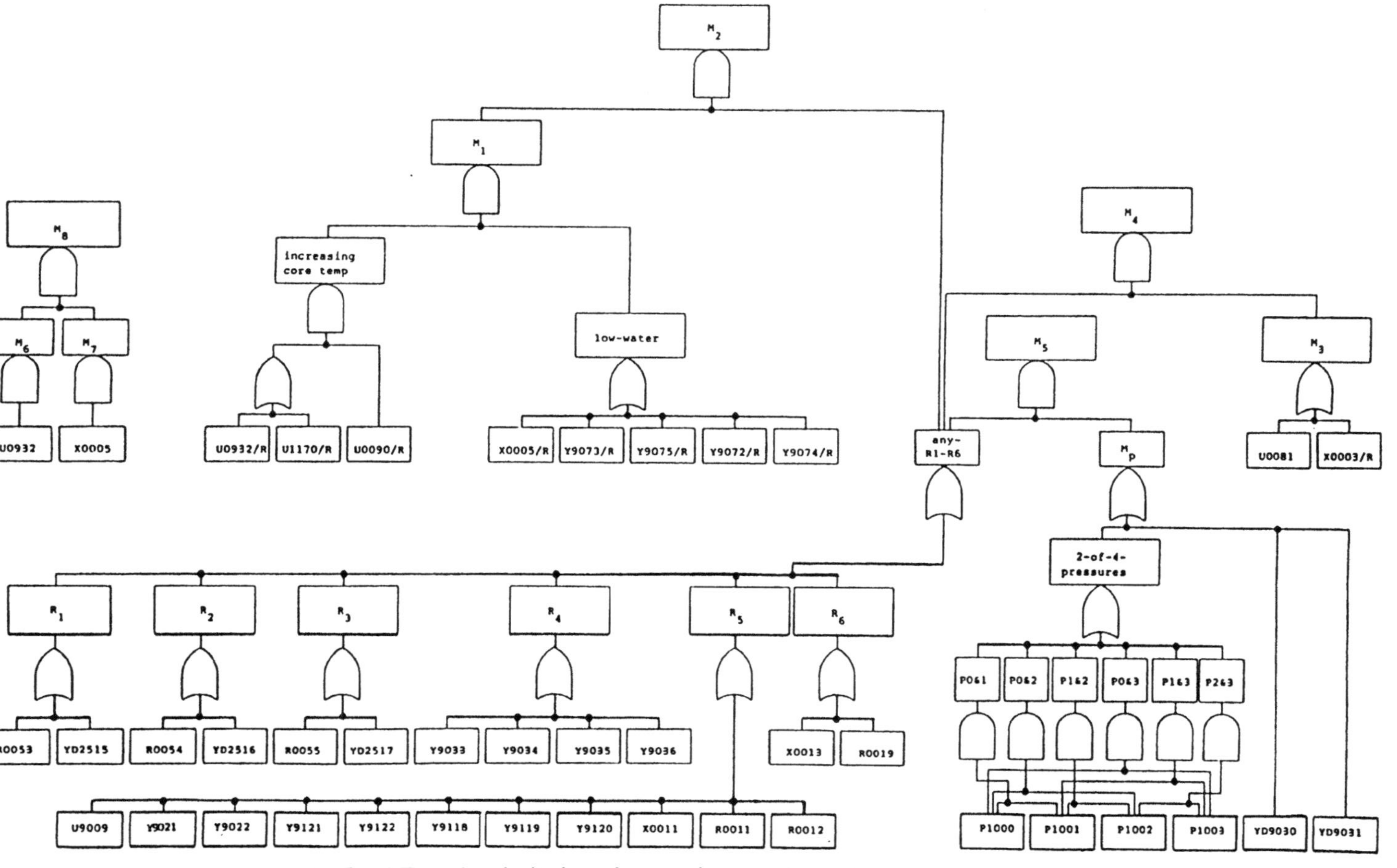

Fig. 3: Goal-Tree for deducing abnormal states from plant sensors information

235

```
(p plant-rad1
        (goal   action process) <goal>
        (fuel-integrity  state degraded) <fuel>
        (rad-level  state high) <rad>    ; rad level in containmn
-->
        (write (crlf)(crlf) | Plant Radiation Levels are high |
               (crlf) | because both the radiation levels |
               (crlf) | in the containment are high and there is |
               (crlf) | evidence that the fuel clad integrity |
               (crlf) | is in a degraded state. | (crlf)))
        (remove <goal> )
        (remove <fuel> )
        (remove <rad> ))

(p cont-pres5
        (goal   action process) <goal>
        (sensors   sens2    <sens2> > 1.0   )
        (sensors   sens4    <sens4> > 1.0   )

-->
        (write (crlf)(crlf) | Both containment pressure sensors |
               (crlf) | P1001 & P1003 are exceeding their |
               (crlf) | setpoint value of 1.0 psig. | (crlf))
        (make con-pres  state high))
        (remove <sens2>)
        (remove <sens4> ))
```

OPS-5 has two kinds of memory: a production memory which holds all
the production-rules, and a working memory which holds the data against
which the premises of the production-rules are matched. Productions can
be entered anywhere in the program. Regardless of where a production is
placed in the program, it will be executed when the conditions defined
in its premise part are satisfied best by the data in the working
memory.

The present system has a trigger-production titled "make-start."
When the working memory contains a working memory element called
"start", this production is fired and the system begins a series of
actions: It first opens two files outside of OPS-5. The first file is
called "signals.dat", and contains the most recent values of the sensors
information. The other file is called "recommendations.dat", and contains recommendations and diagnostic information generated by the
inferencing process. A representative content of this file is shown
here for a test case.

CONCLUSIONS AND RECOMMENDATIONS

The present work demonstrates the usefulness of using a Production-
rule system for plant monitoring purposes. The system is opetational on
a minicomputer and uses a forward-chaining, data-driven mode. This is
favored to an earlier system built on a microcomputer, and operating in
the backward-chaining mode. The nature of the problem of monitoring,
which deals with a large amount of data, that is continuously updated,
favors the present approach. The use of procedural programming simula-

Files Opened

*** reading input from file ***

input accepted

After input was accepted for the
current diagnostic cycle of UNIZE, the
signals data files have been closed.

 The effluent iodine radiation level
in the containment purge exhaust stack
is excessive. (Reading from sensor R0054
exceeds the setpoint of 0.0001 pCi/cc.)

It is definite (100%) that the radiation
level detected by the main steam line monitors
is excessive. (Based on the reading of monitor
RIM-RP58 (sensor Y9033) exceeding the threshold
value of 2.5 mR/hr.)

It is definite (100%) that the radiation
level in the containment is excessive!!
(Inferred from the fact that the value of
sensor X0011 exceeds the threshold of
10.0 mR/hr.)

It is definite that the radiation
level in the containment is excessive!!
(This inference was made on the basis that
the noble gas radiation level from the
lo-range sensor Y9120, exceeds the threshold
value of 10.0 uCi/cc.)

It is definite that the radiation
level in the containment is excessive!!
(Inference reached by the fact that the
#2 hi-range sensor, Y9022, reads over
the threshold value of 10.0 kREM/hr.)

It is definite (100%) that the
radiation level in the containment
vessel is excessive! (This inference
is derived from the high range area
radiation sensor U9009.)

Sensor X0003 indicates that the heatup
or cooldown rate of the average coolant
temperature in degrees F/hr is greater
than 100.0

tions of reactors can provide a more efficient means of identifying the
"desirable" and the "undesirable" plant states. In this operational
mode such simulations provide a source of knowledge about the behavior
of the system and the information obtained from such simulations would
replace the use of setpoints. The use of setpoints may be presently
satisfactory, but they lack the ability to adjust to alternate plant
conditions, particularly during transients. Thus the coupling of symbolic and procedural programming methodologies, and the development of
efficient decision-making and performance-measurement methodologies,
would make the present approach even more powerful in this field of analysis.

REFERENCES

1. L. Brownston, R. Farrell, E. Kant, and N. Martin, "Programming
    Expert Systems in OPS-5: An Introduction to Rule-Based
    Programming," Reading, Massachussetts, Addison Wesley (1985).

2. D. Gvillo, M. Ragheb, M. Parker, and S. Swartz,
    "Situation-Assessment and Decision-Aid Production-Rule Analysis
    System for Nuclear Plant Monitoring and Emergency Preparedness,"
    SPIE Vol. 786, Applications of Artificial Intelligence V, P. 347
    (1987).

3. M. Ragheb and D. Gvillo, "Development of Model-Based Fault
    Identification Systems on Microcomputers," SPIE Vol. 635,
    Applications of Artificial Intelligence, P. 368 (1986).

CHAPTER 5

PROCESS DIAGNOSTICS AND TRANSIENT ADVISOR

A GENERALIZED SYSTEM STATE ANALYZER FOR PLANT SURVEILLANCE*

J. Mott   R. King and W. Radtke

EI International, Inc.   Argonne National Laboratory - West
P. O. Box 736   P. O. Box 2628
Idaho Falls, ID 83401   Idaho Falls, ID 83402

INTRODUCTION

Very large, dynamic, complex systems such as electrical power plants, both fossil and nuclear, are designed using a host of physical parameters that interact with each other on a first-principle basis and via control actions and reactions. Examples of these parameters are pressures, temperatures, flows, voltages, etc. When the plant is built, it is instrumented with thousands of sensors and actuators to both determine and control the design parameters. When the plant is operated, a major problem is ensuring all the parameters are where they ought to be.

A usual solution to this problem is to establish a list of major parameter values, with upper and lower limits as appropriate, for each mode of operation or for each state of the plant. These parameters are examined regularly and checked against their limits in a process referred to as a polling technique. Often the very important parameters will have alarm annunciators to indicate violation of these limits. The problem with this solution is that it is not very accurate in a dynamic sense. Changes in some parameters are accompanied by changes in others. This occurs because the system is closed and many of the parameters are coupled or correlated with each other. For example, if coolant flow increases through a reactor core without a nuclear power increase then the core outlet temperature will decrease. Moreover, the movement of some variables is often a precursor to component or system failure. Such may be the case when a pump motor begins to draw too much power. If the pump power increase is not correlated with a flow increase or a coolant temperature change and corresponding density change, then it may indicate that the shaft is beginning to bind because of foreign material intrusion or that bearings are beginning to fail. In any case, the vast number of correlations are difficult to anticipate and difficult to recognize on a timely basis using a polling technique.

*Work supported by the U. S. Department of Energy under contract No. W-31-109-ENG-38. The submitted manuscript has been authored by a contractor of the U. S. Government under contract No. W-31-109-ENG-38. Accordingly, the U. S. Government retains a nonexclusive, royalty-free license to publish or reproduce the published form of this contribution, or allow others to do so, for U. S. Government purposes.

Perhaps the ultimate solution to the problem is one where the plant is
completely and accurately simulated by a computer program which runs in
real time in parallel with the plant. The program could be kept linked to
the present state of the plant. If any parameters were inconsistent with
the plant model then consequences of the inconsistencies could be diagnosed
and projected into the future via a faster-than-real-time execution of the
simulation. If this solution could be achieved, it would be human-
intensive in developing the modeling, diagnosing, and predicting
methodologies and software; it would be computer-resource-intensive in
application; and it would be specific to a particular plant.

Presently, a solution to the problem exists. It addresses many of the
same areas as the ultimate solution but with none of the drawbacks. It is
a surveillance technique only and does not address diagnosis and
prediction. But it does so without any modeling effort, with a minimum of
computer resources, and in a completely generic fashion that is applicable
to any system. This methodology is based on a learning process which can
range from simple algorithms to advanced artificial intelligence programs
and is referred to as the System State Analyzer (SSA). The remainder of
this paper describes the SSA and then discusses an application at a real
plant. While the methodology is quite general and can also encompass fast
and slow transients, the present work discusses only a pattern-recognition-
based learning process and a steady-state application. In this form, the
methodology is applicable to validating simultaneously hundreds of signal
values and replacing many failed signal values with estimates that depend
only on a dynamic estimate of the state of the system, with both validation
and replacement occurring in near-real-time.

The SSA gives an observer the abilities to easily identify drifting
and failed sensors and to detect the abnormal operation of an entire system
or of any component in a system. These abilities, in turn, can be used for
surveying the plant in general, for controlling the plant, for verifying
that plant startup is passing through the correct sequence of states, for
determining the health and calibration of sensors, and for detecting the
onset of failures in a complex system. The SSA may also be employed for
handling the avalanche of alarms that accompany plant upsets if good
examples of such upsets are available either from real operation or from
computer simulation.

SYSTEM STATE ANALYZER

The SSA is computer software which embodies a mathematical framework
that uses examples of a real system from which to learn, makes a single new
observation of the system, and then uses the learned examples to estimate
the true state of the system. States of the system are represented by
vectors in an n-dimensional mathematical space where the coordinates of the
space are the elements of the vector. Elements of the state vector are
chosen by the SSA user and can range from direct values of the system
parameters to any transformation which produces a scalar value. State
vector elements do not need physical dimensions or units associated with
them nor do they have to be linearly independent. The learning process is
performed according to a set of rules defined by the SSA user.

This methodology is outlined in Fig. 1 where the SSA inference engine
is diagrammed as accepting a set of learned states from the top and a new
observation from the left. After a series of operations denoted by the
function in the SSA inference engine, one result is an estimate of the
learned state which lies "closest" to the new observation and is denoted by
the letter "j" with a subscript "1". The definition of "closest" is not a
geometric one but instead is the state lying closest to the new observation

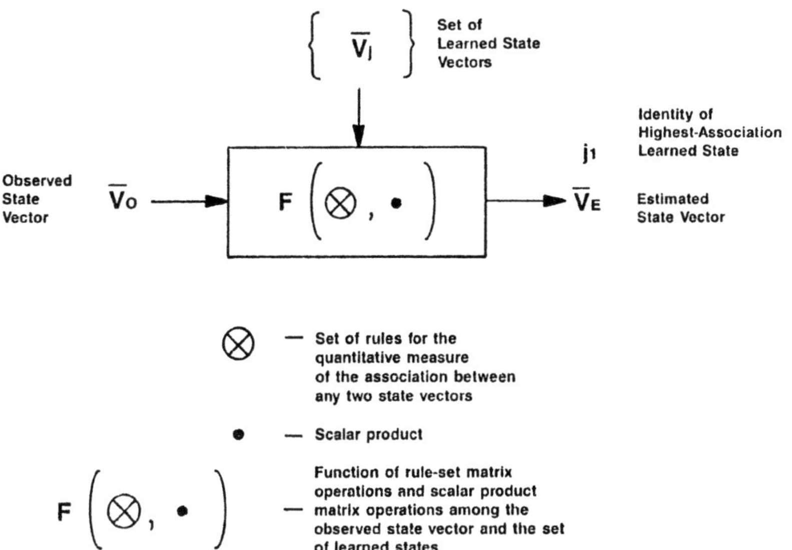

Fig. 1. SSA Inference Engine.

from the point of view of the set of rules that determine the association
of any two vectors. This rule set is denoted in the diagram by a circled
"X". A second result is the estimation of the true value of every element
in the new observation. This estimation takes the form of an estimated
state vector shown on the right of the diagram.

The function in the SSA inference engine diagram of Fig. 1 can be
expressed as a single equation. The operations that the SSA inference
engine performs in this equation are matrix multiplications using the rule
set, normal matrix inversion, and normal matrix multiplication denoted by
the filled dot in the diagram of Fig. 1. First, all pairs of learned
states are associated two at a time using the rule set to create the
elements of a recognition matrix. Next, the new observation is associated
with each learned state using the rule set to produce an overlap vector.
The largest value in this vector identifies the learned state "closest" to
the new observation. Finally, the normal matrix product of this vector
with the recognition matrix produces a set of linear combination
coefficients for combining the learned states into the estimated state
vector. A fundamental theorem about the SSA methodology states that the
processes discussed above, when applied to any true state of the system,
which is an exact linear combination of the learned states, yields a very
close approximation to the true state.

The particular vector elements and rule set that are the subjects of
the present SSA description and application are derived from pattern-
recognition algorithms. Vector elements are simply those numbers which are
directly accessible on the data acquisition system (DAS) without any
transformations. They consist of both direct sensor outputs and values
calculated from the sensor outputs. Often there will be multiple,
redundant sensors with nearly the same output values or different versions
of the same information (e.g., a flowmeter output in mV which is also
present on the DAS as gpm) so that many of the elements of the state vector
are not truly linearly independent. The SSA is robust enough to handle
this situation without any attention from the user and without any
mathematical difficulties.

The present rule set is based on a learned process which compares the
overlap of any two patterns (state vectors) within a fixed framework.
Imagine that you are asked to compare two photographic negatives which are
transparent images on a completely opaque background. You would probably
lay the images over each other, hold them up to a light source, move them
around a little to maximize the light transmission through their common
transparent regions, and come up with an estimate of similarity arrived at
by a judgement of the amount of overlap of the two images. The present SSA
rule set is the mathematical analog of this overlap process.

The present SSA framework is an attempt to give the characteristics of
intelligence to the acts of determining whether the state of a complex
system is consistent with other previous observations of the same system
and, if so, making quantitative estimates of all the variables of the true
state of the system. Figs. 2 and 3 are illustrations of this attempt. In
these figures are displayed a completely hypothetical two-variable system.
Two variables are used because they allow the representation of the system
state vectors on a two-dimensional surface such as a computer screen or
viewgraph. These figures are printouts of an SSA demonstration program
that runs on a PC. When observing these figures, keep in mind that, as
described above, the nucleus of the SSA process depends only on the number
of learned states and not on the number of variables in the system.

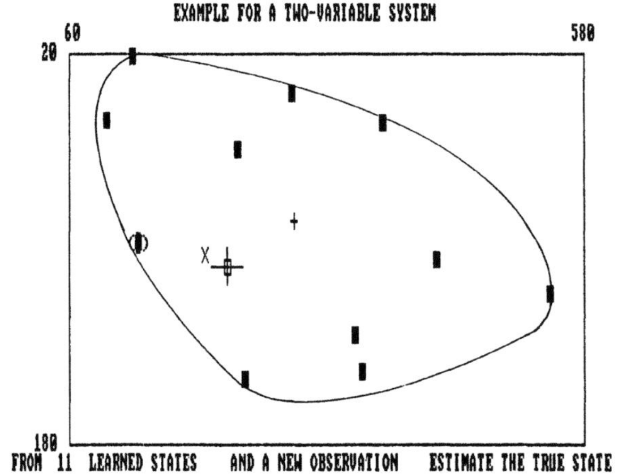

Fig. 2. Example 1 for Two-Variable System.

In Fig. 2 are shown 11 learned states denoted by filled rectangles.
The variables of the system are simply the X and Y coordinates of the
display. These could represent such real variables as temperature and
pressure. The domain of learning is indicated by the solid line
surrounding the learned states. A new observation is made, denoted by the
X symbol. This observation lies inside the domain of learning. The SSA
inference engine immediately determines the "closest" learned state to the
new observation and places a circle around it. Almost as rapidly, the SSA
produces an estimate of the true state of the system and denotes it on the

display by an unfilled rectangle, with uncertainty bars for each of the two
system variables. The uncertainties arise from the average fractional
difference of both system variables between the estimated state and the new
observation. Note that the estimated state lies near to the new
observation within the learned domain and seems to be a reasonable answer
to the question "If I have previously seen this system in states
represented by the filled rectangles, and I now observe it to be in a state
represented by the X symbol, what is my best guess as to the true state of
the system and my best guesses as to the true values of both variables in
the system?". The answer to this question is given by the unfilled
rectangle which simultaneously shows the estimated state position, the
values of both system variables, and the uncertainties in each variable.

Consider now the situation shown in Fig. 3 where the new observation
lies outside the learned domain. The SSA inference engine estimates that
the true state of the system really lies somewhere within its domain of
learning and that both system variables should be considerably changed to
reach a true state of the system. But, in this instance, the SSA is fairly
uncertain about these conclusions because the correlations among the
majority of variables in the new observation are very unlike anything it
has seen before.

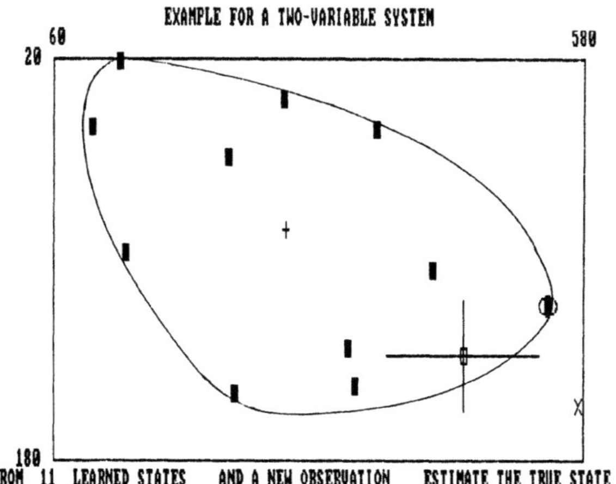

Fig. 3. Example 2 for a Two-Variable System.

APPLICATION TO EBR-II

The Experimental Breeder Reactor - II is a liquid-sodium, metal-
fueled, fast-neutron-spectrum reactor power plant that has been operating
at Argonne National Laboratory - West for the last 23 years. It has a
ducted-flow primary heat transport system that is immersed in a large-
thermal-capacity sodium pool. The radioactive primary sodium is pumped to
an intermediate heat exchanger (IHX), and a secondary heat transport system
moves non-radioactive secondary sodium from the IHX to a sodium-to-water
superheated steam generator. The reactor generates about 62.5 MWt and the

steam generator drives a turbine-generator that outputs about 20 MWe. In
addition to steady-state operation for power generation, the plant is an
experimental facility used for both steady-state and transient experiments
that test fuels, components, and systems. Fig. 4 is a diagram of the
primary cooling system.

Of special interest as far as the SSA is concerned is EBR-II's
advanced DAS. The DAS has over 800 data values available every half
second. These values are continually averaged on five-second, one-minute,
and three-hour bases and are placed in continually updated disk files for
ready availability until they are finally archived on magnetic tape. The
SSA accesses these data over serial links between PCs in the engineering
building and the power plant building. The SSA software has custom
interface programs on the DAS computer to access the disk files and send
them to the main SSA inference engine resident on the PCs. The SSA
currently is set up to accept up to 135 DAS values and can transmit one
such data set to a PC in about ten seconds where the SSA software can
analyze them in about five seconds. These times are appropriate for a
PC-AT with a math coprocessor and SSA with 11 learned states. Eleven
learned states are normally used but up to 30 will fit on a PC with 640 kB
of RAM. The normal mode of operation of this system is to send 11 such
data sets at a time to the PC and then to present graphical information on
the entire data set. The learned data sets and the monitored data sets are
accessed and transmitted from the DAS computer to the PC in almost
identical manners. Their files can be used interchangeably by changing
their names.

Often, signals change by catastrophic failures. If these failures
occur, the SSA inference engine can easily detect them: the new
observations simply go off-scale in the SSA time-dependent comparisons
while the SSA estimations remain on-scale and give approximately the same
values as if the catastrophic failure had not happened, unless such a
failure strongly affects the rest of the plant.

But many failures are much more subtle than catastrophic failures.
Following is a case which involves a signal that drifts very slowly. The
drift is so slow as to be almost unnoticeable, and the signal has larger

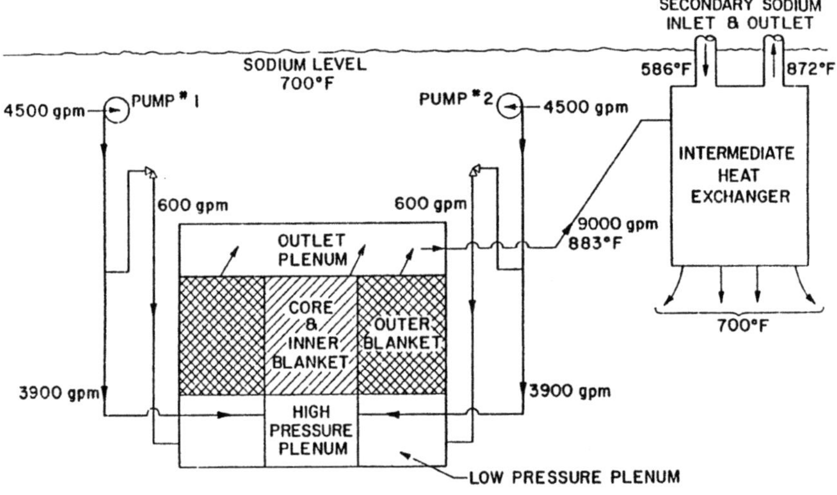

Fig. 4. EBR-II Primary Sodium System Schematic Diagram.

than desired variations over short time periods. The signal is important
because it determines one of the main control actions during plant
operation. This signal is the reactor temperature rise.

During startup of EBR-II, the primary coolant flow is initially set at
its full power value and the entire system is nearly isothermal at 700
degrees Fahrenheit. Control assemblies are inserted to expand the reactor
temperature rise to the full power value of 183 degrees Fahrenheit and
achieve a final mixed mean reactor outlet temperature of 883 degrees
Fahrenheit. During this process, the secondary coolant flow is continually
adjusted to provide a constant reactor inlet temperature equal to 700
degrees Fahrenheit. After full power is achieved, the control assemblies
are periodically adjusted to maintain the full power reactor temperature
rise, and the secondary coolant flow is adjusted to maintain the reactor
inlet temperature at 700 degrees Fahrenheit. The reactor temperature rise
is therefore the key parameter in maintaining reactor power.

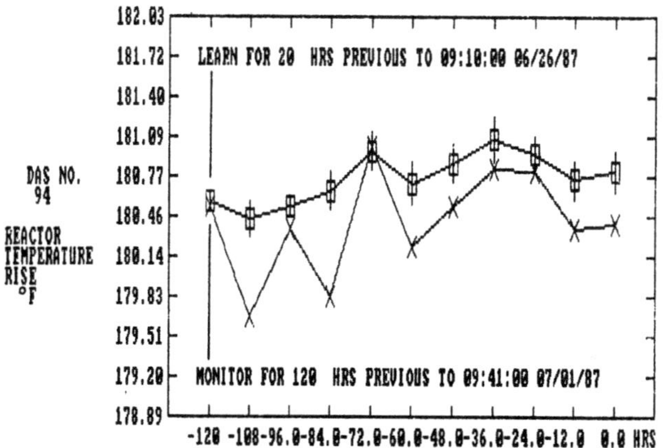

Fig. 5. Typical Appearance for DAS Value 94, as Analyzed by the SSA,
July 1, 1987.

Unfortunately, due to the apparent failure of temperature sensors in
the reactor outlet coolant pipe, the reactor temperature rise is a
difficult parameter to measure accurately. The reactor core has many flow
paths that are affected by core configurations and control assembly
positions. So a host of thermocouples at the outlets of many core
assemblies are necessary to infer the outlet temperature profile. The
usual way of inferring reactor temperature rise thus includes core inlet
thermocouple averages, primary flow, one upper plenum thermocouple, and a
heat balance in the secondary cooling system, all combined in an algorithm
and output as DAS number 94. A typical appearance for this DAS value as
analyzed by the SSA on July 1, 1987, is shown in Fig. 5. Notice that the
general shape of the observations is estimated by the SSA but that the
short-term movements are not strongly coupled to the SSA estimations. The
learn period for this observation is just previous to the monitor period,
just to the left of the vertical line above the -120 hour point on the
horizontal axis. What is evident here is that the average new observation

had already slowly drifted downward with respect to the SSA learn period
five days earlier. Previous to June 5, 1987, the plant had been controlled
by this DAS value. Operators noticed that the plant state would slowly
move out of SSA learned state domains, with many short-term oscillations,
because the plant power was actually being slowly increased to compensate
for the downward drift of the reactor temperature rise as measured by DAS
number 94.

On June 5, 1987, the engineering staff of EBR-II recommended to the
operations staff that the control of the plant be based on a different,
more direct measurement of the reactor temperature rise. This direct
measurement is provided by differences between the average reactor inlet
temperature and a reactor outlet temperature, which is provided by a single
thermocouple at an appropriate location in the primary coolant flow path in
the ducted portion downstream from the reactor core. The direct
measurement is shown as DAS number 95 in Fig. 6. The learn periods and
monitor periods are the same in this figure as in Fig. 5. In Fig. 6, the
plant is being controlled by DAS number 95 and the correlations between the
SSA estimations and the new observations are strikingly good.

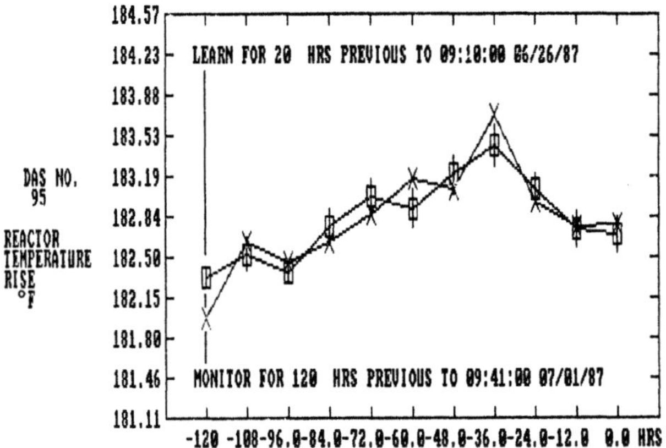

Fig. 6. DAS Number 95 and SSA Estimations.

DAS number 95 had not been used for plant control prior to June 5
because this thermocouple had become suspect over the 23 years of its
operation and had been removed from plant control service several years
ago. During 1986, this thermocouple's long-term performance had been
investigated using the SSA.[1] Studies showed that it could be used
for single plant runs. With the advent of the observed deficiencies
in the performance of DAS number 94, it was fortunate that confidence in
the performance of DAS number 95 had been restored and that it could be
returned to service as a plant control parameter.

CONCLUSIONS

The SSA inference engine has been designed to add information content
to observations of data in a global surveillance application based on a
finite number of previous system state observations. It has been tested
with a variety of simulated data to verify that it could do so in a rapid,
robust manner. The manner in which the SSA does this has the
characteristics of the beginnings of artificial intelligence with the
current pattern-recognition-based rule set for associating system state
vectors. More artificial intelligence characteristics might be expected to
result from an expansion of the rule set into true artificial intelligence
areas.

Application to a real system, EBR-II, offered the opportunity to
verify that real data could also be analyzed by the SSA and to develop
timely, responsive interfaces with the user. In the real application, we
discovered that SSA is easy to use and that the learned domain is
important, easy to capture, and rapidly verifiable via the SSA observations
themselves. If the learned domain is far from the monitored-state domain,
then the SSA simply gives fairly uniform time-dependent estimations for
each state variable with an uncertainty that is characteristic of the
separation between the domains. In effect, the SSA indicates that it knows
only what it has learned, that it will give the best estimates that point
in the direction of what it is being asked to monitor, but that it is
uncertain about these estimates to an extent that the average variable in
the monitored region differs from the average variable in the learned
region. So, in a sense, the SSA is self-diagnosing.

Research further discovered that the real plant problems are not
simply a matter of signals violating rather large operating-limit bands,
which can be handled by straightforward polling techniques. Rather, they
are more subtle and require a fairly high-resolution tool with the data at
least partially transformed into information. The SSA inference engine
discussed in this paper seems to satisfy these real-world needs in a
generic manner that is applicable to a host of steady-state systems such as
nuclear power plants, fossil power plants, or any large process system that
has an adequate data stream available to the SSA.

REFERENCES

1.   J. E. Mott, W. H. Radtke, and R. W. King, EBR-II System Surveillance
     Using Pattern-Recognition Software, in "Proceedings of the ANS/ENS
     Topical Meeting on Operability of Nuclear Power Systems in Normal
     and Adverse Environments", Albuquerque (1986).

DEVELOPING AN EXPERT SYSTEM FOR MONITORING POWER PLANT PROCESSES

Kautto Ari *, Haarla Jyrki **, and Ranta Jukka *

* Imatran Voima Oy, P.O.Box 138
  SF-00101 Helsinki

** Technical Research Centre of Finland
   Electrical Engineering Laboratory
   Otakaari 7 B, SF-02150 Espoo

ABSTRACT

Imatran Voima Oy and the Technical Research Centre of Finland have
collaborated on developing an expert system for monitoring the process
of a nuclear power plant (PWR) and for fault detection. The development
work is supported financially by the Technology Development Centre
(TEKES), Finland. The idea is to build into the expert system a
description of the process and the normal behaviour of the components
so that it can signal an alarm for abnormal events. In addition, the
system will incorporate a certain amount of diagnostics. Completion of
the prototype is scheduled for March, 1988. For the purpose of testing
and obtaining experience on an actual system-plant interface, the
system will be connected to the Loviisa nuclear power plant training
simulator at the end of this year. The system is intended for use by
an operator and it is being implemented with a Symbolics 3620 Lisp
machine using the KEE expert system development environment.

INTRODUCTION

Since the processing of human knowledge is limited, it makes sense
to arrange and group the large amount of information obtained from the
process as well as to model the behaviour of the process in aiming for
successful management of the process. Table 1 gives a list of the main
points required for successful process management. In our opinion a
good information system should be able to provide "real time help" for
each significant event and task/1, 3/. The tabular listing of
requirements does not purport to be exhaustive.

This project concentrates on the main flow system of the nuclear
power plant's secondary circuit, above all on the feedwater system, the
high pressure-preheating system and the turbine bypass. Our project is
limited to the modelling of the energy production function.

Table 1. Main requirements for successful process management

Focus	Implementation in our expert system	
Recognition of process main states	Recognition of cold shutdown, hot shutdown, power run, etc.	Used in process chronological alarm system
Main target state	Different stages of the turbine generator plant's start-up and shutdown	
Principal strategies of controlling the process	Not possible within the framework of our project	
Behaviour of subprocesses, as well as data on how much time is left for performing control actions.	Reference curves with respect to the generator's output. Check of connection states. Prediction of component trip. Supervision of regulating devices.	
Disturbance diagnosis.	During abnormal situations. Provides optional guidance in making a diagnosis.	
Correct use of procedures.	Not provided for.	
Operational organization of information.	Derived information depicted by mimic diagrams. Knowledge concerning planning.	
Efficient information access	Organization of data access actions and minimization of data acquisition chains.	

FUNCTIONAL SPECIFICATION

Monitoring the process

The concepts to be precisely modelled are listed below:

- Main process states; cold shutdown, semi-hot shutdown, hot shutdown, power run.

- Changes in main state and change directions; defined by using the change directions and rate of certain main variables of the process.

- Reference vector of the connection state; contains the preset positions of pumps and valves in certain process situations. The reference vector makes use of changes in the main state and change trends.

-   Target state of the process; at present only two target states are
    used: the target power state when starting up and shutting down
    generators. It would be interesting to find the criteria for other
    process target states too.

-   Readiness for operation; this means the checking of a start-up
    disable for a component as well as the broader concept of checking
    the operating restrictions of the process and subsystems.

-   Productivity; this means functional checking of a subprocess
    (whether the operation of a subprocess fulfils the requirements set
    upon it, which are determined by the generator's output). This also
    embraces the modelling of start-up and stop signals for subprocesses
    as well as the modelling of emergency stop signals/2/. Emergency
    stop means loss of operation due to protection (for example,
    excessively high temperature).

The result of modelling should be a functional alarm system somewhat
like the following example:

If the feedwater flow diverges too much from its reference value (in
proportion to the generator's output), an operational alarm will be
registered. The correct action might be, say, starting up the backup pump if the main pump has tripped and the back-up pump has not
started up at all (insofar as the back-up pump starts up
automatically, the alarm is only at the component level).

Regulating devices

   Regulating devices make it difficult to notice a process
disturbance because their purpose is to keep the main variables of the
process at their set value. If disturbances arise in connection with
the main regulating devices, we often have to limit the plant's output.
For these reasons it is important that we can obtain data on the
correct behaviour of the regulating valves. This information
essentially facilitates the making of a diagnosis. We have used a PI
algorithm, on which most of the standard regulating devices are based,
in order to calculate the position of the regulating valve's shaft. An
alarm is given if the measured position of the shaft diverges
excessively from the calculated value. In addition, we calculate the
valve's aperture and the flow rate through the valve based on the
characteristics of the valves and general laws of flow mechanics. Here
the same idea has been used as was applied by Oivind Berg in his Early
Fault Diagnosis project/4/. These calculated flow values can again be
used in other mass and energy balance computations. These values could
also be used in validating signals but this subject is not dealt with
in the present project.

Deduction of main state of the process

   The process has five different main states, which are described in
the knowledge base: full power, zero power (critical), hot shutdown,
semi-hot shutdown, cold shutdown. The system is able to deduce these
states by means of the following main variables of the process: average
temperature of the primary circuit, generator output, pressure of the
primary circuit, boron content of the primary circuit, reactor output,
reactor tripping signal, position of the control rods. The relation
between these variables and the main states of the plant are stored in
the system's knowledge base.

Problems arise when one or more of the main variables has a value
which is not compatible with the plant's main states. In this case
something is amiss with the process (a component is out of order, there
is a leak in the pipelines), in other words, the availability of the
plant has suffered. The system accordingly deduces the state of the
plant to be the closest possible main state, but it reports at the same
time that the state in question is disturbed.

Determination of the behaviour of subprocesses

The main variables of the subprocesses of the energy production
chain in the plant's power range correlate with the generator's output
(for example, the feedwater flow increases when the generator's output
increases). We have used the training simulator of the Loviisa Nuclear
Power Plant and logged variables which should correlate with generator
output. We have used regression analysis to compute from these
measurements formulas for the relationships between certain process
variables and generator output. Using this data, it is possible to
deduce whether subsystems are behaving properly.

Prediction of component trip

The alarm system is built to incorporate a number of predictive
properties, in particular the prediction of the tripping of the most
important components. Computation of the time for which a given
component will remain running is begun when the process variable
causing a trip diverges sufficiently from the normal value. The
remaining running time is calculated as follows:

$$T(t) = (Xtrip - x\,(t))\, /\, \dot{X}(t)$$

where

Xtrip	is the trip limit of the component
X(t)	is a process variable, for example, temperature, pressure, surface height etc. at the instant t.
$\dot{X}(t)$	is the rate of change of X at the instant t.

Checking of the state of locally controlled valves

There are at the plant a large number of locally controlled
components with local operation functions. The status data for these
components cannot be seen from the control room control panels. For the
most part, such components are drainage and deaerating valves as well
as locally controlled valves which are used during maintenance in order
to isolate a given area.

Monitoring of the status of local components can in some cases be
performed functionally; for example, when the pump is running, there
should occur a flow or pressure rise in the pipeline downstream from
it. If nothing happens or very little happens, the reason can be that
an incorrect state prevails for a locally controlled component or a
drainage or deaerating valve is open. At this stage, there is reason to
stress the difference between reliable and uncertain data in order to
avoid misleading information.

Knowledge acquisition

Heuristic rules are much discussed in connection with expert
systems. We have nevertheless decided to first collect technical and
physical data and knowledge. In this way we can better describe the
dynamics of the process and at the same time we obtain a viewpoint on

where heuristic rules are needed. They may be appropriate for events in
which safety has been jeopardized because in this case the target state
of the process is a cold shutdown. In using heuristic rules there
exists the danger of applying some rule in the wrong context.

The process knowledge at the plant where our system is being
implemented is well documented. To be sure, the utilization of these
documents limits the appropriateness of our system to other plants. We
must find the planning principles from the process knowledge in order
to be able to obtain general knowledge. In addition to process
knowledge, we have had the possibility to exploit the plant's training
simulator, which has been a good source of knowledge. In this way we
have had precollected as well as plant-specific and general knowledge.

In addition, we have run turbine start-ups and shutdowns on the
simulator. From these runs we have collected values for some of the
most important variables, thus permitting us to depict the normal
behaviour of the process.

We have also had the opportunity of discussing these matters with
the automation designer of the plant, the instructor for the training
simulator as well as the experienced shift supervisor. We have tested
the utility of our ideas on them by telling of some events that can be
envisaged and asking what kind of information the operator needs at any
given time and what he should do.

Table 2. Sources and use of knowledge

Source of knowledge	Status data from automation documents	Regulation algorithms and component characteristics from the simulator	Laws of physics	Relationships between variables from the plant and from the simulator
Knowledge	Reference connection state	Operation of regulating devices - main regulating devices	Models of the M&E balance Models for components	Reference curves
Use of knowledge	-Start-up/stop of sub-systems -Start-up/ stop of components -readiness -prediction of trip	Reporting on behaviour deviating from the normal	Diagnosis Explanation: -Connection error -leak -inside pipes -outside pipes (-Instrument fault) (Guidance: -isolation)	Initiating a diagnosis Monitoring help for plant status analysis

INFERENCE MECHANISM

We are trying out two alternative inference mechanisms: a non-
linear static mathematical model and a dynamic simulation model.

Non-linear static mathematical model

A summary of the operation of the inference engine is presented in
Table 3. Process data is updated continuously. Computation and checking
of subprocess functions is modular: on the highest level the most
important variables are checked the whole time. If something abnormal
occurs, the subsystem where the disturbance has been found is subjected
to a rigorous analysis or diagnosis.

The behaviour of regulating devices is monitored continuously.
Some of the most important variables of the process too, for example,
the sum of feedwater flow are checked on a continuous basis. The values
of these variables are compared with precalculated reference values. If
the system detects an excessively large discrepancy between a variable
and its reference value, an alarm is registered and a detailed
computation or diagnosis is initiated in the assumed disturbance area.
If control actions are needed to perform a diagnosis, the system guides
the operator in making them.

Table 3. Operation of the inference engine

	Passive variables	Active variables
Continuously updated data	Process variables	Values of regulating devices
Continuously computed	- Dynamic alarm limits (based on reference curves) - Limit checks - Some derived variables (e.g. rates of change) -Start-up and stop conditions of sub-processes	-positions of the shaft of regulating valves -difference:measured-computed -flows based on the performance curve of the regulating valve and on the laws of flow mechanics
Computed if necessary	-Check of connection of component triggering -Making a diagnosis	
At operator's request	?	

Simulation model

Another approach is based on the use of simulation models together
with the process. Here the dynamic models for pumps, valves and heat
exchangers are used. The computed values for process variables are
compared with the measured values. If discrepancies are detected, the
system lists those components and measurements where a fault is assumed
to exist. Computations are carried out only in the case that a
sufficiently great change has occurred in the process.

Connection state

Determination of the state of the reference connection is based on the following model (the purpose is to maximize plant efficiency). The start-up and stop conditions of subprocesses are based on optimizing heat transfer with a few limitations (for example, critical flow, flashing). In this way a reference state is obtained for a larger part of the motor-controlled valves.

In connection with the start-up and stop conditions for subprocesses, a small dead band is used so that the system does not become too sensitive. This makes possible, up to a certain limit, personal methods of process control. The system checks the connection status of the subprocess when the start-up or stop condition has just been fulfilled or when changes have occurred in the connection state.

MAN-MACHINE INTERFACE

It has been attempted to make the man-machine interface as easy to use and conceptually clear as possible. The display is based on a process diagram, from which the operator can call up other displays using a mouse. The other displays can be, for example, interlock diagrams, a process main state display as well as a correlation curve display. In addition, the display always incorporates an alarm window for outputting messages produced by the system (warnings, alarms).

Alarms and warnings produced by the system are also accompanied by the reason for them. For example:

  08.25.35 RL21D01 pump will trip in 7.20 minutes.
       RL21T010 motor bearing T = 68"C.

Reporting of an imminent trip should be limited (for example, 10 min.) in order to cut down on unessential information.

EXPERIENCES OF THE DEVELOPMENT EQUIPMENT

The development equipment used in the projects is a Symbolics 3620 unit having 8Mb of central memory as well as 380Mb of disc memory. The development environment procured for the equipment was KEE (Knowledge Engineering Environment).

User-friendliness

The user-friendliness of the device is good, at least compared with "old-fashioned" programming. Windowing, object oriented programming as well as a mouse are effective tools for enhancing programming.

Speed

The machine's speed has become one of the greatest problems already in the development phase. Whereas the original intention was to reach a 1-5 second sample-taking interval for 400 measurement quantities, we now have to make do with an approx. 20 second sample-taking interval. This does not, however, prevent us from ascertaining the utility of the system since the system will be connected to a simulator on which the speed of simulation can be adjusted.

FUTURE VIEWS

In future the system could be able to deduce the cause of a
disturbed process main state as well as provide the operator with
instructions on how to restore the situation to normal. This could take
place as follows:

1. On the basis of certain rules the system is able to deduce the
   target main state of the process (for example, in start-up of the
   process the target state would be assumed to be the next hotter
   state of the process if the limiting conditions for this transition
   in state are not in force.

2. On the basis of the target main state, the system creates a listing
   of those subsystems whose state is abnormal or whose state is to be
   changed.

3. By examining these subsystems, the system seeks out the target main
   state of the process and by means of the deductive chain pinpoints
   those components which have an incorrect status. The names of these
   components and the faults affecting them are reported to the
   operator.

Instead of using precomputed static reference curves, the
reference values could be computed dynamically. In this case the
reference value should be calculated so that heat transfer is
maintained continuously. The heat produced in the core would now be an
independent variable and it would be the value referenced. On the basis
of this data, we could calculate, for example, the feedwater flow
required for heat transfer /5/. The area of competence of this kind of
knowledge would be broader than that of static dependent
relationships.

REFERENCES

1. Nelson W. and Blackman H., Response Tree Evaluation: Experimental
   Assessment of an Expert System for Nuclear Reactor Operators. Idaho
   National Engineering Laboratory. NUREG/CR-4272. EGG-2397. 1985. 53
   PP. + app. 5 pp.

2. Gaudio P., Improved On-line Operational Support System Using Success
   Path Monitoring. Combustion Engineering, Windsor, Connecticut,
   U.S.A. TIS- 7732. 11 p.

3. Kautto A. Information Presentation in Power Plant Control Rooms.
   Espoo 1984. Technical Research Centre of Finland, Research Reports
   320. 108 pp. + app. 11 p.

4. Berg O. & al., Early Fault Detection Demonstrated on the NORS
   Feedwater System. OECD Halden Reactor Project. HWR-204. Halden 1987.
   19 p.

5. Miettinen J. and Hämäläinen A., Critical Parameters in a PWR during
   Primary SBLOCA and Secondary LOCA Transients. In Enlarged Halden
   Programme Group Meeting at Gothenburg, Sweden, 3th - 7th June, 1985.
   12 p.

THE DISYS REAL-TIME DIAGNOSTICS/CONTROL SYSTEM AND ITS APPLICATION

TO THE EBRII

Alan M. Christie

Westinghouse Electric Corporation
Technology Enterprises
Expo Mart 335E, 105 Mall Blvd.
Monroeville, PA 15146

Richard W. Lindsay

Argonne National Laboratory
P.O. Box 2528
Idaho Falls, ID 83401

ABSTRACT

DISYS is a real-time diagnostic and control guidance expert system designed to aid plant operators during off-normal events. Currently it is being implemented and tested at the Experimental Breeder Reactor II. DISYS is driven by sensor input and evaluates symptoms and faults associated with components. Component faults in turn allow an evaluation of sub-system, system and plant faults. The hierarchical model of the plant as used by DISYS also allows information regarding system reconfiguration to be generated in the event of a fault. Current testing at EBR II indicates that DISYS can successfully detect component and system anomalies. A model of the Argon Cooling System has been constructed and currently has about 300 rules. Once the diagnostics system for the ACS has been completed, it is intended to implement DISYS on other reactor systems.

INTRODUCTION

DISYS is an expert system which, on a real-time basis, evaluates sensor data, combines this data to determine component symptoms which in turn are evaluated to assess the faultedness of components. The "health" of these components is then used to assess the "health" of sub-systems, systems and so on up to the plant level. This sensor-driven logical hierarchy is seen to be the most appropriate structure upon which to provide real-time the diagnostics and control guidance information. It is also very consistent with the engineering design, and compatible with the human conceptual structure of a complex plant [1].

Because of the inherent uncertainties in the diagnostic process, and also because degrees of faultedness exist, fuzzy logic [2] is used to propagate information up the hierarcial model. Each rule in the model has multiple input pairs and one output pair, each pair consisting of a numerical value for "degree of faultedness" and "credibility of diagnosis". In general diagnosis can only take place within the context of a known system configuration. For example, the diagnosis of a valve in the "closed" position will differ from its diagnosis in the "open" position. For this reason the plant control signals are used in DISYS to determine plant configuration down to the component level. In addition to diagnostics, DISYS is also capable of recommending appropriate control actions to move the plant from an unsafe state to a functionally equivalent safe state or functionally degraded safe state. This feature has not yet been implemented at EBR II.

THE DISYS SYSTEM

The DISYS software is composed of two separate modules - DICON and NOGEN. DICON is the code which is integrated into the real-time plant control system while NOGEN is an off-line code for the development of the physical model and to aid in knowledge engineering. This latter code allows for the rapid development and modification of models which are run on-line by DICON.

In keeping with the operators mental models of the plant, the computer model is based on a logical hierarchy of processes. This hierarchy contains many diverse concepts to adequately describe plant operation. The implementation of these elements within the model must be based on a consistent formulation to allow information transfer between elements of the model. The major elements required for plant modelling are:

- o real-time signal acquisition, validation and processing
- o diagnostic evaluation of components in their various states
- o component, subsystem, system and plant status evaluation
- o generation of control guidance sequences.

These considerations must be addressed in a real world context where uncertainty and ambiguity are bound to arise. Specifically, the following uncertainties may arise:

- o uncertainties due to inconsistencies between redundant sensor readings
- o uncertainties due to discrepancies in the system sets which define faults
- o uncertainties in the faultedness of components, subsystems and systems due to too few sensors.

Even if these uncertainties do not exist, the degree to which the faults are present will depend upon the severity of the symptoms. For these reasons fuzzy logic [2] is used in place of binary (true-false) logic. This allows the degree of faultedness (v) to range from zero to unity, where zero represents a fully faulted condition while unity represents a fully unfaulted condition. The variable v is defined as the status. An additional variable (c), defined as the credibility, represents the degree of confidence in the status, i.e., the degree to which the value of v can be believed. Hence if c=1, the value of v is fully believed; while if c=0, the value of v cannot be believed at all.

The initial status and credibility values are derived form raw sensor data. Each incoming signal (pressure flow etc.) is transformed through a simple mapping to generate the status in the range zero to one. This provides the initial indication of the degree to which the signal is a symptom of some fault. A corresponding credibility (c) is determined from the consistency of multiple readings of the same measurement using a voting technique. Other credibilities, such as to determine the consistency of symptoms with respect to a specific fault may also be generated.

The hierarchical structure can be considered as a nodal network in which lower level nodes propagate status and credibility information to higher level nodes (i.e., it forwards chains). Each node receives multiple status/credibility pairs and uses these to generate an output status/credibility pair. The transformation from input to output depends upon the functional attributes of that node. On the other hand, control and system configuration information flows downwards in a backward chaining manner.

FUNCTIONAL UNITS IN DICON

To adequately represent the diverse logical and physical structures associated with diagnosis and control, DICON has a number of node-types each of which perform a specific function. These node-types can be grouped into five functional categories as briefly discussed below.

o Signal Validation
The validation procedures compares the "distance" of each reading to all other like readings. An inconsistency index for that reading is computed based on the separation of that reading from the rest of the readings in the set. These inconsistency indices are then used to compute a "best" signal value and associated credibility.

o Diagnostics
Having generated validated signals, the fault diagnosis process can proceed. Diagnosis involves two steps: signal transformation and incremental evidence accumulation. Signal transformation was discussed above and involves mapping the signal on to the zero-to-one range to provide non-dimensional status and credibility data to the diagnostic analysis.

Fault diagnosis involves the incremental accumulation of evidence to determine the "health" of components. Evidence is accumulated both for and against the fault; if a piece of significant evidence is missing, the likelihood of that fault is accordingly adjusted. It is in the area of fault diagnosis that expert judgement is used. The diagnostic algorithm is based on Bayes' Theorem of conditional probabilities. In this sense it has the same mathematical foundation as the PROSPECTOR system [3].

o Components

Component nodes serve a central role in DICON since these are the entities upon which fault diagnosis is performed. The condition of higher level systems is fully determined by that of the components. Components can have a variety of operational states such as operating and standby.

Components can also have a variety of fault-related states such as OPERABLE and FAULTED.

All combinations of operational and fault-related states are allowed except that, while in maintenance, the component fault-states are irrelevant.

o System Integration

The top node in the hierarchy is unique in that it ties together the complete logical model. It integrates all the subtrees each one of which represents an operating mode of the plant (e.g., full power, shutdown). Besides this function, it has two other important roles. First it is in a sense, a nerve center since it is the meeting point for the control signals which allow DICON to determine the current operating mode. Secondly, it identifies the allowable state transitions that can be made. For example it a reactor is operating at full power, it might be moved to hot shutdown but not cold shutdown.

The hierarchical organization of these units is summarized schematically in Figure 1. The circular boundary around the Figure shows the conventional information flow between the plant and the operators. DICON evaluates the plant sexes or data in the context of the plant current operating configuration (the dashed line from "CONTROL ACTIONS" implies this context.

IMPLEMENTATION AT EBR II

Subsequent to its development at Westinghouse, DISYS was ported to the EBR-II site for continued implementation and testing under the US. Department of Energy, Man Machine Interface program.

<u>Modelling of the Argon Cooling System</u>

Preparations at EBR-II were made to allow testing by specially instrumenting a small safety related system, the Argon Cooling System (ACS) so that a number of tests could be performed. The ACS provides cooling for subassemblies as they are removed from the reactor and the primary tank and are placed into the Inter Building Cask for transport to the Fuels Examination Facility. In addition, the ACS provides for preheating capability for subassembly installation into the primary tank and reactor. The system consists of two turbine gas blowers, automatically operated valves to route flow, and system purifiers, heat exchangers, etc. to accomplish the system functions. A schematic of the ACS is shown in Figure 2.

The added instrumentation included valve position, system flow, system pressures in more than one location, system temperatures, heater controller status, current to the AC powered turbine blower as well as to the DC powered turbine, turbine vibration, bearing temperatures etc.

A model of the ACS consisting of nearly 300 nodes was generated for DISYS in cooperation with Westinghouse personnel. The program NOGEN was used to construct the model. NOGEN is part of the DISYS package and allows construction of a system model in a reasonable, top down design which permits a logical breakdown of the system. The system model was also diagrammed by Westinghouse using a CAD system for ease in visualizing changes to the model. A typical part of such a model is shown in Figure 3. The graphical representation of the model is important to understanding what DISYS is doing by the operator who may be using the program and it is especially useful for training.

The model created for testing included the operational modes of 1) system shutdown, 2) flow through the Fuel Unloading Machine only, 3) flow through the Fuel Unloading Machine and Fuel Transfer Port, and 4) flow through the Fuel Transfer Port only. DISYS provides for control nodes to be incorporated into the model. The control nodes allow DISYS to determine which mode the system is in. For example, certain valves must be closed with others open when the system is in a particular mode. The ability of the system to identify the operation mode is mandatory to prevent erroneous diagnostics from occurring.

As an example of the diagnostics approach, a description of the model used to determine that the AC turbine is operating properly is provided. A node, "Diagnosis of AC Turbine Failure" was provided as the top diagnostic node. Feeding information to the Diagnosis node are seven nodes which "map" signals to a value between 0 and 1. Below the map nodes are nodes which provide validation (i.e. determines that the signal could be within instrument and system possibilities), and below the validation nodes are the signals from seven sensors. The sensor signals are: AC turbine vibration, Turbine outlet temperature, Turbine outlet pressure, Argon flow, AC turbine supply current phase A, AC turbine supply current phase B, and AC turbine supply current phase C.

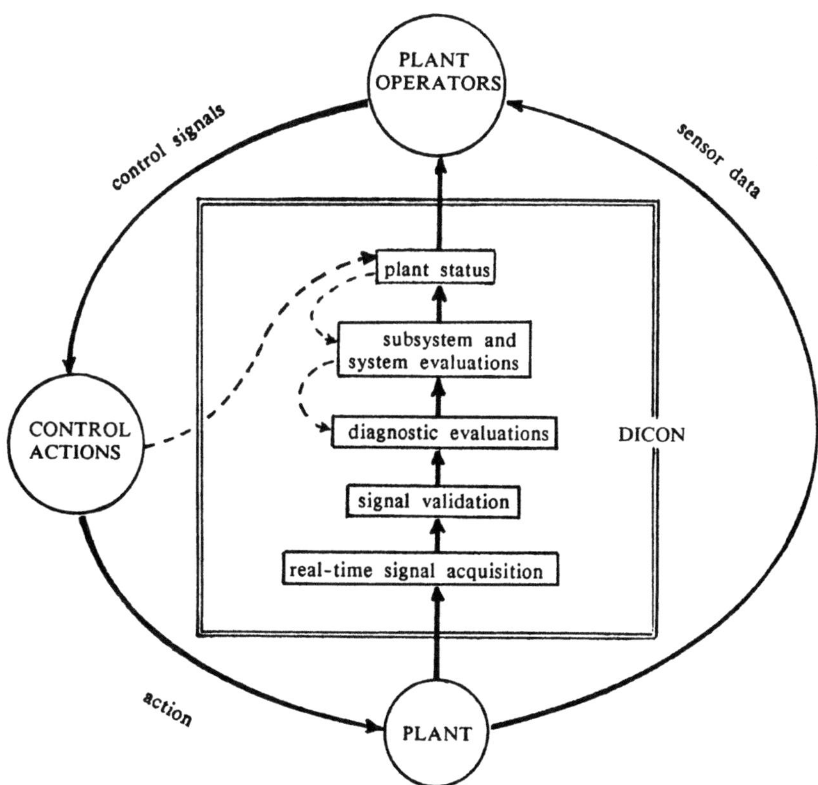

Figure 1   Integration of DICON in the Plant Environment

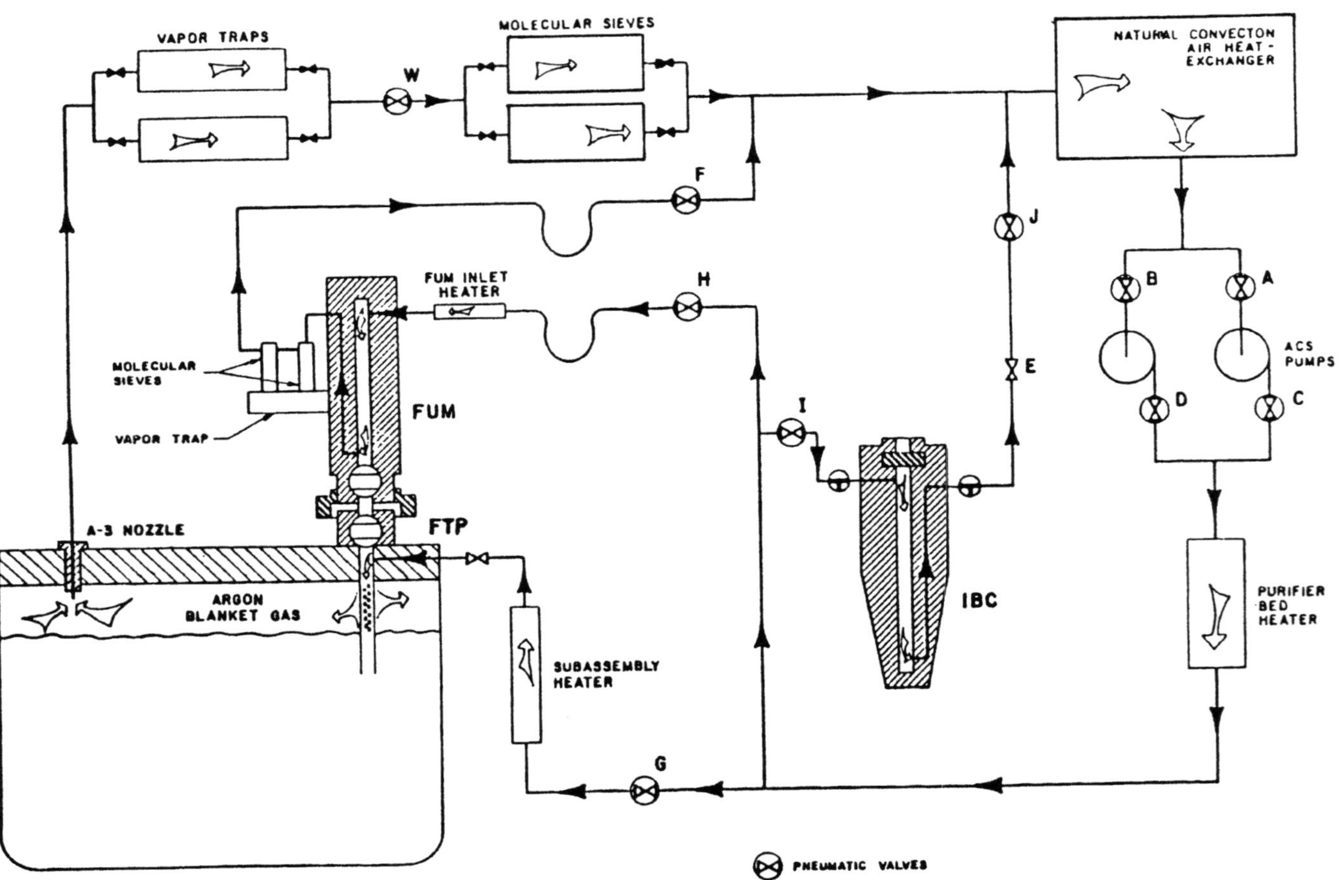

Figure 2   Schematic of the Argon Cooling System

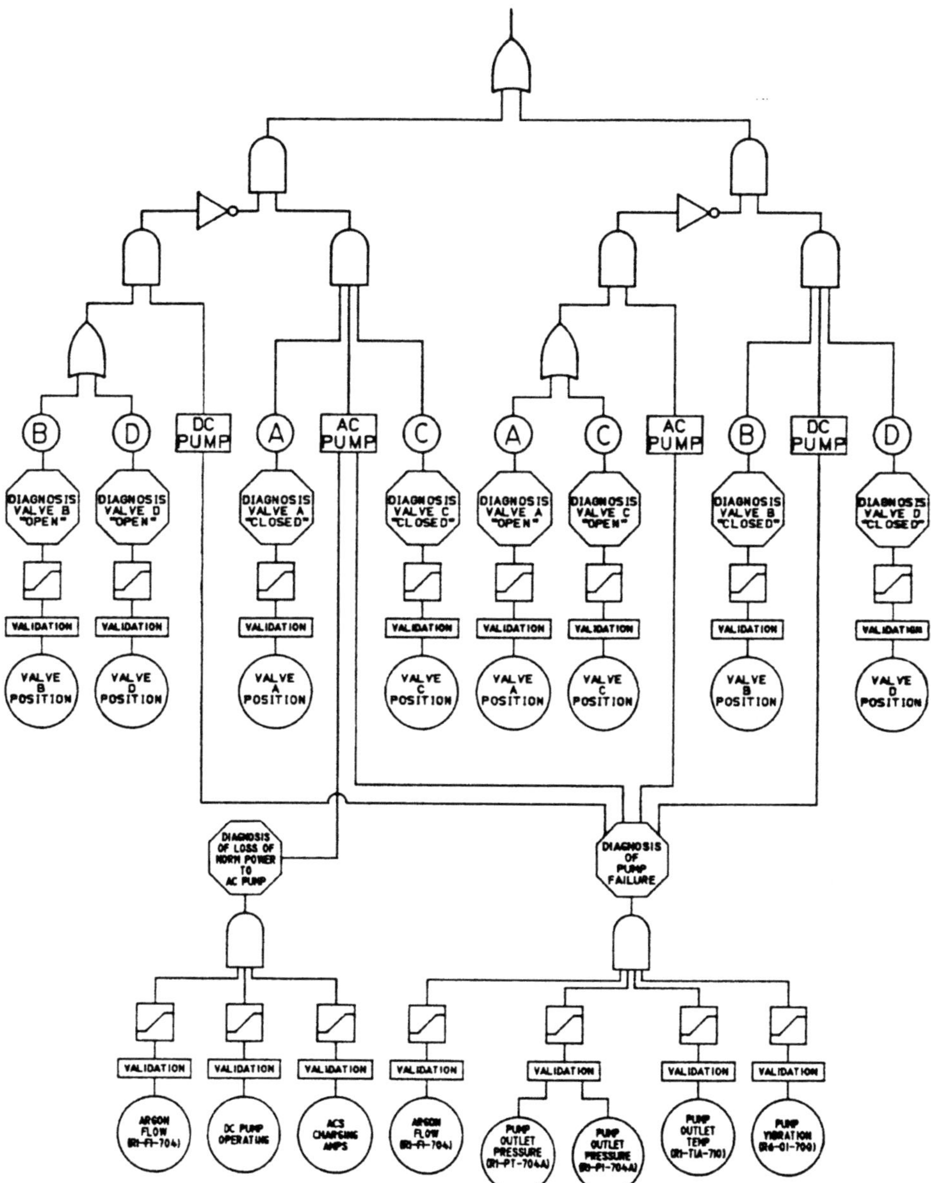

Figure 3 Typical Diagnostic Logic Element for the ACS

The diagnosis node provides the capability to weight the signals so that those signals that are the most indicative of a problem with the AC turbine receive the most weight. For example, if the three current sensors indicate that there is no power to the turbine, and there is no flow, the probability is very high that there is a turbine failure for some reason. A lower probability exists when the other signals are approximately normal but the turbine outlet temperature is high or low.

Because the ACS includes both an AC and a DC powered turbine, and the system operates with only one turbine on-line at a time, the node above the diagnostic nodes for the AC turbine and the DC turbine is a logical OR node. The use of logical nodes combined with control nodes and others including those mentioned, allow a system expert to arrange the analysis of a system such that it can be determined whether the system is operating properly. In the case of the logical OR, DISYS can suggest the operation of the alternate component if it determines that the component picked for operation has failed.

The use of fuzzy logic as previously described, allows for consideration of the cases where it is not possible to uniquely identify that a component is or is not failed. In the case of the AC turbine, an increase in turbine vibration and an increase in the outlet temperature but with no significant change in flow nor turbine current may signal that there could be some degradation in the turbine but is is still performing its design function. DISYS would inform the operator and provide a numerical indication as to whether there had been a failure. In the example given, the possibility of complete failure is considerable less than one.

<u>DISYS Testing and Evaluation</u>

Testing and improving DISYS is ongoing. DISYS testing consists of both formal tests where valves, etc. in the ACS are purposely misaligned, and informal tests where DISYS is used as normal operations are performed by the ACS during Fuel Handling. DISYS has been shown to work as designed, however there are several enhancements that are being made to the system to improve its operation and utility.

As DISYS was originally designed, diagnostics would continue until an anomaly is encountered. The program then stops the on-line diagnostics and prints the diagnostic messages to the operator. The operator must restart the diagnostics. This is obviously not appropriate for operation in a real system environment. Changes are being made to the program to allow DISYS to continue diagnosis even after detection of a system problem. The messages will be written to a file for the use of an operator and will come up to a display automatically in a hierarchical order. The major system function challenged or lost is displayed first, then with the supporting data showing where the failure has occurred.

Further enhancement of DISYS will be the inclusion of graphics so that the system can be displayed. When a problem with the system is encountered, color change, or highlighting the problem area will be provided. The alpha-numeric messages will still be provided so that the needed detail is available to the operator. The addition of graphics requires considerable revision of the DISYS program so that the graphics will improve the interface with the operator rather than detract from it.

Evaluation of DISYS is continuing. One consideration that seems to be emerging from testing is that a program such as DISYS requires considerable computer power. As computer power increases, it is obviously easier to provide diagnostics via computer, however with presently available computers, it probably would not be cost effective to provide diagnostics for every system and component in a large nuclear plant or other complex facility. That fact does not detract from the present use of such a program because it is not typical that all systems and components are fully instrumented in any case. Judicious use of DISYS to monitor systems that are of critical importance will allow enhancement in plant operation and certainly improve the ability of an operator to maintain surveillance of the plant systems and components.

Future Activities

Future additions to DISYS, in addition to including graphics, will include the incorporation of nodes not present or not active in the existing system. For example, a maintenance node that shows that a component is out of service for maintenance and which provides for that consideration in diagnosis is partially included in the DISYS program. The maintenance node is not complete and has not been activated Another node that is contemplated is a prognostic node. The prognostic node has also been included but not activated. Prognosis is visualized as being a prediction based on the rate of change of a parameter or a set of parameters.

As DISYS performs diagnostics and provides messages to the operator, the generation of the specific messages could also be used to trigger the display of either pre-planned procedures or even to provide the input of automatic procedure generators. For the near future, these additions will not be pursued because of manpower limitations and other priority work.

A modelling feature in DISYS allows for the synthetic generation of a parameter based on inputs from other parameters (such as providing a delta pressure from the inlet and outlet pressures). Further exploration of the utility of this node is warranted. In the case of failed instruments and/or lack of instrumentation, a more complex simulation should be possible to provide the desired input. An obvious limitation in the use of this node is that of computer speed limitation.

Conclusion

DISYS development is now complete and its implementation and testing at EBR II are now well along. Further work, primarily in the area of improved graphics, still has to be performed before the system can be considered fully operationally.

Acknowledgments

We would like to thank the U.S. Department of Energy for supporting the work described above under contract DE-AM02-76CH94000.

References

1. J. Rasmussen, "The Human as a System Component" in Human Interaction with Computers, Ch 3, pp. 66-96, Academic Press, New York, NY (1980).

2. L. A. Zedah "A Theory of Approximate Reasoning" in Machine Intelligence Vol 9, Ch 7, pp 149-210, Ellis Horwood (1979).

3. R. O. Duda, et al, "Subjective Baysian Methods for Rule-Based Inference Systems," in Proceedings of the National Computer Conference, 45, pp 1075-1082, American Federation of Information Processing Societies, Montvale, NJ. (1976)

ACCIDENT DIAGNOSIS AND PROGNOSIS AIDE (ADPA)

A. Dale Gunter and Robert A. Touchton

Technology Applications, Inc.
6621 Southpoint Drive North, Suite 310
Jacksonville, FL 32216

This presentation provides a demonstration of a proto-typical expert system developed by Technology Applications, Inc. (TAI) under a contract with the Department of Energy as a part of their Small Business Innovation Research Program. The Accident Diagnosis and Prognosis Aide (ADPA) Demonstration Prototype is a working scale model of a real-time expert system which:

o   Diagnoses an accident situation (as well as a number of underlying failures, events, and conditions deduced along the way).

o   Calculates the change in the likelihood of core damage as a function of the events and failures diagnosed.

o   Dynamically generates a recovery procedure tailored to the specific plant state at hand.

The bulk of the knowledge which supports the prototype was gleaned from existing studies and analyses taken from the mainstream of the nuclear plant community, especially Critical Safety Function Status Trees, Functional Restoration Guidelines (and their associated Background Documents), and Probabilistic Risk Assessments (PRAs). The system to be demonstrated will provide a hands-on interaction with the prototype. Supporting materials will be included which will help the viewer to understand the problems being solved and techniques being used to solve them.

APPLICATION SYNOPSIS

The purpose of ADPA is to improve operator productivity and effectiveness during off-normal or crisis situations. The ADPA Demonstration Prototype was developed to explore the technical feasibility and practicality of chosen AI techniques. The prototype consists of these modules: the Diagnoser, the Predictor, and the Recovery Advisor. The ADPA

environment includes a model of a scaled down nuclear plant
and the ability to simulate a sensor data stream on the hard
disk. Each of the three modules, in turn, has knowledge of a
narrow portion of its relevant domain. The Diagnoser contains
two of the six CSF Status Trees and can diagnose a few dozen
failure, events, and conditions of importance to the other two
modules. The Predictor contains two event trees and three
simplified fault trees and can instantly update and propagate
failure probabilities through the trees based on results
reported by the Diagnoser. The Recovery Advisor contains a
portion of an Inadequate Core Cooling procedure and can
present relevant "Normal Action" statements and "Response Not
Obtained" statements based upon results reported by the
Diagnoser.

<u>Diagnoser</u>

A rule-based paradigm proved to be an appropriate way to
codify the knowledge in the CSF Status Trees along with the
associated information contained in the background documents.
There are many diagnostic inferences that can be drawn from
the heuristics contained in these knowledge sources. The rule-
based paradigm was supplemented by an object-oriented
methodology for applying those heuristics which were tightly
coupled.

The CSF-based knowledge, while extremely useful, is not
sufficient to provide the diagnostic foundation required by
the Predictor Module and Recovery Advisor Module. The
capabilities of the Diagnoser Module had to be extended to
include conditions and events which were not related to
Critical Safety Functions, but which could be inferred from
plant sensors, manually entered observables, and other
conditions and events.

<u>Predictor</u>

A new and unique methodology was developed and tested
which allows for the real-time (re)calculation of the
estimated likelihood of core-melt as a function of plant
status. This methodology uses object-oriented programming
techniques that enable one to codify the event tree and fault
tree logic models and associated probabilities developed in a
PRA study. As soon as the Diagnoser determines that off-normal
conditions exist, the Predictor Module is able to update the
relevant failure probabilities throughout the event tree and
fault tree models by dynamically replacing the "off-the-shelf"
(or prior) probabilities with new probabilities based on the
current situation. The new event probabilities are immediately
propagated through the models (using "demons") and an updated
core-melt probability is calculated. Along the way, the
dominant non-success path of each event tree is determined and
highlighted.

<u>Recovery Advisor</u>

A new and unique methodology was developed and tested
which dynamically generates a recovery procedure tailored to
the current state of the plant. This methodology, which relies
upon object-oriented programming techniques, was found to be
appropriate for representing accident recovery knowledge and

expertise. For example, as soon as the Recovery Advisor is
notified that an Inadequate Core Cooling (ICC) condition
exists, it would begin assessing which of several ICC recovery
techniques is the best one to use in light of current plant
status. If all AC power were lost, the Recovery Advisor would
skip over the primary ICC recovery method because it relies
upon motor-driven pumps which would have now been rendered
useless.

The ability of the Recovery Advisor to access sensored
parameters directly was found to be significant because it
frees the operator from having to manually look up and verify
such information. With this new methodology, procedure steps
aimed at periodically checking to see if the operator still
belongs in the current procedure are continuously monitored by
the system. For example, in the written ICC procedure, the
operator is asked at strategic points throughout the procedure
to check whether the Core Exit Thermocouples have fallen below
700 degrees F (in which case the procedure has been successful
and should be exited). In ADPA, the Recovery Advisor is
constantly checking to see if the ICC condition has subsided,
thus, freeing the operator from having to check manually and
from having to wait until the next strategic point to find out
that the procedure should be exited.

The Recovery Advisor is, thus, able to automatically
generate a procedure on the screen which is focused on just
those steps which are germane to the current situation and
which either involve the verification of non-sensored
parameters or require explicit operator actions.

CONCLUSION

Within the bounds of the prototype scaling factors that
were selected, development of the ADPA Demonstration Prototype
has proven the technical feasibility of using certain AI
programming techniques to provide new and unique crisis
management tools. The nature of the technology chosen for
implementation also sheds favorable light on its economic
feasibility based on the modularity of the codification
techniques (making customization and evolution more
economical) and the ability to take advantage of existing
knowledge sources throughout the industry (such as a completed
Probabilistic Risk Assessment or a Utility Owner's Group set
of procedure background documents).

TAI is convinced that the technology is valuable to the
nuclear industry in terms of improved operator performance and
productivity. The ability of an expert system to quickly and
accurately diagnose an accident situation, provide insight
into how the accident might progress, and provide concrete,
reliable recommendations on the stabilization of the plant
will be of tremendous value to a power plant operator, the
nuclear industry, and the nation as a whole.

SEXTANT: AN EXPERT SYSTEM FOR TRANSIENT ANALYSIS

OF NUCLEAR REACTORS AND INTEGRAL TEST FACILITIES

N. Barbet, M. Dumas, G. Mihelich, Y. Souchet and J.B. Thomas

Département des Etudes Mécaniques et Thermiques
Service d'Etudes des Réacteurs et de Mathématiques Appli-
quées - Centre d'Etudes Nucléaires de Saclay
91191 Gif-sur-Yvette, Cedex France

ABSTRACT

Expert systems provide a new way of dealing with the computer-aided
management of nuclear plants by combining several knowledge bases and
reasoning modes together with a set of numerical models for real-time analysis of transients. New development tools are required together with
metaknowledge bases handling temporal hypothetical reasoning and planning.
They have to be efficient and robust because during a transient, neither
measurements nor models, nor scenarios are hold as absolute references.
SEXTANT is a general purpose physical analyzer intended to provide a
pattern and avoid duplication of general tools and knowledge bases for
similar applications. It combines several knowledge bases concerning measurements, models and qualitative behavior of PWR with a mechanism of
conjecture-refutation and a set of simplified models matching the current
physical state. A prototype is under assessment by dealing with integral
test facility transients. For its development, SEXTANT requires a powerful
shell. SPIRAL is such a toolkit, oriented towards online analysis of
complex processes and already used in several applications.

A. OBJECTIVES

The computer-aided management of complex processes (nuclear reactors,
integral test facilities) requires to simulate the human reasoning mode
for data driven problem solving. Large declarative knowledge bases must be
associated with a library of numerical models. Several expert systems are
under development in this area.

In addition, for real-time applications concerning control assistance,
specific constraints have to be satisfied. The main functions are :
- the estimation of the physical and functional state of the plant ;
- the identification of the normal or abnormal events (control, failures)
  which determine the evolution ;
- the prediction, the planning and revision of the next sequence of
  corrective actions.

During a transient, specific difficulties arise for performing the interpretative tasks :
- the measurements are uncertain and not well adapted to the plant state estimation in every situations ;
- the physical and functional models of the plant are no more reference models with the capability of running several sensitivity computations in real time for the needs of the analysis ; the dynamics of the plant can be highly non linear ;
- complete and consistent sets of diagnosis assumptions, beyond normal operating conditions are not currently available for the interpretation of physical events related to hypothetical failures of the plant, of the measurements, and of the models.

Therefore, the conventional tools of interpretation cannot be used. A new tool for "expert filtering", handling simultaneously in real time the uncertainties concerning the measurements, the models and the plant behavior, must be developed. SEXTANT is designed to reach this goal by developing and combining the knowledge bases and the model libraries.

B. PRINCIPLES

In order to accomplish its analysis, SEXTANT uses three types of informations :
- measurements and signals,
- plant models :
  . numerical models (physical behavior) : measurement and simulation ;
  . logical models (functional behavior) ;
  . operator's and designer's cognitive hypothesises on physical processes, providing qualitative interpretation and order of magnitude of the assumed causes.

The information is structured by SEXTANT in the following manner :
- the state of the system is represented by a state vector $\vec{E}$ (t), which moves in a state space partitioned in a grid of subdomains {Di} ;
- libraries of measurement models Ri and simulation models Pi depending on the subdomain Di are assembled.

The temporal reasoning process used in SEXTANT to determine the transient scenario performs belief management to select the interpretation of the basis of its credibility evolution.

The main principles are :
- time control : the time necessary for the analysis must be monitored and planned by comparison to the real-time evolution of the process ;
- several time scales are defined dynamically depending on the physical situation. For instance, windows are built up around identified events or observed drifts between predicted and estimated evolutions. The goal is to isolate such events and generate a set of local assumptions to be combined to the preexistent scenarios {Sk}. Long term recapitulation is performed on a larger time scale, in order to select scenarios when evidence becomes available, and to avoid proliferation of contradictory beliefs ;
- conjecture/refutation is the main mechanism of the process. It contains heuristics based on a hierarchy of criteria and on efficient simplifications. These simplifications are necessary because, in SEXTANT, neither measurements, nor models, nor scenarios are hold as absolute references, leading to a threat of branching phenomena.

At each time step, SEXTANT provides a best estimate state, $\vec{E}_{be}(t)$, which combines the results of the measurements and of the best prediction during the time step, and provides the set of updated scenarios {Sk(t)}.

C. CONTENTS

1. SEXTANT is a multilevel and modular system which contains several libraries of numerical models and declarative knowledge bases devoted to the previously described functions.

The induced specifications for the shell are :
- deep representation of the functional logic of the plant (frames) ;
- flexibility of the inference engine, combining forward and backward chaining ;
- communication with libraries of numerical models ;
- time modeling, task scheduling (agenda mechanism), interruption management for real-time communication with the physical process ;
- temporal belief management, with contexts or "viewpoints" capabilities ;
- general mechanism for planning.

This shell design matches advanced industrial applications involving complex processes and real-time problem solving. Corresponding implementations and cognitive methodologies are still under development.

2. Library of physical models. In a first step, SEXTANT has been oriented towards analysis and intelligent report of transients leading from normal operation to potential core uncovery in a post-trip situation.
- Measurement models : they work for the reconstitution of the system state.
- Predictive models for analysis : they are fast, the running time must be easy to predict, they must be condensed and adapted to the detail level of information available from the instrumentation and the detail level required for predicting key parameters.

At the present time, SEXTANT uses three analysis models for computing the primary system and the steam generator.

The models are generic. They are developed and assessed by comparison with existing computer codes and several systems (pressurizer, primary system, steam generator secondary side) in situations corresponding to their availability domain. The physical control parameters determining the evolution are represented by general terms and the object representation by schemata allows the definition of the precise controls (injection systems, leaks) depending on the system whose behavior is analyzed.

The parameters associated with the controls (for instance, flowrate and specific enthalpy) are determined by using demons with several options (fixed values, default values, computation aiming at a specified goal).

The basic model is a two-volume computation of mass and enthalpy balance and of pressure evolution, taking into account thermal desequilibrium.

Depending on the estimated state of the system, by using available measurements and previous computations, the fluid circulation in the systems is computed separately, with simplified, specialized models (pumps on, natural circulation under one or two-phase flow, condensation and reflux).

The computation modules (basic or simplified, equilibrium models) are represented as goals in the logic programming language, eventually called as variables whose instantiation is determined in the course of reasoning, according to the situation and the needs of the analysis.

3. Functional representation. It describes the main components and their functional logic. It will be gradually completed.

4. Diagnosis knowledge bases. Three types of diagnosis bases are used (figure 2) : two for the initialisation of the analysis, and one for the current time interpretation. At the present time, they are oriented towards post-trip transients analysis.

An early diagnosis rule base (D1) works after crossing threshold and aims at determining the ruling physical causes in terms of mass and energy balance, together with their order of magnitude, by comparing the slopes at the departure from the normal operation. The existence of a complete set of measurements and of definite physical states are beneficial to the early analysis.

A delayed diagnosis rule base (D2) takes place after stabilization of the effects of the safety systems. It is drawn from existing diagnostic logic charts. The conclusions are compared with the early interpretation. Multiple fault diagnostic is allowed and leads to several competing scenarios at the beginning of the interpretation. This step of diagnostic takes advantage of the significant differentiation that can be reached together with the stabilization of the effects of control.

A current time diagnosis rule base (D3) is under development. It uses an object representation based on schemata with classes and instances to describe the hierarchical structure of the assumptions. At the class level, they are linked to the identified ruling symptoms (concerning the mass balance, for instance). At the instance level, they are linked to the detailed causes.

The structure of a part from the steam generator rule base concerning mass balance symptoms provided by level and pressure measurement analysis is sketched in figure 1.

In some cases, parameters have fixed values, in other cases, they have to be determined by computation (ruling parameter fitting or multi-parameter optimization). The actions are automatically performed by demons.

5. Supervision knowledge bases. The rules performing the analysis time management and those performing the conjecture-refutation mechanism are assembled in a test version.

The principles of the temporal reasoning process are sketched in figures 2 and 3.

The interpretation can be initialized :
- from the departure from normal operation, where the early diagnostic takes place ;
- from any time in the transient by launching the process of current time interpretation.

There are primarily three time scales in the current time interpretation :
- a short time (depending on the slope of the measured and predicted evolution and other parameters) loop of estimation and best-estimated prediction ;
- according to a set of criteria, the comparison between the estimated and predicted states leads to close analysis windows around "events", to generate the assumptions linked to the ruling symptoms, to go backward, check by sensitivity computation, select a subset of assumptions and combine them consistently and economically with existing scenarios ;
- a recapitulation is performed on a larger time scale, by using a Reason Maintenance mechanism[2].

Objects
Assumptions
Assumptions /SG
Assumptions /models
Assumptions /system
Assumptions /measurements
Mass
Energy
(ruling symptoms)
excess of
Lack of
increasing flow in
decreasing flow out
increasing flow out
decreasing flow in
SG auxiliary feed
SGT Rupture (p1>p2)
Decreasing vapor flow
Increasing vapor flow
SG auxiliary feed
SGT Rupture (p1>p2)
tripped on ; increasing
tripped off ; decreasing

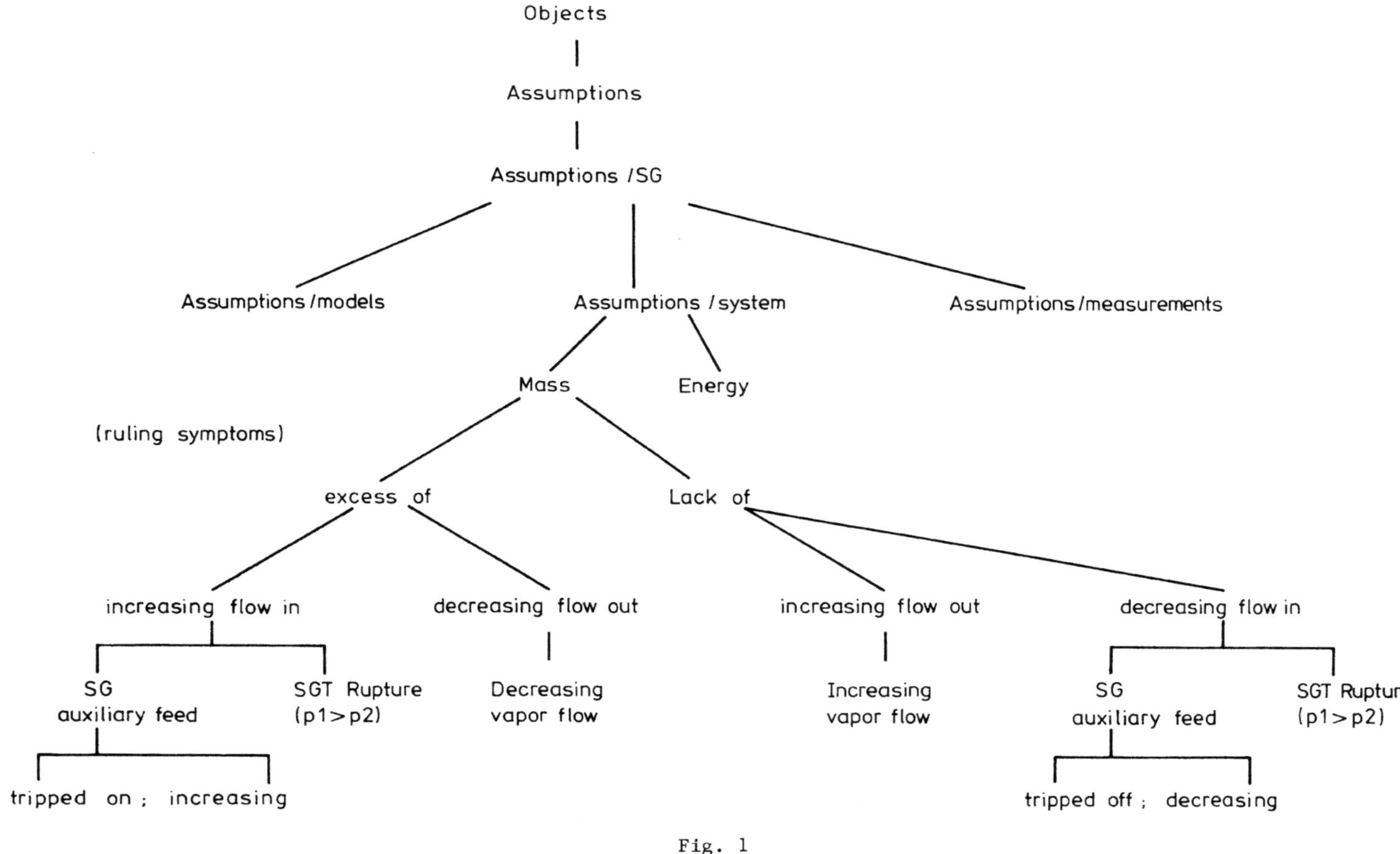

Fig. 1

E
(D1)
Δt_s (D2)
(D3)
t

Fig. 2

E
Sjo
Sj
Ee
H1
Sko
Sk
Δtc
Ebp
Sj1
t
Wn-1
Wn
Wn+1
Wn+2
Wn+3
Rℓ
Sk1

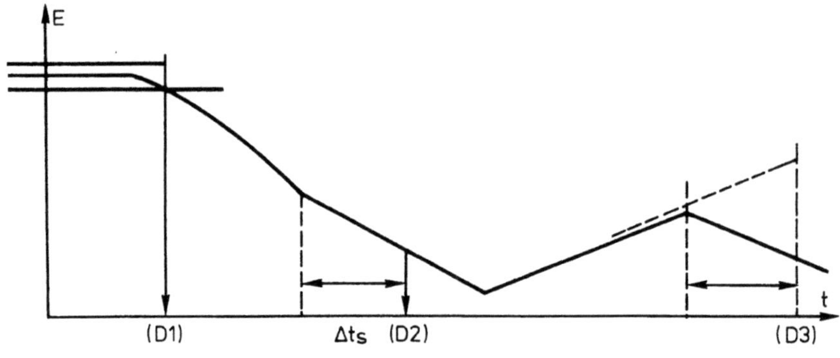

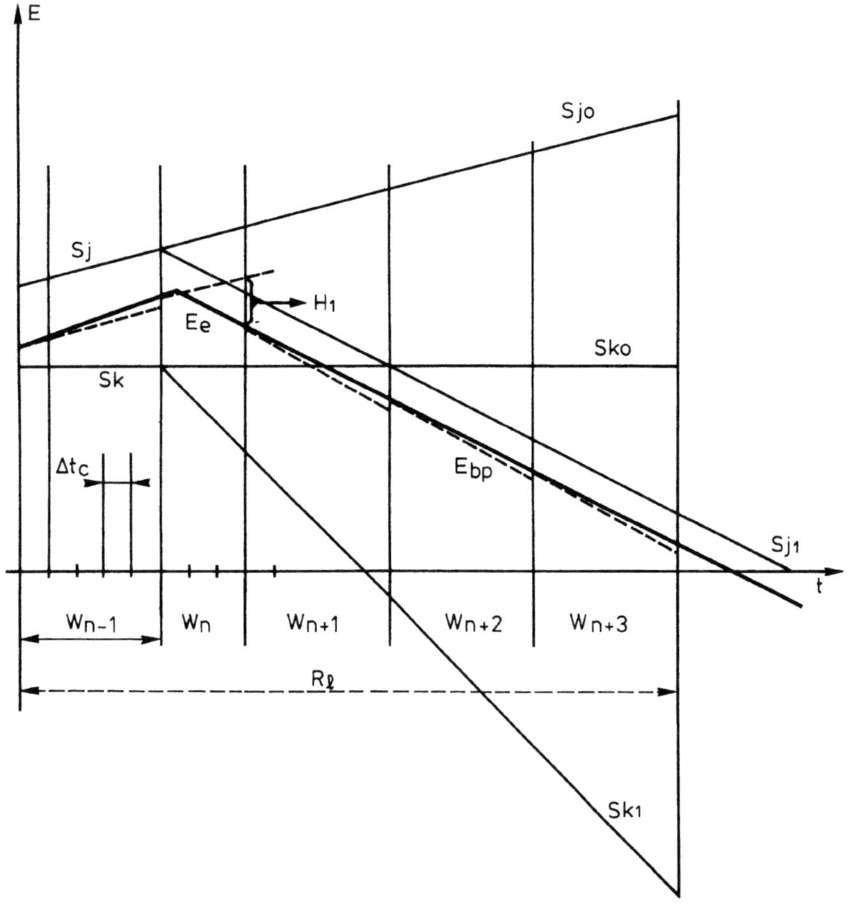

Fig. 3

D. ASSESSMENT, DEMONSTRATION

Two sides of SEXTANT have to be assessed :
- the ability of the models to follow a real transient ;
- the efficiency of the hypothetical reasoning in the case of complex
  transients.

They are under test on LOFT small break LOCAs with good preliminary
results.

A small break transient is analyzed from the beginning. The break is
identified and a first evaluation of the area is performed. The analysis
is going on with looking for break uncovery and the estimation of long term
primary system mass and energy balance. On the steam generator secondary
side, the system is able to identify and evaluate leaks at the steam valve.
Tests are performed by withdrawing informations from the system and
watching the interpretation process going on.

The models are assessed separately, the diagnosis rule bases,
according to their type, are checked whether analytically (the class
structure improves significantly the soundness of the construction) or by
integral tests.

The temporal reasoning process constitutes an autonomous level of the
system. The assessment is performed by tests on formal problems.

E. SPIRAL, A SHELL FOR THE DEVELOPMENT OF EXPERT SYSTEMS

SPIRAL is a shell for multipurpose expert systems developed in
CEA/DEMT. It provides cooperation between several modes of knowledge
representation :
- rules and facts in first order logic ;
- object-centered representation describing the system by classes ane class
  instances ;
- description of local reasoning mechanisms with rules.

The main inference mechanism of SPIRAL is a backward-chaining engine,
working by resolution, unification, backtracking, with extended reasoning
control possibilities. An agenda mechanism has been recently implemented.

This product also supplies a programm environment providing maintenance of large knowledge bases, handling files and using external numerical
models. A context handling mechanism and a forward-chaining engine are
under implementation in SPIRAL.

F. CONCLUSION

SEXTANT is designed for several application areas :
- analysis and report of test facility experiments ;
- computer-aided code assessment using an experimental date base ;
- computer-aided reactor control, either in normal or accidental transients ;
- interpretation and decision simulation for human factor and procedure
  analysis.

It will therefore be necessary to link complementary knowledge bases
specific for each application and developed in another connection. The
multipurpose use of such knowledge bases represents simultaneously an
advantage and a methodological challenge for the implementation of
industrial expert systems.

REFERENCES

1. J. L. Laurière, "Intelligence Artificielle", Eyroles (1985).
2. J. Doyle, A thruth Maintenance System, <u>Artificial Intelligence</u>, 12 (1979).

DEEP KNOWLEDGE EXPERT SYSTEM FOR DIAGNOSIS OF MULTIPLE-FAILURE

SEVERE TRANSIENTS IN NUCLEAR POWER PLANT

Robert P. Martin and Dr. B. Nassersharif

Department of Nuclear Engineering
Texas A&M University
College Station, Tx   77843-3133

ABSTRACT

TAMUS (Transient Analysis of MUltiple-failure Simulations) is a prototype expert system which is the result of a project investigating and implementing event confidence-levels (used by reactor safety experts in reactor transient analysis) in the form of an expert system. Currently, TAMUS is designed to diagnose reactor transients by analyzing simulated sensor and plant thermal hydraulic information from a system simulation. TAMUS uses a knowledge base of existing emergency nuclear plant operating guidelines and detailed thermal-hydraulic calculation results correlated to confidence-levels. TAMUS can diagnose a number of reactor transients (for example, loss-of-coolant accidents, steam-generator-tube ruptures, loss-of-offsite power, etc.). Future work includes the expansion of the knowledge base and improvement of the "deep-knowledge" qualitative models.

I. INTRODUCTION

Inherent to the control of complex systems such as nuclear power plants is the inability to systematically or mechanistically define methods for analysis, diagnosis, and mitigation of unanticipated transients. The application of established methods do not sufficiently emphasize the likelihood of lost or invalid data, the potential for multiple failure, and the need for quick response to the transient(s). Computer diagnostic and transient mitigation systems have yet to be installed at a plant in a position other than personnel training.

The state-of-the-art in Artificial Intelligence (AI) and Expert System technology has matured to a degree that the potential development of a computer aided/automated diagnostic and transient mitigation system can be considered. Since traditional methods cannot handle complex systems efficiently, AI techniques provide a means to "emulate" an expert reactor operator rather than follow mechanistic methods.

TAMUS (Transient Analysis of MUltiple-failure Simulation) is a confidence-level based expert system written in InterLISP on the Xerox 1108 computer system with new versions being developed in Common LISP for the SYMBOLICS 3640 computer system, the Texas Instruments (TI) Explorer, and the Sun microsystems machines. TAMUS analyzes incoming data from a nuclear power plant (or from a power-plant simulation), interprets the thermal-hydraulic data over consecutive time increments, determines the transient sequence, allows the operator to add relevant information asynchronously, and returns diagnosis and suggested remedy responses (taken from nuclear-plant emergency operating guidelines)[1]. On the Xerox

1108, the program takes advantage of the capabilities offered by KEE
(Knowledge Engineering Environment[2]). KEE's main role within TAMUS is to
assist in defining confidence levels to be used by the confidence-level
based inference engine written in LISP and provide graphics and user-
interface support. On the Symbolics,TI,and SUN systems, TAMUS is
completely LISP supported.

TAMUS was designed for the analysis of nuclear power systems, but it
is not limited to such systems. TAMUS uses a knowledge base of component
thermal-hydraulic conditions and transient knowledge base of component
thermal-hydraulic conditions to infer diagnosis and operating actions.
Component models and the knowledge base can be modified to model any
system.

Normal operation of a nuclear power plant is controlled by the plant
trip and control system. When an emergency occurs, an accurate and
thorough understanding of the state of the reactor is necessary to
diagnose and mitigate the transient because consequences could be severe.
For example, during the first few minutes of an event, as many as 100
annunciators can alarm in the control room. Reaction to this information
overload and the mechanical failure of certain sensors can lead to a
misinterpretation of the situation; consequently, a plant can be
permanently damaged and safety of the public may be at risk.

For a proper evaluation of a potential accident, it is necessary to
closely monitor the many components of a nuclear reactor system. Ideally,
an operator would want to know the exact state of the reactor at all
times. In cases of multiple-failure transients, diagnosis is more
difficult because the existing operating guidelines do not sufficiently
cover cases of multiple-failure transients. Often the failure of one
component causes the failure of another or interfere with mitigation of
the transient. To remedy this problem, a LISP machine could be interfaced
with the existing plant computer which collects plant information. An
expert system such as TAMUS could analyze many transients in real time.
Analysis of thermal hydraulic data is correlated against emergency
operator guidelines and a corresponding mitigating response is arrived at.
In complex systems such as nuclear power plants, allowances must be made
for lost or invalid sensor data. Therefore, special models are necessary
to validate sensor data.

A. LITERATURE REVIEW

REACTOR[3], a rule-based expert system also designed for analysis of
nuclear reactors, lacks the real-time and multiple-failure analysis
feature TAMUS has been programmed to perform. A similar expert system for
improving operator diagnostic ability has been developed by Wells and
Underwood[4,5]. The U.S. Nuclear Regulatory Commission (NRC)[6] has under
development an expert system called The Reactor Safety Assessment System.
Given parametric values, known operator actions, and the time sequence
information in the data to generate conclusions for assessed situations.
Also, much research in developing a sensor validation system is being
conducted at Ohio State[7].

A number of other applications of expert systems to plant operations
have been described. These include a program for probabilistic risk
assessment[8], operation analysis of the Savannah River reactors[9], and
automated monitoring of plant performance for the Oak Ridge National
Laboratory High Flux Intensity Reactor[10].

III. DESCRIPTION

The purpose of TAMUS is to monitor a "simulated" nuclear reactor
facility, detect deviations from normal operation as they happen, respond
with diagnosis of what is happening in the system, and recommend operator
actions for the transient.

Analysis in TAMUS is done by considering the thermal hydraulic conditions or physical status of reactor components. The program "understands" that missing or unknown information does not imply that it is not important. Events that may have a significant impact on the analysis at a future transient time are also included in such analysis.

TAMUS is a highly interactive and asynchronous program, the user can add relevant information to the knowledge base at any time (by moving the mouse to a mouse sensitive window on the screen). The main program calls a subprogram that backchains through each condition of each component searching for unknowns. The main program evaluates a new set of confidence-levels for the newly learned information and repeats the analysis for an "improved" diagnosis.

During multiple-failure transients, the occurrence of a second transient can override the conditions that identify a first transient that may have been present for some time. Such cases pose a particular problem for the program to accurately identify both transients. To solve this problem, a history monitor was developed. The monitor simply records the frequency of a given transient diagnosis. When the second transient appears, the monitor is called to judge if the first transient is being accurately described. If not, the history monitor overrides the decision made by the main program.

TAMUS in its present state is a preliminary program incorporating most of the presented ideas along with the addition of graphical display of data.

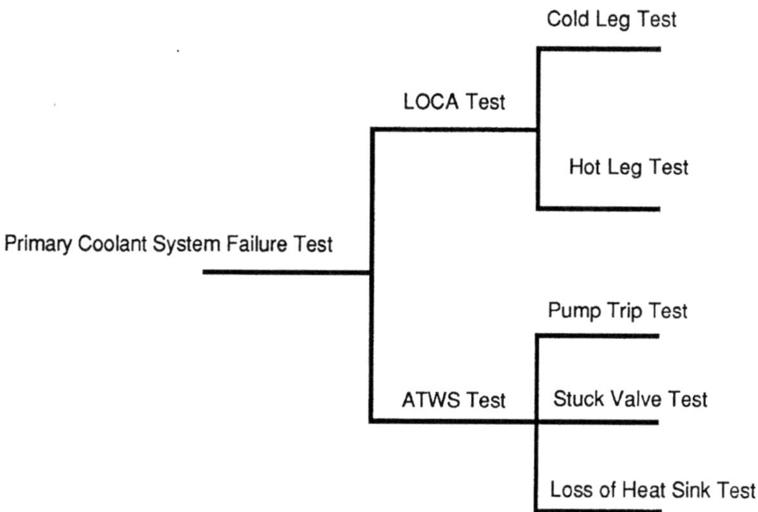

Figure 1. Example tree structure of transient knowledge base

A. KNOWLEDGE BASE STRUCTURE

TAMUS manipulates both a knowledge base of transient identifier patterns and a knowledge base containing a qualitative model of a nuclear power plant. The inference engine used by TAMUS uses the information stored in both knowledge bases to arrive at confidence level values that are used to infer particular plant states.

In the Common LISP version, the transient knowledge base is tree structured; while on the Xerox the transient knowledge base defines only one level of transient failure. At the top level of the tree structure general system failures are defined; these are classified as: coolant system failure, steam generator related failures, feedwater system failures, emergency systems failure, etc. The intermediate level of the tree structure defines an accident identification such as loss-of-coolant-accident (LOCA). The basic levels define specifics on the source of failure and give accident evaluation. This might included information such as the failed component and the best-estimate location of the failure. Inherent to each knowledge level are the thermal-hydraulic pattern that describe the particular event. Also included in the transient knowledge base are rule definitions necessary to distinguish fine details at the basic levels. Figure 1 shows an example of the tree-structured knowledge base used by the recent versions of TAMUS.

The component knowledge base is a qualitative model of the nuclear power plant system. Information given along with the component are the components that are immediately adjacent to the main component and the relevant thermal-hydraulic conditions attributed to that component during both transient and normal operation. During a transient, the inference engine uses the transient thermal-hydraulic conditions and knowledge base to make a diagnosis.

B. CONFIDENCE LEVEL ASSESSMENT

A confidence-level is the normalized confidence (0-1) that an "expert" expresses in a specific transient pattern describing the actual transient. The confidence-level value for each transient is determined by examining the states of every relevant system in a nuclear reactor. When a transient occurs, it can be identified by an "expert" from the pattern made by the thermal-hydraulic states on reactor systems. In otherwords, in a given reactor system similar thermal-hydraulic sequences will occur every time a particular transient occurs; thus, the problem of transient analysis is one of pattern recognition. For the confidence-level assessment method, a list of all reactor systems and their states are made for a particular transient. Since transients do not follow the exact path every time they occur, system analysis must concentrate on general transient trends, rather than exact numbers. Therefore, states are given in symbolic form that describe the general trend of the system. For example, when a LOCA occurs, the pressure gradient in the reactor cooling system (RCS) is always DECREASING; thus the "system-state" pair is [Pressure.Gradient RCS DECREASING].

Assigned to each "system-state" pair in the transient list is a number that represents a weighing factor to be used with other weighing factors in the transient list to give an overall confidence level in the transient diagnosis. These factors are determined by applying a fuzzy set theory similar to Bayesian theory[11]. Given the list of conditions that apply to a particular transients, a value judgement is made on the importance of a certain condition in the diagnosis of the transient by extensively analyzing actual plant and code-calculated data (e.g., from TRAC[12]). By comparing detailed transient simulations[13-15] against heuristic results, estimates of the confidence-levels can be obtained. Interviewing experts in reactor safety could result in improved values for the confidence-levels, provide additional parameters, and improve our qualitative models.[16] When a single transient occurs, the sum of all the weighing factors is equal to 100%. Accurately determining individual weighing factors is, therefore, the most important part in developing a successful transient analysis program by this method. The physical meaning of the weighing factor is the percentage that the pattern represented by a transient is satisfied by a given "system-state" pair. In estimating the weighing factor value, a thorough understanding of the transient is necessary. For example, if one considers a LOCA transient, relevant system-states that identify a LOCA are the following: Reactor Coolant System (RCS) pressure, RCS pressure gradient, RCS pressure rate of change, RCS temperature, RCS temperature gradient, and containment vessel pressure and radiation (of course, there are others that are relevant, but for this example a simple model will suffice). The data from the individual systems must now be ranked according to usefulness in diagnosing the transient. In this case RCS pressure and RCS pressure gradient data are the most valuable because the occurrence of a LOCA immediately affects these two quantities. From this point these two

quantities can be given an initial weighing factor value. The other
system-state pairs can also be given values relative to each other once
the "rank of usefulness" is determined. Factors that determine the
usefulness of system data are: response time to a transient, availability,
and uniqueness to a transient.

It is entirely possible that a component in a certain state can
contribute to the identification of more than one transient.
Consequently, greater information on conditions of other parameters may be
necessary to identify the transient. If all of the conditions that
identify the transient occur, a confidence-level of 100% is assigned to
that diagnosis. If certain conditions are unknown to the system, the
confidence-levels are calculated so that weighing factors are normalized.
In this way, unknowns are not completely ignored, they are just removed
from the list describing a transient. Confidence levels associated with
each parameter condition at a certain time are added together to arrive at
an overall confidence-level for a particular transient.

C. INFORMATION REDUCTION

During the operation of nuclear power plants or their numerical
simulation, much data is produced describing current conditions of every
subsystem and component inherent to the plant. Information is generated
in the form of thermal-hydraulic data (pressures, temperature, etc),
annunciator status (*i.e.*, on/off status, etc.), and sensor data (radiation
levels, power levels, etc.). When a transient occurs, an operator
experiences an overload of information that can be difficult to diagnose.
An experienced operator or analyst can diagnose a transient effectively by
examining the "proper" gauges and annunciators that will give him the
information essential for diagnosis and mitigation of the transient.
Likewise, for an expert system designed to diagnose reactor transients, it
must know what essential data to reference. The information load is great
even for the computer; therefore, the expert system must be designed for
sequential deduction, similar to the operator.

The strategy for maximizing the information from a minimum number of
data sources is to concentrate on data that is unique to a transient and
is most often available. By analyzing transient calculation data, such as
that produced by TRAC, an understanding of the essential data needed for
diagnosis can be established. Since the expert system is designed to be
real-time, it can increase the amount of information that can be learned
from plant data by analyzing change over a length of time. Once the
essential data sources are determined, the method of sequential deduction
can be followed. This method starts by addressing a set of "primary"
systems data, when a transient occurs, the "trouble area" is diagnosed
further by addressing a set of "secondary" systems data; this can continue
until an adequate diagnosis has been made. By using this method, the
objective of diagnosing both transient and failure location can be
accomplished simultaneously.

D. DIAGNOSIS AND MITIGATION

TAMUS uses object oriented programming to perform diagnosis. In
analyzing transient conditions, a qualitative model of the nuclear
reactor's thermal hydraulics is developed. By using a first order
approximation to parameter gradients over transient time, symbolic data in
the form of HIGH, LOW, DECREASING, INCREASING, NORMAL, STABLE, ON, or OFF
is inserted into a blackboard in the component knowledge base associated
with both the component and it's thermal-hydraulic characteristic
(pressure, temperature, pressure gradient, etc.) or operation status.
Once all or a subset of the conditions over an interval are inserted into
their specific slots, analysis of the transient begins. For pressure and
temperature data from the coolant system, a least squares fit is applied
to the last three successive data points. Then new pressure and
temperature data are extrapolated from the least squares data. All
numerical data from simulation is converted into a symbolic form. For
example, if pressure in the primary coolant system (PCS) drops from 2250
psia to 2235 psia in one second, the program converts the data to a

pressure that is LOW (with respect to the normal pressure of 2250 psia), a pressure gradient that is DECREASING, and a pressure rate that is QUICKLY. These values are then inserted into their respective slots in the PCS subclass of the class of REACTOR COMPONENTS. Given all the relevant component data, top level transient analysis is performed by confidence-level-based inference engine. If the conditions for a failure exceed a confidence level of 70% (a number that seems to work well), it is reported on the screen.

Figure 2 shows a typical screen display. If certain sensor information is not available, TAMUS uses existing information for diagnosis. In such cases, the result of the analysis may report that there is more than one possible transient or it produces a lower confidence-level in the diagnosed transient. TAMUS can then "ask" for the "critical" information necessary to increase the confidence-level in the final diagnosis. Mitigation is performed when a diagnosis has been made. Given the diagnostic conclusion, mitigation procedure is given from a rule-based knowledge base currently condensed from operator guidelines.

IV. ASSESSMENT SIMULATIONS

Three simulations have been tested to demonstrate TAMUS's abilities. The first simulation described a simple single-failure transient tripping a second at a certain time during the simulation. In this case a loss-of-coolant-accident (LOCA) occurred. The break size was defined by rules specifying it as one of three: SMALL, MEDIUM, or LARGE. In this case it was a MEDIUM LOCA. As expected, the plot of pressure in the RCS dropped and the analysis reported that a MEDIUM LOCA was occurring. When the pressure dropped below 15.16 MPa (2000 psia), a REACTOR TRIP occurred (physically, this was the insertion of the control rods into the reactor core to terminate the chain reaction). At this time, TAMUS successfully reported the REACTOR TRIP, while at the same time reported that the MEDIUM LOCA still existed. This simulation demonstrates TAMUS's ability to analyze two overlapping transients for the sample case and the ability to remember previous significant events (typically, at the time of the reactor trip, the reactor pressure resembles one in the state of a large-break LOCA, rather than the medium-break LOCA).

The second simulation exhibited TAMUS's ability to handle many unknowns. Given only primary-system pressure and temperature data, many transients, depending on the state of the reactor, were possible. In the final analysis, TAMUS correctly reported that the transient could be any combination of a loss-of-offsite-power (LOSP), steam-generator-tube-rupture (SGTR), or loss-of-feedwater (LOFW). All of which were possible transient conditions with the given data. Additional information was necessary to resolve the conflict.

The third simulation involved a combination of the previous simulations. States of certain components were unknown, while significant events had to be remembered. Generally, this simulation showed TAMUS's ability to handle a greater variety of transient situations. In this case, the initiation of the high-pressure-injection system (used to inject emergency cooling water into the RCS) showed up as a notch in the smooth curve of the RCS pressure. TAMUS had to recognize this to report a correct analysis. The result of this simulation was a SGTR (in contrast to a small-break LOCA having similar characteristic on the thermal-hydraulic level, but significantly different).

VI. PLANS

TAMUS is only a prototype. Presently, a XEROX 1108 is required to perform both the signal processing and the transient analysis. The signal processing (analysis of thermal-hydraulic data) is CPU intensive. This problem is planned to be remedied by interfacing a second computer (see figure 3) to run a reactor simulation and convert the numeric parameters to symbolic parameters. Numerical or symbolic data will be directly mapped to specific memory locations or the respective knowledge base memory slots in a LISP machine running TAMUS. A third computer will be added to act as an expert knowledge base for mitigating operator actions. This computer will take the results from TAMUS and reactor parameters from the simulation computer and determine operator actions from it's rule-based knowledge base to remedy an accident and relay the result to the operator. The advantage of detailed simulation is that it provides an automated testing and verification system for the diagnosis program. Therefore, the diagnosis program can "learn" from the detailed physical model and expand its knowledge base and improve the confidence-level in the final diagnosis. The history monitor will also be expanded to handle other situations that might benefit from it's abilities. Immediate plans include interviewing and working with industry and government nuclear reactor safety experts to expand the nuclear reactor qualitative model and transient knowledge base and to begin the design of the rule-based knowledge base for mitigating operator actions.

VII. CONCLUSIONS

The application of AI to engineering problems is a virtually new field with great potential. By producing an innovative AI technique for nuclear power plant operation and control the state-of-the-art of theoretical artificial intelligence (AI) concepts has been advanced by applying it to improve the operation and control of complex engineering systems such as nuclear power plants.

Application of a real-time diagnostic expert system for nuclear reactor operations represents a significant application of artificial intelligence techniques. Real-time analysis along with the integration of event-oriented diagnostic strategies provides a powerful combination for mitigating transients in nuclear power systems. Since rule-based systems are typically CPU intensive, real-time analysis is difficult, if not impossible. By using a confidence-level assessment method, real-time analysis is possible.

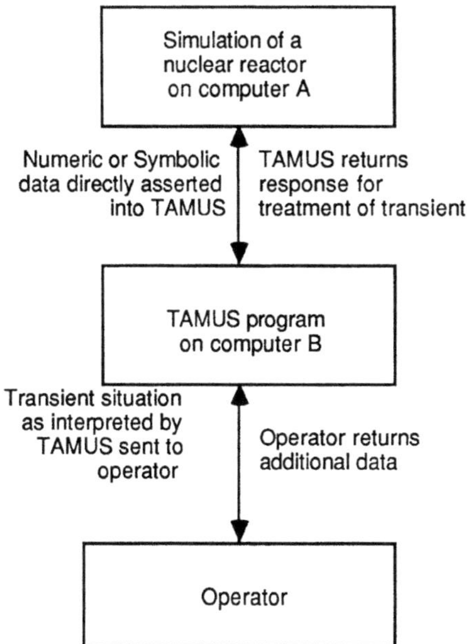

Figure 2. An improved TAMUS design

The use of a confidence-level assessment method for transient diagnosis has shown to be quite capable of handling most transient situations. Confidence-levels are derived from event-oriented descriptions of each transient in a nuclear reactor model. These descriptions consist of a list of thermal-hydraulic parameters, the state of each parameter,

and a confidence-level that this condition alone identifies the transient.
By adding individual confidence-levels, an overall confidence-level is
acquired. This value is compared against a minimum confidence-level, and
the transient most likely occurring in the system at that time is
diagnosed.

The development of this system presents an innovative approach to
nuclear plant operations. A real-time computer automated diagnostic aid
could potentially mitigate incorrect diagnosis of transients. It could
also serve in operator training or college level courses demonstrating
diagnostic reasonings or strategies, plant-operator interactions, and the
behavior of nuclear systems under transient conditions.

In summary, TAMUS is a prototype that applies expert system
technology to the operation of nuclear power plants and shows great
potential for automated real-time accident diagnosis.

REFERENCES

1. Westinghouse Owners Group, _Emergency Response Guideline Seminar_,
    Westinghouse Electric Corp., Sept. 1981.
2. _Knowledge Engineering Environment_, Intellicorp, ver. 2.1, 1985.
3. W. R. Nelson, _Reactor: An Expert System for Diagnosis and Treatment
    of Nuclear Reactor Accidents_, AAAI, August 1982.
4. A. H. Wells and W. E. Underwood, "Knowledge Structures for a Nuclear
    Power Plant Consultant," Trans. Am. Nuc. Soc. _41_, 41 (1982).
5. A. H. Wells and W. E. Underwood, "Reactor Operator Diagnostic Ability
    Training Using Knowledge-Based Computer-Aided Instruction," Trans.
    Am. Nuc. Soc. _46_, 42 (1984).
6. M. A. Bray, D. E. Sebo, and B. W. Dixon, "Reactor Safety Assessment
    System--A Situation Assessment Aid for USNRC Emergency Response,"
    Expert Systems in Government Symposium, edited by Kamal N. Karna
    (IEEE Computer Society, 1985).
7. D. W. Miller, B. Chandrasekaran, and D. Sharma, "Intelligent Process
    Control Operator Aid -- An Artificial Intelligence Approach",
    _Proceedings of the Sixth Power Plant Dynamics, Control and Testing
    Symposium_, April 1986.
8. B. W. Dixon and M. F. Hinton, "Reviewing the Development of an
    Artificial Intelligence Based Risk Program," Trans. Am. Nuc. Soc. _50_,
    291 (1985).
9. J. E. Suich, "Logic Programming for Operational Analysis of the
    Savannah River Reactors," Trans. Am. Nuc. Soc. _50_, 293 (1985).
10. P. J. Otaduy, "Demonstration of Expert Systems in Automated
    Monitoring," Trans. Am. Nuc. Soc. _50_, 298 (1985).
11. G. Shafer, A. Tversky, "Languages and Designs for Probability
    Judgment", _Cognitive Science_, 9, pp. 309-339, 1985.
12. Safety Code Development Group, "TRAC-PF1/MOD1, An Advanced Best-
    Estimate Computer Program for Pressurized Water Reactor Thermal-
    Hydraulic Analysis," Los Alamos National Laboratory report, LA-10157-
    MS (NUREG/CR-3858).
13. J. F. Dearing, R. J. Henninger, and B. Nassersharif, _Dominant
    Accident Sequences in Oconee-1 Pressurized Water Reactor_, NUREG/CR-
    4140, LA-10351-MS, April 1985.
14. B. Nassersharif, _Alternate Steam Generator Tube Rupture Mitigation
    Strategies for the Three Mile Island Unit 1 During a Loss-of-Offsite
    Power_, LA-UR-85-182, 1985.
15. N. S. DeMuth, D. Dobranich, and R. J. Henninger, _Loss-of-Feedwater
    Transients for the Zion-1 Pressurized Water Reactor_, NUREG/CR-2656,
    LA-9296-MS, May 1982.
16. M. A. Meyer and J. M. Booker, _Sources of Correlation Between Experts:
    Empirical Results from Two Extremes_, NUREG/CR-4814, LA-10918-MS,
    April 1987.

DIGITAL DATA MONITORING DISPLAY AND LOGGING -

AN APPROACH TO INTELLIGENT DATA DISPLAY AND CONTROL

Edward P. Ficaro and David K. Wehe

Department of Nuclear Engineering
University of Michigan
Ann Arbor, MI   48109

INTRODUCTION

A digital data acquisition system for monitoring plant variables has been designed and implemented at the University of Michigan's Ford Nuclear Reactor (FNR), a 2 Megawatt, open-pool, research reactor. The digital data provided by this system will be useful for :

* closed loop control,
* real time experimental calculations,
* advanced simulation-as-knowledge techniques,
* improved operator training, and
* expert system applications.

The purpose of this paper is to discuss the transition to the digital data world and the anticipated applications and benefits.

THE FNR AND THE MOTIVATION FOR THIS WORK

The Ford Nuclear Reactor is one of the oldest research reactors in the U.S. and is still heavily used for the production of radiation for experiments, production of irradiated materials to the commercial sector, and student and reactor operator training. In this latter capacity, the reactor serves as a testbed for both students and the commercial reactor operators to study reactor plant kinetics and control.

As part of a major DOE-University initiative, the University of Michigan has been working on developing and applying theoretical advances in artificial intelligence to reactor plant diagnosis. This work has involved using simulation as a form of physical knowledge to diagnose possible event causes. It is recognized that it would be advantageous to test these techniques in an actual environment. A research reactor offers this potential under simpler conditions than would a commercial power plant. For this reason, and for the numerous tangential benefits, we are implementing a digital data acquisition system at the FNR.

Table 1. Selected Analog Signals

Parameter	No.	Sensor	Signal Type	Requirements of Interface Module
Temperatures (F)	8	RTD	Floating, 1 - 5 VDC	Grounded Buffers, and Filtering
Shim and Control Rod Positions (in.)	4	Variable Resistor	Floating, 0-0.5 VDC	Gnded Buffers, Filtering and Amplification x 10
Flow Rates (gpm)	3	dp Cells	Grounded, 1 - 5 VDC	Buffers and Filtering
Radiations (mr/hr)	7	NaI, G-M and a GAD	Grounded, 0-50 mVDC	Buffers, Filtering, and Amplification x 100
Power (MW) and Period (s)	2 1	Comp. Ion Chambers	Grounded, 0-10 mVDC	Buffers, Filtering, and Amplification x 500
Neutron Flux	2	Fission Chambers	Grounded, 0-10 mVDC	Buffers, Filtering, and Amplification x 500
Conductivity (mhos)	3	Conduct. Cell	Grounded, 0-10 mVDC	Buffers, Filtering, and Amplification x 500

THE SIGNALS CONSIDERED

With visions of an eventual expert system, it was decided to monitor all of the major plant information that is available to the reactor operator. This information can be split into two categories, analog and digital. The analog system measures the dynamic responses of the continuous reactor variables, such as primary and secondary flow rates, heat exchanger inlet and outlet (primary and secondary) temperatures, bulk pool temperature, positions of control rods, flux and power levels, and building radiation values. Table 1 lists the analog signals available for acquisition. The digital system displays the status of the discrete parameters, such as rod magnet contact, automatic control on/off, and rundown and scram occurrences. Table 2 gives a partial list of the 24 digital signals incorporated into the acquisition system.

With the conversion of these signals to digital information acceptable to the computer, one can easily devise an overall plant status picture (i.e., a computer monitor display) and ideally have the computer rationally control system plant variables as an operator would. On a limited budget of about $5000, the transition to the digital world can be difficult. In this paper, we show, for the signals presented above, how to accomplish this and some of the pitfalls which can be (and were) encountered.

Table 2. Partial List of Selected Digitals

Control Switches	Annunciators	Scram Signals
Cooling Tower Fans(3)	Automatic Control On	High Building Radiation
Primary Pump	Automatic Rundown	High Power, No Flow
Secondary Pump	Reactor Scram	Shim Range Defeated
Hot Demineralizer Pump	Pool Level	

THE DATA ACQUISITION SYSTEM (EDAC)

The analog data monitoring system developed for the FNR
consists of an IBM AT compatible computer, an analog-to-digital
(A/D) converter, two 16-channel multiplexing cards, and an
isolation /signal conditioning module for each signal. The
primary objective in selecting and designing the components for
this system was to acquire the fastest and most accurate system
for under $5000.

The computer used in EDAC has a 16-bit microprocessor
running at 8 MHz with 1 Mbyte of RAM and a 30 Mbyte hard disk.
Since data would be acquired over long periods of time at a
fairly quick rate, a machine that could process the data and
write it reliably to disk in the time allotted was a major
concern. For this reason, a personal computer with a large
amount of RAM was needed for the storage and manipulation of
data, and a large hard disk with a fast seek time (35 μs) to
save the data expeditiously.

The A/D converter selected is the Metrabyte Corporation's
Model Das-8.   It is an 8 channel, 12 bit successive
approximation A/D converter with an average conversion time of
30 μs. This provides the capability for sampling at rates
approaching 33 kHz. With a full input scale of $\pm$5 volts, a
resolution of 2.44 mV is achieved. This plug-in, half-slot, PC
board is also equipped with three countdown counter/timers, one
of which is connected to the CPU clock in the computer. When
functioning as timers, they can be used in conjunction with the
interrupt handlers provided on the card to generate interrupts
signaling the start of an A/D conversion. If used as counters,
they can accurately be used as scalers for such instruments as
gas filled detectors and fission counters. In addition to the
above, seven bits of TTL digital I/O are available, composed of
4 bits of output and one input port of 3 bits.

With the use of the Metrabyte Model EXP-16 expansion
multiplexer, each of the Das-8 channels can be multiplexed to
accept 16 differential channels. This provides the capability
to monitor 128 different signals which are selected by using the
4 bit digital output port on the Das-8 card. The Exp-16 cards
also provide cold-junction compensation for thermocouples,
signal amplification, and low pass filtering. However, the
amplification is selectable for the entire card not for each
individual signal, and the filtering is modest. For this reason
and the fact that the Exp-16 card did not suitably isolate
(there were potential feedback paths) itself from the FNR
system, special isolation/conditioning modules had to be
designed for each individual analog signal.

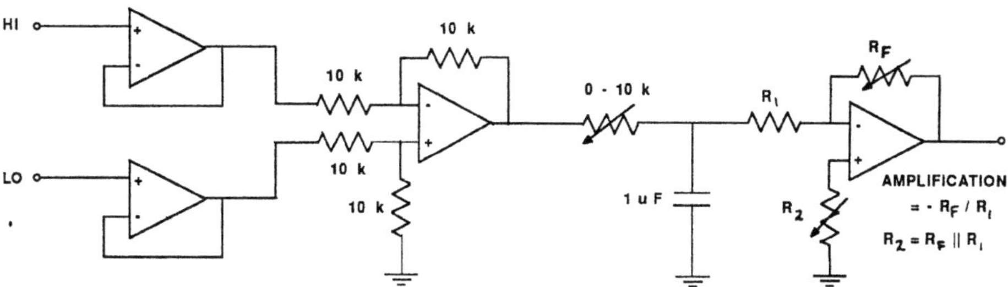

Figure 1. Isolation/Conditioning Schematic for Grounded Signals

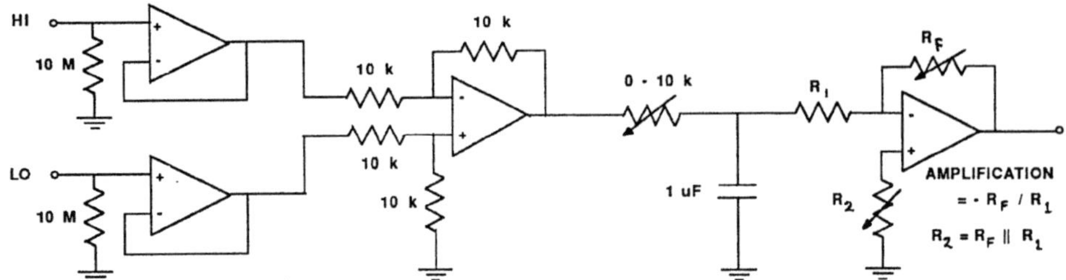

Figure 2. Isolation/Conditioning Schematic for Floating Signals

Our most important concern in interfacing a computer to a sensitive safety system was to insure we did not affect the system's signal recorders by attenuation or feedback. Therefore, isolators in the form of operational amplifier (op amp) voltage followers were used to isolate EDAC from the reactor system. Figures 1 and 2 show the schematic diagrams for the isolation/conditioning modules needed for the different signals (cf. Table 1) present in the FNR measurement system.

Since grounded signals possess a common ground with the computer system (assuming the computer has a common ground with the reactor instrumentation), they can be inputted directly into the buffers (voltage followers). The floating signals, however, did not have a common ground with the FNR instrumentation; therefore, a resistor had to be added to each buffer to establish a ground reference. In order to limit the effect of the pathway developed from these referencing resistors, 10 megaohm resistors were used. This resulted in the desired negligible feedback and attenuation. From that point, both circuits work identically. The differential voltage from the followers passes through a unity gain differential amplifier producing an inverted, single-ended, grounded signal. This signal is filtered to reduce the noise (>60 Hz) and amplified into the 0 to 5 volt range where the full resolution of the A/D conversion can be obtained.

Our experience indicates that three factors are particularly important in designing the computer acquisition system interface: ground loops, shielding and op amp specifications. Grounding problems are most prevalent with floating signals, but if the computer system and power supplies do not have a common ground with the measurement recorder system, a ground loop between the two systems occurs, and inaccurate results arise.

The second concern, shielding, is most important for signal clarity. When cables are run over long distances, the cables should be shielded to prevent RF interference and crosstalk between adjoining cables. By properly grounding the shielding, one can significantly improve the clarity of the signal as shown in Figure 4.

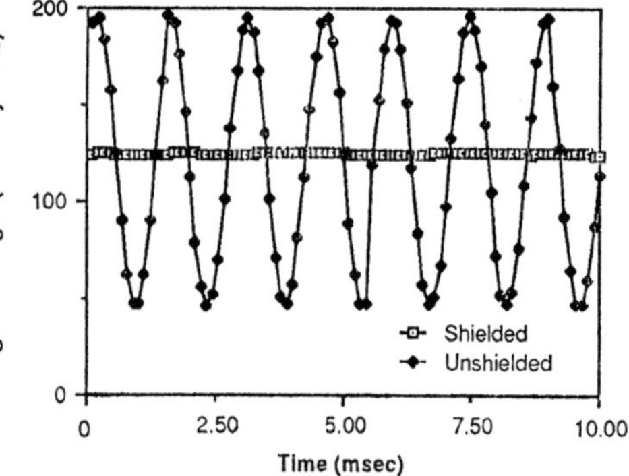

Figure 3. Effect of Shielding

Lastly, special care must be taken when choosing the proper op amps to provide accurate results. High input impedances are needed to insure negligible current is drawn from the monitored signal and hence prevent attenuation. Offset voltages and drift must also be minimized, especially when the signals of interest are of the same order of magnitude as the offsets. This problem was most significant in the power signals which are in the millivolt range. Since offsets are typically 2-5 mV, non-linear amplification occurred as a result of the buffer stage. For this reason, a new design which we are currently developing uses ultra-low offset (< 5 μV) op amps, and hence a very precise linear response.

To test the precision of the system, a shutdown was monitored and compared to the data (Figure 4) from one of the power meter strip recorders. This particular meter reads the fraction of the power set by a range selector. During a shutdown, the range selector starts at 2 MW and cycles down in the order 1 MW, 500 kW, 200 kW, 100 kW, 50 kW and 20 kW. The curves in Figure 5 are identical in both shape and magnitude, and thus indicates that this part of the system operates as desired.

THE DIGITAL SENSOR SYSTEM

The system used for monitoring system status annunciators and scram signals consists of Metrabyte's Model PIO12 24 bit, parallel, digital I/O card and rectifying modules for each signal. Modern annunciators are typically LED displays with 0-5 VDC inputs which could be inputted directly to the Model PIO12 card. However, the FNR, like many older reactors, uses 120 VAC annunciator signals. For this reason, conversion units were designed to interface these signals to the digital I/O port. The schematic for this conversion is shown in Figure 5. The input voltage (120 VAC) to this circuit is stepped down to 12 VAC and half rectified by the diode. The low pass filter produces a "choppy" DC waveform which is fed into a comparator set at a level of 1 volt. With this circuit connected to the "condition normal" signal, an abnormal condition such as a scram

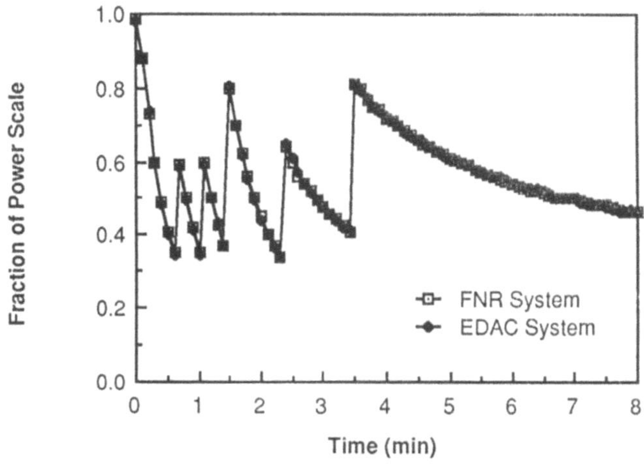

Figure 4. Reactor Shutdown from 2 MW. Meter reads fraction of power set by a range selector switch (2MW,1MW,500kW,200kW,100kW,50kW,20kW).

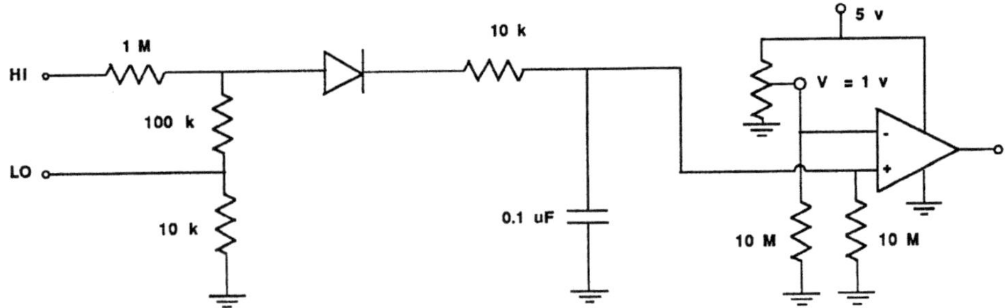

Figure 5. Digital Sensor Rectifying Schematic

would drop the voltage seen at the comparator below 1 volt and hence produces an output voltage of 5 VDC to the Model PIO12 card.

It must be noted that there will be a delay between the actual condition and the time it is sensed. This lag time is a result of the low pass filter which must discharge when the voltage is removed by the abnormal condition. Since the circuit time constant is τ = RC = 1 ms; it will take 2.5 ms to drop from 12 volts to the comparator level of 1 volt. However, as will be explained in the following section, this problem can be corrected in the software.

THE ACQUISITION CODE (TriAc)

In order to do any meaningful analysis of normal plant life characteristics, data has to be acquired at a modest rate (every minute) over each cycle (10 days). Form this data, one is able to see long term system changes (e.g. decreases in heat exchanger efficiency due to sludge build-up, pump degradations from vibration). It is also important, from a control point of view, to monitor the plant variables during power excursions such as startups/shutdowns, irradiation experiments, and most importantly reactor scrams. For this reason, a three-level acquisition code written in Turbo Pascal, TriAc, was developed.

TriAc is designed to monitor all of the analog signals (30) and the digitals (24) at a rate of 50 times a second. This is accomplished in a two-stage operation. Running as a background task, an interrupt routine, invoked by an interrupt produced by the A/D timer every 20 ms, starts and stores to memory the A/D conversions for each signal. It also polls the digitals and notes any change. While this is occurring in the background, TriAc stores the data in three arrays labelled TRANSIENT, EXCURSION, and STEADYSTATE according to the following procedure. TRANSIENT stores the data acquired 50 times a second for a period of 60 seconds (3000 records). Every second the latest data from TRANSIENT is copied into the second array, EXCURSION, which holds 20 minutes of data (1200 records). The last array, STEADYSTATE, holds the data acquired at every minute for an 8 hour period (480 records). In addition to the data stored in the three arrays, the date and time at the beginning of each refreshed array is also stored in RAM.

Under normal operation, only STEADYSTATE is written to disk while the other arrays are continually overwritten.

However, if the reactor goes out of automatic control (signalled by the monitored digital AUTO ON/OFF or AUTOMATIC RUNDOWN), the record position is flagged and the data in array EXCURSION will begin to be written to disk starting 5 records previous to the digital change. The reason for backtracking 5 records is to compensate for the delay time of the digital conversion circuitry mentioned previously. Data from EXCURSION is continually written to disk until the reactor is placed back in automatic control. The time of occurrence can be calculated from the flagged record position and the date and time the array was started.

If a scram occurs, both TRANSIENT and EXCURSION are written to files on disk along with the flagged record position and the time and date the arrays were started so that the time of occurrences can be calculated. TRANSIENT will stop being written when the scram signal has cleared (at least 3000 records will be written, regardless of how fast the signal is cleared), while EXCURSION will continue to be written to disk until steady-state operation is achieved. Regardless of the condition, NORMAL is always written to file on the hard disk. Collecting and writing data in this manner insures that all power excursions and scrams are accurately mapped and show all pertinent information in reasonable time steps.

In order to analyze the above data, an analysis code was written that converts the A/D conversions to appropriate units. It presents this data in three different manners: table form, a strip recorder and as a plot. The strip recorder has the advantage of being able to show many different signals simultaneously so that correlations can be made. It can also present this data on time scales of 8 seconds to 2 weeks, allowing one to compare the data with the FNR strip recorder data. The plot routine allows single and double ordinate plots for producing fine-tuned presentation slides.

THE STEP INTO THE AI WORLD

The practical difficulties which were encountered in acquiring real data signals have yielded a greater understanding of the plant's electronics and signals. The knowledge gained will enable us to better diagnose hardware problems using signal validation and fault tolerant techniques since we know the characteristic of each signal. In addition, the access to reliable and realistic plant data enables us to begin applying and testing advanced control techniques in a setting which is constrained by reality.

The development of an expert system holds the most intriguing and innovative possibilities. The reactor plant dynamic displays can be translated into an object-oriented framework with message passing used to animate component characteristics. The monitoring software will be expanded into an expert system using basic physical laws, plant procedures and specifications, and the reactor operator expertise. The diagnostic techniques developed, which use faster-than real-time-simulations, can be tested online using the reactor data. Techniques for reasoning and learning, using plant data and basic physical knowledge, will also be developed and applied. Ultimately, we hope to have a system so powerful and complete that we can move to closed plant operation.

The development and installation of the data acquisition and monitoring system represents a major advance in the technology at the Ford Nuclear Reactor. But perhaps more importantly, it provides a practical focus to begin applying artificial techniques in a more realistic environment.

ACKNOWLEDGEMENTS

The authors would like to acknowledge the contributions of **Chris Brannon** for his instrumental help in designing and interfacing EDAC to the FNR and **John Allen Stamos** for his expertise in helping develop and program the acquisition and analysis codes.

REFERENCES

Hassberger, J. A. and Lee, J.C., "Macroscopic Mass and Energy Balance for Nuclear Plant Diagnostics Using FUZZY Logic", ANS-Artificial Intelligence Conference, Aug.31-Sept 2, 1987.

MOAS: A REAL-TIME OPERATOR ADVISORY SYSTEM

I. S. Kim and M. Modarres

Department of Chemical and Nuclear Engineering
University of Maryland, College Park, MD 20742

INTRODUCTION

A real-time knowledge-based operator advisory system (Maryland Operator Advisory System; MOAS) is being developed at the University of Maryland using PICON (real-time expert system shell for process control). MOAS is an expert system designed to assist the plant operators during transient conditions initiated by the Condensate and Feed Water System (CFWS) of a pressurized water reactor. The MOAS performs 1) intelligent prealarming and alarming, 2) real-time sensor validation and sensor conflict resolution, 3) real-time hardware failure diagnosis, and 4) real-time corrective measure synthesis. The core of the MOAS knowledge base is constructed using an integrated deep-knowledge approach called Goal Tree-SuccessTree (GTST). The process domain knowledge of the CFWS is completely and rigorously modeled in the GTST format, which is then used to build up the knowledge base of the expert system. This paper will briefly describe the GTST model, design of the current MOAS and our ongoing and future efforts.

GOAL TREE-SUCCESS TREE MODEL

Goal Tree-Success Tree[1-6] (or Composite Trees) consists of a goal tree and success trees. Goal trees can completely, rigorously, and hierarchically describe a process plant and its operations. Goal trees incorporate a structured approach that shows how a specific objective in a plant is achieved. This is done by defining the objective and partitioning it into a series of related sub-objectives or goals; these goals are in turn broken down into subgoals or functions. The partitioning continues until the description of function cannot be made without referring to plant hardware or process conditions.

As soon as hardware or a process condition is explicitly mentioned, the tree can be viewed as a success tree which describes success paths (i.e., ways that a subgoal can be achieved), not goals. The success tree represents the logical relationship among hardware success, process conditions, and human actions that are necessary to achieve a subgoal. Various modes of operation, if existing, can also be incorporated in the success tree. The structure of a typical GTST is illustrated in Fig. 1.

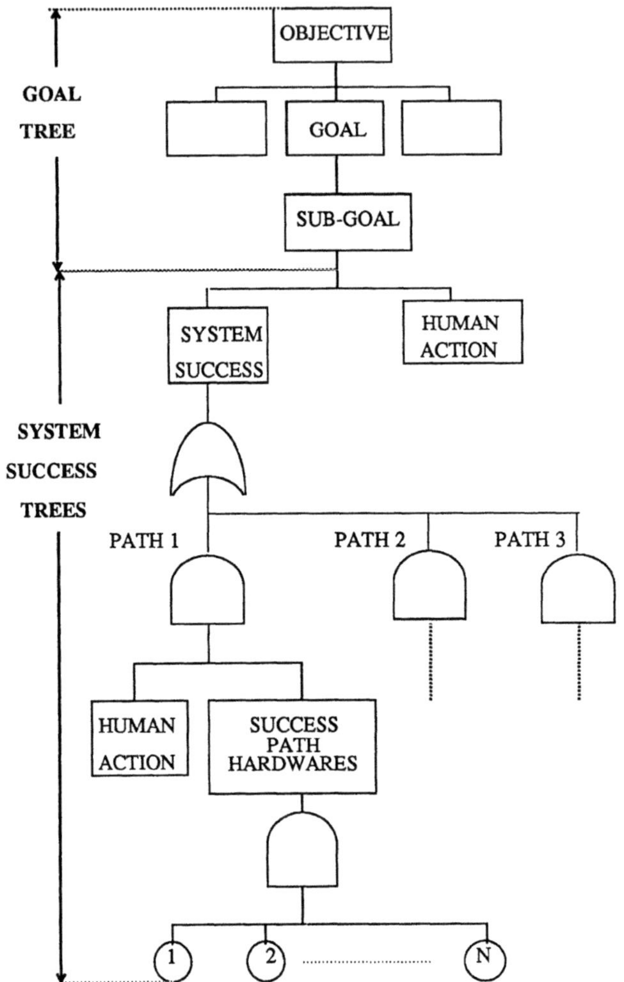

Fig. 1 Typical Structure of a Goal Tree-Success Tree

298

A REAL-TIME EXPERT SYSTEM FOR OPERATOR ASSISTANCE

Based on the GTST model, a real-time expert system[7-9] for operator
assistance is being developed at the University of Maryland, using PICON
shell on the LMI computer[8-9]. This expert system is an on-line expert
advisor that can assist plant operators in coping with process upsets in the
CFWS of a pressurized water reactor.

The process domain knowledge of the CFWS operation is first modeled in
the Goal Tree-Success Tree format, and the GTST is then used to construct the
knowledge base of the real-time operator advisory system. The development
of the GTST and its ultimate conversion to a knowledge base is briefly described in the remainder of this section.

The partial loss of the CFWS has been the most dominant contributor to
the nuclear plant outages[10]. However, the partial loss of this system does
not completely disable the heat removal capability of the CFWS, provided that
remedial operator actions be performed in a timely fashion. This is because
the CFWS can provide several discrete levels of heat removal capability; each
level corresponding to a known reactor power level. Table 1 shows typical
discrete power levels for a PWR.

By the timely and proper response of the operator, the reactor power can
be reduced to a level corresponding to the heat removal capability of the
CFWS when necessary. The reduction in the number of plant outages will improve not only the plant economy but also the plant safety by minimizing the
likelihood of an unwanted plant trip. If the plant power, however, cannot be
maintained at any of the discrete success levels of the CFWS, then the operator should be quickly advised to manually shutdown the plant.

Knowledge Base

Process domain knowledge of the CFWS operation was completely and rigorously represented in the Goal Tree-Success Tree with the top objective of
"Cycle Equivalent Availability Maximized," and with the most relevant goal of
"Operability of the Condensate and Feed Water System Maximized." A good
example of the GTST, although in much smaller scale, can be found in Ref. 6.

PICON accepts the domain knowledge in the form of production rules
(i.e., [If ..., then ...]). For the knowledge base of the MOAS expert
system, the rules are derived from the GTST. The logical structure of the

Table 1. Success Criteria for CFWS

| Success of | Plant Power Levels | | | |
Equipments	100%	75%	50%	30%
Condensers	6/6	5/6	4/6	3/6
Condensate Pumps	3/3	2/3	1/3	1/3
Booster Pumps	2/3	2/3	1/3	1/3
Main Feed Pumps	2/2	2/2	1/2	1/2

GTST provides the ability to encode each goal and subgoals in production rule
formats. For instance, for the following case

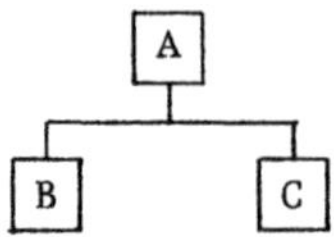

the rule, [If B and C, then A], can be derived. This kind of straightforward
mapping programming of the entire GRST may be used in certain cases, providing
that the knowledge base is built in such a way that it can be effectively
used by the inference engine for problem solving. In almost all cases, however, the knowledge engineer needs to devise the most effective way of encoding a GTST by first carefully examining the tree, and organize the rules
for achieving computational efficiency and reducing the search time. The
well-organized knowledge structure of the GTST helps the knowledge engineer
find better search routines and thereby develop a more efficient real-time
expert system.

Real-Time Inference Engine of PICON

     PICON has an efficient real-time inference engine designed specifically
for process control applications. Both forward and backward inference chain-
ings can be employed. Although frames are used in PICON to represent the
attributes of each rule (e.g., scan and alert intervals, category, or priority of a rule) and of each variable (e.g., currency interval of a sensor
value), PICON is essentially a production rule system.

     The attributes are used by the inference engine to effectively deal with
dynamic large-size data in time-critical situations. For example, alert
intervals and categories of rules can be used for relevant rule sets to be
invoked and for potential problem areas to be focused, when a potentially
significant event occurs.

System Status Flag in MOAS

     In order to control the real-time inference process and message display
while multiple process anomalies are occurring in the CFWS, it was necessary
to adopt a system status flag (SSF) consisting of four imaginary colors:
     1) Green: Whenever the MOAS does not see any abnormal operating situa-
tion in the CFWS, the color of the SSF remains green.
     2) Red: If a certain significant event is detected in the CFWS, then the
SSF color will be changed to red and stay red until the problem is completely
solved by the MOAS. While the SSF is red, no message will be provided to the
operator.
     3) Yellow: If the process problem is completely solved by MOAS, the SSF
color is changed from red to yellow and the inference results are displayed
to the operator.
     4) Black: If the conclusion that the plant should be shutdown is reached
by the MOAS, then the message will be given to the operator and the SSF color
permanently changed to black.

PERFORMANCE OF MOAS

     The MOAS expert system is capable of performing basic system function
monitoring for a given power level and success path. If an abnormal condition is detected in the CFWS, it will provide the operator with the inference
results consisting of one or several of the following:

1) Fault detection (FD): The expert system informs the operator about
the malfunctioning of an operating equipment when it is detected.

2) Process or component status (ST): Statuses of components or processes
which are important at a given time instant and for the specific plant operation mode are displayed to the operator. For instance, if a drain dump control valve of a feedwater heat exchanger is in manual mode when the heater
condensate level is high, then the recognition of the status of the control
valve is important. Therefore, a message like "CV127 in manual mode" will
be given to the operator with other appropriate messages in this case. As an
example of process status, the predicted time margin to the steam generator
feed pump auto trip due to low-low suction pressure may be calculated in
real-time with consideration of time delay, if any, and displayed to the
operator.

3) Operational aid (OA): Appropriate operational aid is provided to the
operator whenever possible. The aid may be recommendation of a proper control action or of proper verification points. For example, in case the feed
water regulating valve fails as is due to the pneumatic pressure loss, the
operator will be given a message like "Try to maintain normal water level in
steam generator 12 with the differential pressure controller 422 in manual
mode and ....."

4) Prealarming (PA): Prealarming is provided to the operator before
something important occurs. As an example, if there is a possibility of low
pressure feedwater heat exchanger 11 level becoming high-high, then the operator wil be warned of a potential automatic turbine trip by the plant protective system.

5) Corrective measure (CM): Upon partial loss of CFWS, the expert
system will seek the maximum achievable power level and its corresponding
optimum success path. In doing so, the availabilities of standby components,
if necessary, and the possibility of putting the available components on line
in due time will be checked by the expert system. When the expert system
reaches a final decision, a corrective measure will be provided to the operator to quickly reconfigure the CFWS.

Test runs of MOAS have been carried out using PICON simulation facility
which permits simulation of each sensor value. Hardware failures or process
faults are simulated by manipulating time-dependent sensor values such that
they are indicative of the failures or faults. The tests have shown that
MOAS provides the operator with a course of reliable inference results.

The MOAS messages are displayed to the operator with one of the above
five message classifications specified in front of every MOAS message. This
is an ergonomic feature to promote quicker acceptance of the MOAS messages by
the operator.

ONGOING EFFORTS TO IMPROVE MOAS

The real-time inference performed by the current version of MOAS is
based on the on-line data transmitted by "perfect" sensors and component
status indicators. In other words, the integrity of the plant instrumentation system is not questioned. But, malfunctions in the plant instrumentation system may corrupt the inference performed by the MOAS.

A methodology for real-time sensor validation or real-time sensor failure diagnosis is being developed, which is based on the real-time inference
from 1) comparison of redundant sensor data (if any), 2) validation of sensor
data using engineering principles such as mass or energy balance, or 3) validation of sensor data using pump characteristic curve. The idea here is that
the deep knowledge such as conservation principles or pump curve can be used
to formulate sensor-value constraints. The constraints will then be used to
detect inconsistency or conflict among the sensor values in real-time environment by MOAS.

However, the violation of a sensor-value constraint does not necessarily mean that there is a sensor failure. The reason is as follows: Satisfaction of a sensor-value constraint is an indication that something is normal. (For example, satisfaction of a mass-balance constraint on two flow-sensor values, upstream and downstream, is an indication not only that the two sensors are working fine but also that there is no leakage between the two measurement points.) But violation of a sensor-value constraint does not necessarily imply a sensor failure. The violation might also indicate a possibility of a certain hardware malfunction. In case of a constraint using a pump curve, for instance, not only sensor failure but also pump failure (such as damaged pump impeller) may lead to the violation of the pump-curve constraint.

Therefore, it is necessary to perform hardware failure diagnosis in parallel with the sensor data validation. Failure diagnosis also has been an integral part of our current research[11].

ACKNOWLEDGMENTS

The authors wish to thank Mr. R. N. Hunt of Baltimore Gas and Electric Company for providing us with plant operating insight and valuable comments. We also acknowledge the support of the University of Maryland's Systems Research Center provided through an NSF Grant.

REFERENCES

1. R. N. Hunt and M. Modarres, Integrated Economic Risk Management in a Nuclear Power Plant, Annual Meeting of the Society for Risk Analysis, Knoxville, TN (1984).
2. M. L. Roush, M. Modarres, and R. N. Hunt, Application of Goal Trees to Evaluation of the Impact of Information upon Plant Availability, Proceedings of the ANS/ENS Topical Meeting on Probabilistic Safety Methods and Applications, San Francisco, CA (1985).
3   M. Modarres, M. L. Roush, and R. N. Hunt, Application of Goal Trees in Reliability Allocation for Systems and Components of Nuclear Power Plants, Proceedings of the Twelfth International Reliability Availability Maintainability Conference for the Electric Power Industry, Baltimore, MD (1985).
4. M. Modarres, M. L. Roush, and R. N. Hunt, Application of Goal Trees for Nuclear Power Plant Hardware Protection, Proceedings of the Eighth International Conference on Structural Mechanics in Reactor Technology, Brussels, Belgium (1985).
5. M. Modarres and T. Cadman, A Method of Alarm System Analysis in Process Plants with the Aid of an Expert Computer System, Computers and Chemical Engineering, Vol. 10, No. 6, 557-565 (1986).
6. I. S. Kim and M. Modarres, Application of Goal Tree-Success Tree Model as the Knowledge Base of Operator Advisory Systems, Nuclear Engineering and Design, Vol. 103 (1987).
7. P. A. Sachs, A. M. Paterson, and M. H. M. Turner, Escort--An Expert system for Complex Operations in Real Time, Expert Systems, Vol. 3, No. 1 (1986).
8. R. L. Moore, L. B. Hawkinson, C. G. Knickerbocker, and L. M. Churchman, A Real-Time Expert System for Process Control, Proceedings of the First Conference on Artificial Intelligence Applications, IEEE Computer Society (1984).
9. R. L. Moore, L. B. Hawkinson, C. G. Knickerbocker, and Michael E. Levin, Expert Systems in Real-Time Environments, International Topical Meet-

ing on Computer Applications for Nuclear Power Plant Operation and Control, Tri-cities, Washington (1985).
10. C. H. Meijer and B. Frogner, On-Line Power Plant Alarm and Disturbance Analysis System, EPRI Report, NP-1379 (1980).
11. D. T. Chung and M. Modarres, Diagnosing Faults through the GTST Deep-Knowledge Model, AICHE National Meeting, New York City (Nov. 1987).

ARTIFICIAL INTELLIGENCE EXPERT SYSTEM

FOR NUCLEAR POWER PLANT CONTROL AND MAINTENANCE

George B. Westrom

Odetics, Inc.
1515 S. Manchester Avenue
Anaheim, CA   92802

A major concern in emergency response in a nuclear power plant is the lack of real experience with a catastrophic condition. Experts do more than just follow a set of rules. They have experience and insight into problems and are able to use their professional judgment.

The objectives of this expert system are:

- Detect subtle off-normal trends in a reactor power plant system.
- Alert the operator with color graphics and synthesized voice commands.
- Identify the source of the problem and advise the operator on steps to restore the plant to normal operation.
- Provide an explanation facility and database to supply detailed information on component history and reasoning process.

The RHRS (Residual Heat Removal System) was selected for the initial expert systems development because it is extremely important during refueling and cold shutdown operations.

During cold shutdown operations, much of the backup safety equipment is disabled for operational or maintenance reasons. The operators get little indication of some abnormal events, and events can progress quickly if decay heat has not fallen off. Continued operation of the decay heat removal systems is essential to maintain core cooling and to prevent core uncovery and resultant fuel damage. A number of diverse protection systems are reduced during cold shutdown operations; many instruments are either not available or are operating at reduced accuracy. Additionally, some of the cold shutdown monitoring information may not be available in an ideal location on the control board, but in fact may require secondary panel observations by the operators. Computer hardware which can host this system are the Micro Vax II, Sun 3 Workstation, TI Explorer and Symbolics.

<u>Application Description - Decay Heat Removal System Advisor</u>

The expert system application test vehicle chosen for the Phase I implementation is a Residual Heat Removal system advisor - a subset of conditions to be modeled during cold shutdown operations. The RHR system is comprised of two identical parallel 100% capacity trains consisting of circulating pumps and heat exchangers to dissipate the decay heat loads.

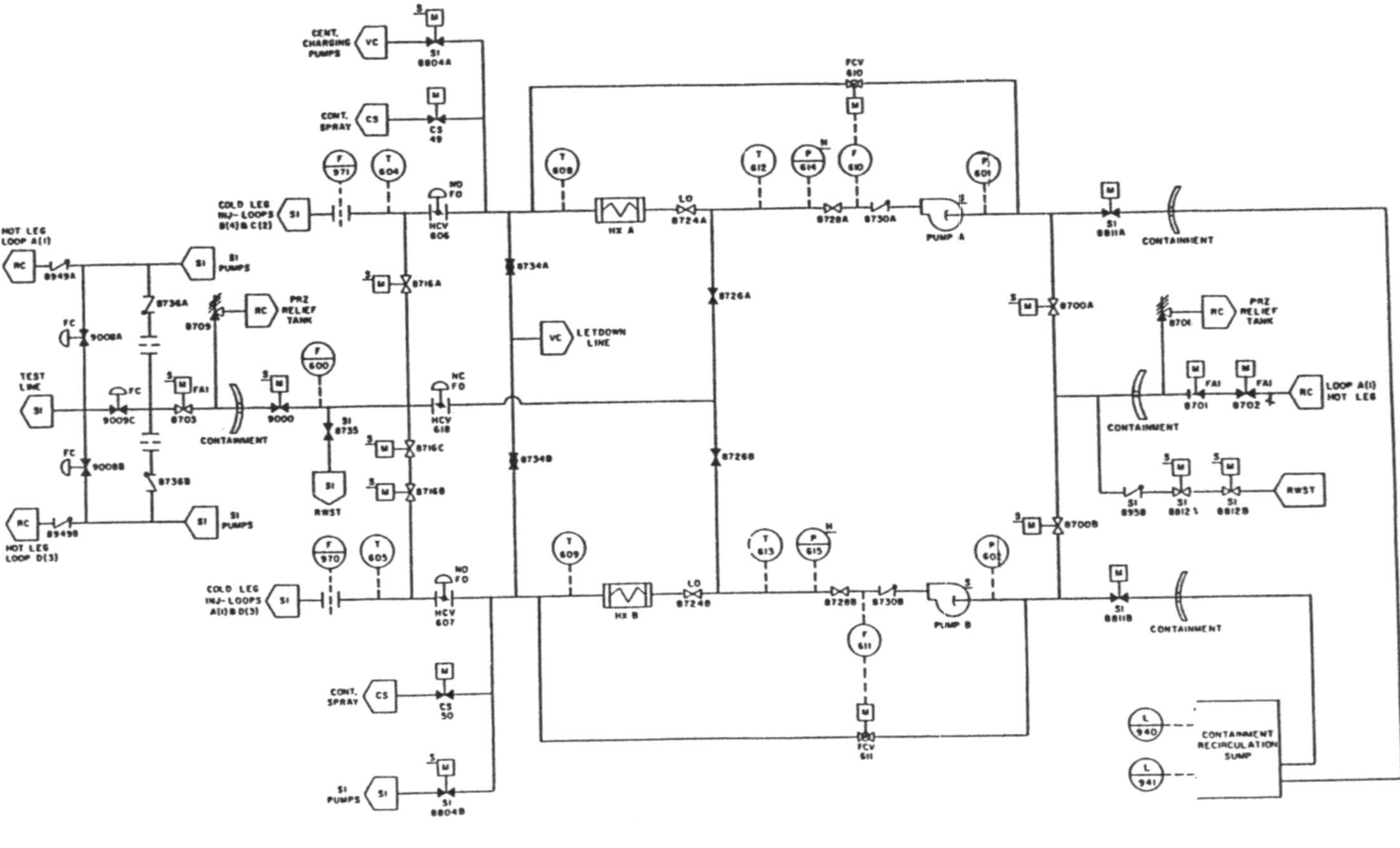

Figure 1   RHR System

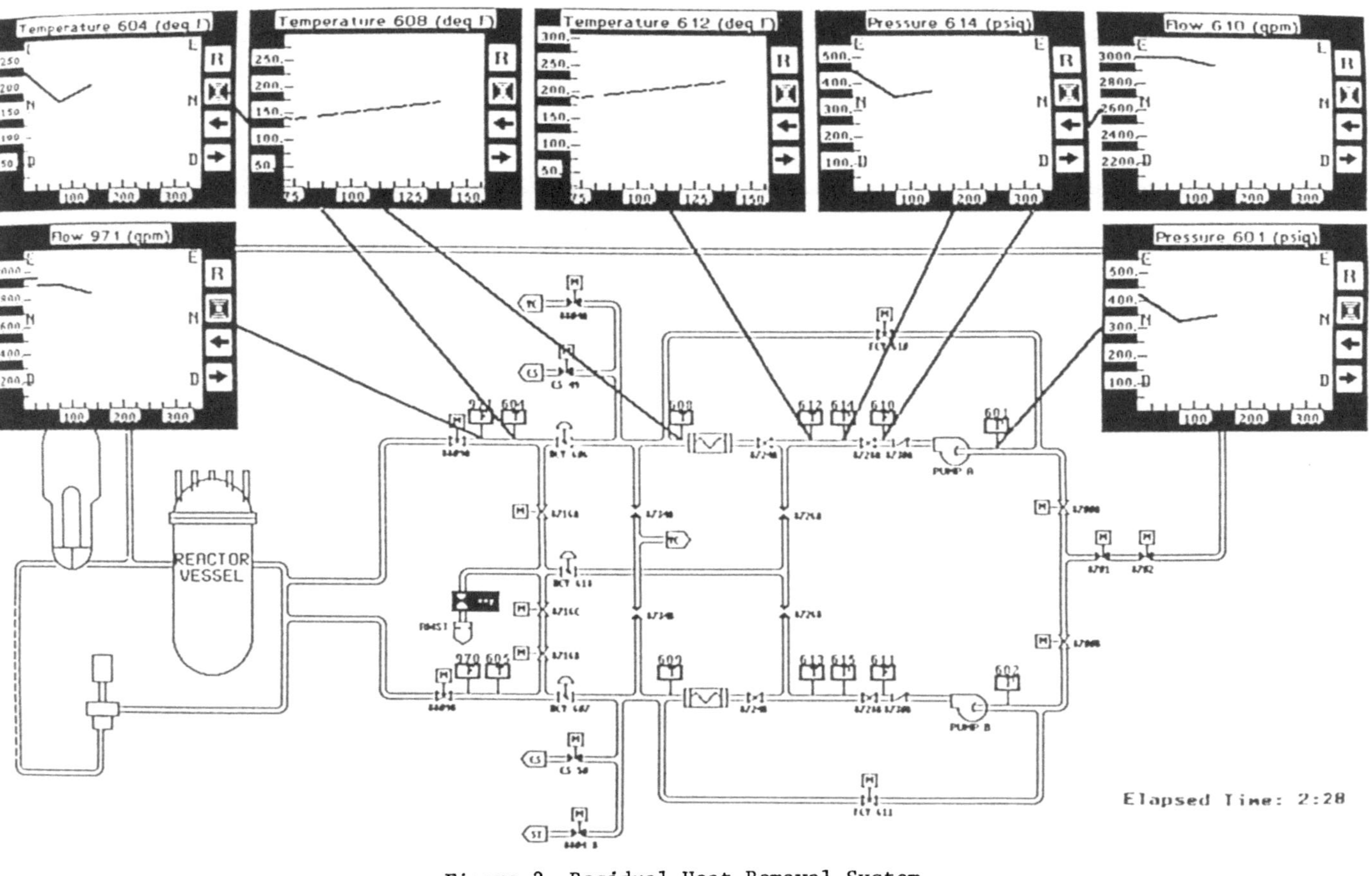

Figure 2   Residual Heat Removal System

The Zion plant as described in NSAC/84 Report, "Zion Nuclear Plant
Residual Heat Removal PRA", by Prichard, Lowe and Garrick, Inc. was
selected as the model for our system. The sensors (temperature, pressure
and flow meters) and all the valves important for this operation are shown
in Figure 1.

The Residual Heat Removal System advisor which we have developed for
Phase I is a prototype expert system that interacts with the user and the
process. If a problem is indicated during refueling and cold shutdown, it
advises the crew of the steps to be taken which will lead to a stable
plant. The expert system advisor provides a substantial safety support
system for monitoring key functions during decay heat removal and other
functions during cold shutdown. The benefits for a utility for this type
of advisory system lies in the fact that it provides a means of improving
the safety of the nuclear plant operation while also providing significant
economic benefit. Economic benefits arise from early detection and
diagnosis of faulty components before equipment or plant damage would
result, and reducing plant downtime.

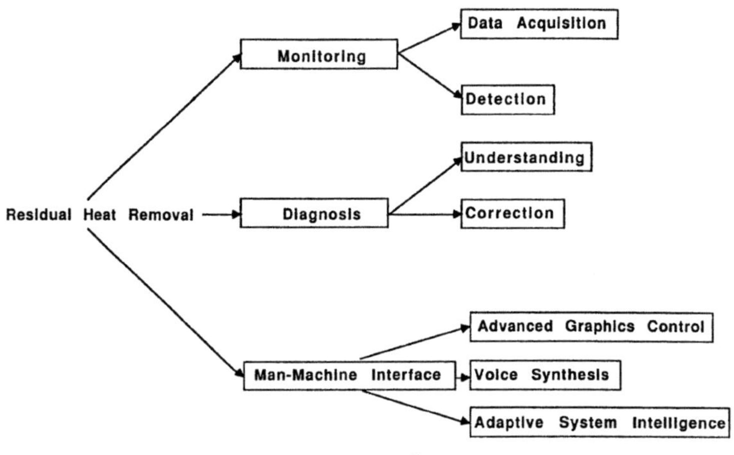

Figure 3

The expert system has an advanced graphics man-machine interface so
that the user, a control room plant operator, need not be a programmer to
understand this advisory system. A computerized graphic representation of
the Zion plant RHRS operational components was used to interface with the
user as shown in Figure 2. Strip chart graphic windows can be displayed
by the operator for any of the sensors selected for observation. When a
fault is indicated, a voice synthesized message to the operator is
automatically annunciated.

EXPERT SYSTEM SOFTWARE IMPLEMENTATION

The Automated Reasoning Tool (ART) developed by Inference Corporation
was the tool we selected to use to develop this system. The computer
hardware used for this purpose was a Symbolics 3675 system.

A system design consisting of the data, facts and rules for the system operation was developed.   A block diagram of the system design is shown in Figure 3.

A monitoring function was developed which reads in simulated data from the plant sensors (i.e., temperature, pressure and flow meters) and displays the operational valve settings for the RHR operation.   The monitoring function consists of sensor data acquisition and rules to detect any off-normal conditions during the operation.

A diagnostic rule-base built on the knowledge acquired from experts was developed to indicate the possible faults and to guide the user to take the appropriate steps to restore the plant to a stable condition.

CHAPTER 6

HUMAN-MACHINE INTERFACE

AN EXPERT SYSTEM FOR MODELLING OPERATORS' BEHAVIOUR IN CONTROL
OF A STEAM GENERATOR

Pietro Carlo Cacciabue[1], Giovanni Guida[2], and
Alberto Pace[2]

[1]Joint Research Centre, Commission of the
European Communities, Ispra, Italy

[2]Progetto di Intelligenza Artificiale
Dipartimento di Elettronica
Politecnico di Milano, Italy

INTRODUCTION

Modelling the mental processes of an operator in charge of controlling
a complex industrial plant is a challenging issue currently tackled by
several research projects both in the area of artificial intelligence and
cognitive psychology. Progress in this field could greatly contribute not
only to a deeper understanding of operator's behaviour, but also to the
design of intelligent operator support systems (Hollnagel, Mancini and
Woods, 1986). In this paper we report the preliminary results of an experimental research effort devoted to model the behaviour of a plant operator by
means of Knowledge-based techniques.

The main standpoints of our work is that the cognitive processes
underlying operator's behaviour can be of three main different types, according to the actual situation where the operator works. In normal situations,
or during training sessions, the operator is free to develop deep reasoning,
using knowledge about plant structure and function and relying on the first
physical principles that govern its behaviour. In dynamic operational
situations the operator can resort to empirical reasoning, based on shallow
knowledge which directly relates plant behaviour to possible interventions.
In emergency conditions, the operator must take the appropriate interventions in a very short time and often operates under stress; thus he must
rely on a kind of immediate reasoning, mostly based on remembering ready-
for-use solutions to the current problem acquired in the past.

The work reported in this paper is concerned with the design and
implementation of an experimental system where the three levels of behaviour
above introduced can be appropriately modelled and tested. The knowledge
based approach has been assumed as a fundamental design choice, and different
knowledge representation and reasoning techniques have been adopted for each
one of the three levels of mental processes considered, namely: qualitative
modelling for deep reasoning (Michie, 1982; Chandrasekaran and Milne, 1985;
Guida, 1985); production rules for empirical reasoning; and simple transition rules for immediate reasoning. The experimental system developed has
been focused on a practical case study concerning the control of a steam

generator of a nuclear power plant. It has been implemented using KEE 3.0,
Lisp and Fortran on Symbolics 3640 and SUN 3-75 workstations, and is
currently used for experimental activity. In particular it is utilised for
research on the different types of knowledge used for implementing the three
levels of behaviour considered, with special attention to the deep mental
model of the plant and to the shallow knowledge used for empirical and
immediate reasoning.

The paper is organised in the following way: Section 2 introduces the
concepts of deep, empirical and immediate reasoning, and illustrates the
overall architecture of the proposed model. Section 3 presents the case
study of the steam generator, and illustrates the main modules of the experimental system developed, namely: the quantitative simulator, qualitative
plant model, the empirical knowledge base, and the immediate knowledge base.
Section 4 outlines the major points of the experimental activity done on the
system. Finally, section 5 presents some conclusive remarks, and discusses
promising directions for future research.

MODELLING OPERATOR'S BEHAVIOUR

<u>Levels of Knowledge and Reasoning</u>

In analysing the behaviour of an operator employed in the control of a
complex industrial plant at least three different levels of mental processes
can be identified, each one involving specific knowledge and reasoning mechanisms:

[1]   First level: deep reasoning
      In normal situation a plant undergoes slow and regular variations and
      it is either controlled manually by the operator or automatically by a
      control system. In this case, as well as during training at simulator
      sessions, the operator is free to develop *deep reasoning* using first physi-
      cal principles, a mental (mostly qualitative) model of the plant and
      direct observations on plant behaviour. Deep reasoning allows the
      operator to assess his knowledge on the behaviour, function and
      operation of the plant, and to acquire new knowledge. Such knowledge
      is is compiled into appropriate mental representations and it is
      stored for future use at empirical or immediate reasoning levels.

[2]   Second level: empirical reasoning
      In dynamic operational situations characterised by moderate time con-
      straints, typical of the operational transients such as the star-up or
      the shut-down of the plant, the operator can resort to an *empirical reason-
      ing* activity, based on shallow < situation → action > associations
      between observations of plant behaviour, internal states and available
      intervention mechanisms, or on simple mental models of the task at
      hand. Empirical reasoning typical involves deduction capabilities,
      and is focused on identifying appropriate control actions to settle
      the transient situation occurred. Also empirical reasoning allows the
      operator to acquire new knowledge possibly relevant to the immediate
      reasoning level.

[3]   Third level: immediate reasoning
      In emergency or critical situation the automatic control system is
      disconnected from the plant and manual control has to be performed.
      In this case the operator is under great stress and has a restricted
      time to find an appropriate solution to the problem at hand. There-
      fore, no sophisticated reasoning activity nor shallow empirical rea-
      soning can be performed, and the operator must resort to *immediate reason-
      ing* which basically relies on remembering previous experience. Immedi-
      ate reasoning uses direct < problem → solution > associations

acquired during the previous levels of reasoning and relies on a mechanism of similarity matching between the current situation and the past experiences.

Overall architecture of the model

The general configuration of the experimental model of operator's behaviour developed is illustrated in Figure 1. It includes two parts connected by an interface. On one side there is a simulator of the plant and an event generator, both based on quantitative techniques. On the other side the three different levels of operator's behaviour considered are represented, using a qualitative approach. The interface allows conversion from quantitative to qualitative representations and vice versa.

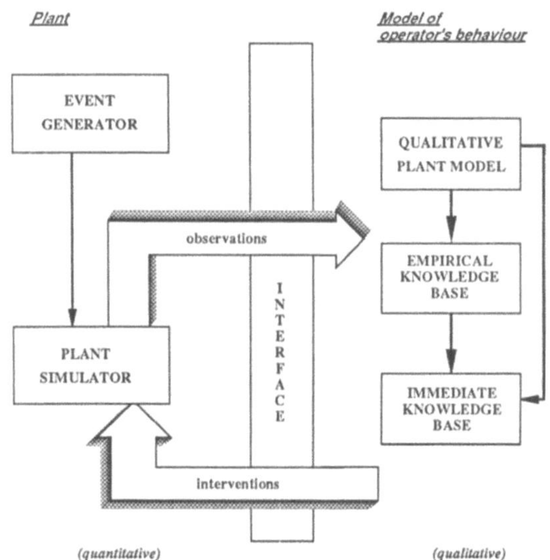

Fig. 1. Architecture of the experimental model.

The quantitative plant simulator and the event generator are devoted to simulate the real plant, which of course is not available for experimentation with the model of operator's behaviour. The plant simulator is designed using traditional system modelling techniques; the main principles on which it is based will be briefly described, with reference to the case study of the steam generator, in the following section. The event generator is employed for driving the plant simulator and , in particular, for simulating possible incidental sequences together with the relevant control problems. The model of operator's behaviour includes three subsystems:

- a *qualitative plant model*, which is assumed as an adequate representation of the mental model of the plant used for deep reasoning;

- an *empirical knowledge base*, which contains a shallow model of the plant control task utilised in empirical reasoning;

- an *immediate knowledge base*, which stores ready-for-use knowledge about plant control used for immediate reasoning.

These subsystems will be illustrated in the following sections, with reference to the case study concerning the operator in charge of controlling a steam generator during abnormal events.

The connections among qualitative plant model, empirical knowledge base and immediate knowledge base shown in Figure 1, denote the processes of knowledge acquisition which are supposed to take place in the mind of the operator during deep and empirical reasoning.

THE CASE STUDY: CONTROLLING THE STEAM GENERATOR OF A NUCLEAR POWER PLANT

In the following the modelling approach discussed in the previous sections is applied to the operator in charge of controlling the steam generator of a nuclear power plant during transient conditions. The steam generator is one of the most significant components of the whole energy production process and is controlled during normal operation by an automatic feedback regulator. When an abnormal event occurs, as well as during operational transients, the regulator is cut out and the operator must directly control the behaviour of the steam generator taking the appropriate interventions to face the current problem. Therefore the steam generator presents a variety of operational situations which seem appropriate to study the various levels of mental processes which characterise the operator's behaviour outlined in the previous section.

The quantitative model

In a steam generator heat is transferred from a primary (hot) fluid to a secondary (cold) fluid which is vaporised. Primary and secondary fluids flow in separate circuits and do not intermix. Under the assumption that the power exchanged between primary and secondary circuit is known, only the phenomena occurring in the secondary side of the steam generator will be considered. Now, supposing to have at each time instant complete mixture between vapour and liquid inside the steam generator, the overall mass and energy conservation laws can be applied obtaining (through a few steps) the following equations:

$$\frac{dh}{dt} = \frac{1}{V_{tot}\left(\frac{\partial v}{\partial P} + v\frac{\partial v}{\partial h}\right)} \left[v\frac{\partial v}{\partial P} Q_{eff} + v^3(w_{out} - w_{in}) \right] \tag{1}$$

$$\frac{dP}{dt} = \frac{1}{V_{tot}\left(\frac{\partial v}{\partial P} + v\frac{\partial v}{\partial h}\right)} \left[-v\frac{\partial v}{\partial h} Q_{eff} + v^2(w_{out} - w_{in}) \right] \tag{2}$$

P (pressure) and h (enthalpy) are the state variables, and Q (exchanged power), h_{in} (input enthalpy) and w_{in} (input water flow) are the inputs of the system. Moreover, V_{tot} denotes the total volume, v the specific volume, w_{out} the output steam flow, h_{out} the output enthalpy, and Q_{eff} the net exchanged power (given by: $Q_{eff} = Q + w_{in}(h_{in} - h) - w_{out}(h_{out} - h)$).

The system of the two non-linear differential equations (1) and (2) governs the dynamic behaviour of the steam generator, which can be easily obtained through intergration. From these equations, applying Euler explicit integration method, a quantitative simulator of the steam generator has been developed in Fortran and is presently running on a SUN 3-75 workstation.

The qualitative plant model

On the basis of equations (1) and (2) a qualitative model of the steam generator may be obtained following the approaches proposed by De Kleer and Brown [1984 and 1986], or Iwasaki and Simon [1986a and 1986b], or Kuipers [1985 and 1986]. However, these approaches suffer of major problems, mainly because their qualitative arithmetics can lead to ambiguous results. A tentative remedy to this problem is to branch on each ambiguity found, but the result is that lots of solutions are generated, the majority of which are not acceptable. Similar results would also be obtained using the QSIM algorithm proposed by Kuipers (1985 and 1986). Therefore, these approaches are too weak to cope with such a complex case as the steam generator.

The main idea that characterises the approach to qualitative modelling we propose is to rely in the definition of the model not only on the static form of the equations that govern the behaviour of the system at hand (as all the above mentioned approaches do), but also to elicit further knowledge about the real system performance by analysing the actual dynamic variations (during operations) of the involved variables. Using the quantitative simulator it is indeed possible to evaluate the quantitative behaviour of the system in a very large number of situations, and to discover through statistical analysis new qualitative properties of system behaviour. This information is then combined with the arithmetics of integer numbers in order to solve the major problem encountered in traditional qualitative simulation. For this purpose, it has been necessary, however, to modify the definition of a qualitative variable. If a is a generic quantity, Λa denotes the corresponding qualitative variable. The symbol Λ is to be viewed just as a *qualitative operator* (like ∂ in the theories of De Kleer and Brown (1984)). The value of Λa may be any integer number, and it represents the qualitative value of the corresponding quantity a. $\Lambda a = 0$ denotes that the value of a coincides with its *landmark*, i.e. with a nominal value chosen according to the specific characteristics of the system considered. For example, it can be the steady-state value of the pressure or the normal value of the level in the steam generator. $\Lambda = -1$ represents a value that is generically lower than the landmark, while $\Lambda = +1$ denotes a generically higher value. These values are to a large extent arbitrary and can be chosen according to the characteristics of the system under study. However the symbol Λa is always referred to its landmark, i.e. its nominal value, and it represents a variation.

According to the approach introduced above, the system of equations (1) and (2) representing the behaviour of the steam generator is first transformed into qualitative expressions, keeping the sign of the partial derivatives, and then, using the quantitative simulator, the actual contribution of each individual derivative is evaluated. These contributions are represented in the qualitative equations through appropriate integer coefficients. In this way the following qualitative model is obtained:

$$\Lambda \frac{dh}{dt} = \Lambda Q + \Lambda h_{in} - \Lambda w_{in} \, , \tag{3}$$

$$\Lambda \frac{dP}{dt} = 2 \Lambda Q + 2 \Lambda h_{in} - \Lambda w_{in} - \Lambda P \, . \tag{4}$$

Finally, the qualitative prediction of the system behaviour is easily obtained by integrating equations (3) and (4) by the Euler integration method which implies that:

$$\Lambda h(\tau+1) = \Lambda h(\tau) + \Lambda \frac{dh}{dt}(\tau) \tag{5}$$

$$\Lambda P(\tau+1) = \Lambda P(\tau) + \Lambda \frac{dP}{dt}(\tau) , \tag{6}$$

where $\Lambda h(\tau+1)$ and $\Lambda P(\tau+1)$ represent the future state, $\Lambda h(\tau)$ and $\Lambda P(\tau)$ the current state, and $\Lambda \frac{dh}{dt}(\tau)$ and $\Lambda \frac{dP}{dt}(\tau)$ are computed according to equations (3) and (4).

In this way a qualitative model of the steam generator has been constructed which is more informed than the traditional ones. Its main characteristics are:

- the model is qualitative and causal but not ambiguous (i.e., not weaker — at least from this viewpoint — than a quantitative one);

- the model is much simpler than both quantitative and traditional qualitative ones;

- the model is transparent to the operator and appropriate to support deep reasoning (Chi, Feltovich and Glaser, 1982; Williams, Holan and Stevens, 1983).

<u>The empirical knowledge base</u>

The empirical knowledge contains qualitative production rules concerning the regulation skills employed for empirical reasoning. This knowledge has been derived from the design principles of the regulator of the steam generator which automatically controls its behaviour during normal operations. As the major concern of the control task is to keep the water level inside the steam generator steady through an appropriate regulation of inlet flow, three different actions can be considered: proportional derivative and integral with respect to the deviation of the current level from its normal value. The following qualitative equations represent the kernel of the empirical knowledge base:

$$\Lambda w_{in}(\tau) = -\frac{2}{3}\Lambda z(\tau) + \frac{1}{3}\Lambda z(\tau-1) - \frac{1}{9}\Lambda z_{acc}(\tau) , \tag{7}$$

where z denotes the water level and z_{acc} denotes the sum of the values of the water levels over the past time instants (τ denotes a discrete time instant). The intuitive meaning of equation (7) is straightforward. When the operator is controlling the steam generator, he can read the measure of the actual water level on the instrumentation panel. According to his observation he can decide to increase, decrease, keep constant the inlet flow. The three effects outlined above can influence his decision: the first effect (proportional) appears whenever the level is not at its normal value and must be corrected; the second effect (derivative) predominates when the level is rapidly moving in one direction; and the third effect (integral) prevails when the level is far from its normal value for a long time, notwithstanding previous interventions. The empirical knowledge base is processed by a traditional rule interpreter. This inference engine, using an appropriate search strategy, can concatenate rules together and form the rules together and form the deduction chains needed for solving the control problem at hand.

The immediate knowledge base

The immediate knowledge base is a collection of qualitative transition rules concerning changes in the state of the steam generator obtained through application of appropriate inputs. Rules have the following structure: < current-state, inputs $\rightarrow$ future-state > and mean that, if the steam generator is presently in the current-state, and inputs are applied to it for one time step, it will reach the future-state in the next time instant.

This knowledge has been mostly gathered from experience, and is used for immediate reasoning. A similarity matching algorithm looks in the immediate knowledge base for the rule whose current-state best fits the present state of the steam generator and whose future-state is the desired normal value, and, if one is found, the suggested inputs are considered an appropriate intervention for facing the transient situation at hand.

EXPERIMENTAL ACTIVITY

The experimental system outlined in the previous section has been developed using a composite programming environment: the quantitative simulator and the event generator are implemented in Fortran on a Sun 3-75 workstation, while the qualitative plant model and the empirical and immediate knowledge bases, together with the relevant inference mechanisms, are implemented using KEE 3.0 and Lisp on a Symbolics 3640 machine.

The system has been used for a variety of experimental activities, which include, among others:

-   the validation and testing of the qualitative plant model simulation activities;

-   the validation and testing of the empirical and immediate knowledge bases in control and regulation tasks;

-   the automatic acquisition (compilation) of immediate knowledge as a side-effect of deep and empirical reasoning activities;

-   the automatic incremental extension and refinement of the content of the immediate knowledge base through the new automatically acquired immediate knowledge.

The last two points listed above have involved the design of a specific algorithm for automatic knowledge acquisition and refinement. This algorithm is able to generate new transition rules from experience, and to extend and refine the content of the immediate knowledge base whenever a new rule is discovered which can bring a real and valuable addition to its current content.

CONCLUSION

In this paper the first preliminary results of a research effort devoted to the experimental study of the behaviour of an operator of a complex industrial plant have been described. In particular, a model of operator's behaviour has been proposed, which can be used for experimenting several cognitive processes.

Two issues have been partially addressed in the research reported in the paper and will be further investigated in the future. The first issue concerns the use of the proposed model for the analysis of the operator's performance during an accident: this can be studied using this model in combination with an event generator of components failure, based on reliability techniques. In this case the use of an explicit operator model represents an interesting alternative approach to the merely behaviouristic methodologies currently in use for human reliability analysis. The second issue concerns the general topic of knowledge acquisition, with particular attention to the origin of knowledge and learning mechanisms. The experience so far developed in the automatic compilation of immediate knowledge is only a first step in this direction. Among others, the following aspects seem specially worth investigating:

- the origin and use of qualitative knowledge about physical systems in novices and experts;

- the validation and refinement of qualitative models through experience;

- the automatic acquisition of empirical knowledge from deep reasoning and experience;

- the automatic acquisition of immediate knowledge from deep and empirical reasoning and experience;

- the automatic extension, refinement and validation of empirical and immediate knowledge bases, with particular attention to the problems of quality (consistency, completeness, correctness, goodness, etc.), dimensions and usability.

A last point which will be considered in the future research is the evaluation of the cognitive adequacy of the model proposed as a candidate for representing a mental model of operator's behaviour.

REFERENCES

Chandrasekaran B., and Milne R. 1985, Reasoning about structure, behavior and function. *ACM Sigart Newsletter* 93, 4-9.
Chi M.T.H., Feltovich P.J., and Glaser R. 1982, Categorisation and representation of physics problems by experts and novices. *Cognitive Science* 5, 121-152.
De Kleer J., and Brown J.S. 1984, A qualitative physics based on confluences. *Artificial Intelligence* 24, 7-83.
De Kleer J., and Brown J.S. 1986, Theories of causal ordering. *Artificial Intelligence* 29, 33-61.
Guida G. 1985, Reasoning about physical systems: Shallow versus deep models. In *Expert Systems and Optimisation in Process Control,* J. Efstathiou, A. Mamdani (Eds.), Gower Technical Press, Aldershot, UK, 135-159.
Hollnagel E., Mancini G., and Woods D. D. (Eds.) 1986, *Intelligent Decision Support in Process Environments*, NATO ASI Series, Springer-Verlag, Berlin, FRG.
Kuipers B. 1985, The limit of qualitative simulation. *Proc. 9th Int. Joint Conf. on Artificial Intelligence,* Los Angeles, CA, 128-136.
Kuipers B. 1986, Qualitative simulation. *Artificial Intelligence* 29, 289-338.
Iwasaki Y., and Simon H.A. 1986a, Causality in device behavior. *Artificial Intelligence* 29, 3-32.
Iwasaki Y., and Simon H.A. 1986b, Theories of causal ordering: Reply to De Kleer and Brown. *Artificial Intelligence* 29, 63-72.
Michie D. 1982, High-road and low-road programs. *AI Magazine* 3(1), 21-22.
Williams D.M., Hollan J.D., and Stevens A.L. 1983, Human reasoning about a simple physical system. *Mental Models.* D. Gentner and A. Stevens (Eds.), Lawrence Erlbaum, Hillsade, N.Y, 131-153.

KNOWLEDGE-BASED DIALOGUE IN INTELLIGENT DECISION SUPPORT SYSTEMS

Erik Hollnagel

Advanced Information Processing Division
Computer Resources International
Vesterbrogade 1A, DK-1620 Copenhagen V, Denmark

ABSTRACT

The overall goal for the design of Intelligent Decision Support Systems (IDSS) is to enhance understanding of the process under all operating conditions. For an IDSS to be effective, it must: (1) select or generate the right information, (2) produce reliable and consistent information, (3) allow flexible and effective operator interaction, (4) relate information presentation to current plant status and problems, and (5) make the presentation at the right time. Several ongoing R&D programs try to design and build IDSSs. A particular example is the ESPRIT project Graphics and Knowledge Based Dialogue for Dynamic Systems (GRADIENT) which addresses the problems of: (1) knowledge-based alarm handling with real-time state and fault identification, (2) prevention of incidents through monitoring of operator actions, and (3) increased flexibility of the graphics interface through abandoning the restrictions of pre-defined displays. The GRADIENT project regards the interaction between man and machine as a whole and integrates several specialised support functions through the common concept of a knowledge-based, graphical dialogue. The starting point is the analysis of the functionality needed for this purpose. The knowledge-based systems provides the content of the interaction, graphical expert systems provides the form, and the whole is controlled by a dialogue system. There are two categories of users for the GRADIENT system: the designer of a particular application, and the operator who uses it to control the process. In both cases the GRADIENT project considers the functions that are needed to support these user categories.

INTRODUCTION

A recent article about the next generation of nuclear power plants argued that "...artificial intelligence approaches are essential to the revival of the commercial nuclear power industry and a key force that will help us advance the next generation of nuclear power" (Uhrig, 1986, p. 30). We are all aware of the serious accidents that have occurred in NPPs in the 1970es and 1980es, and of the impact this has had on the future for NPPs - in both public and industry. Uhrig also listed four problems that should be solved before plant construction could be resumed:

o   What is the plant going to look like?
o   What is the plant going to cost?
o   When is it going to be finished?

o   How it can be operated after it is finished?

In the long term AI may well contribute to a solution to most of these problems - because knowledge-based systems (KBS) gradually are being used to solve financial and administrative problems, in addition to the traditional applications. In the short term, however, we should only expect significant advances with regard to the problem of operation. In the following I will try to give an outline of some of the developments that are taking place within the European Strategic Programme for Research in Information Technology - also known as ESPRIT.

SYSTEM COMPLEXITY AND SYSTEM COMPREHENSION

The technological development has forced its own changes on the way NPPs look and function. Critics of NPPs have claimed that they are inherently unsafe because they are tightly coupled systems with complex interactions (Perrow, 1984). The use of KBS, however, offers a solution to the handling of the complexity which may render Perrow's arguments invalid.

Computerisation has led to a minituarisation of information presentation and the increased capabilities of computer graphics has tempted many to move away from the large mimic boards. Computerisation has also provided the possibility of including additional information. This has lead to a difficult situation, because there are few established principles on how to manage the dialogue between the computer and the operator. The overall is goal is to improve understanding of the system under operating conditions, i.e., in real-time, both in adverse circumstances and during normal operation (start-up, maintenance, refuelling, technical specifications monitoring, diagnosis, etc.), as well as to increase the grasp under more analytical circumstances (procedure analysis, post-trip analysis, PRA). I will primarily consider the former (operating conditions) but under that also include the whole problem of designing the man-machine interface.

AI can be used in NPPs not only for operator support in the control room, but in practically all phases of the fuel cycle (Kretzschmar, 1986). Basically, AI and KBS can be used to handle complexity - to reduce complexity to within the levels manageable by the human - and complexity is what characterises NPPs, during construction, operation, and maintenance. Complexity can be reduced and understanding enhanced by:

o   **Selecting or producing the right information:** The right information depends on a situation analysis, and the goal is basically to reduce the information to what is needed - no more and no less.

o   **Presenting the information in the right way:** The right way means that it should fit into the situation, i.e. not be controlled by rigid display standards or limitations, but rather be reshaped to fit into whatever the operator is observing at the moment. Maintaining coherence and cognitive momentum is essential (Woods, 1984). This means that the system must understand what is going on and use that intelligently, e.g. in a dialogue controller.

o   **Presenting the information at the right time:** The right timing is crucial for smooth operation. Consequently, the systems must either be fast enough to honour any time demand, or restrict themselves to the type of answers they can deliver in time. AI is no universal solution, but should be judiciously combined with other means of signal processing. Timing is also important with respect to the demands of human decision making. A simple command can be given immediately before it has to be executed, but a piece of advice or a set of alternatives may need some time before they can be acted upon. Again, this can be achieved if a model of the operator's characteristic way of thinking is included in the system.

The use of computers have produced many examples of operator support systems, and the need for them is constantly emphasised by the, seemingly unavoidable, accidents that do occur. The type of activity that is mainly in focus is decision making, and the interest is therefore in the design and implementation of Intelligent Decision Support Systems (IDSS).

ISSUES IN INTELLIGENT DECISION SUPPORT SYSTEMS

There are many formal theories of decision making but few of these are sufficiently developed to serve as a basis for actually designing decision support systems. That is because they generally consider decision making under idealised rather than real circumstances, hence cope with only part of the complexity. Some of the unsolved problems refer to the design of artificial reasoning mechanisms, the structure and representation of knowledge, and the use of information across the man-machine interface.

The purpose of an IDSS is to enhance the quality of information for the decision maker. This involves the problems of determining WHAT information to present (contents, meaning), HOW it should be presented (format, context, and receiver characteristics), and WHEN it should be presented (timing in relation to the decision, whether it should be presented automatically or on request), cf. above. In order to solve these problems, a number of detailed questions must be asked, relating to the problem areas described in the following (Hollnagel, 1988):

One problem area concerns the conditions under which decision support is needed. The classical situation is **information overflow** where the operator gets more information than he can process. This is the situation typically seen in NPP control rooms during accidents, when several hundred alarms may be active at the same time. There are, however, two other situations that may be just as important. One is the case of **incomplete information** where the situation is under-specified, either ill-defined or vague. The operator consequently has to complete the information by his own knowledge, typically falling into the trap of the strong-but-wrong responses (Reason, 1988). The other case is that of **insufficient information** where the information is potentially there, but where cannot be retrieved or produced in time. The varying conditions obviously require different types of decision support.

Another problem area is providing a model or reference description of the decision making process. One need is to find a good behavioural classification, particularly to be able to distinguish between various levels of decision making - e.g. skill-based, rule-based or knowledge-based (Rasmussen, 1986). Another question is whether reasoning should be modelled as time or event based (non-monotonic or monotonic). In particular, the model should be able to answer questions about the locus of errors in decision making (i.e. at which step or stage in the process), the selection of decision strategies, the scheduling of decision making in parallel to other activities, the control of execution, and the handling of conflicts and interrupts.

A third problem area concerns operator modelling as a representation of the decision maker. Here we again find the question of a proper language or formalism (Stassen, 1986), as well as questions of whether the model should be synchronised by emulation or by pattern matching, and whether the operator should be considered as a rational decision maker or a human being guided by expedient heuristics (e.g. Reason, 1988; Tversky & Kahneman, 1974).

A final, and very important problem area, concerns the use of information by the IDSS. Here questions must be asked of whether one should support a global view or rather focus attention on details, of whether one should prefer prediction (hence prevention) to diagnosis, of whether the system should be based on a principle driven design or aim at accident avoidance (not repeating the last accident, which seems to be the basic philosophy, perhaps for funding reasons), and whether the IDSS should be active or passive.

The Top-down Approach To IDSS Development

IDSS can include several things. One is highly processed information that list the alternatives and provide sufficient background information for the operator to accept the assessments. Another is the step beyond that, being the actual recommendations for what should be done, again as advice rather than commands. A third is the implementation of the decision, i.e. checking that the appropriate actions are taken in the appropriate manner. The latter is particularly important, because decision making does not stop with making the decision. Actually implementing it is important, and itself an area where AI and KBS can be of significant value (EOPs, rules and regulations, etc.).

Developing an IDSS can either be done bottom up, trying to solve a specific set of problems, or top down, trying to achieve a goal that is relatively independent of a specific problem. The former is more often the case, but the latter is essential to find and develop alternate solutions. From the top-down point of view the fundamental criteria for an IDSS are the following:

o   the information produced must be reliable and consistent,
o   the IDSS must allow flexible and effective operator interaction,
o   information presentation must relate to current plant status, and current problem,
o   the information must truly support the operators needs, rather than force him to comply with the rigid functioning of the system (IDSS) itself.

It is probably no surprise that many of these problems still are under research. There are accordingly not precise answers to all of them, but in most cases experience provides good guidelines to start from. Even though a completely satisfactory solution may not be found, it is important that none of the problem areas are neglected in the design considerations.

THE GRAPHICAL DIALOGUE ENVIRONMENT

There are several R&D programs which try to design and build IDSSs. Many of them are to be found outside the nuclear field because they have a general aim, e.g. in the European ESPRIT programme. They are, however, highly relevant for the application of IDSSs in NPPs. A particular ESPRIT project is called GRADIENT, a five year ESPRIT project to design and develop a graphics and knowledge based dialogue interface for industrial supervision and control (S&C) systems.

Background

Current S&C systems provide least support for the operator at those moments when it is most needed and the dialogue between the system and the operator is usually inflexible and non-adaptive. If graphics display systems are used, they function by assembling pre-packaged pictures to display the information the display designer has anticipated (Alty et al., 1985). The GRADIENT project investigates the use of KBS to support operator decision making during normal and adverse conditions of system operation, using techniques from knowledge processing and user-system modelling, to support a flexible, dynamic dialogue system. This support will be supplied by a set of interacting expert systems which will provide:

o   Real-time state and fault identification and alarm handling.
o   Prevention of incidents through monitoring of operator actions.
o   Increased flexibility of the graphics interface through abandoning the restrictions of pre-defined displays.

There are several other research projects, in ESPRIT and elsewhere, that consider particular aspects of the use of advanced information technology in process control applications. Most of these, however, focus on a particular type of support, for instance alarm filtering, diagnosis, emergency operations, procedure following, etc., and implement that more or less as a stand-alone system. The unique nature of the GRADIENT project is that it considers the interaction between man and machine as a whole and integrates the several specialised support functions through the common concept of a knowledge-based, graphical dialogue.

The starting point for the GRADIENT project is therefore the analysis of the functionality needed for this purpose. For the actual working situation in industrial process control the crucial point is to present the **right** information in the **right** way at the **right** time. In GRADIENT the KBS will provide the content of the interaction, the graphical expert systems will provide the form, and the whole will be controlled by a dialogue system. There are, accordingly, two categories of users for the GRADIENT system. One is the designer of a particular application, the other is the operator who uses it to control the process. In both cases the GRADIENT project considers the functions that are needed to support these user categories.

Designer Support

The designer has to handle the complexity of the man-machine system when it is specified. One problem is that of compliance and consistency with guidelines, design (Human Factors) knowledge, technical specifications and regulations, etc. This can obviously be supported through a KBS. Another problem relates to the efficiency of the dialogue between the process of the operator. If it is defined too rigidly, it is doomed to fail at the critical moments - simply because they have not been anticipated. It is a basic assumption of the GRADIENT project that it is not feasible for the designer to specify the actual interaction in every little detail or to account for every possible contingency. The designer should rather specify the overall strategies and principles of dialogue control that are to be applied, and the guiding principles that the system should use to minimise operator workload and maximise comprehensibility. There are basic principles that guide that, but they have to be applied dynamically. For this purpose he needs a number of powerful tools - for dialogue specification, for knowledge representation, and for graphical design - which enable him to develop and implement a specific application from the general functionality inherent in these tools, and to structure the form and content of the dialogue. GRADIENT, in particular, will provide the designer with high-level dialogue specification and graphical design techniques.

Operator Support

For the operator, knowledge-based support can be provided in a number of different ways, some of which have been mentioned above. In the GRADIENT project the emphasis has been put on facilitating the response to a disturbance through intelligent, real-time alarm handling and provision of recommendations, and on reducing the occurrence of 'human errors' through a situation dependent supervision of actions (Hollnagel, 1987). The output of the specific knowledge-based systems constitute the main content of the dialogue. This is presented to the operator according to the specifications provided by the designer, as implemented by the functionality in the dialogue system and the graphical expert system. Several systems, that address one or the other of these problems, develop stand alone solutions which often lead to an increase in the workload for the operator, hence do not function as planned. Very few systems consider several of these functions together. GRADIENT views each of these functions as a KBS support, and integrates it all under the view of the dialogue control system. Consequently, one of the centers of the GRADIENT system is the dialogue system, which takes care of the dialogue control and routes information between the different 'objects' in the system.

GRADIENT System Components

The GRADIENT system contains several subsystems, as shown in Figure 1. Each of these are outlined below. Further details can be found in the reports issued by the project, as listed in the references.

o   **Quick Response Expert System (QRES):** The purpose of QRES is to support the operator during a system failure. In order to accomplish this QRES must identify system failures and system failure states, inform the operator of their existence, evaluate their severity, and recommend the proper corrective action. QRES processes data from the S&C system and sends the results to the Dialogue System (DIS). Since QRES is a time critical system it must respond fast enough to satisfy the demands of the process dynamics (Belleli et al., 1986).

o   **Response Evaluation System (RESQ):** RESQ is an advisory system that takes the operator's actions, relayed through the Dialogue System, as input. RESQ monitors the actions to detect situations where the action pattern is inconsistent or incorrect. Such situations may occur when the operator forgets actions that were performed some time ago or whose execution has been delayed, when several lines of action are mixed due to simultaneous or overlapping execution, in the transition between work shifts, or due to the 'normal' slips and mistakes in human action.

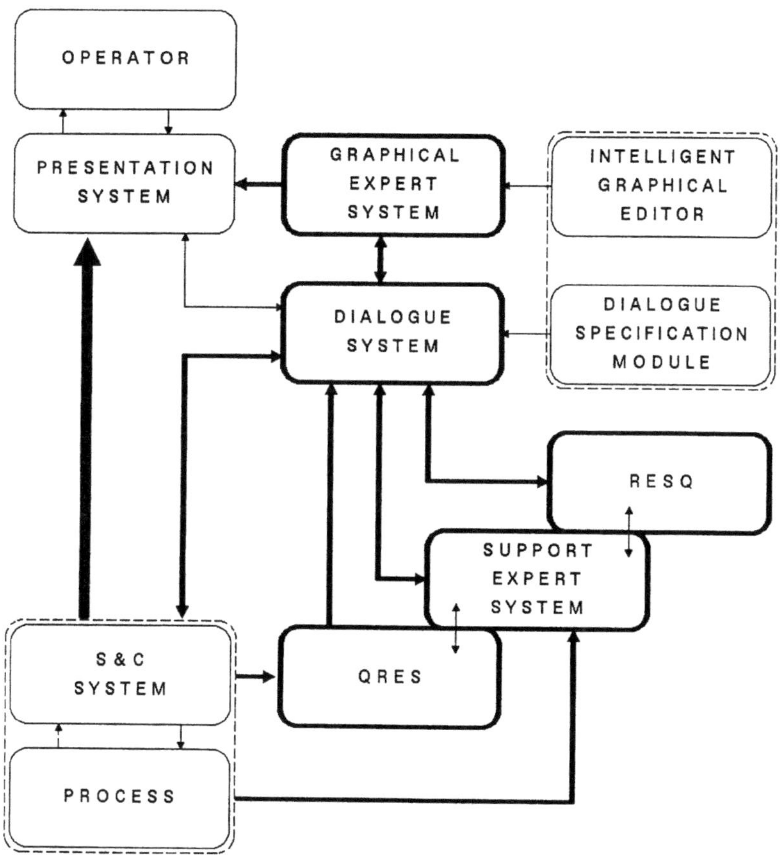

Figure 1   ESPRIT #P857 - GRADIENT
General System Architecture

o   **Support Expert System (SES):** The main function of SES is to support the operator with knowledge and information in addition to what is provided by the other expert systems - QRES, RESQ, DIS, and GES. To this end, SES will use and maintain a knowledge base that contains fundamental (design) knowledge of the process, cumulated operating experience and a full user description model. SES is a specialised form of the "conventional" expert system, with full knowledge storage and search capabilities but a restricted inference system. Together with QRES and RESQ it makes up a complex of dynamically interacting expert systems which is expected to perform significantly faster than present solutions.

o   **Dialogue System (DIS):** The Dialogue System takes care of the communication between the operator and both the Process and Support Expert System (SES). It does not handle 'front-end' dialogue, which, in the form of input / output, is the job of the Presentation System. Process information is monitored by the Dialogue System but is routed directly to the Presentation System for display (Alty et al., 1986; Neilson, 1987a&b).

o   **Graphical Expert System (GES):** GES is intended to support the DIS by handling the main stream of measurement and control information from the S&C system. GES will dynamically compose pictures and picture sequences, using knowledge of the process (cf. Borys et al., 1986), the user model, graphical representation techniques and dialogue techniques. GES will use several different types of knowledge base. One may contain data about how the system should present information to the user, e.g. picture composition, coding techniques, the use of shape, colours, windows, frames, etc. Another may contain knowledge about the behaviour and requirements of the human operator, and be used to select the most effective presentation technique. A third may contain application specific technological knowledge (Borchers et al., 1986). GES will be preceded by several phases of Intelligent Graphical Editors (IGE).

The basic idea in the GRADIENT project is not simply the use of AI, but rather the use of multiple KBS to support the operator. The goal is to provide a tool - for the designer and the operator - rather than a prostheses or a replacement of either. By using this approach, and considering each KBS as an object in its own right, the whole system becomes easy to modify, maintain, and extend. These are necessary virtues since the industrial application of AI is still in its infancy.

ACKNOWLEDGEMENTS

The work reported here was carried out under ESPRIT Project #857 in a consortium consisting of CRI, BBC (Heidelberg, F. R. Germany), University of Kassel (Kassel, F. R. Germany), University of Strathclyde (Glasgow, Scotland), and University of Leuven (Leuven, Belgium). The support of the CEC and the contributions of the other members of the project consortium is gratefully acknowledged.

REFERENCES

Alty, J. L., Elzer, P., Holst, O., Johannsen, G. & Savory, S. (1985). **Literature and user survey of issues related to man-machine interfaces for supervision and control systems.** Copenhagen, Denmark: Computer Resources International.

Alty, J. L., Mullin, J. & Weir, G. (1986). **Survey of dialogue systems and literature on dialogue design** (P857-WP3.1/.2-UST-001). Glasgow, UK: University of Strathclyde.

Belleli, T., Christensen, B., Hollnagel, E., Hovgaard, H.-O. & Jepsen, M. (1986). **Design and architecture of QRES1** (P857-WP5-CRI-033). Copenhagen, Denmark: Computer Resources International.

Borchers, H. W., Elzer, P., Siebert, H., Weisang, C., Borys, B.-B., Fejes, L. & Johannsen, G. (1986). **Survey of existing sets of picture elements and editors.** (P857-WP3-BBC-012). Heidelberg, F. R. Germany: Brown, Boveri & Cie AG.

Borys, B.-B., Hansel, H.-G., Johannsen, G. & Schmidt, J. (1986). **Task and knowledge analysis. Methodology and application in power plants** (P857-WP2.1-UKS-007). Kassel, F. R. Germany: Gesamthochschule Kassel.

Hollnagel, E. (1987). **Plan recognition in modelling of users.** Invited paper for the International Post-SMiRT Seminar on Accident Sequence Modelling, München, F. R. Germany, August 24-25.

Hollnagel, E. (1988). Information and reasoning in intelligent decision support systems. In G. Mancini, D. D. Woods & E. Hollnagel (Eds.), **Cognitive engineering in dynamic environments.** London: Academic Press.

Kretzschmar, J. G. (1986). **Expert systems in the nuclear sector.** (In Second International Expert Systems Conference, London, 30 September - 2 October 1986.) Oxford, England: Learned Information.

Neilson, I. (1987a). **Techniques for specifying dialogue** (P857-WP3-UST-017). Glasgow, UK: University of Strathclyde.

Neilson, I. (1987b). **Dialogue specification: A select bibliography** (P857-WP3-UST-018). Glasgow, UK: University of Strathclyde.

Perrow, C. (1984). **Normal accidents.** New York: Basic Books.

Rasmussen, J. (1986). **Information processing and human-machine interaction.** New York: North-Holland.

Reason, J. (1988). Cognitive aids in process environments: Prostheses or tools? In G. Mancini, D. D. Woods & E. Hollnagel (Eds.), **Cognitive engineering in dynamic environments.** London: Academic Press.

Stassen, H. (1986). Decision demands and task requirements in work environments: What can be learnt from human operator modelling. In E. Hollnagel, G. Mancini & D. D. Woods (Eds.), **Intelligent decision support in process environments.** Berlin: Springer Verlag.

Tversky, A. & Kahneman, D. (1974). Judgment under uncertainty: Heuristics and biases. **Science, 185,** 1124-1131.

Uhrig, R. E. (1986). Toward the next generation of nuclear power plants. **Forum for Applied Research and Public Policy,** (Fall), 20-31.

Woods, D. D. (1984). Visual momentum: A concept to improve the cognitive coupling of person and computer. **International Journal of Man-Machine Studies, 21,** 229-244.

A MAP MULTILEVEL SYSTEM AS A MAN-MACHINE INTERFACE FOR EMERGENCIES MANAGEMENT

F. Argentesi and N.M. Avouris

Commission of the European Communities
Joint Research Centre, AI Laboratory
I-21020 Ispra (Va), Italy

Abstract

In order to create a valuable man-machine interface for the interactive emergency management systems we are developing in our laboratory a "new" paradigm, that of hypermaps or multilevel maps which is presented in this paper. Within this concept the user interaction with the emergency system is supported by a set of linked maps spanning from a very high level, like an overview map of the area under threat, down to a very low level like detailed technical drawings of the machinery concerned. A rich network of navigation routes is allowed by the HyperMap system permitting rapid changes in the conceptual levels of the problem solving task. The map support system permits, within each level, access to a variety of active tasks such as local expert systems, data bases of various kinds and simulators. This interface is designed for an emergency management system called ChEM (Chemical Emergency Manager), developed at JRC.

1.Introduction

Emergencies management is an area of increasing importance for nuclear, non-nuclear industry, government authorities and institutions and even international organizations faced with handling natural or technological disasters.

The classical approach to emergencies management has been that of preparing emergency plans before the actual start of the event. These emergency plans are usually limited to a small number of emergency scenarios, i.e. the scenarios that can be reasonably foreseen a priori. This approach could result in low level of preparedness, inflexibility or wrong reactions during the emergency.

Computerized interactive emergency management systems can implement more complex and flexible models and therefore cope with a wider spectrum of emergency events. The design and implementation of computerized emergency management systems is a very complex task involving deep interaction between computer science problems and the difficulties linked to the modelling of emergency processes.

In a complex situation, like that immediately after a nuclear accident, the emergency handlers are called to take into consideration a series of technological, economical, environmental and sociopolitical factors and to decide, usually under time pressure, on the safety and health of a great number of people. An easy to operate and efficient computerized integrated emergency management system which would supply all the required information and suggest measures based on expert's knowledge, could be proved an invaluable tool for the decision maker.

In our laboratory at JRC Ispra we are developing emergency management systems for both chemical and nuclear industry. The architecture of these systems is quite complex involving the design and implementation of a variety of models : knowledge processors, simulation models and data models for information management.

In order to create a valuable man-machine interface for this type of systems, which we consider as one of the most important features of a successful design, we are developing ideas and specific technical implementations based on intelligent spatial data organization principles. The result of these efforts has been the creation of a "new" paradigm, that of *hypermaps* or *multilevel maps*.

2. Investigating space organization

The idea under investigation during the design of the emergency management system interface is that of using the metaphor of a navigable space as the basis and common reference point of the integrated system. This method of data and processes organization has been discussed in various contexts[2].
Space order is a powerful organizational factor, something easily observed in our everyday life. For example libraries organized in the traditional way give considerable thought to the spatial arrangement of information. Books on the same subject are always collected together. Stacks are arranged so that the visitor could walk (navigate) between them, reaching up for any desired subject.
 In the field of emergencies management software, spatial order is of prominent importance. The effectiveness and efficiency of the rescue and evacuation as well as mitigation and contingency plans must be based on geographic and time specific data. A good picture of the situation and the spatial distribution of available human and material resources can be shown only on dedicated maps of the accident and the surrounding space. It is obvious then that a decision support system should exploit in the maximum degree the ideas of spatial data organization.
This has lead us to use maps, which are traditionally a familiar way of depicting spatial distribution of significant properties and characteristics of the earths surface, as a central point of our interface.
This supplies the emergency commander with the modern version of the traditional weapons that the field commanders during the crucial battles have always used in order to keep track of events and their forces.
Finally the use of maps is required in most official guidelines about nuclear emergency response plans [8], in relation to various spatial characteristics like topography, evacuation routes, maps of sampling and monitoring points, shelter areas, demographic maps etc.
 In what form however, these maps should be organized and integrated with the emergency management software? the choice of existing solutions offered by the computerized mapping industry is vast. It is ranging from small scale, presentation graphics based, mapping systems dedicated to assist sales managers, up to the highly sophisticated "Geographical Information Systems" (GIS), which can serve in a great number of applications like urban and regional planning, environmental and natural resources management etc.

3. Mapping Systems

A first fundamental question any map system designer has to answer is whether to treat a map as a picture or a database.
The first approach has lead to a series of optical-device-based systems. The American National Ocean Service is proposing a two-sided videodisc that will contain approximately 108,000 images of maps, charts and satellite and aerial photographs, while a system developed by LTV Vought Corp to serve as a flight simulator for the American Navy contains a visual database of the earth's surface stored on 16 videodiscs for a total of 864,000 frames. The information provided by these vast pictorial databases however does not go very far beyond the traditional paper maps, severely limiting what can be done with the fundamental geographic relationships stored on maps; more complex database representations yield the map a much more flexible tool for the decision maker.
The second trend with the accent on spatial databases and spatial analysis techniques has lead to the integrated GIS systems which constitute ".. general sets of tools for collecting, storing, retrieving, transforming and displaying of spatial and attribute data" [6]. The GISs usually run on sophisticated computing environments with expensive graphic workstations. Their development requires skills of software engineering, cartography, computer aided design and computer graphics, data base management theory and even remote sensing and image processing technology.
The four components of conventional GISs are: (i) Data input subsystem that collects and processes spatial and attribute data derived from maps, remote sensors and other sources; (ii) data storage and retrieval subsystem, which contains the database management system; (iii) data manipulation and analysis subsystem that consists of simulation models etc and (iv) map display and reporting subsystem for output of the data on the CRT or the plotter in the form of maps, or the printer in the form of tables.
A GIS system of the plant and its surrounding area can be very useful for emergencies management. In a system like this the base map of the geographical area can be digitized, manipulated and reproduced on a graphic computer terminal. Various data themes like water resources of the area, distribution of urban communities, flora and fauna patterns, land use, weather characteristics, road and transport networks, maps of infrastructure utilities like water, gas, electricity etc. can be overlaid on the map. Spatial relations among the various data themes can be examined and specialized maps can be developed. So if the emergency manager would like, for example, to identify milk animals which are in potential danger of contamination in the plume exposure pathway, he should overlay the thematic map of dairy farms and the polygon of the zone under threat and create a new map which would contain the details of the farms to be given the warning.

Existing GIS systems which have been developed by using conventional computer science techniques have however limitations[4,5] which can be summarized in: (a) Due to their complicated nature GISs require familiarity and the expert knowledge of geographers and experts of other related subjects for their full use and operation. (b) The subjects covered by a GIS , planned to be used for emergency management are diverse, so the knowledge which relates to these subject domains consists of technical information, simulation models, experts opinions etc , becoming so large that it is rarely, if ever, used in its entirety for emergency management decision making. (c) Existing GISs are mainly used for planning, resources management etc, so the user interface is designed without consideration to   friendliness and speed, which are prime requirements for an emergency management tool.

4. The HyperMap approach to Intelligent geographic systems

Some of the problems and limitations of conventional map systems have been considered and an attempt has been made to solve them designing an HyperMap emergency management interface system. The most important design features of HyperMap are discussed in this paragraph:

(1) The HyperMap consists, apart of the four components of a conventional GIS, of a knowledge base which contains, in the form of rules, the interpretation of events and relations among data items. There can be various knowledge base modules associated with different subjects ranging from technical information concerning plants, to knowledge about the ecology of the area etc. This should be combined with the fundamental characteristic of the HyperMap system : its full integration with the rest of the emergency management software. The map support system should permit,within each level, access to a variety of active tasks such as local expert systems, other data bases of various kinds (including pictorial data bases) and simulators. The higher level of the map navigation system is capable of top view analysis where the required cooperation among the underneath tasks is possible.

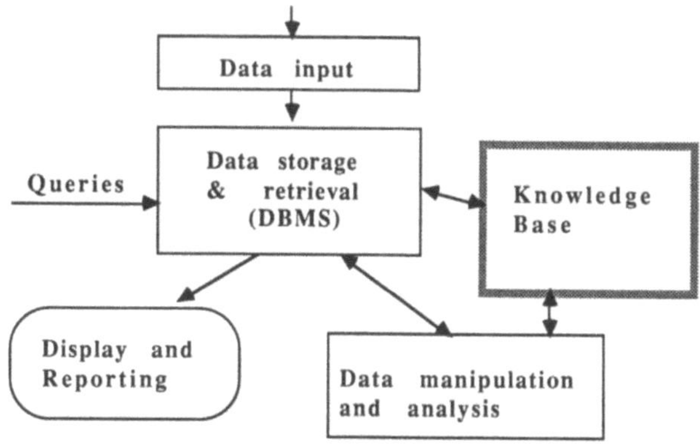

Fig. 1 The Subsystems of an Intelligent map system

(2) The nature and complexity of emergencies demand a "multilevel" approach. The emergency handlers during different stages of their effort and according to the relative position and distance from the centre of action require a different level of information. So the spatial style involved deals not just with a single data surface or information space but an entire set of them. The transition points between different information spaces are called Zooming Points. The result is a hierarchical set of information spaces through which the user plunges, pursuing some topic from plane to plane. So a characteristic of HyperMap is that it consists of a system of linked maps and not just a single detailed map like in most geographical information systems. A navigation through the HyperMap results therefore in moving between different maps through the relevant Zooming Points permitting rapid changes in the conceptual levels of the problem solving task. This differs considerably from the graphical zooming

capabilities of most mapping systems which just change the scale of their presented image, using map generalization algorithms and techniques.

(3) The contents of the maps could vary, so a HyperMap can contain:
 (a) Raster images of the earth's surface of the type produced by processed remote sensing data. These can supply an alternative view of the area concerned and serve in observing certain changing patterns on the earth's surface.
 (b) Conventional maps i.e. graphical representation of sets of points, lines and areas defined by their location in space with reference to a coordinate system, their topological relation and their non-spatial attributes. A great range of general topographical maps supported by specialized "thematic" maps can be used. A vast range of digitized maps already exist in most planning departments of central and local government and specialized map digitizing companies can supply HyperMap with material. That is why every attempt has been made to conform with existing standards.
 (c) In the lower level, HyperMap can contain detailed technical drawings of the machinery and other equipment concerned as well as plans of the buildings, architectural designs, sections etc which can be of great use for better understanding and quick identification of the danger. The use of any existing digitized drawings in the plant infrastructure department where a growing number of CAD tools are applied could save a lot of effort.
 (d) Text. All the above types of maps can be combined with explanatory or advisory text or documents like accident reports etc. Text is considered as an active element of the HyperMap structure and therefore it can be structured in various levels in the form of HyperText[7], containing Zooming Points into various other maps etc.

(4) A series of thematic maps which can be overlaid in various combinations and can be superimposed on various levels, provide information about the spatial distribution of significant properties of the area involved. These can be for example : demographic maps, municipal, jurisdictional or country boundaries of the area under threat, various exposure pathways with zones of responsibility, emergency service stations like ambulance, hospital, fire-fighting units, police stations, plans of evacuation routes and areas, plans of shelter areas, land use maps, water systems, plan of food processing plants and dairy farms in the surroundings of the plant, main and secondary transportation network maps etc.

(5) Another important feature of the HyperMap is the speed of navigation between various levels, which is not one of the prime considerations in the analysis oriented traditional GISs. In HyperMap the quick synthesis of the new image in every level is very important, since speed is of high importance in emergency management applications.

(6) An important additional feature is the ability of the system to access optical storage devices like optical videodiscs, cd-rom etc which can contain a great number of frames and can reproduce a realistic image of a plant, a frame sequence which simulates navigation in interactively chosen routes [3], still photos etc. The New Media group of our laboratory is developing new ideas on the navigation of pictorial data bases and their application on HyperMap should provide additional components for a realistic reconstruction of the scene of an accident.

5. ChEM: a prototype of a HyperMap decision support system

The principles of HyperMap design have been applied on a Chemical Emergencies Management system (thus called ChEM), which is under development in our laboratory at JRC Ispra[1].
The HyperMap prototype system to support ChEM is written in C, developed under the Metawindow graphic environment which guaranties transportability of the system in a variety of machines.
The HyperMap Manager (HM) is the most important module of the system. Other modules and utilities are concerned with map creation, editing, downloading map files into the HyperMap format, maintaining the map-symbol library etc.
The HyperMap system operates as follows: The HM maintains a directory of all the maps currently in the HyperMap along with their position in the map-hierarchical system. When the system starts, the HM identifies the top level map which is retrieved, initialized and given the status of the "current map". Information about child-maps and child-tasks is also loaded in a list which is constantly associated with the current map. A table of all the map-sheets which can be overlaid on the current map is also maintained along with their status (i.e. whether they are activated or not). This information about overlay- sheet status is saved even if the current map changes. So it can be guarantied that if for example the user is investigating the primary evacuation routes overlaid on a certain area and wishes to inspect temporarily some weather report, when he returns at the level he was working, his evacuation routes will be undisturbed.
The user interaction with the HyperMap is via mouse and pop- up menus. Using the mouse for instance the user can select a zooming point in order to inspect a detail of the current map. This

results in the change of the current map status. The HM identifies the selected map in the map directory, retrieves the related information, calls the map drawing module, which visualizes the map on the screen, and updates the current map lists and tables. So at the end of the operation the ex child-map has become the new current map, while the ex current map is now called "parent-map" and can be reached by a zooming- up operation.

An attributes data base containing information about the elements of the different maps is associated with the HyperMap. It should be mentioned that the available information for a certain element can differ considerably, between different levels of the map structure. So there is different degree of detail supplied about a certain industry if the information is extracted from the map of the region, instead of the map of the industrial site itself.

A series of maps have been created for the ChEM system. The maps were related to industrial areas of the north Italy where serious chemical accident have occurred. The maps were produced by using a technique combining a vision system and graphic digitizers.

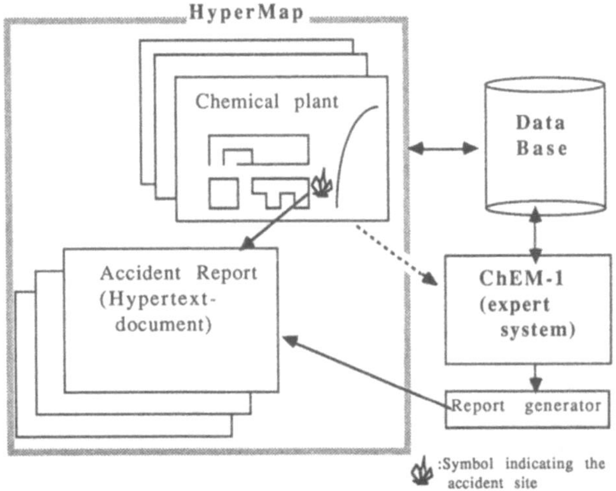

Fig. 2 The structure of ChEM-1 HyperMap system prototype

ChEM has been designed as a decision support system for chemical emergencies caused by the accidental release in the environment of polyhalogenated aromatic compounds (e.g. PCB, PCDF and PCDD), a class of supertoxic chemicals which in recent years have caused serious accidents like the one in Seveso, Italy etc. The system is being implemented in independent modules which perform different tasks like identification and classification of the threat, analysis and evaluation, response and neutralization, post- emergency actions etc. From these only the first module has been implemented and tested so far. It is an expert system developed on the shell Savoir which is available for a wide variety of computing environments. Savoir is capable of linking a knowledge base with external programmes for numerical computation and data input. ChEM is therefore designed as a coupled system which can be integrated with a mathematical simulator for the release and diffusion of the toxic material into the soil and other media. An important component of this module is the knowledge base which contains the expert knowledge on the field expressed through more than 1500 rules. ChEM-1 is activated by the user by indicating the chemical plant in which the accident occurred in the corresponding map. All the information related to the specific plant, as seen from the particular level of the HyperMap system is fed to the module which also defines the context of the accident by asking the user all the necessary questions. ChEM-1 can produce an estimation of the level of the threat according to an association network implemented in a knowledge base.

The output of ChEM contains also general information about the incident (release, explosion, discovery of disposal sites etc), qualitative information about toxic levels, estimate of the damage to the environment, information about the dynamics of the current state and suggestions about the urgent actions to be taken by the decision maker. This is all contained in a report produced by the Report Generator (RG) which can be dynamically incorporated in the HyperMap structure as a Hypertext document, it can therefore be accessed through a relevant Zooming Point created at the accident site.

The output parameters can be fed into the map management system which can produce a graphic image of the levels of threat and their spatial distribution in the area superimposed on the corresponding map. The system can also indicate and graphically display communities in immediate or potential threat, zones where the fauna and flora are in real or potential danger of various degrees and other environmental and ecological hazards. Using the available overlay map-sheets about infrastructure utilities and services it can suggest immediate and long term countermeasures, display evacuation routes, gathering points, rescue services and estimate the response time of the responsible teams.

It can also be used to feed data into mathematical simulators which using mathematical models and information about weather conditions, territory topology and other data can simulate the event sequence and display graphically the various effects it has on the emergency indicators.

6. Conclusions

The possibility of using new-media-based navigable pictorial data bases, investigation of various spatial data base models and the possibility of using the system on-line in combination with constant monitoring for periodic reevaluation of the situation and forecast of future occurrences of emergencies, are some of the new ideas we are investigating during the course of this research project.

The ideas presented in this paper about the design of an intelligent emergency management interface which:

(i) would take advantage of the experts' knowledge on the various factors of the crisis;

(ii) supply the emergency manager with an easy to understand picture of the situation , graphically depicted in relation to familiar realistic plans and maps of the area;

(iii) permit navigation through different parts of the plant by zooming into a particular area or having an overview of the situation by zooming out ;

(iv)provide the possibility of studying particular aspects of the crisis through "thematic" maps, we believe that   transform the decision support software into an invaluable tool for those handling nuclear or chemical emergencies.

7. References

1. F.Argentesi, L.Bollini, S.Facchetti, G.Nobile, W.Tumiatti,G. Belli, S.Ratti, S.Cerlesi, G.U.Fortunati, V.La Porta, "ChEM: An Expert System for the management of chemical accidents involving halogenated aromatic compounds", World Chemical Accidents Conference, Rome ,July 1987, pp 227- 230.

2. R.A. Bolt, "The human Interface, where people and computers meet", Lifetime learning Publ, 1984.

3. A. Lippman, "Movie-map an application of the optical videodisc to computer Graphics", ACM 0-89791- 021, April 1980.

4. S. Kubo, "The basic scheme of Trinity, A GIS with Intelligence", 2nd Int. Symb. on Spatial Data Handling, Washington, July 1986, pp 363- 374.

5. R.N. Coulson, L.J. Folse, D.K. Loh, " Artificial Intelligence and Natural Resource Management", Science, Vol 237, pp 262-267 (1987).

6. P.A. Burrough, "Principles of Geographical Information Systems for Land Resources Assessment", Clarendon Press, Oxford 1986.

7. J. Conklin, "Hypertext: an Introduction and Survey", IEEE Computer, September 1987, pp. 17-41.

8. US N.R.C. , F.E.M.A. , "Criteria for Preparation and Evaluation of Radiological Emergency Response Plans and Preparedness in Support of Nuclear Power Plants",Rev 1, Nov 1980.

A MODEL-BASED DISPLAY

L. Beltracchi

U.S. Nuclear Regulatory Commission
Washington, DC   20555

ABSTRACT

A model-based display is identified, discussed, and illustrated. The model
used in the display is based upon the Rankine Cycle, a heat engine cycle.
Plant process data from the loss of main and auxiliary feedwater event at
the Davis-Besse Plant on June 9, 1985 is used to illustrate the display.

The model used in the display fuses individual process variables into process functions. It also serves as a medium to communicate status of the
process to human users. The human user may evaluate the goals of operation
from the displayed process functions. Because of these display features,
the user's cognitive workload is minimized. The opinions expressed herein
are the author's personal ones and do not necessarily reflect criteria, requirements, and guidelines of the U.S. Nuclear Regulatory Commission.

INTRODUCTION

The individual control panels within a control room of a nuclear power plant
contain meters, indicators, gauges, control stations, and switches, etc. An
operator's evaluation of a process function within the heat engine cycle requires the evaluation of one/several instruments that display measured process variables. Unfortunately, this may require an operator to move among
several panels, store the value of the individual process variables within
short-term memory, and then process the collected data to evaluate the process function.

Computer-driven cathode-ray tubes (CRTs) are also used in control rooms to
integrate and display related process data. For example, the typical Safety
Parameter Display System (SPDS) in nuclear power plants consists of a
computer-driven CRT. Many display formats within the SPDS consist of sets
of process variables and their numerical values that may be used to evaluate
a safety function. Stored logic within the computer is used to monitor the
individual process variables and to display perceptual cues when a violation
is detected.

The language of the above-described interfaces consist of text and numbers
for individual process variables. The operator must read and evaluate each
process variable displayed to fully assess the process functions and goals
of operation. Furthermore, the operator must understand the meaning of each

item of text to interpret individual process variables as well as to evaluate the relationship among the process variables.

Norman (Ref. 1) identifies and discusses a new interface language: an iconographic language. The basic syntax of an iconographic language consists of graphic primitives. The primitives, such as vectors, are used to form graphic segments. The graphic segments are then used to form an icon. In an iconographic language, the form of the expression is related to its meaning. However, the user must know the semantic meaning of the icon for it to serve as a useful interface. Moreover, because humans learn from birth to understand images before text, icons serve as a powerful language by which to communicate data and information.

An example of an iconographic language for potential use in control rooms of power plants is identified and discussed next. The model used as the basis of the iconic display is the Rankine Cycle, a heat engine cycle. An image of the Rankine Cycle is computer generated from measured process variables of the coolant water to form the icon. Because the displayed image integrates related process variables and models the process, the human operator may directly evaluate functions within the process. In summary, the icon is an image of the process that integrates related process variables. The process icon serves as the user's mental model of the heat engine cycle.

RANKINE CYCLE

The Rankine Cycle is a heat engine cycle. The basic Rankine Cycle is composed of four process functions consisting of: 1) heat addition to a working fluid, which results in a phase change from liquid to vapor; 2) an isentropic, reversible expansion of the fluid, which does work upon a turbine; 3) a reversible, heat rejection process of condensation, which results in a phase change from vapor to liquid; and 4) an isentropic compression of the liquid by pump work. In real world performance, the basic Rankine Cycle is unachievable. However, each of these process functions discussed above is related to the ideal cycle by an efficiency factor. The measured properties of the coolant within the cycle and the efficiency factors are used to construct a model of the actual cycle in the process. Additional details on the Rankine Cycle are found in Obert (Ref. 2).

The process within the secondary coolant loop of a pressurized water reactor (PWR) is directly related to the Rankine Cycle. The process experienced by the coolant water within a cycle for a boiling water reactor (BWR) is also directly related to the Rankine Cycle. The measured properties of water, such as temperature and pressure at various points within the process cycle, are used to formulate the Rankine Cycle.

A MODEL OF THE PROCESS

Reference 3 identifies a method for converting measured temperature, pressure, and water level into a process icon. The display format for the process icon consists of temperature versus entropy coordinates of water. These coordinates are one of the forms in which a Rankine Cycle may be expressed. While the entropy coordinate is used to form the icon, it is unnecessary to display the entropy values. However, the display of a temperature grid is useful to the user.

The process icon in Reference 3 illustrates portions of the TMI-2 accident and Ginna's steam generator tube rupture event. The display format in Reference 3 is used herein to illustrate a model-based interface. Details for coding of data into a process icon are in Reference 3.

On June 9, 1985, Toledo Edison Company's Davis-Besse Nuclear Power Plant, a two-loop pressurized water reactor, experienced a partial loss of feedwater while operating at 90% power (Ref. 4). Following a reactor trip, a loss of all feedwater occurred. While operators acted to restart the safety-related auxiliary feedwater system, operator actions outside the control room were also taken to place a non-safety-related, electric motor-driven startup pump in service. Meanwhile, the plant's two steam generators had essentially boiled dry before feedwater from any source became available to them. Nevertheless, operators were successful in re-establishing feedwater and bringing the plant to a stable shutdown condition.

The plant's process computer was operating and recorded process and system data during the event. A copy of the process and systems data is in the Nuclear Regulatory Commission's Public Document Room. The author obtained a portion of this data. The data describe operations between 01:34:00 AM and 01:59:00 AM. The loss of the two main feedwater pumps occurred at 01:35:00 AM, which was followed by a reactor trip. The display formats and discussion that follow are based on an analysis of and processing of this data.

Figure 1 contains data that models operations in the primary coolant system and the steam generators at 01:34:00 AM. At that time, the plant was operating at 91.4% of design power. The display format consists of temperature and entropy coordinates of water, for temperatures between 400 F and 750 F, and for entropy between 0.5 BTU/#m F to 1.5 BTU/#m F.

The bell-shaped curve in the center of the figure describes the saturation state of liquid water and steam vapor. The portion of the curve to the left of the apex is the saturated water line. The portion of the curve to the right of the apex is the saturated steam line. The apex of the curve is known as the critical point, where liquid and vapor are indistinguishable from one another. In terms of an iconographic language, the area bounded by the curve is known as a graphic segment. As an aid for interpreting the data, a temperature grid is also presented. While the use of entropy is important to locate data within the format, it is generally unnecessary for interpreting the data. Thus, no entropy grid is used after Figure 1.

The states of water as a function of temperature and entropy (Figure 1) must be known by a user. To the left of the saturated water line, the water is in a liquid state and subcooled. To the right of the saturated steam line, water is in a vapor state and superheated. Between the saturated water line and the saturated steam line, saturated water and saturated steam co-exist with one another. This knowledge must be known by a user to evaluate the data in the display.

The horizontal line in the two-phase region at 648 F is known as the pressure bar. The pressure bar represents the average pressure in the primary coolant system, which at the time of the event was about 2182 psia. The saturation temperature of water at this pressure is 648 F. The pressurizer in the primary coolant system contains saturated steam and saturated water when at steady state operation. Thus, it is logical to locate the pressure bar in the two-phase region.

The water level in the pressurizer is also coded on the pressure bar. The graphic segment used to code the water level is the little square on the line. The pressurizer water level, as shown, is 61.9% full. If the square were on the saturated water line, it would indicate that the pressurizer were full of water. If the square were on the saturated steam line, it would indicate that the pressurizer were full of steam. A linear scale is used between these saturation lines to code water level.

The hot leg temperatures in the primary coolant system are 605 F. With an
operating pressure of 2182 psia, this means that hot leg water is 43 F subcooled. This information is coded into the display by a horizontal line
that originates at 605 F on the saturated water line and extends to the left
(into the subcooled region of water) by a distance proportional to the
amount of subcooling. Although this method of coding subcooling is arbitrary and is not in total agreement with thermodynamics, it does convey information to a user in a clear, legible manner.

The cold leg temperatures in the primary coolant system are 561 F. With an
operating pressure of 2182 psia, this means that cold leg water is 87 F subcooled. This subcooling is coded in the same manner as the hot leg temperature. The area between the hot leg and cold leg temperatures represents the
range of coolant temperatures of the primary coolant exclusive of the pressurizer. This mass of subcooled primary coolant is encoded as the polygon
shown in Figure 1. Note that the amount of subcooling increases with distance from the pressure bar. Finally, the temperature rise across the reactor, 44 F, is also coded into this graphical segment of the display as the
vertical height of the polygon.

The steam generator process line is the horizontal line located below the
polygon and in the two-phase region. This process line represents the boiling of saturated water to saturated steam in the secondary side of the steam
generator. The saturation temperature in the steam generator is 527 F. The
water level in each steam generator is also coded in the same way as was
pressurizer water level. The subcooling of the feedwater is coded in the
same way as in the primary coolant loop.

The saturated steam in the steam generator is superheated before it leaves
the steam generator. The saturated steam of 527 F is heated to 596 F, which
represents 69 F of superheat. The expansion of the superheated steam in the
high pressure turbine is partially shown as the vertical line that originates from the 596 F point shown in Figure 1. The display segments for the
secondary coolant represent a portion of the Rankine Cycle discussed earlier.

The bulk of the heat transfer from the primary coolant system to the secondary coolant system is used to convert saturated water to saturated steam.
The heat transfer process is a function of the temperature difference (56 F)
between the average primary coolant temperature (583 F) and the saturation
temperature of the secondary coolant (527 F). It is also a function of the
overall heat transfer coefficient, which is basically determined from nucleate boiling. The superheated steam is a function of the primary hot leg
temperature and the film boiling heat transfer coefficient, which is much
lower that the nucleate boiling coefficient. The superheated steam is only
9 F colder than the hot leg temperature of the primary coolant. It is important that superheated steam enters the high pressure turbine, because
wet steam will result in damage to the turbine blades.

The process data integrated into Figure 1 form an image of the process and
is thus a process icon. The basic heat transfer functions within the process may be evaluated from the information encoded within the graphic segments that compose the icon. Thus, the goals of operation, such as superheated steam from the steam generator, may be monitored by a user of the
icon. Also, the superheat is an important factor that influences the efficiency of the heat engine cycle, which determines the amount of electrical
energy produced. Furthermore, malfunctions within systems that result in
signatures on process variables may also be detected from the process icon.

Figure 2 contains a process icon for operations at 01:38:30 AM. The primary
coolant temperatures are about isothermal at 557 F and subcooled by 77 F.
These temperatures are close to the heat sink temperatures of the steam generators, which are about 545 F. The small temperature difference between

the primary coolant and the secondary coolant indicates the systems are
coupled and heat is being transferred. It also indicates that the steam
generators have enough feedwater to serve as a heat sink. The steam leaving the steam generator is superheated to a temperature that is higher than
the indicated primary coolant temperature. This could be the result of a
lag in the temperature sensor in conjunction with the low heat transfer
coefficient associated with steam vapor. The superheated steam leaving the
steam generator is throttled as it flows through the turbine bypass valve.
Part of the throttling process is illustrated in Figure 2.

Figure 3 contains a process icon for operations at 01:44:30 AM. At this
time, all feedwater flow to the steam generators is lost. Subcooled feedwater has been deleted from the icon. Also, the primary coolant is heating
up (562 F) as a result of the loss of the steam generator as a heat sink.
This is evidenced by the increased temperature difference between the primary coolant and the secondary coolant in conjunction with the increased
water level in the pressurizer.

Figure 4 contains a process icon for operations at 01:47:30 AM. The primary
coolant continues to heatup (572 F) because of the lack of a heat sink in
the steam generators. The temperature difference between the primary coolant and the steam in the steam generators is about 30 F. The pressurizer is
about full of water. The pressure in the primary coolant system is high as
seen from the location of the pressure bar and the saturation temperature in
the primary coolant system of 662 F.

Figure 5 contains a process icon for operations at 01:50:00 AM. The primary
coolant continues to heatup (580 F) because it is unable to transfer heat to
the secondary coolant. The temperature difference between the primary coolant and the steam in the steam generators is 42 F. However, the primary
coolant remains in a liquid state and is subcooled by 85 F. The pressurizer
is about full of water. Feedwater flow to the steam generators has not been
established.

Figure 6 contains a process icon for operations at 01:53:00 AM. The pressurizer is full of water and the primary coolant temperature is 592 F, which
is about the maximum for the transient as feedwater flow is re-established
shortly after this time. The primary coolant remains in a liquid state and
is 60 F subcooled. The steam generators remain ineffective heat sinks as a
result of the lack of feedwater.

Figure 7 contains a process icon for operations at 01:56:00 AM. Feedwater
has been re-established, and the primary system is thermodynamically re-
coupled to the steam generators. The primary coolant temperature is 568 F,
which is much cooler than it was three minutes earlier. The colder primary
coolant has greater density, thus the smaller volume of the coolant accounts
for the drop in pressurizer water level. The steam leaving the steam generator is superheated to a higher temperature than the primary coolant temperature, but is less than 592 F.

Figure 8 contains a process icon for operations at 01:59:00 AM. The main
bulk of the primary coolant is subcooled and isothermal at 542 F. The
saturation temperature in steam generator one is 539 F and in steam generator two is 533 F. The steam leaving the steam generators is superheated.
The steam leaving steam generator one is at 561 F and from two is at 555 F.
Again, these temperatures are greater than the primary coolant temperature
in the steam generators. However, the data in Figure 10 indicate that the
primary system is thermodynamically coupled to the steam generators and that
the heat cycle is intact.

Figures 1 to 8 describe a major safety function: the cooling of the reactor
core. The data within these figures describe this function for normal oper-

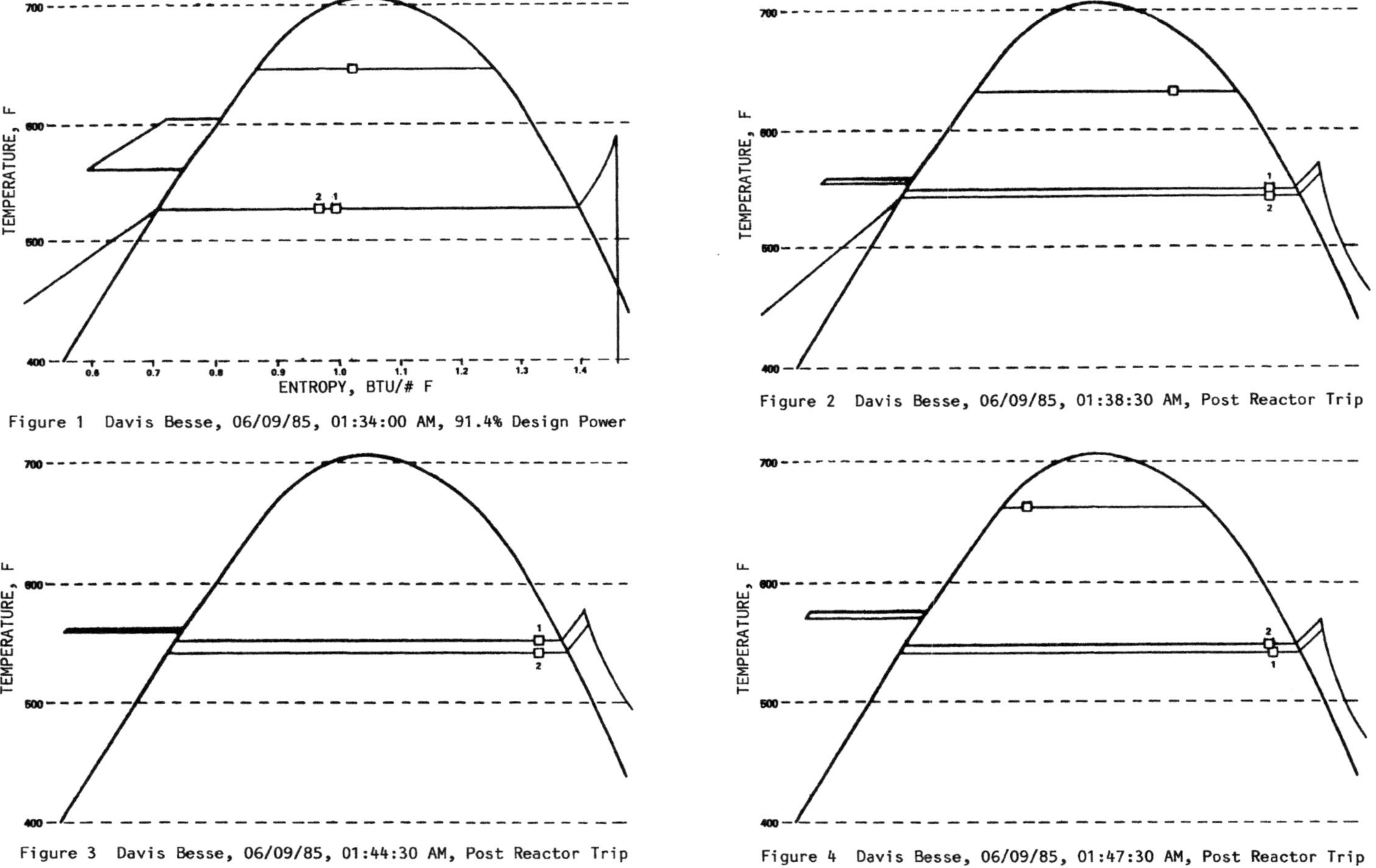

Figure 1 Davis Besse, 06/09/85, 01:34:00 AM, 91.4% Design Power

Figure 2 Davis Besse, 06/09/85, 01:38:30 AM, Post Reactor Trip

Figure 3 Davis Besse, 06/09/85, 01:44:30 AM, Post Reactor Trip

Figure 4 Davis Besse, 06/09/85, 01:47:30 AM, Post Reactor Trip

342

700
600
500
400
TEMPERATURE, F
Figure 5 Davis Besse, 06/09/85, 01:50:00 AM, Post Reactor Trip
700
600
500
400
TEMPERATURE, F
Figure 6 Davis Besse, 06/09/85, 01:53:00 AM, Post Reactor Trip
700
600
500
400
TEMPERATURE, F
Figure 7 Davis Besse, 06/09/85, 01:56:00 AM, Post Reactor Trip
700
600
500
400
TEMPERATURE, F
Figure 8 Davis Besse, 06/09/85, 01:59:00 AM, Post Reactor Trip

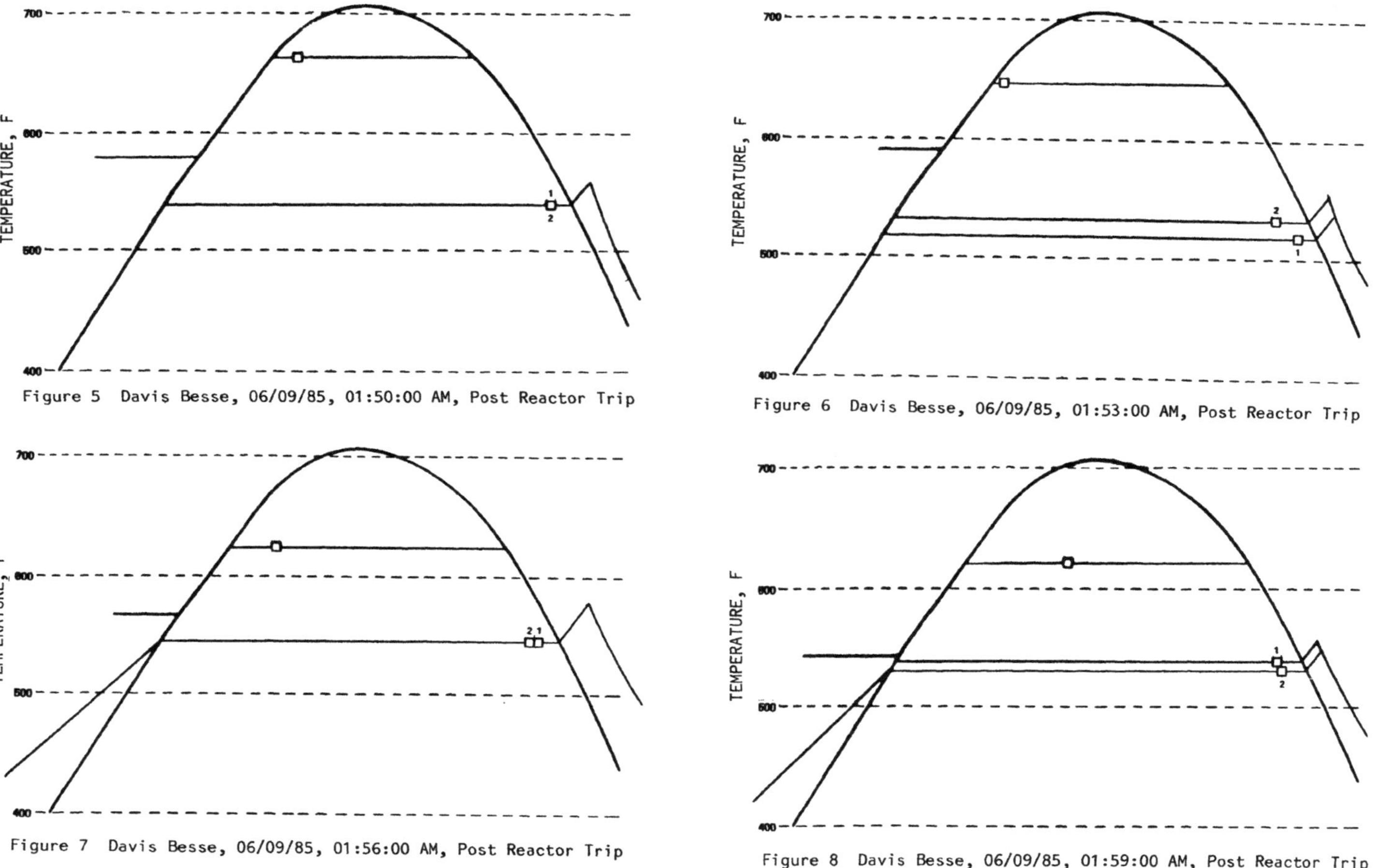

343

ation as well as abnormal operation. The sequence of figures present a
trend of this function, which is a powerful tool for monitoring plant opera-
ations. Finally, the thermodynamic goals of operation may easily be monitored. These goals are: 1) maintain liquid coolant inventories within the
process; and 2) sustain the heat engine cycle from heat source, the reactor,
to the ultimate heat sink, the environment. The figures herein describe
only the heat engine cycle to the steam generators, but with the appropriate
data, it would be a simple matter to extend the process icon to cover the
total cycle.

CONCLUSIONS

A model-based display was identified, discussed, and illustrated. The model
used in the display is based upon the Rankine Cycle, a heat engine cycle.
The format for the display consists of temperature and entropy coordinates
for the coolant water. Plant data from the loss of main and auxiliary feedwater event at the Davis-Besse Plant on June 9, 1985 was used to illustrate
the display. The data from the event was used to construct a process icon
for various points of time during the event.

A process icon may be used to evaluate process functions such as the heat
transfer coupling between the primary coolant system and the secondary coolant system for normal and abnormal operations. Furthermore, the process
icon serves to integrate the measured process variables into a image that
may be rapidly evaluated by a user. Because of these features, the user's
cognitive workload is minimized and human error should be reduced. Finally,
the user's training is continually reinforced because the model is based on
the "first principles" of the process.

REFERENCES

1.   D.A. Norman, J.D. Hollan, and E.L. Hutchins, "Direct Manipulation In-
     terfaces," ICS Report 8503, May 1985, NPRCD-UCSD Intelligent Systems
     Group, Institute for Cognitive Science, University of California, San
     Diego.

2.   E.F. Obert, Elements of Thermodynamics and Heat Transfer, Mc Graw Hill
     Book Company, 1949.

3.   L. Beltracchi, "A Process/Engineered Safeguards Iconic Display," Pro-
     ceedings of the Symposium on New Technology in Nuclear Power Plant
     Instrumentation and Control, Washington, DC, November 28-30, 1984.

4.   U.S. Nuclear Regulatory Commission, NUREG-1154, "Loss of Main and Aux-
     iliary Feedwater Event at the Davis Besse Plant on June 9, 1985," July
     1985, available from the National Technical Information Service,
     Springfield, VA 22161.

FOCUSING ON THE HUMAN FACTOR IN FUTURE EXPERT SYSTEMS

Sallie E. Gordon

Department of Psychology
University of Idaho
Moscow, Idaho

INTRODUCTION

Technological advances in the area of artificial intelligence have
produced expert systems that hold much promise for the design, operation,
and maintenance of complex systems such as nuclear power plants. Such
systems have been designed and implemented in a wide variety of task
settings (see papers in this volume).

In spite of the gains that have been made in the application of
expert systems, there are still several difficult problems which have yet
to be resolved. One of these problems is a frequently noted lack of user
acceptance of newly fielded intelligent systems (e.g., Rohm, this volume).
This lack of acceptance can be attributed to a variety of factors,
including unfamiliarity with computer technology, difficulty in adjusting
to interface mechanisms, fear that the system was designed to replace
the human operator, and a feeling that the human can perform the job
better than the system. Some of the problems may be related to the fact
that expert system design is essentially in it's infancy; much as we
may not like to admit the fact, sometimes systems are built where the
person simply can do a better or faster job on the task.

However, most of the problems with user acceptance of expert systems
are derived more from difficulties in developing an integrated human-
machine system. That is, by focusing predominantly on the task itself,
and developing an intelligent system to perform that task, expert system
designers neglect many of the factors that will affect not only Total
System performance, but also user satisfaction. Optimizing the integration of a system with the human user is the province of human factors,
and for this reason, human factors principles and methods should play a
major role in expert system design.

The primary goals of this paper are to discuss the role of human
factors in the design process, and to examine some of the key factors
that tend to affect Total System performance and user satisfaction with
the expert system. Finally, several suggestions will be made regarding
characteristics of future expert systems that would optimize Total System

performance as well as user acceptance and use of the expert system.

HUMAN FACTORS AND INTELLIGENT SYSTEMS

While there are many areas of overlap between traditional mechanical systems and advanced intelligent systems (i.e., information displays, control devices, etc.), unique features of advanced systems are presenting problems that require new techniques and guidelines. For example, one of the most important features is that the system _overtly_ communicates with the user in various forms. Among other functions, computers ask questions of the user, present the user with a number of action choices, provide feedback about what has occurred in the (intelligent) system, provide supporting information, recommend decision strategies, and ask for additional information. As technology advances, we will see many new functions added to this list. The added dimension of overt communication necessitates consideration of the human-machine interaction at a cognitive level as well as at a more fundamental physical level.

Another difference between traditional and advanced systems is the people who are designing them. Because much of the design involves computer programming, the designers are predominantly software engineers rather than, for example, electrical or mechanical engineers. Because human factors engineering has traditionally been associated with hardware systems, software designers may be less familiar with the field of human factors and the need for consideration of human factors design principles.

Looking back over recent years, one can see an analogy between the development of traditional mechanical systems and the development of many advanced computer systems. Software designers are highly trained specialists motivated to provide the best product their skills allow. However, their repertoire of skills often leads to the design of very complex systems that may be difficult for the user to learn and use, or present information and judgments in a manner that the user does not find helpful or trustworthy. The result of such a design process is often a devotion of a large amount of resources (both time and money) to an expert system that is basically unacceptable from the users point of view.

Many problems of user acceptance can be avoided if proper human factors principles are followed in the design and evaluation of the system. It is surprising and discomforting to note that in the major texts on expert system design (no names mentioned), the primary concern is with designing the internal mechanisms of the system itself. The sections on problem definition and delineation of the subtasks are written as there were no user at all; the system is performing some function in a social vacuum. _Nowhere_ are there sections on analyses of the human-machine system as a whole. When the texts discuss evaluation of the system, it is again an internal evaluation of the expert system, not an evaluation to determine system usability and total human-machine system performance. It is suggested that this neglect in the design process of user needs is one of the main reasons few systems reach utility in the real world.

The application of human factors principles to the design of expert systems is an even newer endeavor than the design of the expert systems themselves. As such, there are no hard and fast rules or guidelines. However, there are some general guidelines, principles, and insights that can be extrapolated from the human factors profession and applied to the development of any intelligent system that is being designed to supplement (rather than replace) a human operator. The guidelines revolve around a number of central issues in the design of expert systems. Some of these issues will be addressed in the following sections along with implications for desirable qualities in future expert systems.

Computer support systems can provide several different forms of assistance. It is becoming increasingly popular to classify types of support (e.g., design, diagnosis, planning, etc.) and develop generic intelligent system approaches or algorithms for each system type (Clancey, 1985; Cohen, Greenberg, & DeLisio, 1987). One rudimentary way to classify intelligent systems is to simply dichotomize them into two categories. One form of assistance is the provision of information to aid users in performing the task themselves (i.e., sensory information aggregation, probability estimation, etc.). Another form of assistance is the provision of advice or a judgment based on a comparison between situational characteristics and the system knowledge base. Most expert systems are designed to provide the latter type of support, and the user is left free to decide whether he/she wishes to use the judgment. However, consideration of the total human-computer system suggests that this may not be the most appropriate approach. A decision to build an intelligent system to perform the task _for_ the user may sometimes be premature.

As background for this argument, we must back up and consider the human factors design process. In designing an intelligent system, the human factors designer would focus attention on the following steps:

1. _Problem Specification_. Similar to the problem specification process in traditional engineering, this step involves a detailed specification of the problem, a statement of the objectives of the system (including both human and machine), specification of constraints, etc.

2. _Task Analysis_. This step consists of identifying the physical and psychological tasks and their associated subtasks which must be completed by either the human or the machine in order to meet system performance requirements.

3. _Task Allocation_. A critical step is the analysis to determine which subtasks are best performed by the human and which are best performed by the machine.

4. _Determine Information Needs_. Designers must identify the informational needs of the user in order to accomplish those subtasks allocated to the human (and also for tasks the human may wish to perform along with the expert system).

5. _Apply Human Factors Principles to the Interface Design_. Finally, the human-machine interface is optimized at the sensory-motor level and at the cognitive level by following well-established, basic human factors principles.

It has become apparent that many expert systems have been designed and fielded without having had the benefit of many of these procedures. The evidence comes in the form of field tests where the product user expresses dissatisfaction with the system, and sometimes simply refuses to use the system. The reasons often cited for this dissatisfaction strongly imply a lack of consideration of these essential steps in the design process. Some of these reasons are discussed below.

SOME FACTORS AFFECTING TOTAL SYSTEM PERFORMANCE AND USER SATISFACTION

Problem Specification

The problem specification may be conducted by design engineers and/or

supervisors with little or no input from the user. This can result in
expert systems that cover too narrow a portion of the field, a portion
of the field that the user feels is inappropriate, or other similar
problems. This is a fairly common problem because technology is not at
the point where expert systems can be feasibly developed with a large
knowledge base. Because the scope or breadth of the expert system must
be defined, this task should be done after extensive, systematic interviewing of the end users.

<u>Task Allocation</u>

     In addition, designers often circumvent the subtask allocation
process described above, and implicitly or explicitly make the decision
to build an expert system to perform the entire task sequence. This
puts the user into a difficult "either or" position. Either the human
performs the problem solving task or the computer performs tha task and
the human must decide whether or not to use the judgment. This causes
a variety of unanticipated problems, many of which ultimately result
in users being dissatisfied with the product.

     The most obvious problem is that the user does not want the system
to act as a replacement. The user may enjoy performing the task, and in
fact, may feel that he/she is being replaced by a machine that "won't do
as good a job anyway". It is not uncommon for management to have an
expert system developed for just those tasks that the user finds enjoyable
(although somewhat difficult) and leave to the user those tasks that are
tedious or boring.

     A related problem with the replacement approach is indicated by
difficulties currently being faced in airline cockpit design. By replacing
the human operators with automated systems, designers run the very real
risk of an "out-of-the-loop" problem (Wickens, 1984). That is, when a
human is largely replaced and forced to sit back and passively monitor
the system, their performance may degrade in many unforeseen ways. These
include an increased latency and reduced accuracy of failure detection
when the human is out of the system loop (Ephrath & Young, 1981; Kessel
& Wickens, 1982), and a loss of familiarity with the system over time
(Wickens, 1984).

     Other problems can ensue from building expert systems to perform an
entire task. One is based on the fact that the current state of technology cannot deliver expert systems capable of solving problems under
all circumstances. We are left with a human who must either perform the
task him/herself, or determine the conditions under which he/she will use
and trust the judgment of the machine. There are some interesting
problems inherent in this type of situation. First, the user must decide
whether to perform the task without consulting the expert system. Several
studies have shown that for easy to moderately difficult tasks, people
generally prefer to perform the task themselves (Gordon, 1986; Jenkins,
1984). However, at the "difficult" end of the task continuum, users
will allocate the task to the expert system (Fitter & Cruikshank, 1983;
Gordon, 1986). For example, physicians given the opportunity to use the
expert system MYCIN consulted the system only on very difficult cases
(Fitter & Cruikshank, 1983). Thus, the particular nature of the task as
well as the difficulty of any <u>particular</u> problem will determine the use
of the expert system. A problem with this use pattern is that, by
definition, the problems where the expert system is consulted are exactly
where the system will have the most difficulty performing the task. The

result is that, in general, the user is experiencing the expert system at
its worst. This, in turn, cuases the user to have an inaccurately low
estimate of the system capability. To make matters worse, this is coupled
with a standard psychological tendency for very "negative" instances to have
a particularly strong weight in impression formation (Wyer, 1974). In light
of this type of psychological analysis, it is not so surprising that designers promise great achievements, and the user is often left disappointed.

An important conclusion can be drawn from this analysis. If the problem
cannot be broken down into component parts with optimal allocation of subtasks, then an attempt should be made to dtermine the circumstances where the
machine will perform the task and where the user should perform the task.
Finally, if the user must decide when to use the expert advice, he/she should
be given enough information support to enable the user to perform the task
along with the expert system (e.g., a list of all plausible hypotheses for
diagnostic problem solving), as well as adequate feedback about the reasoning
used by the expert system in making the inference.

Information Needs of the User

Following from the discussion above, we can conclude that informational
needs of the user are of at least three types. The first is data for performing any subtasks allocated to the user. This would include aggregation
of sensor data, probability estimates, various calculations, and so forth.
The second type of information is decision support data so that the user can
attempt to perform the task concurrently with the expert system. The third
type is an explanation of the reasoning behind the expert system judgment.
Expert systems are typically designed to provide an explanation of their
reasoning by retracing the rules used to determine the judgment and then
listing those rules for the user. Unfortunately, this type of explanation
is of little benefit to the user. It does not provide information concerning
the reasoning behind the choice of that particular rule set. As several
researchers have pointed out (Kidd & Cooper, 1985; Swartout, 1983), the
reasoning is implicit in the rule chains, and as such cannot be provided to
the user. This leaves the user in the position of not really understanding
the basis for the decision. Ultimately, the user will have difficulty
becoming comfortable with a system he/she does not understand, a system that
does not really have the capacity to explain itself.

Several analyses have shown that to adequately provide an explanation
of their reasoning, expert systems will need to include some type of declarative knowledge base (Clancey, 1983; Swartout, 1983). That is, the knowledge
underlying the development of the expert system will have to be explicitly
built into the system.

Design of the Interface

Mapping Psychological to Physical Variables. Every user has in mind
some overall goal as well as subgoals in interacting with a system. One
major task facing the user of an expert system is determine what actions
are required to accomplish the subgoals. Ideally, it should be quite easy
for the user to identify what actions are needed. Thus, input devices should
be both psychologically and physically easy to use (Norman, 1986).

In addition to mapping psychological goals and intentions onto the
physical system, users must understand what the system is doing. The user's
"internal model" of how the system functions is created by the system displays. For this reason, messages should be designed to present as clear and

accurate a picture of the system states and functioning as possible.

    <u>Attention and Workload Considerations</u>. Some of the tasks expert system designers are attempting to augment take place within an environment that consists of a large amount of stimulus elements and complexity (e.g., nuclear plant control rooms). Because of this, the expert system should be designed in conjunction with other technologies rather than in isolation, and in such a way that the end result is not an increase in operator mental workload. This will undoubtedly mean making use of aggregate data displays (such as the integrated displays discussed by Wickens, 1984) and simple, easily comprehended judgments and explanations. Consideration of mental workload necessitates the cooperation between expert system designers and designers of any associated information display systems.

    Research in "cognitive automation" has shown that increased training on a system can decrease the amount of mental effort required to interact with the system (Schneider & Shiffrin, 1977). Because of this effect, some means should be found for incouraging the operator to use the expert system frequently, even in instances where the user does not necessarily feel the need for this consultation.

SYSTEM USABILITY

    In addition to being involved in the design process, human factors has a special role in the test and evaluation of a system. Because systems are frequently poorly received by end users, test and evaluation of system usability is critical. A complete evaluation of system usability includes both objective and subjective measures. Objective measures should include standard tests such as error analysis, time on task, expert system use patterns and other measures of Total System performance. Subjective measures are necessary for determining the users perception and attitudes toward the system. Shortliffe (1983) suggested six questions implicitly asked by users that are worth repeating here:

    1) Is the system's performance reliable?
    2) Do I need this system?
    3) Is it fast and easy to use?
    4) Does it help me without being dogmatic?
    5) Does it justify its recommendation so that I can decide what to do?
    6) Is it designed to make me feel comfortable when I use it?

    The best means to ensure acceptance of a system are to be very complete in preliminary interviews with potential users, and during the design process have users give responses to questions such as those listed above (as well as more specific questions about system functioning).

SOME SUGGESTIONS FOR FUTURE SYSTEMS

    In summary, certain system characteristics are desirable to ensure the use and acceptance of an expert system. First, as Norman (1984) pointed out, the user must view the expert system as a <u>tool</u> rather than as a replacement. And further, the system should be a tool that makes the user's task easier and more enjoyable.

    Second, users are not prone towards adapting themselves greatly to meet the needs of the system. Thus, it is the system which must be <u>flexible</u> in adapting to the user and the current situation. This will require several types of flexibility: (a) adaptation to the expertise level of the user; (b) adaptation to the strategy preferences of the user; (c) a knowledge base designed to be flexible and able to handle a wide range of

problems; and (d) flexible explanation facilities that provide short,
simple explanations with the capability of providing further information
upon demand.

Third, the system should be built in such a way that it reduces rather
than increases mental workload. Psychological goals and intentions should
map easily onto actions to control the system. The system states and
functioning should be displayed in a manner which is easily interpreted
by the user.

Fourth, although not a part of the expert system design per se, the
issue of training and experience with the system is an important one.
Experience with the system under conditions of easy to moderately difficult
tasks will cause the user to have had more positive interactions with
the system. This will help to cognitively counteract the negative experiences that will undoubtedly occur for difficult task conditions. (Obviously, every effort should be made to minimize system errors under difficult
task conditions.) Another reason for training and familiarization with the
system is to reduce workload by increasing the "automatization" of the
interaction process.

Finally, it is suggested that an important feature for future expert
systems will be the ability to learn.   Because intelligent systems are
difficult to build that can handle all possible situations, they are usually
defined specifically in terms of breadth and depth. This quality may be
difficult for the user to accept, particularly if the capabilities are
focused on areas that the user perceives to be inappropriate or unhelpful.
Also, the limitations of the expert system will cause difficulties for the
user if those limitations are not easily distinguished. For these reasons,
it will be desirable to create an expert system that can expand its own
knowledge base as a function of experience. This will give the operator
an incentive to interact with the system frequently, and will alleviate
the expansions and modifications that have to be performed by software
engineers. It is recognized that the design of systems capable of learning
from experience is still in a preliminary stage of research, however the
capability is an important one to keep in mind.

In summary, there are several qualities that will enhance the usability
and acceptability of expert systems. Some of those qualities are feasible
with current technologies. The characteristics more difficult to implement
should be approached to the extent possible, and certainly kept in mind for
future expert systems.

REFERENCES

Clancey, W.J., 1983, The epistemology of a rule-based expert system:
     A framework for explanation, Artificial Intelligence, 20, 215-251.
Cohen, P.R., Greenberg, M., and DeLisio, J., 1987, MU: A development
     environment for prospective reasoning systems, in Proceedings,
     (AAAI) Sixth National Conference on Artificial Intelligence,
     Morgan Kaufmann, Los Altos.
Ephrath, A.R., and Young, L.R., 1981, Monitoring vs. man-in-the-
     loop detection of aircraft control failures, in: "Human Detection
     and Diagnosis of System Failures," J. Rasmussen and W.B. Rouse,
     eds., Plenum Press, New York.
Fitter, M.J., and Cruikshank, P.J., 1983, Doctors using computers:
     A case study, in: "Designing for Human-Computer Interaction,"
     Academic Press, London.
Gordon, S.E., 1986, Building a structural model of computer-aided
     decision making, in: Proceedings of the 30th Annual Meeting
     of the Human Factors Society, Human Factors Society, Santa Monica.

Jenkins, J.P., 1984, An application of an expert system to problem solving in process control displays in: "Human-Computer Interactions," G. Salvendy, ed., Elsevier, Amsterdam.

Kessel, C.J., and Wickens, C.D., 1982, The transfer of failure-detection skills between monitoring and controlling dynamic systems, Human Factors, 24, 49-60.

Kidd, A.L., and Cooper, M.B., 1985, Man-machine interface issues in the construction and use of an expert system. International Journal of Man-Machine Studies, 22, 91-102.

Norman, D.A., 1986, Cognitive engineering, in: "User Centered System Design: New Perspectives on Human-Computer Interaction," D.A. Norman and S.W. Draper, eds., Lawrence Erlbaum Associates, Hillsdale.

Schneider, W., and Shiffrin, R.M., 1977, Controlled and automatic human information processing I: Detection, search, and attention, Psychological Review, 84, 1-66.

Shortliffe, E.J., 1983, Medical consultation systems: Designing for doctors, in: "Designing for Human-Computer Communication," M.E. Simes and M.J. Coombs, eds., Academic Press, New York.

Swartout, W.R., 1983, EXPLAIN: A system for creating and explaining expert consulting programs, Artificial Intelligence, 21, 285-325.

Wickens, C.D., 1984, "Engineering Psychology and Human Performance," Bell and Howell Company, Columbus, Ohio.

Wyer, R.S., 1974, "Cognitive Organization and Change: An Information Processing Approach," Lawrence Erlbaum Associates, Potomac.

OPERATOR PERFORMANCE WITH THE HALO II ADVANCED ALARM SYSTEM FOR

NUCLEAR POWER PLANTS - A COMPARATIVE STUDY

Edward Marshall, Craig Reiersen and Fridtjov Øwre

Institutt for Energiteknikk
OECD Halden Reactor Project
1750 Halden, Norway

INTRODUCTION

In recent years, the development of an increasingly capable and reliable computer technology has resulted in a trend towards the installation of computer-based information displays in the nuclear and process plant control room. The presentation of alarm information is a potentially fruitful area for the application of this technology. In the modern control room, hundreds of alarms may become active within minutes of a process transient. Unstructured, this information may be rather unintelligible to the operator, especially if he must interpret and act upon it within a short time to prevent plant shutdown. A system which is able to process and present alarms so as to minimise the cognitive demands made upon the operator and direct him towards the source of a disturbance might be expected to offer real advantages over a conventional tile-based annunciator display.

The Halden Project has been working with computer-based alarm systems for a number of years. Experimental interest has focussed on two issues - whether a reduction in the number of alarms presented to the operator helps him to interpret and handle a process transient, and how best to use the flexibility and potential of a computer-based system in order to display alarm information. The HALO (Handling Alarms using LOgic) system, developed at the Project, uses logic to reduce the number of active alarms during a process transient. Processed alarm information is subsequently presented to the operator on colour CRT's in a hierarchical display structure. The HALO system has been implemented on the Project's own full-scale NORS simulation of a PWR nuclear power plant and in 1984 an experiment was carried out comparing operator performance with three computer-based alarm systems. This experiment was described at a previous ANS/ENS meeting (Marshall and Øwre 1986, Øwre and Marshall 1986).

The results of this study, in which two man crews operated the simulator during a series of transients, revealed little by way of systematic performance differences although operators expressed a clear preference for the HALO symbolic system with its three-level display hierarchy of graphically presented alarm information. However it was apparent that, at times, operators found the three tier system somewhat cumbersome and time consuming to use. The display structure has consequently been modified and this paper reports a second study in which comparisons were made between an enhanced NORS/HALO system and a conventional tile based alternative.

METHOD

<u>Experimental Design</u>

Two groups of five subjects were observed coping with a complex series
of feedwater transients using either an enhanced NORS/HALO alarm system or a
conventional annunciator display. The transients were administered in the
context of a realistic control room task which involved running up a turbine
and which lasted for over four hours. A set of computer-presented procedures
was provided to assist with this task.

<u>The Enhanced NORS/HALO Alarm System</u>

The modified NORS/HALO system presented alarms at only two levels of
detail. At the top level, a permanently displayed mimic overview of the
plant was implemented using a full graphics workstation developed at Halden
(Figure 1). All process alarms were filtered on-line to remove redundant
information and alarm data were then presented on the overview in two modes:
1) as colour changes to alphanumeric text, bar graphs and the process mimic,
and 2) as a direct reference to lower level process formats. If a variable
entered an alarm condition, the alphanumeric process code for the format on
which that variable was located started to blink on the overview. The colour
of the blink depended upon the priority of the alarm - red for high priority
and yellow for low priority. Alarms were accepted by using a standard touch
panel to call up the appropriate process format. When alarms were accepted,
the blinking changed to a steady light. If the operator already had the
appropriate format on his CRT screen when a new alarm was indicated, he was
still obliged to call up that format with the touch panel in order to accept
the alarm. If the value of a variable changed such that there were no longer
any active alarms on a format, the blinking of the corresponding code on the
overview was extinguished. At the second level, individual alarms were embedded in the ordinary process formats where they appeared as colour changes
to displayed parameters and symbols.

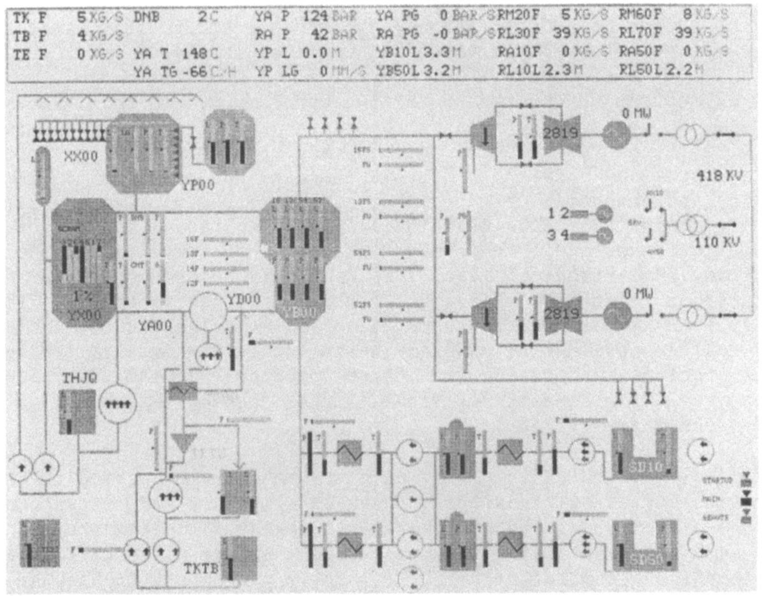

Figure 1. NORS/HALO Alarm Overview

<u>The Conventional Alarm System</u>

The conventional system consisted of fifteen alarm blocks, each
comprising ten alarm tiles - sufficient to instrument one feedwater train.
The alarm blocks were mounted in a purpose-built frame and were arranged so
as to maintain some correspondence with process flow. Each individual tile
displayed a single alarm status for a single process variable. The tiles
blinked red for high priority alarms and yellow for low priority alarms.
Alarms could be accepted by pressing a green button located at the bottom of
each block. When an alarm was accepted, the blink changed to a steady light.
If several alarms were blinking on the same block, they could all be
accepted by a single push of the accept button. If a variable ceased to be
in alarm condition, the light behind the appropriate tile went out. In the
conventional alarm condition no alarm information was presented on the
process formats.

<u>The Experimental Scenario</u>

In many simulator experiments which strive to achieve a rigorous and
ecologically valid evaluation of different systems, the potential importance
of "task", or conceptual, fidelity to the operational situation is often
overshadowed in a drive for the more readily perceptible physical fidelity.
This focus on the need for physical realism may obscure a tendency to
manipulate other factors for reasons of expediency rather than validity, and
may lead to systems being tested under conditions which are fundamentally
different from those occurring in the control room.

The scenario devised for this experiment sought to reproduce those
aspects of the operator's normal control room work which might be expected
to influence his use of an alarm system. Rather than present the operator
with a series of transients in an artificially short time period, the
transients were embedded in a "background" task in which the subject was
required to take one of the two turbines from hot shut down to 92% power.
This task was adapted to give the subjects long periods of passive monitoring interspersed with short intervals of activity - typical of the
operator's role in modern highly automated plant. The main intention of this
scenario was to avoid subjects waiting for something to happen and unnat-
urally focussing attention on the alarm system. The scenario lasted for over
four hours and a set of computer-presented turbine run-up procedures was
developed to help the subjects in this task. The procedures were generated
using the "Microtext" authoring system (Allman, 1984) implemented on an IBM
PC.

<u>The Transients</u>

Two transient sequences were inserted during the experiment. The first
transient sequence comprised three independent faults, the effects of which
were overlapping:

FAULT I
Control system fault leading to erroneous shutdown of a condenser ejector.
This caused a slow rise in condenser pressure until alarms for high pressure
and temperature were activated. These symptoms referred forwards to cause
alarms in the condensate transfer system.

FAULT II
Overheating bearing in a feedwater pump. Initially this prompted a high
temperature warning and subsequently caused a pump trip which affected the
flow of feedwater to the steam generators and raised levels in the feedwater
supply tanks.

FAULT III
Pump trip in the high pressure pre-heaters. This had little immediate effect
and caused only one alarm, for the pump trip itself. The fault was inserted
towards the end of the transient sequence as the subject started to recover
the plant.

The second transient sequence included a single fault and occurred
shortly before the subject achieved the required power level on the turbine:

FAULT IV
Erroneous closing of a main condensate isolation valve. An imbalance in the
water levels on either side of the valve prompted a number of alarms in
several plant subsystems.

Subjects

Ten subjects took part in the experiment. These included four experienced operators from the Halden boiling water reactor, and six volunteers
from Halden project staff. All the subjects had participated in previous
experiments with the simulator.

Training

Each subject was trained individually. Training was carried out the day
before the subject was tested and lasted for about four hours. Training
could be restricted to such a brief period because of the subjects'
familiarity with the process. The first part comprised a re-introduction to
that part of the process which the subject would be required to manipulate
and monitor during the current experiment. The second part consisted of an
introduction to the instrumentation used in the experiment, and practice in
using the computer-presented procedures for running up the turbine.

The Experimental Layout

The subject had access to two identical work stations consisting of a
CRT screen, touch panel and tracker-ball. He was able to call up process
formats and manipulate the plant from either station. In both experimental
conditions the alarm displays were located between the two work stations.
The IBM PC used to display the turbine run-up procedures was positioned at
the side of the right hand work station. A Supervisor's desk was positioned
some four metres behind the subject's control desk. It included one work
station plus a CRT dedicated to a text alarm list display. The supervisor
was responsible for running the primary side of the plant and was also
available to answer general queries which the subject might have about plant
operation.

Procedure

The ten subjects were assigned at random to the two experimental
conditions, with the restriction that each condition included two experienced operators from the Halden reactor, and three members of Halden project
research staff. Each subject was trained and tested individually. During the
experiment's lengthy periods of relative inactivity - for example, when
monitoring the automatics manipulating the process - the subject was allowed
to read a newspaper and chat with the supervisor. He was also supplied with
coffee as often as he wished. The first transient sequence was inserted
about two hours into the scenario, and the second after about two hours

forty five minutes. During the two transient sequences the help and advice
available to the subject was curtailed so as not to prejudice the outcome of
the experiment.

<u>Measurement of Subject Performance</u>

During each trial, video and audio recordings were made of operator
activity, together with detailed simulator event logs which showed which
alarms were activated and when, and logged the actions which the subject
carried out on the plant. A 15 item questionnaire immediately after the
experiment explored subjects' attitudes to the alarm systems they had used
and to the experimental scenario and conditions.

RESULTS

Although efficient use of an alarm system undoubtedly contributes
significantly towards <u>diagnostic</u> success, a comparison of more immediate and
direct measures of subject performance with the different alarm systems was
made in this study. The experimental data were analysed with reference to
three criteria - the rapidity with which subjects were able to <u>detect</u> the
first alarm for each fault, <u>select</u> the appropriate process format for more
detailed investigation of the disturbance and <u>locate</u> the disturbed variable
on the format.

Data relating to the first two of these criteria are analysed quantitatively; subject performance in locating disturbed variables on the formats
is described qualitatively.

<u>Detections</u>

Table 1 describes the mean times (secs) taken by subjects in the two
experimental conditions to detect the first alarm caused by each of the
four faults used in the experiment.

A relatively crude, yet informative, analysis was performed on the
data. Each subject's detection time for the first symptom of each fault was
ranked as "fast" ($<$10 secs), "medium" (10-120 secs) or "slow" ($>$120 secs). A
chi-squared analysis on the ranked data showed no significant difference in
the distribution of ranked detection times between the two experimental
conditions (χ^2=3.37, df=6, p$>$.10). This is perhaps not too surprising given
the restricted section of plant used in the experiment and the use of an
audible warning to support the visible alarm. Analysis of the video recordings indicated that the discrepancy on the second fault can be explained by
differences in the way in which the two alarm systems supported format
selection (see below). All 5 NORS/HALO subjects rapidly moved to the correct
format and located the disturbed variable. They subsequently concentrated on
this problem and failed to attend to further new alarms until the appropriate remedial actions had successfully been implemented. In contrast,
subjects using the conventional tiles failed to find the right format and
were seeking more information when the first alarm for the second fault was
triggered. Their response times to this alarm consequently were relatively
rapid, and both alarms were initially perceived as having a common cause.
This observation illustrates some of the problems which operators may face
when dealing with parallel and independent faults.

Table 1. Mean Times (secs) to Detect the First Alarm from
each Fault.

	Fault 1	Fault 2	Fault 3	Fault 4
NORS/HALO	5	152	15	5
CONVENTIONAL	19	6	12	5

Selections

Table 2 summarises the mean times (secs) taken by subjects in the two
experimental conditions to select the appropriate process format for inspection, following detection of the relevant alarm.

Subjects' format selection times for each fault were ranked as above,
and a chi-squared analysis was subsequently performed on the ranked data
(x^2=17.19, df=6, p<.001). The overview, which contains a direct cue to the
process format in the shape of a blinking alphanumeric code, clearly
supported more efficient selection of the process format. In two faults no
subject using the conventional tiles managed to select the correct format
within three minutes of detecting the relevant alarm. In each case this can
be explained by an inconsistency between the process code displayed on the
alarm tile and the code for the relevant format. This inconsistency arises
because alarms, like other plant instruments, are identified by plant
subsystem and the various subsystems themselves are not always represented
on a single format. So the operator selecting a process format bearing the
same code as the alarm tile is likely to find himself on the main format for
that subsystem whereas the alarmed variable may in fact be located on a
different format.

Such confusion may be avoided by providing additional labelling on the
alarm tiles. However, this would further increase the demands upon the
operator - particularly when dealing with the whole plant. We would argue
that performance with the focussed CRT- based overview would not deteriorate
to the same extent as plant size, and the potential number of alarms, is
increased.

Table 2. Mean Times (secs) to Select the Appropriate Format
for each Fault.

	Fault 1	Fault 2	Fault 3	Fault 4
NORS/HALO	10	15	31	5
CONVENTIONAL	*	65	*	10

* no subject selected the appropriate format within a 3 minute time limit.

<u>Variable Location</u>

Analysis of the video recordings, and subjects' responses to a post-experimental questionnaire, indicated that colour was a powerful aid to the location of disturbed variables on the formats. This was especially true when the subjects were trying to find several alarmed variables represented on a number of formats. In order to locate these variables, subjects in the conventional condition often addressed a series of different indications, and showed signs of forgetting the name of the alarmed variable by looking repeatedly to the alarm tiles. In contrast, subjects using the NORS/HALO system utilised the colour cue to locate disturbed variables very rapidly.

<u>Questionnaire Responses</u>

The questionnaire data generally were consistent with the observed performance data; although subjects in both conditions found it easy to detect new alarms on the two displays, NORS/HALO subjects found it easier to move down to the appropriate process format and to locate disturbed variables on the format. It appears that the attempt to achieve an ecologically valid experimental scenario was fairly successful; the subjects approved of the scenario and its reality was rated quite highly.

DISCUSSION AND CONCLUSIONS

When alarm information was clear and unambiguous with regard to the identity of the disturbed plant system and the process format on which the disturbed variable is located, the conventional alarm system was used effectively. However, when there was a conflict between the identity of the process format and the plant system the conventional system was less efficiently used. Moreover, important reservations about such alarm systems may be raised when large numbers of alarms are active. On a full-scale simulation in which the entire plant was equipped with alarm tiles, the tiles would necessarily be distributed over a wider physical area. Transients may be envisaged in which hundreds of alarms are active. In such circumstances, a degradation in the success with which the operator could detect, identify and interpret new alarms might be expected, since the spatial distribution of the alarm panel would require him to move repeatedly between the panel, to identify new alarms, and the control desk to investigate the relevant process format.

In contrast, it can be argued that an increase in the number of alarms is unlikely to affect significantly the NORS/HALO operator's use of the alarm system. The alarm overview is focussed on a single CRT screen, so the operator is not required to move from his central control desk despite an increase in the number of alarms. Consequently, he is unlikely to encounter such difficulties in detecting new alarms and identifying the process format on which they are located. Moreover, as the overview presents the codes for the process formats on which disturbed variables are located, a proliferation of alarm signals is not matched by a similar proliferation in the number of alarms on the overview. This might be expected to help the operator interpret alarm information, and move to the disturbed formats more easily. It is important to note that, although the overview reduces the amount of alarm information presented on the conventional tiles, this represents not a loss of information but a restructured presentation, with all the alarms available to the operator on the process formats.

A number of general conclusions about the comparative value and pract-
icability of computer-based alarm systems can be drawn from this study:

- The feasibility of installing a computer-based alarm system which
  includes sophisticated logical filtering and an advanced colour graphics
  presentation system on a scale commensurate with a real power plant has
  been demonstrated.

- If a computer-based alarm system is to be installed, its application
  should not be limited simply to text lists of alarm messages. The
  computer's capability for logical processing and graphic presentation
  should be fully exploited. Subjects in this study showed a clear
  preference for symbols and colour in information representation.

- In a computer-based control room, the alarm system should be integrated
  with the system for process information display. It would be disadvan-
  tageous to use a separated conventional alarm system.

- In mixed control rooms the alarm system should be part of the computer-
  based operating system. There was no experimental evidence to support
  the building of conventional annunciator panels.

REFERENCES

Allman, G., 1984, "Microtext - User Guide". Transdata Ltd.

Marshall, E. and Øwre, F., The Experimental Evaluation of an Advanced
Alarm System. in "Proceedings of the International ANS/ENS Topical
Meeting on Advances in Human Factors in Nuclear Power Systems", Knoxville,
Tennessee, USA, April 1986.

Øwre, F. and Marshall, E.C., HALO - Handling of Alarms using Logic:
Background, Status and Future Plans. in "Proceedings of the International
ANS/ENS Topical Meeting on Advances in Human Factors in Nuclear Power
Systems", Knoxville, Tennessee, USA, April 1986.

CHAPTER 7

INTELLIGENT DATA BASE DEVELOPMENT

REACTOR EMERGENCY ACTION LEVEL MONITOR:

KNOWLEDGE ACQUISITION EXPERIENCES

Robert A. Touchton

Technology Applications, Inc.
6621 Southpoint Drive North, Suite 310
Jacksonville, FL 32216

This paper presents the Knowledge Acquisition experiences
in developing the Reactor Emergency Action Level Monitor
(REALM) Expert System Prototype. REALM is an expert system
which interprets plant sensor data and provides advice on the
proper emergency classification. The REALM project is being
funded by the Electric Power Research Institute, Consolidated
Edison is serving as the host utility, and the effort is being
conducted by Technology Applications, Inc.

REALM is being designed to provide expert assistance in
the identification of a nuclear power plant emergency
situation and the determination of its severity, ultimately
operating in a real-time, on-line processing environment.
REALM embodies a hybrid architecture utilizing both rule-based
reasoning and object-oriented programming techniques borrowed
from the Artificial Intelligence discipline of Computer
Sciences. The rule base consists of two general classes:
event-based rules, which codify the logic embodied in the
current EAL tables, and symptom-based rules, which draw
inferences even when no specific problem events can be
identified. The symptom-based rules go beyond the current EAL
structure to address the more problematic scenarios and entail
a more symbolic representation of the plant information. As
such, a great deal of knowledge about nuclear plant behavior
and operations has been captured within REALM.

The paper will discuss briefly the direct knowledge
acquisition techniques used by the project team (who are
themselves power industry engineers), to extract relevant
knowledge from plant specifications and procedures. More in-
depth dimensions will be provided in describing the Knowledge
Engineering efforts conducted with plant personnel and
industry experts. Four specific techniques will be defined and
how each was used will be discussed.

KNOWLEDGE ENGINEERING METHODOLOGY

Knowledge engineering can be defined as the transfer of problem-solving expertise from some knowledge source to a computer program. It is the foundation upon which an expert system is built, but looms as perhaps the most difficult task. Several general approaches to knowledge engineering exist: 1) allowing the human expert to enter knowledge directly into the computer by means of some software product; 2) acquisition of knowledge by the computer directly from non-human sources, such as written texts or databases; 3) using a knowledge engineer as an intermediary between the human expert and the computer. Of these options, the last is the most commonly used and was, in fact, the primary method used for developing the REALM knowledge base and rule base.

Knowledge consists of a set of facts, beliefs, and heuristics for a given domain. Knowledge engineers are given the task of retrieving information from all possible sources and rearranging it in such a fashion that a computer can use it effectively (we refer to this as "codification").

Knowledge sources can be divided into two categories. First there are the conventional knowledge sources which include textbooks, journals, databases, etc., and which provide a wealth of general knowledge and facts. Expert knowledge sources comprise the second category and include domain experts and personal experience. Domain experts are used for their heuristic knowledge and/or beliefs in the way problems should be solved. Both categories of knowledge sources were used in the creation of REALM.

The development of an artificially intelligent emergency classification system requires domain expertise in a number of closely related disciplines including emergency planning, accident analysis, accident assessment (including available instrumentation) and probabilistic risk assessment. The development process also requires knowledge of plant design and operating procedures. If expertise is defined as a combination of skills, facts, beliefs and heuristics, then sources of expertise may include people, documents, and even computer codes.

In the following sections, the problem domain has been divided into its various dimensions and the potential sources of knowledge identified.

<u>Emergency Planning and Emergency Classification</u>

This aspect of the problem domain is the umbrella under which all other aspects reside. The other dimensions were explored only to the degree to which they would aid in solving the emergency classification problem at hand. This dimension requires understanding of the regulatory basis of the current EALs and requires the ability to make reliable judgements on what approaches for augmenting the current EALs will be successful. This area also requires an understanding of the technical and non-technical basis of a utility's current EALs.

The majority of the knowledge in this area was provided by the principal investigator. This includes representation

and interpretation of the plant observables, codification and
augmentation of the current EALs, and the development of
heuristic rules in general. Additional knowledge of their
basis and way-of-doing-business was provided by various plant
personnel.

Other sources of knowledge for this domain aspect
included the plant-specific Emergency Plan and Procedures,
including NUREG-0654 (<u>1</u>), the EALs (<u>2</u>, <u>3</u>), and NUREG-0396 (<u>4</u>).

<u>Accident and Degraded Core Behavior</u>

This dimension of the problem domain relates to the
identification and even prediction of certain key plant
threats and intermediate states (e.g. "LOCA" or "inadequate
core cooling"). This aspect requires an understanding of the
event sequences (and their symptoms) and plant states which
can lead to a significant release of radionuclides to the
environment. This involves such topics as internal threat
recognition, critical safety functions, fission product
barriers, and degraded core accident progression.

Once again, the principal investigator was the primary
source of expertise. In addition, supplemental expertise was
brought to bear in the areas of unusual accident and degraded
core behaviors and data acquisition and interpretations during
accident conditions. Two consultants from Volian Enterprises,
Inc. provided expert consultation in the area of severe
accident behavior, assisting the project team in developing
rules designed to interpret unexpected plant responses during
a severe accident and interpret situations involving missing
or ambiguous sensor data.

Other sources of knowledge for this dimension of the
problem included the plant-specific Emergency Operating
Procedures (<u>5</u>, <u>6</u>) and their background documents (<u>7</u>) and the
Indian Point Probabilistic Safety Study (<u>8</u>). NUREG/CR-4151 (<u>9</u>)
provided guidance on the mapping of critical safety function
degradation into fission product barrier challenges.

<u>Plant Design and Operation</u>

Since the plant's physical and operational
characteristics are represented within REALM, some effort was
allocated to this dimension of the problem domain. Our
interest in modeling the plant is strictly limited to those
objects necessary to properly classify an emergency. This
reduces the scope of required expertise significantly,
focusing on general plant design and operation, conduct of
operation and plant-specific tech specs. (For example,
detailed aspects of normal operation not closely related to
emergency classification, such as water chemistry or
performance monitoring, were not addressed.)

The primary source of knowledge for this aspect of the
problem domain was provided by the lead knowledge engineer's
extraction and condensation of the plant information contained
in the plant specific System Descriptions (<u>10</u>) and Tech Specs
(<u>11</u>). These sources were augmented by input from the principal
investigator and from plant personnel in the engineering,
emergency planning, operations, and training departments.

REALM KNOWLEDGE ENGINEERING PROCESS

Traditionally during the domain definition phase, the knowledge engineer must become familiar with the application domain prior to beginning the knowledge acquisition process. In order to efficiently gather information from the domain expert, the knowledge engineer must be able to communicate with the expert using the language of the expert's domain. The knowledge engineer must, therefore, develop a glossary of terms and acronyms used in the knowledge domain. After becoming comfortable with the knowledge domain, the knowledge engineer can begin characterizing the problem and identifying experts. However, in the case of the REALM knowledge engineering effort, the lead knowledge engineer was already quite familiar with the application domain, since he is an experienced, degreed nuclear engineer. His engineering background greatly improved the quality and efficiency of the knowledge engineering process for REALM.

We found the REALM knowledge acquisition process to be iterative in nature. As knowledge structures and concepts were developed, other information was needed to complete open concepts. Therefore, it was necessary to loop back to the process many times.

Because of the large size and interdependencies of the REALM Rule Base, the development team found it necessary to develop a knowledge engineering aid referred to as REALMNET, a utility program which takes as input the name of a rule class (or subclass) and builds a diagram showing the AND and OR linkage among the various rule antecedents and consequents. This REALMNET diagram was invaluable in refining the decision-making logic used by REALM.

KNOWLEDGE ENGINEERING TECHNIQUES

There are many good techniques that the knowledge engineer can use when acquiring and codifying knowledge. Four of these techniques which were used to varying degrees in the REALM development process are pure reclassification, goal decomposition, forward scenario simulation, and procedural simulation.

Pure reclassification is the technique of directly codifying problem observables into objects and activities that fit into a given style of solution. This technique was used in the codification of knowledge contained in plant documentation where the project team could extract concepts, designs, and decision rules directly.

Use of the goal decomposition technique requires the knowledge engineer to generate an initial general question. This question is then posed to the expert. The knowledge engineer continues the line of questioning with "Why, What if, How" questions until a solution is reached. This method was used most often during formal knowledge engineering sessions involving an outside expert. Interviews with outside experts were recorded on either video or audio tape for review at a later date. Recording the interview provides a permanent record which can be studied by the knowledge engineer. After

careful study, the lead knowledge engineer extracted concepts
from the interview session(s) and created or updated the
knowledge base and rule base accordingly.

The forward scenario simulation technique uses a scenario
for the expert to work with. The knowledge engineer and expert
walk through the scenario step by step and note the reasoning
necessary to achieve the goal. This technique was used during
the later stage of development, with an actual Indian Point 2
Emergency Exercise Scenario being used to fine-tune and debug
the REALM reasoning process.

In the case of procedural simulation, a general problem
is posed to the expert. The expert is asked to "think out
loud" while solving the problem. The knowledge engineer then
just observes the problem solving technique. This technique
was used (in spirit) when the project team was invited to
observe the Indian Point 2 Emergency Exercise. In this case,
the "problem" was a full scale simulation of a severe accident
at the plant and the "expert" was actually a significant
number of Indian Point 2 staff members who had assumed their
emergency management responsibilities.

REFERENCES

1. Criteria for Preparation and Evaluation of Radiological
    Emergency Response Plans and Preparedness in Support of
    Nuclear Power Plants (NUREG-0654), Nuclear Regulatory
    Commission, January 1980.

2. Emergency Plan for Indian Point Unit Nos. 1 and 2.
    Consolidated Edison Company of New York, Inc., version in
    effect July 22, 1985.

3. Emergency Procedures Document for Indian Point Unit Nos.
    1 and 2. Consolidated Edison Company of New York, Inc.,
    version in effect October 2, 1985.

4. USNRC, 1978. Planning Basis for the Development of State
    and Local Government Radiological Emergency Response
    Plans in Support of Light Water Nuclear Power Plants.
    NUREG-0396, EPA 520/1-78-016. Washington, DC.

5. Indian Point No. 2 Emergency Operating Procedures, E-Set
    Procedures. Consolidated Edison Company of New York,
    Inc., Revision 0, January 25, 1985.

6. Indian Point No. 2 Emergency Operating Procedures, F-Set
    Procedures. Consolidated Edison Company of New York,
    Inc., Revision 0, January 25, 1985.

7. Indian Point No. 2 Emergency Operating Procedures,
    Background Documents. Consolidated Edison Company of New
    York, Inc., Revision 0, January 1, 1985.

8. Power Authority of the State of New York and Consolidated
    Edison Company of New York, Inc. Indian Point
    Probabilistic Safety Study. Amendment 2, December 1983.

9.   D. W. Faletti and J. D. Jamison. <u>Integration of Emergency Action Levels with Combustion Engineering Emergency Operating Procedures</u>, September 1985. NUREG/CR-4151, PNL-5392.

10.   <u>Indian Point Station Unit No. 2 System Descriptions</u>. Consolidated Edison Company of New York, Inc., version in effect February 1981.

11.   <u>Appendix A to Facility Operating License DPR-26, Technical Specifications and Bases</u> for Consolidated Edison Company of New York, Inc., Indian Point Nuclear Generating Unit No. 2, Docket No. 50-247, Amendment #93.

A CONCEPTUAL VIEW DATA MODEL OF A DISTRIBUTED CONTROL AND MONITORING
SYSTEM FOR A NUCLEAR POWER PLANT

L. E. Fennern

GE Nuclear Energy
175 Curtner Avenue
San Jose, California

INTRODUCTION

How one views data in an advanced nuclear power plant will impact the
automation of plant operations and influence the design of controls and
instrumentation equipment and user interfaces. Three views of data are
possible[1]: the conceptual view, a superset of what is sometimes termed
the "global schema", which describes data as it exists in the plant; the
logical view, a superset of what is sometimes termed the "external schema",
which describes the view of data required by applications; and the storage
view, a superset of what is sometimes termed the "internal schema", which
describes how the data elements are represented in a database management
system.*

Through careful attention to data base management, closer consistency
between these three views of data can be achieved. Such consistency is
desirable, in order to fully realize the potential for greater data accessibility and improved plant availability made possible through application
of digital microelectronic and optical data transmission technologies which
have been proposed for distributed control and monitoring systems of some
advanced nuclear power plants[3].

---

*Alternatively, these have been described as the integrated view, user view
and implementation view, respectively[2].

DATAFLOW

Defining the conceptual view process data flow is an important first
step in developing the overall control structure and configuration of a
software intensive, realtime distributed control and monitoring system[4,5].
The data flow can be conceptually described by "bubble diagrams", which show
processes, data input/output flow between processes, and data terminators.

The data flow diagram for a distributed control and monitoring system
might show the processing tasks (data transformations) to be performed by
the hardware and software of the system, and the data signals (input/output)
needed to support these tasks. Each process entity and signal entity can
be given a name, associated with abstract or specific types, and provided
a control data network address where the process transformation is performed or the input/output signal is terminated, respectively.

DATABASE

The conceptual view data flow can be represented logically by the entity-relationship (ER) model. This model provides a framework into which
knowledge about entities and their relationships can be deposited[6], and
has been suggested as a possible approach to an integrated database for
operational plant control[7]. Extended entity-relationship (EER) approaches[8],
the collection of entities into classes, and the use of frame-based representation, provide further enhancements to the data model representation[9].

Entity-relationship databases can be implemented using a first generation database model, such as a relational database, or alternatively, using
the inherent data structures provided by "logic" programming languages[10].
In the first approach, meaning can be built into the database system using
an external module to interprete data interrelationships and inheritance.
In the second approach, interpretation can take place as part of the database management system itself.

For productive system design and improved operation and maintenance,
the database containing the data flow for a distributed control and monitoring system should serve to verify and automate the design of the system
during development phases, and provide powerful diagnostics and operator
aids later during system operation. This is more easily achieved when the
logical and storage views of data are less distinct from the conceptual
view.

APPLICATION

A conceptual view of data in the entity-relationship format for the
distributed control and monitoring system is shown in Figure 1. Three key
entities are defined in the center column of the figure: (1) "processes"
(data transformations); (2) "signals" (input/output); and (3) "MUX units"
(intelligent multiplexing data stations). Each process may require several input signals, and a given signal may be used in several processes.
Furthermore, each process or signal is related to a MUX unit, representating
a data station address, where the data transformation or input/output termination occurs, respectively.

 During the design of the system, process, signal and MUX unit entities
are related to other entities such as control & instrumentation configurations (e.g., remote motor operated valve, remote operated pump, etc.). The
left side of the database then provides productivity and verification aids.
As an example of a productivity aid, signal routing (transmission sources
and destinations) for the control data network and distribution of process
transformations can be determined automatically from algorithmic and heuristic procedures providing functional allocation. As an example of a
verification aid, the input and output signals associated with a process
can be checked for consistency with the related control and instrumentation configuration.

As the design is implemented in actual hardware and software, and the
interfaces of the system to actual field devices becomes known, the general
information in the left column of Figure 1 is supplemented with specific
information as shown in the right column. This can be thought of as the
"physical side" complement to the "conceptual side", as described in
reference 10. When completed, knowledge of the system in its operation
can be quickly acquired. As an example, if a certain card in a multiplexing
unit is removed for maintenance, a complete report of equipment and signals
which will be unavailable, along with other pertinent information, can be
provided.

CONCLUSIONS

This paper provides an integrated, conceptual view of data for a distributed control and monitoring system, and shows that design and operational advantages can be obtained when the conceptual view of the data is

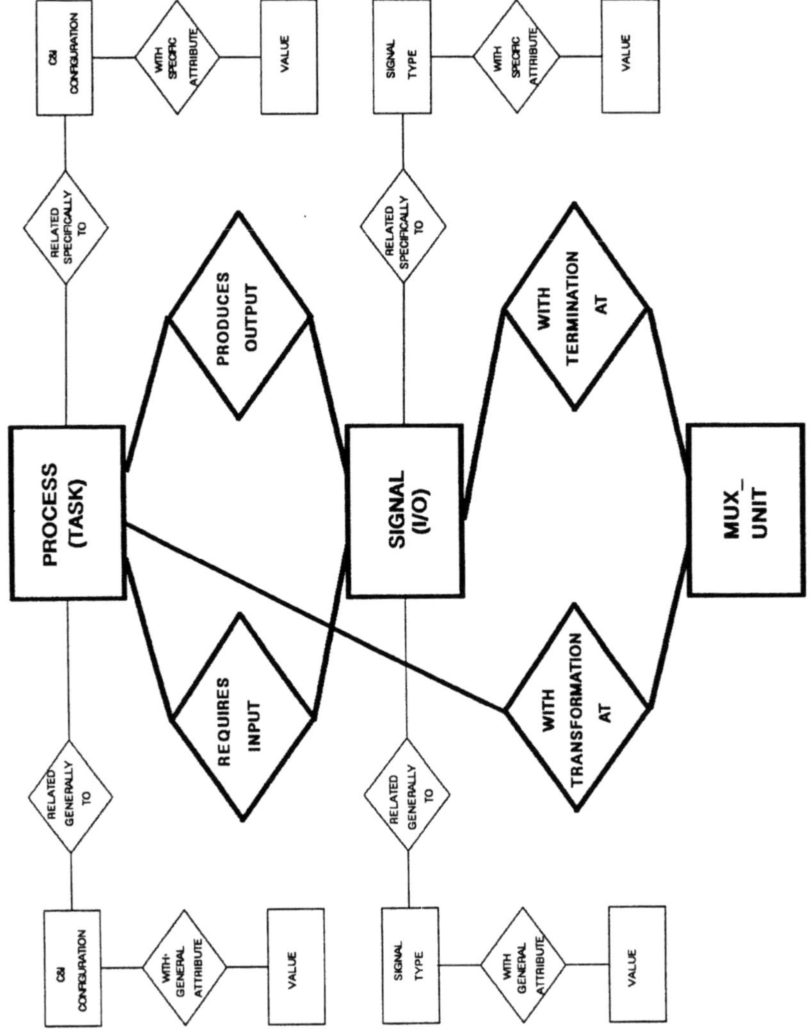

Figure 1. Distributed Control and Monitoring System Conceptual View Data Model

preserved in the logical and storage views. The model described is currently being applied to rapidly confirm the satisfaction of functional and performance requirements during system design for advanced nuclear power plants.

REFERENCES

1. Jerome M. Fox, J. Carlos Goti, Claude R. Miller, and James M.
    Sagawa, "Implementing a self-defining entity/relationship
    model to hold conceptual view information", in _Entity-
    Relationship Approach to Information Modeling and Analysis_,
    P. P. Chen, ed., Elsevier Science Publishers B. V. (North-
    Holland), ER Institute, 1983.
2. Mary E. S. Loomis, "Logical Data Modeling -- A Step Toward Inte-
    gration", COMPINT '85, Montreal.
3. D. R. Wilkins, T. Seko, S. Sugino, and H. Hashimoto, "Advanced
    BWR: design improvements build on proven technology",
    Nuclear Engineering International, June, 1986.
4. Ward and Mellor, _Structured Development for Real-Time Systems_,
    Yourdon, 1985.
5. Derek J. Hatley, "A structed Analysis Method for Real-Time Sys-
    tems", Seminar at the Fall DECUS U.S. Symposium, December,
    1985.
6. PETER PIN-SHAN CHEN, "The Entity-Relationship Model -- Toward a
    Unified View of Data", ACM Transactions on Database Sys-
    tems, Vol. 1, No. 1, pp. 9-36, March, 1986.
7. Robert W. Colley, Stephen E. Seeman, Devin E. Smith, John Gabriel,
    Ewing L. Lusk and Ross A. Overbeek, "An Entity-Relationship
    Model for Nuclear Power Plants", ANS Topical Meeting on
    Computer Applications for Nuclear Power Plant Operation
    and Control, Tri-Cities (Pasco), Washington, September
    8-12, 1985.
8. Toby J. Teorey, Dongqing Yang and James P. Fry, "A Logical Design
    Methodology for Relational Databases Using the Extended
    Entity-Relationship Model", Computing Surveys, pp. 197-222,
    Vol. 18, No. 2, June, 1986.
9. David J. Hartzband and Fred J. Maryanski, "Enhancing Knowledge
    Representation in Engineering Databases", IEEE Computer,
    pp. 39-48, September, 1985.
10. Ewing L. Lusk and Ross A. Overbeek, "Databases and Automated
    Reasoning", ANS Topical Meeting on Computer Applications
    for Nuclear Power Plant Operation and Control, Tri-Cities
    (Pasco), Washington, September 8-12, 1985.

AN APPROACH TO BUILD A KNOWLEDGE BASE

FOR REACTOR ACCIDENT DIAGNOSTIC EXPERT SYSTEM

Kazuo Yoshida*, Toshihiko Aoyagi**, Yasuhiro Hirota***,
Minoru Fujii* , Kazuo Fujiki*, Masao Yokobayashi*,
and Atsuo Kohsaka*

*Department of Reactor Safety Research
 Japan Atomic Energy Research Institute
 Tokai, Ibaraki, Japan
 **Management Administration Department
 Kyushu Railway Company
 Fukuoka, Japan
 ***Research and Development Department
 CSK Corp.
 Tokyo, Japan

INTRODUCTION

In the development of a rule based expert system, one of key issues is how to acquire knowledge and to build knowledge base (KB). On building the KB of DISKET[1],[2] , which is an expert system for nuclear reactor accident diagnosis developed in JAERI, several problems have been experienced as follows. To write rules is a time consuming task, and it is difficult to keep the objectivity and consistency of rules as the number of rules increase. Further, certainty factors (CFs) must be often determined according to engineering judgement, i.e. empirically or intuitively. A systematic approach was attempted to handle these difficulties and to build an objective KB efficiently.

The approach described in this paper is based on the concept that a prototype KB, colloquially speaking "an initial guess", should first be generated in a systematic way and then is to be modified and/or improved by human experts for practical use.

Statistical methods, principally Factor Analysis, were used as the systematic way to build a prototype KB for the DISKET using a PWR plant simulator data. The source information is a number of data obtained from the simulation of transients, such as the status of components and annunciators etc., and major process parameters like pressures, temperatures and so on.

BRIEF DESCRIPTION OF DISKET SYSTEM

The system organization of DISKET is illustrated in Fig. 1. The system consists of an inference engine and a knowledge base as in usual expert systems. The inference engine IERIAS[1],[2] , which has been developed at JAERI, identifies the cause and the type of an accident with the KB and plant process data as shown by black arrow signs. The KB is

made from the relevant information about the plant such as actuating
conditions of components or signals etc., and characteristics of transient
responses obtained from analytical studies as well as plant transient
records as shown by white arrow signs. In the developmental stage, a
plant simulator installed at JAERI has been used in stead of an actual
power plant.

In the DISKET system, dynamic phenomena are taken into consideration
by rule representation, and partitioning KB into several knowledge units
is successfully made for effective search of the rules.

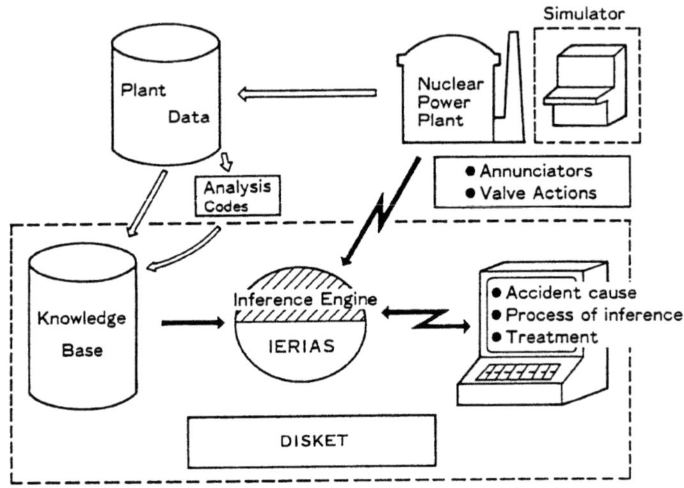

Fig.1 Concept of DISKET system

PROCEDURE OF BUILDING KNOWLEDGE BASE

The procedure of building prototype KB using the statistical methods
consists of three steps. In the first step (data acquisition step), a
number of transient data obtained from the simulator are transformed into
matrix named "Transient Data Matrix". In the next step (factor analysis
step), the factor analysis, which is one of the methods used in multivariate analysis, is applied to the transient data matrix to classify the
types of transients into several groups. In the last step (rule
extraction step), the correlation coefficients between alarms and groups
of transients are calculated to extract IF-THEN rules and to determine the
CF of each rule. The each step is described in more detail below.

Data Acquisition Step

The transient data obtained from the plant simulator are categorized
into two types. The first one is ON/OFF type data such as annunciators
actuations, status of pumps or valves and so on. Another one is numerical
data such as pressures, temperatures and so on. In the Data Acquisition
Step, which is illustrated in Fig. 2, data of a transient are digitalized
keeping the characteristics of the transient according to the following
criteria. For ON/OFF type data, 1/0 is given corresponding to the on/off
status, and as to numerical data, 1/0/-1 is given corresponding to the
trends of increase/unchange/decrease, respectively. In the present study,
early changes during transients are only focused on. For convenience, any
data type is generally called alarm and these digitalized data are called
alarm data hereafter. The transient data matrix is obtained by arraying

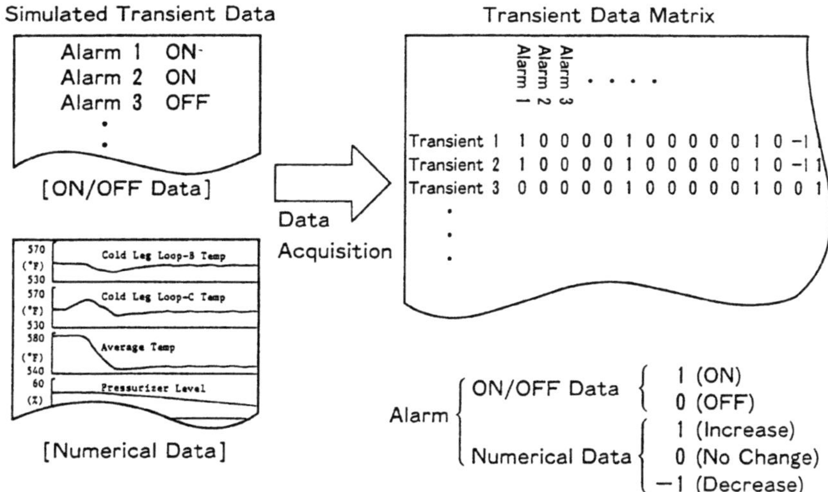

Fig.2   Data Acquisition Step

these data for each transient. For example, Aij, an element of the
matrix, takes the value of either 1/0/-1 and indicates the status of alarm
j in transient i. If Aij is 1, it means that alarm j is turned on or
increase in the early stage of transient i. The alarm data vector, (Ail—
-AiJ), can be considered to characterize transient i and is the basis in
the following analysis.

Factor Analysis Step

The Factor Analysis is then applied to the matrix to find out so
called "factors", for each of which a group of transients are correlated,
being separated from other groups. This grouping may be interpreted as
extraction of similarities among patterns of alarm data vectors. The
factor loading matrix is obtained as a result of the factor analysis.
From this result, it is possible to distinguish several dominant factors
and to classify the transients into groups corresponding to those factors.
Fig.3 shows an example of this step. An element of the factor loading
matrix, saying Bij, indicates the intensity of correlation between
transient i and factor j. A transient which shows strong correlation to
factor j can be categorized into group j. Thus transient group j is
defined. For example, transients 1, 10 and 8 are categorized into group
1, because the matrix elements for these transient and factor 1 have
bigger values than the others, as indicated by square in the figure.

Rule Extraction Step

In the Rule Extraction Step, the correlation coefficients between
alarms and factors are calculated to extract rules and determine their
certainty factors as shown in Fig. 4. An element of this matrix, saying
Cij, indicates the intensity of correlation between alarm i and transient
group j. For a Cij whose magnitude is relatively large, larger than 0.4
in the present study, the following rule can be extracted.

$$F(\text{alarm } i, T) \rightarrow H(\text{group } j, (Cij)^2)$$

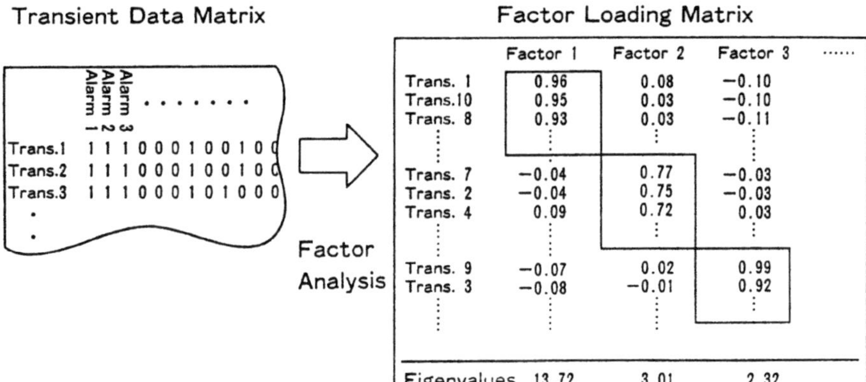

• Identification of dominant factors

• Grouping of transients by the factors

Fig.3 Factor Analysis Step

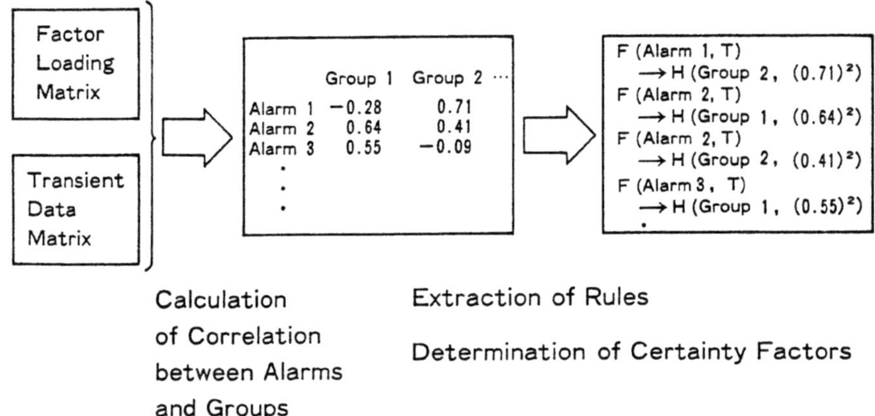

Calculation
of Correlation
between Alarms
and Groups

Extraction of Rules

Determination of Certainty Factors

Fig.4 Rule Extraction Step

Here F and H denote finding (process data) and hypothesis, respectively. In this extraction, it is assumed that the square of Cij, which is known as the determination coefficient in statistics can be directly used as the CF. The above rule is interpreted as "When alarm i is ON, an accident in group j may have occurred with CF of $(Cij)^2$."

VERIFICATION OF THE APPROACH

The verification studies of this approach were carried out using a compact PWR simulator installed at JAERI as a model of an actual nuclear plant. This simulator which is named SRS (Safety Research Simulator) is designed to simulate an existing 822 MWe PWR type power plant with three loops. Its simulation capability covers a wide range of the plant operation modes from cold shut down to full power operation, and various transients caused by malfunctions of components or systems.

Data Acquisition

For the present study, 34 types of transients initiated from normal operating condition with rated power were selected to make the transient data matrix.

Results of Factor Analysis

ANALYST code package[3)] was used for factor analysis. From the factor analysis firstly applied to this matrix, 34 transients were clearly divided into two groups, one is a group of transients with reactor scram and another group is without scram. Then further factor analyses were performed individually for each transient data matrix corresponding to each group of transients. By these second analyses, the former group was divided into three sub-groups and the latter group was divided into four sub-groups. These results are shown in Table 1. As also shown in the tables, it is interesting that physical explanation of "factors", in other words group of transients, can be given.

Table 1   Results of Factor Analysis

Transients with scram

	Simulated Transients	Physical Explanation
Group B1 (Factor 1)	ATWS with T. Trip 1 CRD Drops ADV Leakage. Turbine CV Fails Close	Power-Cooling Mismatch
Group B2 (Factor 2)	SG Tube Rupture Uncontrolled CRD W/D VCT Level Cont. Fails Low Turbine CV Fails Open Failure of Turbine Runback SG Lo-Level Mis-detect.	Power Cooling Mismatch (Mild)
Group B3 (Factor 3)	Small Leak in RCS Uncontrolled CRD Insertion Hi-Temp Mis-signal in C. Leg VCT Level Cont. Fails High Loss of Charging Flow	Loss of Primary Coolant Inventory
Group B4 (Factor 4)	P.Loss of Condenser Circ. Water HP Heater Drain Pump Trip	Failure of Feedwater System

Transients w/o Scram

	Simulated Transients	Physical Explanation
Group A1 (Factor 1)	PORV Sticks Open Spray Valve Fails Open PRZ Hi-P Mis Detect. PRZ Lo-P Mis Detect. Mis-demand on CRD W/D	Failure of Pressure Control
Group A2 (Factor 2)	Turbine Trip Generator Trip Loss of Cond. Circ. Water SG Hi-Level Mis Detect. P. Loss of FW Total Loss of FW	Failure of Secondary Heat Removal
Group A3 (Factor 3)	1 RCP trips 2 RCPs trip 3 RCPs trip 1 RCP sticks Steam Line Leakage	Failure of Reactor Core Heat Removal

Rule Extraction and Building of Knowledge Base

In the DISKET system, each transient and each group of transients are treated in KB as hypotheses to be identified, which are classified into a hierarchical structure named "taxonomy" with several levels such as global, intermediate and fine or 1st, 2nd, and 3rd levels. For building a prototype KB of DISKET, the results of previous step were directly applied. That is, the taxonomy in the prototype KB with two 1st level and seven 2nd level hypotheses were given, corresponding to the groups/sub-groups defined in the previous step, respectively, and 34 3rd level hypotheses were classified under the 2nd level hypotheses. The example of taxonomy is partially shown in Fig. 5.

The rules for 1st and 2nd level hypotheses were extracted from the results of the calculation of correlation coefficients between alarms and factors described in the previous section. The rules for 3rd level hypotheses were extracted by identification of any distinctive or representative alarms among transient data associated with each sub-group as shown in Fig. 6. The table in Fig.6 which is a portion of the transient data matrix shows the status of alarms in the transients classified into group A3. "4 KV Breaker Auto Trip" and "Main Steam Isolation Valve Close" are "ON" only in the transient 3 and 5 respectively. Therefore, two rules can be extracted as shown in Fig.6. In the inference of DISKET, a hypothesis is identified with the satisfaction of several individual rules, and the final CF for a hypothesis is the summation of CFs of satisfied rules based on Basian rules. Therefore, "0.2" is a proper value for CF in each rule, as about five individual rules are expected to be used to identify one hypothesis.

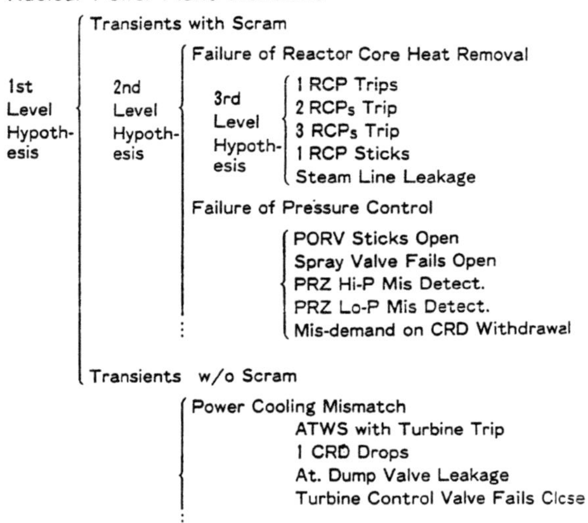

Fig.5   Hierarchical Structure of Hypothesis

Alarm	Group A3				
	Trans. 1	Trans. 2	Trans. 3	Trans. 4	Trans. 5
4 kV Breaker Auto Trip	0	0	[1]	0	0
Main Steam Isolation Valve Close	0	0	0	0	[1]
Main Steam Flow High	0	1	1	1	1
Main Steam Pressure Low	1	0	0	0	1

F(4 kV Breaker Auto Trip, T) →H (Transient 3, 0.2)

F (MSIV Close, T) →H (Transient 5, 0.2)

CF = 0.2 : Number of applicable rules to identify

each transient ≃ 5

Fig.6   Rule Extraction for 3rd Level Hypothesis

<u>Diagnostic Results</u>

The evaluation of performance of the prototype KB thus built was carried out by making diagnoses for several transients. An example of diagnostic result is shown in Fig. 7. The transient used for diagnosis is all three reactor coolant pumps trip due to breakers open by excessive load followed by reactor coolant flow reduction and reactor scram. As to the 1st and 2nd levels of taxonomy, the correct hypotheses are identified with high CFs, 0.83 and 0.92 respectively. As to the 3rd level hypotheses, one of which is the final target of diagnosis, similar type of transients are all listed up, placing the target transient at the top, though not with enough high CFs compared to other levels of taxonomy. Hence the performance of the prototype KB is fairly satisfactory. The refinement of prototype KB by experts is necessary to improve the accuracy of diagnosis for the 3rd level hypotheses.

The inference executed at 5 seconds after the accident

1st Level	C. F.	Hypothesis
	0.83	Transient with Scram
	0.20	Transient w/o Scram
2nd Level	C. F.	Hypothesis
	0.92	Failure of Reactor Heat Removal
	0.48	Failure of Secondary Heat Removal
	0.44	Failure of Pressure Control
3rd Level	C. F.	Hypothesis
	0.51	All(3) RCPs Trip
	0.23	2 RCPs Trip
	0.23	Steam Line Leakage
	0.23	1 RCP Sticks
	0.23	1 RCP Trips

Fig.7   Example of Diagnostic Results for All Reactor Coolant Pumps Trip

CONCLUSION

From this study, it is concluded that the statistical method, Factor Analysis, is powerful for building a prototype of knowledge base of an expert system for reactor accident diagnosis like DISKET.

REFERENCE

1) Yoshida,K., Yokobayashi,M., Ayoyagi,T., Shinohara,Y. and Kohsaka,A. : Development and Verification of an Accident   Diagnostic System for Nuclear Power Plant by using a   Simulator. ANS Topical Meeting on Computer Applications for   Nuclear Power Plant Operation and Control. Pasco, USA September (1985)

2) Yokobayashi,M., Yoshida,K., Kohsaka,A. and Yamamoto,M. : Development of Reactor Accident Diagnostic System DISKET using   Knowledge Engineering Technique, J.Nucl.Sci.Tech., 23[4], (1986)

3) Fujitsu Co. Ltd. : Software package ANALYST for Statistical   Data Processing, FUJITSU 70AR-0800-1 (1982)

INCORPORATING "FUZZY" DATA AND LOGICAL RELATIONS
IN THE DESIGN OF EXPERT SYSTEMS FOR NUCLEAR REACTORS

Michael A. S. Guth[*]

Instrumentation and Controls Division
Oak Ridge National Laboratory[†]
Oak Ridge, Tennessee 37831-6008

INTRODUCTION

This paper applies the method of assigning probability in Dempster-Shafer Theory (DST) to the components of rule-based expert systems used in the control of nuclear reactors. Probabilities are assigned to premises, consequences, and rules themselves. This paper considers how uncertainty can propagate through a system of Boolean equations, such as fault trees or expert systems. The probability masses assigned to primary initiating events in the expert system can be derived from observing a nuclear reactor in operation or based on engineering knowledge of the reactor parts. Use of DST mass assignments offers greater flexibility to the construction of expert systems in two important respects.

First, DST mass assignments have the advantage over classical probability methods of accommodating when necessary uncommitted probability assignments. Thus the DST probability framework can incorporate expert system inputs from imprecise or "fuzzy" data. Second, DST applied to the Boolean rules themselves leads to a probabilistic logic, where a given rule may be valid with probability less than unity: "fuzzy" logical rules.

When knowledge bases contain rules which may not always hold, or rules that occasionally must be operated upon with imprecise information, the DST mass assignments are shown to be a rigorous methodology for calculating probability assignments throughout the system.

The method of distributing probability to subsets of possible values of a random variable, rather than to individual sample points, was introduced by Dempster (1967), and was subsequently extended to the theory of belief functions by Shafer (1976, 1982, 1985, 1986).

---

[*]Subcontractor, this research was sponsored by appointment to the U.S. Department of Energy Laboratory Cooperative Postgraduate Research Training Program administered by Oak Ridge Associated Universities. Special thanks go to Bonnie Brummitt and the Publication and Information Processing Center staff for preparing this text.
[†]Operated by Martin Marietta Energy Systems, Inc., under contract DE-AC05-84OR21400 with the U.S. Department of Energy.

Since 1985, DST has grown in popularity among AI researchers studying knowledge representation and reasoning under uncertainty, so that it now rivals Fuzzy Set Theory (FST) as a major uncertainty paradigm [Prade 1985]. The article by Lee, Grize, and Dehnad (1987) provides an excellent comparison of the theoretical and computational requirements of DST, FST, and classical probability theory.

Surprisingly, little work has been completed to date extending DST to truth-valued logic. Using DST, we will present a framework in which an analyst can represent the fact that he cannot determine whether some event has occurred or not. The event can be either true or false, but not both.

In classical two-valued logic, a statement or proposition is either true or false; similarly an event either occurred or did not occur. In three-valued logic, there exists a separate term denoting "unknown" or uncertain knowledge about the actual event.

Systems of Boolean equations form both fault trees and rule-based expert systems. This paper illustrates a simple set of Boolean equations depicting cause-consequence relations for a nuclear reactor or nuclear power plant to show how uncertainty over initiating events, subsequent results, or rules themselves can affect inferences with such models.

BACKGROUND FOR DST

Define the <u>probability space</u> (Ω, F, P), where Ω is a nonempty set of possible outcomes, F is a σ-field of Ω, and P is a probability measure defined on F. An element of Ω is called a <u>sample point</u>, while a subset of Ω is called an <u>event</u>. In DST probability is assigned to events, while in classical probability theory, probability is distributed among the sample points. Since a sample point is also a subset of Ω with one element, it can be treated as an event and have probability assigned to it in DST. This is how DST subsumes classical probability theory and generalizes it.

Mass Distribution Function

Consider the state space, Ω, with elements $\omega_i \in \Omega$. In classical probability theory, probability can be assigned to individual elements, ω_i, such that $\sum_i P(\omega_i) = 1$. The innovation of DST was to look to the <u>power set</u> 2^Ω, defined as the set of all possible subsets of Ω, and assign probability mass, $m_\Omega(\omega)$, to exclusive-OR (XOR) transformations of all $\omega \subset \Omega$. The mass distribution function of DST is a Borel measure satisfying the following two conditions: $0 \leq m_\Omega(\omega) \leq 1$ and $\sum m_\Omega(\omega) = 1$. In the conventional DST notation ω_i is an element of Ω, while ω is both a subset of Ω and thereby an element of 2^Ω .

A third condition for $m_\Omega(\omega)$ is unique to DST. Implicit in the DST mass summation expression $\sum m_\Omega(\omega) = 1$ is the term $m(\phi)$, which denotes mass assigned to the null set. In DST, $m(\phi)$ is set equal to zero as a premise.

Three-Valued Logic

This paper considers a state space of $\Omega = \{$True, False$\}$ to evaluate events and the validity of propositions. The conventional selection of a state space in DST, also called the frame of discernment, has been chosen as the set of propositions (about a random variable's outcome) not the set of truth values. Therefore, this paper extends DST to a new domain, which we believe can bridge the gap between artificial intelligence and reliability analysis.

For our state space the power set $2^\Omega = \{\phi, T, F, (T,F)\}$ contains a term (T,F) denoting an exclusive-OR for an observation that could be either true or false, but not both. The (T,F) term comes from the DST uncommitted belief term in the power set. After renormalizing for ϕ, the effective power set becomes $2^\Omega = \{T, F, (T,F)\}$.

<u>Modus Ponens</u>

<u>Definition 1</u>: <u>Material implication</u> between A and B is expressed "If A, then B" and written "A → B." <u>Definition 2</u>: The <u>inverse</u> of the logical relation "A → B" is "A → ~ B." <u>Definition 3</u>: <u>Modus ponens</u> of the logical relation "A → B" is "A ∧ (A → B) → B." This paper determines the impact on <u>modus ponens</u> when a researcher receives information, that can be quantified into a probability, on the rule (A → B) itself. In binary logic, if a premise occurs, then the consequence is known as well. In designing expert systems, AI researchers use this logic and refer to "sure-fire" rules. If the premise occurs with some probability less than one, then the consequences would be expected to occur with the same probability. But what can be said if the rule itself is not strictly valid with probability one? The answer to this question can be found by first comparing the truth values emerging for the consequence, B, for given values of A and the rule itself (A → B).

Due to space limitations, we cannot provide here a formal justification of Table 1 based on DST. However, a formal derivation can be found in Guth (1987). Truth tables can also be constructed for the Boolean OR and AND operators that incorporate the three-valued logic. Again to conserve space, we will utilize the logic from these tables published elsewhere in the literature, e.g., Zimmerman (1985, p. 137), without reprinting them here.

<u>Mass Assignment Equations</u>

Denote the mass distribution for A as $m(A) = (a, b, c)$, where "a" denotes the mass assigned to the True element, b to the False element, and c to the XOR(T,F) element. Therefore, $a + b + c = 1$. Similarly denote the mass assignment for B by $m(B) = (x, y, z)$. Then we obtain the mass assignments for the Boolean AND and OR operations as follows:

$$m(A \wedge B) = (ax, b + y - by, az + cz + cx) \quad , \qquad (1)$$

$$m(A \vee B) = (a + x - ax, by, bz + cy + cz) \quad . \qquad (2)$$

From Table 1 if we let the mass on the rule (A → B) denoted by R and $m(R) = (r_1, r_2, r_3)$ then

$$m(B) = (ar_1, ar_2, 1 - ar_1 - ar_2) \quad . \qquad (3)$$

Table 1. <u>Modus ponens</u> truth table

A → B

		T	F	(T,F)
	B			
	T	T	F	(T,F)
A	F	(T,F)	(T,F)	(T,F)
	(T,F)	(T,F)	(T,F)	(T,F)

To see the relevance of these two forms of uncertainty - uncommitted mass and probabilistic rules - on expert systems for nuclear reactors, consider the following example. Our objective is to conclude the probability that the top event will occur, and determine how uncertainty propagates through the model. Let a hypothetical expert system for a pressurized water reactor be summarized by the following six Boolean equations:

$$\text{Pressure Balanced} = \text{Power Source Normal} \wedge \text{Power Sink Normal} \tag{4}$$

$$\text{Power Sink Normal} = \text{Sec. Flow Normal} \wedge \text{Sec. Temp. Normal} \tag{5}$$

$$\text{Sec. Temp. Normal} = \text{Sec. Pumps Operating} \wedge \text{Condenser Vacuum Normal} \tag{6}$$

$$\text{Power Source Normal} = \text{Temp. Pri. Normal} \wedge \text{Pri. Flow Normal} \tag{7}$$

$$\text{Pri. Flow Normal} = \text{Pri. Pump Operating} \vee \text{Backup Pump On} \tag{8}$$

$$\text{Pri. Pump Operating} = \text{No Mechanical Failure} \wedge \text{Power Normal} \tag{9}$$

where $\wedge$ denotes the Boolean AND operator and $\vee$ denotes the Boolean OR operator, the abbreviation "Pri." refers to Primary, "Sec." to Secondary, and "Temp." to Temperature.

Each of the six Boolean equations can be interpreted as an "If, then" rule of a rule-based expert system. For example, Eq. (6) states "if the secondary pumps are operating AND the condenser vacuum is normal, then the secondary temperature is normal." Equation (8) states "If the primary pump is operating OR the backup pump is on then primary flow will be normal."

The equal sign in Eqs. (4) through (9) implies <u>logical equivalence</u> and could be replaced by the symbol <===>. Logical equivalence implies that the material implication ($\rightarrow$) works in both directions. Thus, e.g., Eq. (6) also states "If the secondary temperature is normal, then the secondary pumps are operating AND the condenser vacuum is normal."

To facilitate review of these equations, it will be convenient to introduce some short-hand notation for the events. Let the initiating events be denoted by letters A through F with intermediate events I_1 through I_5 leading to the top event of interest: Pres. Therefore, let Eqs. (4) through (9) be represented by

$$\text{Pres.} = I_1 \wedge I_3 \tag{4'}$$

$$I_1 = A \wedge I_2 \tag{5'}$$

$$I_2 = B \wedge C \tag{6'}$$

$$I_3 = D \wedge I_4 \tag{7'}$$

$$I_4 = E \vee I_5 \tag{8'}$$

$$I_5 = F \wedge G \tag{9'}$$

Thus the top event, Pres. represents "Pressure Balanced," and D represents "Temp. Pri. Normal," etc.

Let R_i denote rule i for i = 4, 5, 6, 7, 8, 9, and let the individual R_i correspond to Eqs. (4') through (9'), respectively. We will assign mass to the rules in the same way that mass is assigned to the initiating events. The R_i will be used as parameters for a study of _modus ponens_. The expert systems rules Eqs. (4') through (9'), coupled with individual R_i for the validity of each rule, are graphically depicted in Fig. 1.

The fact we have written Eqs. (4') through (9') with logical equivalence (=) rather than material implication (→) has nontrivial consequences for analysis with the tree in Figure 1. The equal sign implies that one can study effects propagating up or down the tree. With material implication, the logic would work in only the direction of the implication, e.g., only up or only down the tree.

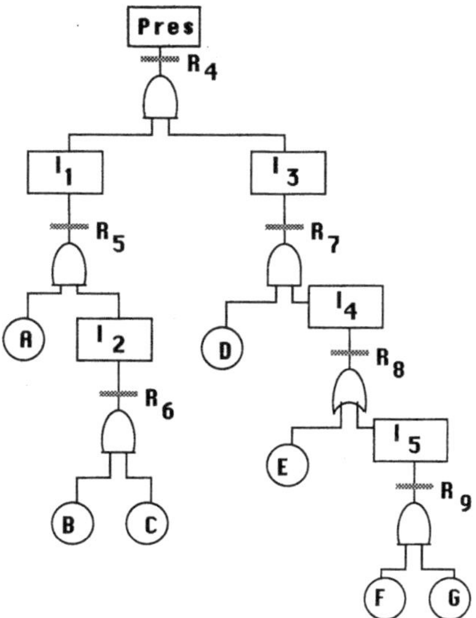

Fig. 1.   Expert system rule tree.

Three different cases for the mass assignments on the initiating events and on the rules are shown in Table 2. For the first case the researcher can clearly observe whether the components A through G occur, but he faces some uncertainty over whether the rules are always valid. In Case II the researcher places uncommitted probability on both the components and the rules.

In Case III the researcher has precise information on both the components and the rules. Case III represents a conventional binary logic

$$\text{Table 2.} \quad \text{Mass assignments for alternative cases}$$
$$\text{(True, False, True-False)}$$

Variable	Case I	Case II	Case III
A	(.9, .1, 0)	(.9, 0, .1)	(.9, .1, 0)
B	(.9, .1, 0)	(.9, 0, .1)	(.9, .1, 0)
C	(.7, .3, 0)	(.7, 0, .3)	(.7, .3, 0)
D	(.8, .2, 0)	(.8, 0, .2)	(.8, .2, 0)
E	(.7, .3, 0)	(.7, 0, .3)	(.7, .3, 0)
F	(.9, .1, 0)	(.9, 0, .1)	(.9, .1, 0)
G	(.8, .2, 0)	(.8, 0, .2)	(.8, .2, 0)
r_4	(.8, 0, .2)	(.8, 0, .2)	(.8, .2, 0)
r_5	(.9, 0, .1)	(.9, 0, .1)	(.9, .1, 0)
r_6	(.6, 0, .4)	(.6, 0, .4)	(.6, .4, 0)
r_7	(.8, 0, .2)	(.8, 0, .2)	(.8, .2, 0)
r_8	(.7, 0, .3)	(.7, 0, .3)	(.7, .3, 0)
r_9	(.8, 0, .2)	(.8, 0, .2)	(.8, .2, 0)

assignment, and one might expect the results from this case to eliminate mass assignments on the uncommitted (T,F) term. However, the application of _modus ponens_ leads to uncommitted mass assignments on the consequence even in the binary mass assignment case.

The results for Cases I and II are identical and shown in Table 3. No mass is assigned to the false elements of the consequences in either of these two cases due to the nature of the _modus ponens_ operator. Mass can only be assigned to the false element of the consequence if there exists positive mass on both the true element of the premise and on the false element of the rule. For Cases I and II no mass is assigned to the false element of the rule.

The results for Case III, given in Table 4, reveal the same mass assignment for the true element of the consequences; however, part of the mass previously assigned to the (T,F) element has been redistributed to the false elements. The probabilistic interpretations of initiating events shown in Case III have often been employed by the risk analysis community and form the basis for probabilistic risk assessment studies of nuclear reactor safety and exposure to radiation. Yet Case III also considers a probabilistic interpretation for rules, which thereby leads to the three-valued logic mass assignments for the consequences given in Table 4.

The mass assignments shown in Tables 3 and 4 were obtained using Eqs. (1), (2), and (3) with local computations for each intermediate event up the tree. See Shenoy and Shafer (1986) and Shafer, Shenoy, and

Table 3. Equilibrium mass assignments for
Cases I and II

Event	True	False	True or false
Pres.	0.0958	0.0000	0.9042
I_1	0.3062	0.0000	0.6938
I_2	0.3780	0.0000	0.6220
I_3	0.3910	0.0000	0.6090
I_4	0.6110	0.0000	0.3890
I_5	0.5760	0.0000	0.4240

Table 4. Equilibrium mass assignments for
Case III

Event	True	False	True or false
Pres.	0.0958	0.0239	0.8803
I_1	0.3062	0.0340	0.6598
I_2	0.3780	0.2520	0.3700
I_3	0.3910	0.0978	0.5112
I_4	0.6110	0.2618	0.1272
I_5	0.5760	0.1440	0.2800

Mellouli (1986) for a discussion on when local computation techniques can
be employed.

For many rule-based expert systems the existence of possible nonunity
mass assignments on initiating events or rules has been an important
caveat preventing further inference. In DST mass assignments a rigorous
methodology has been proposed to accommodate these probabilistic mass
assignments and explain their impact on the mass assigned to the rules
themselves.

The numerical results show that for some consequential events, e.g.,
I_4 and Pres., the majority of mass is assigned to (T,F). Moreover,
assigning mass directly to the false element of the rule, as opposed to
leaving "non-true" mass on the uncommitted belief term (T,F), does result
in a redistribution of mass to the false element of the consequences.

For those rule-based systems which contain probabilistic or frequency
interpretations of their components, DST mass assignments coupled with a
probabilistic interpretation of rules represents a comprehensive framework
for inference with imprecise information. Our experience with expert
systems in robotics and nuclear reactor control confirms that many of the

expert systems extant use "decision rules" rather than strictly logical material implication rules. Thus we feel confident that the researcher interested in studying the effects of uncertainty in his expert system will find these methodologies useful.

DISCUSSION OF RELATED LITERATURE AND CONCLUSIONS

Dubois and Prade (1987) as well as Chatalic, Dubois, and Prade (1986) have written on DST extensions to expert systems. Both of these works employ the traditional set of propositions as a frame of discernment. Moreover, the latter paper sketches the expert system rules in the form of a "dependency tree" that resembles more of a semantic network than a fault tree with Boolean AND and OR gates.

A conference sponsored by Electric Power Research Institute (EPRI) in May 1987 had two sessions related to plant control: (1) Expert Systems Applications in Nuclear Power Plants and (2) Expert Systems Applications in Plant Diagnostics. One of the EPRI conference papers by Neuschaefer et al. (1987) reported development of a general expert system shell for plant diagnosis where the user supplies the knowledge base in the form of cause-consequence relations depicted in a fault tree. Their system proposes to account for uncertainty in instrument readings through conventional signal validation and filtering.

Skatteboe, Tangen, and Berge (1987) describe the use of alternative mathematical models in the construction of expert systems. These authors distinguish between "shallow" versus "deep" information, where shallow information corresponds to "rules-of-thumb" built into expert systems and deep information derives from actual statistical frequencies. The rules-of-thumb appear to be analogous to rules in our models that hold with a high probability, e.g., 0.9. Their same work also discusses qualitative reasoning based on imprecise quantitative measures.

Finally, DST has been incorporated into expert systems for structural damage assessment (Ishizuka, Fu, and Yao, 1981), and in a medical genetics diagnostic system (Gouvernet, Ayme, and Sanchez, 1982).

In conclusion, this paper has shown how uncertainty over the logical relations embodied in expert system rules can be mathematically represented and propagated through a system of Boolean equations. We have borrowed the concept of _modus_ _ponens_ to admit a probabilistic logic for the validity of rules, where some given rule may be valid with a probability less than one.

This paper has also demonstrated an application of the uncommitted belief term in DST to the possibility set {True, False}. The uncommitted term can represent component failures that may or may not have occurred, or the observer cannot tell whether it occurred. To the extent expert system researchers use "decision rules" and imprecise input data, the forms of uncertainty explored here should prove insightful to future modeling efforts.

REFERENCES

Chatalic, Philippe, Didier Dubois, and Henri Prade, (1986), "An Approach to Approximate Reasoning Based on Dempster's Rule of Combination," paper presented at RAI/IPAR '86 Conference on Robotics and Artificial Intelligence, June 18-20, 1986.

Dempster, Arthur P., (1967), "Upper and Lower Probabilities Induced by a Multi-Valued Mapping," _Annals of Mathematical Statistics_, Vol. 38, pp. 325-339.

Dubois, Didier, and Henri Prade, (1987), "On the Combination of Uncertain or Imprecise Information," L.S.I. Report #263, Universite Paul Sabatier, 118 route de Narbonne, 31062 Toulouse Cedex, France, submitted to _Inter. Journal of Approximate Reasoning_.

Gouvernet, J., S. Ayme, and E. Sanchez, (1982), "Approximate Reasoning in Medical Genetics," in _Fuzzy Set and Possibility Theory: Recent Developments_, R. R. Yager (Ed.), (New York: Pergamon Press).

Guth, Michael A. S., (1987), "Uncertainty Analysis of Rule-Based Expert Systems Using Dempster-Shafer Mass Assignments," _Inter. Journal of Intelligent Systems_, forthcoming.

Ishizuka, M., K. S. Fu, and J. T. P. Yao, (1981), "SPERIL-1: Computer-based Structural Damage Assessment Systems," CE-STR-81-36, School of Civil Engineering, Purdue University, West Lafayetta, IN, November 1981.

Lee, Newton S., Yves L. Grize, and Khosrow Dehnad, (1987), "Quantitative Models for Reasoning under Uncertainty in Knowledge-Based Expert Systems," _Inter. Journal of Intelligent Systems_, Vol. II, pp. 15-38.

Neuschaefer, Carl H., Peter W. Rzasa, Eugene Filshtein, Richard L. Burrington, and Robert Donais, "Application of C-E's Generic Diagnostic System to Power Plant Diagnostics," paper presented at EPRI conference on Expert Systems Applications in Power Plants, Boston, MA, May 1987.

Prade, H., (1985), "A Computational Approach to Approximate and Plausible Reasoning with Applications to Expert Systems," _IEEE Trans. Pattern Anal. Mach. Intel._, Vol. PAMI-7.

Shafer, Glenn, (1976), _A Mathematical Theory of Evidence_, (Princeton, N.J.: Princeton University Press).

(1982), "Bayes's Two Arguments for the Rule of Conditioning," _Ann. Statistics_, Vol. 10, pp. 1075-1089.

(1985), "Conditional Probability," International Statistical Review, Vol. 53, #3, pp. 261-277.

(1986), "The Combination of Evidence," _Internat. J. Intelligent Systems_, Vol. 1, No. 3, pp. 155-179.

(1986) with P. P. Shenoy, and K. Mellouli, "Propagating Belief Functions in Qualitative Markhov Trees," Working Paper No. 186, School of Business, University of Kansas, Lawrence.

Shenoy, P. P. and G. Shafer, "Propagating Belief Functions Using Local Computations," _IEEE Expert_, Vol. 1, No. 3, pp. 43-52.

Skatteboe, Rolf, and Grethe Tangen, and Kaj Berge, (1987), "Models Applied in Knowledge Based Diagnosis," paper presented at EPRI conference on Expert Systems Applications in Power Plants," Boston, Mass.

Zimmerman, H.-J., (1985), _Fuzzy Set Theory - and Its Applications_, (Boston: Kluwer-Nihoff Publishing Co. 1985).

A NUCLEAR FACILITY SECURITY ANALYZER WRITTEN IN PROLOG

B. D. Zimmerman

Westinghouse Hanford Company
P.O. Box 1970   W/C-86
Richland, Washington

INTRODUCTION

The Security Analyzer project was undertaken to use the Prolog "artificial intelligence" programming language and Entity-Relationship database construction techniques to produce an intelligent database computer program capable of analyzing the effectiveness of a nuclear facility's security systems. The Security Analyzer program can search through a facility to find all possible surreptitious entry paths that meet various user-selected time and detection probability criteria. The program can also respond to user-formulated queries concerning the database information. The intelligent database approach allows the program to perform a more comprehensive path search than other programs that only find a single "optimal" path. The program also is more flexible in that the database, once constructed, can be interrogated and used for purposes independent of the searching function.

THE PROLOG LANGUAGE

In order to understand the functioning of the Security Analyzer program, it is helpful to understand a little about how Prolog itself functions.

A Prolog program is made up of two parts: a database of "facts," and the "rules" that operate on those facts. Processing consists of attempts to use the available facts to satisfy the requirements of the rules and thereby provide solutions to questions that the programmer has posed.

A very simple example of a Prolog-type rule is as follows:

X is a Mother if X has a child and X is female

Prolog facts might contain information such as:

> John has a child
> John is male
> Mary has a child
> Mary is female

When presented with the question "who is a mother?", Prolog will activate
its rule about mothers and try to satisfy its requirements. It will find
that John satisfies the first requirement but does not satisfy the second.
Prolog will then automatically backtrack and try to find a different solution to the first requirement: Mary in this case. When it is found that
Mary satisfies the second requirement as well as the first, Prolog will
state that it has found a solution to the original question. If so instructed, Prolog will continue using the rules in this searching and backtracking fashion until it has found all solutions to the original
question. This searching and backtracking results in a "depth first search"
mode of operation.

The example given here is a very simple one. The mechanical process
of performing this simple type of inference is known as unit resolution.
In reality, much more complicated statements (which can be thought of as
containing and's and or's) are normally used, and rules can refer to other
rules as well as directly to database facts. One of the main rules in
the Security Analyzer program has the general form:

> a legal path step has been taken if ___ and if ___
> ...and if ___.

The "if" clauses are structured so as to direct Prolog to search the
appropriate database sections concerning geographical layout, intrusion
sensor locations, and detection probability and path transit time data,
as needed, to find step-by-step paths that intruders could take to reach
the target location. A legal path step has been found when Prolog is
able to use available database information to simultaneously satisfy all
the "if" clauses. A complete path has been found when Prolog has discovered a sequence of legal steps from the specified start point to the specified destination.

ENTITY-RELATIONSHIP MODELING

The Entity-Relationship modeling approach was chosen to structure the
database of facts for the Security Analyzer because: 1) it requires that
the relationships be specified in a way that follows very naturally from
the "real world" problem and 2) it results in a form of the information
similar to that which Prolog requires.

Figure 1 presents the formal Entity-Relationship diagram created for
the Security Analyzer application. The rectangles, called "entity sets,"
represent sets of listable objects, such as "facility components" (rooms,
doors, walls, sensors etc.-a very broad category here). Diamonds, called
relationship sets, represent relationships between the entities, such as
"connects" (for example, a particular room is connected to another room).
Ovals, called "value sets," represent lists of values (either numbers or
"logical values" represented by text strings) that can be associated with
entity sets or relationship sets. An example of a value set is the Transit
Time set, which represents the set of time data that specifies how long
adversaries of various types require to overcome the different kinds of
barriers in the problem or to traverse various rooms, hallways etc.

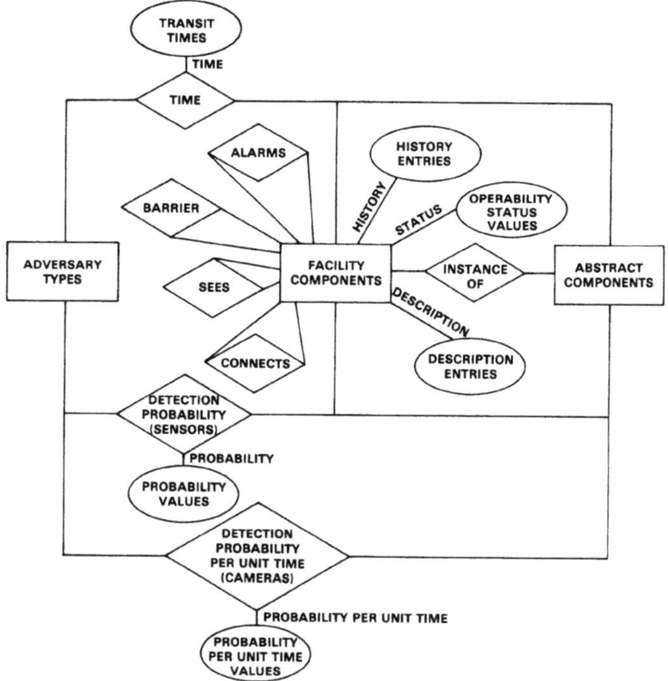

Figure 1. Security Program Entity Relationship Diagram

For the Security Analyzer application, the Entity Relationship diagram
contains the following components (see Figure 1):

1. The FACILITY COMPONENTS entity set includes all the schematic
diagram components, such as halls, doors, fences, sensors, courtyards
etc. Typical entities in this entity set would be "nl_door3" (night-latch
door #3) or "s12" (intrusion sensor #12).

2. The ABSTRACT COMPONENTS entity set includes generic names that can
be associated with specific entities in the FACILITY COMPONENTS entity set.
For example, the entity "door3" from the FACILITY COMPONENTS entity set is
associated with the abstract component "standard building door." If, during
execution of the Security program, the Prolog logic is unable to discover
needed facts about door3, it can attempt to satisfy its needs by searching
for facts concerning the generic category standard building door.

3. The ADVERSARY TYPES entity set allows for the existence of several
classes of intruders with different capabilities and characteristics.
Currently, two types of intruders have been implemented: class one is slow
moving and is unable to go through walls, class two is faster moving and
is able to break through walls.

4. The relationship sets labeled CONNECTS, ALARMS, SEES, and BARRIER
represent relationships between the various FACILITY COMPONENTS entities.
For example, a specific camera sees a specific hall, a specific sensor
alarms when a specific door is opened or a specific area is crossed, a
specific courtyard connects to a specific hallway, or a specific barrier
(perhaps a door) lies between a specific courtyard and a specific

hallway. A fact statement associated with the SEES relationship set would
look like: "sees(camera3,door7)," indicating that camera3 sees door7.

5. The relationship set INSTANCE OF indicates that a specific entity
in the FACILITY COMPONENTS entity set is an "instance of" a more abstract
component in the ABSTRACT COMPONENTS entity set. For example,
"ins_of(door3,std_bldg_door)," indicates that door3 is an "instance of" a
standard building door. A particular class of intruder may require a
certain amount of time to break through a standard building door, rather
than having a particular amount of time associated with door3 itself.

6. The relationship sets DETECTION PROBABILITY and DETECTION PROBA-
BILITY PER UNIT TIME indicate that a particular class of adversary will have
a particular probability of detection if the adversary trips a particular
sensor (or one of a particular class of sensors), or will have a particular
probability per unit time of detection if the adversary is visible with a
particular camera. Value sets associated with each of these relationship
sets indicate that there is a specific set of probability or probability
per unit time values from which to choose. A typical detection probability data record would look like "prob(s12,adv1,0.8)": adversary type 1
has an 80% chance of being detected when encountering sensor s12.

7. The relationship set TIME indicates that a particular class of
adversary will require a particular amount of time to cross a particular
area or negotiate a particular barrier (door, fence, or wall). As before,
time values may need to be found through the use of the ABSTRACT COMPONENTS
generic classifications for walls, doors etc. The value set TRANSIT TIMES
indicates that there is a particular set of time values from which to
choose. TIME data records look like "time(door3,adv1,5)": an adversary
of class 1 will require 5 minutes to defeat door3.

8. Value sets labeled OPERABILITY STATUS, DESCRIPTIONS, and HISTORY
contain individual values (symbols or text strings) for these attributes.
These attributes are all associated with entities in the entity set FACILITY
COMPONENTS.

The Entity-Relationship diagram of Figure 1 organizes the "real world"
problem into a tractable form and acts as a "template" for construction of
the actual Security Analyzer database.

FUNCTIONS OF THE SECURITY ANALYZER PROGRAM

Program Control and Database Retrieval Functions

These functions are provided primarily as auxiliaries to the path
searching function.

1. A main menu that allows selection of the function to be
performed.

2. Rules that search the database for any available information
concerning the description and location of a specified sensor.

3. Rules that produce a listing of the current operational status of
all sensors and cameras known to the database.

4. Rules that allow the operator to change the current operability
status of any camera or sensor known to the database. Changes to the

operability status information are accomplished by the use of Prolog assert
and retract statements.

   5. Rules that allow the addition, listing, or deletion of records
giving the history of any specified component in the database. Prolog
assert and retract statements are employed to allow this capability.

Path Searching Function

   The path searching function is the heart of the Security Analyzer
program. As stated earlier, the purpose of the searching capability is to
allow the program to find any and all paths, from a specified location to
a specified location, that will result in less than a specified probability
that the adversary will be detected and that will require less than the
specified amount of time for the adversary to transit, for a selected
adversary type. From a security point of view, this amounts to finding
the weakest paths in the security system.

   The searching function operates in the following way. A maximum
detection probability value, a maximum transit time value, a starting
location, a destination location, and an adversary type are given to the
program in response to prompts. The program then takes a single step
toward the destination by searching the database to see where it can reach
from the beginning location in one step. When a tentative step is identified, the program searches the database to determine if any operable sensors
will trip by entering the new area, whether any operable cameras will see
the adversary located in the new area, whether any barrier must be negoti-
ated to enter the new area, and how much time it takes the specified class
of adversary to traverse the new area.

   If a barrier is found, another search is made to determine if there
are any operable sensors associated with that barrier, whether any operable cameras will see the adversary negotiate the barrier, and how much
time it will take the adversary to negotiate the barrier.

   If any operable sensors, either associated with the new area or a
barrier, are discovered, the detection probability associated with those
sensors and the specified adversary type is found from the database and
is statistically added to the detection probability accumulation register.
If the search determines that one or more cameras will see the adversary
traverse the new area or negotiate an associated barrier, the detection
probability per unit time is found from the database for that camera and
that adversary type, it is multiplied by the time needed for the adversary
to traverse the area or overcome the barrier, and the resulting probability
increment is statistically added to the detection probability accumulation
register.

   In this way, the total detection probability increment associated with
the tentative path step is found and added to the detection probability
already accumulated from previous path steps (if any). The total time
associated with the current and previous path steps is accumulated in a
similar way. A test is then made to determine if either the total detection
probability or the total transit time exceed the user-specified values.
If either does, then the tentative path step is discarded. If both do
not exceed the specified values, the tentative path step becomes the new
beginning location, a next tentative step is selected from those available
in the database, and the process is repeated.

   If the tentative path step results in exceeding the detection probability or transit-time-limiting value entered by the user, Prolog

automatically discards that choice and goes to the database to make another
choice. If all possible choices have been exhausted, Prolog will automatically backtrack as many steps as needed until a previously untried
step can be found. If no previously untried steps can be found, then the
program concludes that no paths from the original beginning point to the
specified destination point have detection probabilities and transit times
less than the specified values.

Circular Paths and Heuristics

The path finding algorithm described previously could lead to the
program "going in circles," that is, endlessly repeating a sequence of
path steps. In order to prevent this, the program includes a test to
ensure that the same physical location (a room, courtyard, etc.) is not
used more than once as part of a particular path. The program also includes
certain "heuristic" rules to make processing more efficient. For example,
there is a "directness test" rule, which checks to see if two locations
that appear in a path list separated by several other locations are in
fact immediately adjacent to each other in the physical layout. If they
are adjacent such that one could have been reached from the other in a
single step, then the intervening path steps that the program has taken
are superfluous (unless the roundabout path avoids a barrier, a camera,
or a sensor; this situation is also checked). The Security Analyzer with
fully implemented heuristics runs much faster than without heuristics
(how much faster depends on the complexity of the problem).

TEST CASE RESULTS

The Security Analyzer program was used to analyze a hypothetical multi-
story nuclear facility. Approximately two man-weeks of effort were required
to create a database consisting of over 100 rooms and areas, several sets
of stairs and elevators, approximately 200 intrusion sensors and TV cameras,
and surrounding triple security fence and gate system. Because of the
"combinatorial explosion" of possible paths for a problem of this size, the
computations were performed in ten steps: five steps started outside the
facility found paths to five possible building entry locations, the remaining five runs found paths from these entry points to the target location
inside the building.

For the particular (hypothetical) time and detection-probability upper-
limit constraints chosen, the program investigated 25,595 possible paths
and found 140 solutions (paths to the target location requiring less than
the specified time and less than the specified detection probability).
Many of these were quite devious and might not be obvious to a human
analyst. Figure 2 presents a typical printout that the Security Analyzer
might produce when it has found a path from "cyl" (courtyard #1) to "rm426"
(room #426). The first line specifies the intruder detection probability
and the estimated transit time for the path (all values are purely hypothetical). The second line specifies the path (symbols such as "dll"
stand for doorll). The following lines provide a listing of the sensors
or cameras that the intruder would have encountered along this path.

The set of ten runs performed for this analysis required approximately
100 minutes of VAX®cpu time. While this processing time is longer than
might be desired, one security analyst with whom this test case was performed estimated that the total time to perform a Security Analyzer analysis
(including construction of the database) is about one-fourth of the time
required by his current analysis methods, which consist of a combination
of manual and computerized steps.

PATH WITH 0.985868 Detection Probability Found. Estimated time: 2.94

```
rm426   d409   rm424   rm415   d408   rm441   rm440   rm321   d11   rm327
  rm1   d13    cy1
```

THE FOLLOWING SENSORS WERE TRIPPED:

```
c408        camera
s451        inf_sensor
s452        inf_sensor
s453        inf_sensor
s409        mag_sensor
c408        camera
c403        camera
s408        mag_sensor
c403        camera
s11         mag_sensor
c117        camera
```

Figure 2. A Typical Security Analyzer Path Printout

CONCLUSIONS

The Security Analyzer is a successful application of the Prolog symbol
processing language and the Entity-Relationship modeling technique. It
performs several database retrieval and realtime updating functions, but
its "search" function is its most important capability. There are two
general areas of application for this search function:

1. It can provide a CAD capability for designers of a security
system. The Prolog database structure of the program allows for easy
addition or deletion of sensors, fences, etc. A designer can quickly
create various security system configurations and have the program find
the vulnerable points in the design.

2. It can provide an on-line capability to alert security personnel
in charge of a facility to unexpected degradations of the system. For
example, if a given sensor or camera is discovered to be malfunctioning,
that information can be entered into the program and the program instructed
to perform a search analysis to identify the weakest paths into a restricted
area. The program will use the current operability status of the sensors
and cameras to determine the detection probabilities. In this way, a
comprehensive analysis of the current state of facility security can be
performed at any time, and unexpected degradations of security caused by
specific combinations of inoperable sensors and/or cameras in complex
security systems can be detected.

The Security Analyzer was originally written in Quintus Prolog® for
execution on a VAX. Versions currently exist that run on the Sun 3/160
Workstation® (also under Quintus Prolog), and the IBM-PC® using
TurboProlog®).

AN OPS5-ENGLISH ENGLISH-OPS5 PARSER AND LANGUAGE GENERATOR

B. Baird and S. Wertheimer
Connecticut College, New London, CT

N. Gove and P. Otaduy
Oak Ridge National Laboratory, Oak Ridge, TN

INTRODUCTION

The artificial intelligence community has, for at least 50
years, been concerned with the problem of machine understanding
of natural language. Another problem, of concern to programmers, is human comprehension of computer programs. The work
undertaken here may be thought of as an attack on the conjunction of these two problems in a very restricted domain: the
computer language to be made comprehensible to the human is OPS5
while the computer is being asked to understand sentences which
give rise to parts of OPS5 programs. This work translates OPS5
productions to and from English and other languages.

The translation process does not go directly from source
language (either OPS5 or English) to the target language; there
is a collection of intermediate structures which contain all of
the information contained in the original OPS5 production and
which also make the translation job easier. Thus the translation proceeds as

    OPS5 <----> intermediate semantic structures <----> English
         |
         --> OPS83, French, German, other styles of
             English

where the double arrow means one can go in either direction.
Parsing and translation are done using the programming language
Prolog [2]; this choice was made because of its powerful parsing
capabilities. Both OPS5 and the restricted subset of English
used can be defined by definite clause grammars [4], the rules
of which are directly expressible in Prolog. In addition,
starting with the intermediate semantic structures, translations
to languages other than English can be done. We have developed
these for OPS83 (a programming language which is more general
than OPS5), French, German and several other styles of English.
The different English style translations are meant for different end users, such as OPS5 programmers, or those with some
programming background but no knowledge of OPS5.

INTRODUCTION TO OPS5

OPS5 [1] is one of several rule-based languages used in the
nuclear industry for monitoring and advising systems. It comprises a collection of "productions," a set of "condition elements" which make up "working memory," and a "conflict resolution strategy." A production is a rule of the form IF (conditions) THEN (actions), where the conditions can be tested by
looking at the contents of working memory, and the actions can
either perform some communication function with the external
environment (input, output, file access) or change working
memory by removing, adding, or modifying working memory elements. Only one production is active at a time; the active one
is selected by the conflict resolution strategy. The OPS5 rule
interpreter tests the conditions in the IF part of all productions to see if they are satisfied, and then chooses from among
those which succeed by using a predetermined strategy. After
the actions of the successful production are performed, the
process repeats until either it is told to stop or halts because
no production has satisfiable conditions.

INTERMEDIATE SEMANTIC STRUCTURES AND PROLOG

All translations of OPS5 productions, whether originally
written in English or OPS5, pass through an intermediate phase
where the production is translated to a collection of Prolog
facts. These facts are called the intermediate semantic structures (ISS).

There are several reasons why this method was found to be
preferable to a direct translation from source language to
target language:

(1) The ISS are easier for the Prolog program to work with
since they are Prolog facts. It was difficult to translate
directly, say, from OPS5 to English. In order to perform this
translation it was necessary to hold on to some of the OPS5
information before making a decision about the form of the final
English translation. This meant that much of the information
about the production needed to be asserted as Prolog facts. It
was then a short step to put all of the production information
into Prolog facts.

(2) The ISS could be handled both coming and going: in
other words, all translations, whether from OPS5 to English, or
English to OPS5, or OPS5 to another language, stop at this
common meeting ground. This makes for greater modularity,
greater flexibility of movement within the program, and allows
comparison of different versions of productions. Translations
can proceed in steps and thus utilize other work. For example,
in order to translate an OPS production to English, the program
first translates the production to the ISS and then translates
the structures to English. In order to move from OPS5 to
French, the program utilizes the same intermediate structures
and translates them to French.

The ISS contain information such as the production name,
the type and order of condition elements, the type and order of
actions, information that pertains to actions, and information
about variables. All of these pieces of information are in sep-

arate Prolog facts, which are cross-referenced with each other.
All of the OPS5 information is contained in these structures.
An example is given in the Appendix.

OPS5 TO ENGLISH

As mentioned in the last section, the translation of an
OPS5 production to English first involves an extraction of the
ISS and then involves a translation of these Prolog facts to
English. Several factors guided the decisions about the choice
of English phrases and about just what information should be
included in the English version.

The choice of the kind of English was guided by the examples found in the OPS5 book [1] and by a desire to produce a
translation that sounded, as far as possible, natural. The
English version of the production contains only the information
about the OPS5 production that is necessary to understand the
operation of the production; some of the information contained
in the original OPS5 production may not appear in the final
English version. The ISS will contain this information, but not
the English version. For example, no explicit mention is made
of variable names. If the OPS5 production uses a variable name
only once, then no reference to that variable name is ever made
in the English. If a variable name is used more than once, then
the English version makes reference to the place to which the
variable refers, but no explicit mention is made of the variable
name. In those cases where the variable names are the only
means of distinguishing elements the English version uses the
phrase "instance 1," "instance 2," etc. Likewise, no mention of
BIND and CBIND actions are given; the results of these actions
are automatically correctly annotated. Variable names, bind
actions, and cbind actions are internal to OPS5 and not an
integral part of the production. Thus mention of them can
safely be omitted. An example of a translation from an OPS5
production to its English version is given in the Appendix.

OPS5 to OPS83

OPS83 [3] is a more recent computer language, which combines a rule-based production system (like OPS5) with procedural
features (like Pascal). Any procedure can be used as the right-
hand-side of an OPS83 rule. The production part of OPS83 is
similar to OPS5 but with different terminology and syntax rules.
OPS83 is more powerful and versatile than OPS5 and so there may
be some interest in converting OPS5 programs to OPS83 ones.

The first phase in the translation of an OPS5 production to
the (approximately) corresponding OPS83 rule was to translate
the OPS5 production to the ISS. The second phase reads the ISS
and then constructs an OPS83 rule.

The left-hand-side, or conditions part, of an OPS5 production is similar in OPS83, except for differences in punctuation
and naming patterns. On the right-hand-side, or actions part,
some of the action verbs (e.g., MAKE, REMOVE, MODIFY) require
little translation. Others can not be translated precisely as
there is no exact counterpart in OPS83 (e.g., BUILD, CBIND,
SUBSTR, LITVAL, ACCEPTLINE) although the same effects can usual-

ly be obtained somehow in OPS83. For these cases, a partial
translation is given with comments. For some cases, (e.g.,
READ, OPEN, CLOSE), local variables are created by the translation program to fit the OPS83 syntax. The format options in
OPS5 are TABTO(n) and RJUST(n). The TABTO form can be converted
to the RJUST form if the lengths of the print fields are known
in advance. The translation does this where possible; otherwise an approximate OPS83 format is given.

In general, the OPS83 version would require context-
specific editing before running. The present translation program may provide a way to automate most, but not all, of the
editing changes necessary to provide a functionally equivalent
OPS83 rule. For example, the BUILD option in OPS5 can cause the
creation of new productions during execution. This capability
does not exist in OPS83; however, any reasonable number of OPS83
rules could be prepared with variable parameters in both conditions and actions to simulate most, if not all, practical uses
of BUILD.

ENGLISH TO OPS5

In translating from OPS5 to English, the source language
syntax admits only one form (up to trivial rearrangements) to
convey a desired meaning. The resulting English form closely
parallels the OPS5 form and uses, in essence, the same syntax
expressed in a human rather than computer language. The first
objective in understanding an English description of an OPS5
production was to express this restricted syntax as a formal
grammar, necessarily similar to the syntax of OPS5. A recog-
nizer was then built using Prolog, and then, using the ideas of
definite clause grammars (DCG) [4], the ISS were produced.
This left two problems: the design of a program to convert the
ISS to OPS5, and reworking the DCG so that it would recognize
and understand more natural English descriptions of OPS5 productions.

The first problem turned out to be rather straightforward
to solve. Although the ISS were originally designed to enable
easy translation to English, they also proved to be easy to use
to go back to OPS5.

The ultimate solution to the second problem would be a
program which would understand any English description of a
valid OPS5 production. The work described here was incremental;
it built upon the DCG which understood only one version of a
production. This evolving DCG was gradually made less restric-
tive, while keeping its generation of semantic structures capabilities constant. At the time of this writing the program
accepts English which is quite close to the English comments one
often finds as documentation to OPS5 productions [1].

APPLICATIONS TO THE NUCLEAR INDUSTRY

This work has direct application to improving communication
between nuclear reactor control room operators and expert-system
based reactor control programs. Proposed monitoring programs
for the High Flux Isotope Reactor (HFIR) at the Oak Ridge
National Laboratory are written in OPS5; however, the reactor

operators and their supervisors are unfamiliar with this relatively obscure language. By using the OPS5 to English translator, those people unfamiliar with the OPS5 syntax can read and
understand the logical content of OPS5 productions and thereby
detect programming errors, suggest modifications, and also
better understand the operation of the control program. On the
other hand, either a programmer or non-programmer can make use
of the English to OPS5 translator to add new productions to an
existing program, or create new programs from scratch.

REFERENCES

1.   Brownston, L., Farrell, R., Kant, E., and Martin, N.,
      "Programming Expert Systems in OPS5," Addison-Wesley,
      Reading, MA (1985).
2.   Clocksin, W. and Mellish, C., "Programming in Prolog,"
      Springer-Verlag, New York (1984).
3.   Forgy, C., "The OPS83 Users Manual," Production Systems
      Technologies, Inc., (1985).
4.   Pereira, F. and Warren, D., "Definite Clause Grammars
      for Language Analysis," Artificial Intelligence 13, pp.
      231-78 (1980).

APPENDIX

OPS5 production input by user:
```
(p test           ;Remove duplicate responses among annunciators
   {(alarm ^name <name1> ^response <act> ^priority <prior1>
           ^response-occurances <occurances>) <original>}
   {(alarm ^name {<name2> <> <name1>} ^response <act>
           ^priority { >= <prior1>}) <duplicate>}
  -->
   (modify <original> ^response-occurrences (compute
       <occurrences> + 1) ^other-annuns <name2>)
   (ops-remove <duplicate>)).
```

English version input by user:
The production's name is test;
If there is an element "alarm" (1);
and if there is also an element "alarm" (2) whose "name" is not
   the same as the value in "name" of "alarm" (1) and whose
   attribute "response" is the value in "response" of "alarm"
   (1) and whose attribute "priority" is greater than or equal
   to the value in "priority" of element "alarm" (1) ;
Then
Modify "alarm" (1) to make "response-occurrences" equal to
   the sum of the value in "response-occurrences" of element
   "alarm" (1) and 1 , to set "other-annuns" equal to the value
   in "name" of element "alarm" (2) ;
Remove from working memory the element "alarm" (2).

Intermediate semantic structures obtained from the English
   (virtually identical to the ISS obtained from the OPS5
   production):
```
     prname(test).
     cename(act(1),alarm, modify,'').
     cename(ce(1),alarm,+,instance(1)).
     cename(ce(2),alarm,+,instance(2)).
     cevar(cevar_6,ce(1)).
```

```
            cevar(cevar_7,ce(2)).
            vars([var_23,ce(1)],[name,1]).
            vars([var_24,ce(1)],[response,1]).
            vars([var_25,ce(1)],[priority,1]).
            vars([var_26,ce(1)],['response-occurrences',1]).
            vars([var_27,ce(2)],[name,1]).
            field(2,priority,[geq(vbl(var_25))]).
            field(2,response,[vbl(var_24)]).
            field(2,name,[neq(vbl(var_23))]).
            rfield(1,'other-annuns',[vbl(var_27)]).
            rfield(1,'response-occurrences',
               [compute(sum(vbl(var_26),1))]).
            action(1,modify,cevar_6).
            action(2,remove,remwme(cevar_7)).
```

OPS5 production translated from the intermediate semantic
   structures:

```
(p test
   {<cevar_6> (alarm  ^name <var_23> ^response <var_24>
        ^priority <var_25> ^response-occurrences <var_26>)}
   {<cevar_7> (alarm ^name { <var_27> <> <var_23>}
        ^priority >= <var_25> ^response <var_24> )}
 -->
   (modify <cevar_6> ^other-annuns  <var_27>
        ^response-occurrences  (compute <var_26> + 1 ) )
   (remove <cevar_7> )).
```

English translation of the intermediate semantic structures:
The name of the production is test;
IF there is a WME of class "alarm" (instance 1) ;
AND if there is also a WME of class "alarm" (instance 2)
   whose attribute "name" is not equal to the value in
   attribute "name" of WME of class "alarm" (instance 1) ,
   whose attribute "response" is the value in attribute
   "response" of WME of class "alarm" (instance 1) , whose
   attribute "priority" is greater than or equal to the
   value in attribute "priority" of WME of class "alarm"
   (instance 1) ;
THEN
MODIFY the WME of class "alarm" (instance 1) to set
   attribute "response-occurrences" equal to the sum of the
   value in attribute "response-occurrences" of WME of class
   "alarm" (instance 1) and 1 , to set attribute "other-
   annuns" equal to the value in attribute "name" of WME of
   class "alarm" (instance 2) ;
REMOVE from WM the element of class "alarm" (instance 2) .

OPS83 translation of the intermediate semantic structures:

```
rule test
      {
        &1   (alarm );
        &2   (alarm        name <> &1.name ;      response =
             &1.response ;      priority >= &1.priority );
        -->
        modify &1 (   response_occurrences = &1.response-
             occurrences + 1 ;  other_annuns = &2.name    );
        remove &2 ;
        };
```

PROLGRAF-B: A KNOWLEDGE-BASED SYSTEM FOR THE AUTOMATED

CONSTRUCTION OF NUCLEAR PLANT DIAGNOSTIC MODELS

Sergio B. Guarro [*]

Department of Mechanical, Aerospace & Nuclear Engineering
University of California, Los Angeles

INTRODUCTION

Automated diagnosis of undesired operational transients in a nuclear power
plant process has been an active subject of research for the last ten
years. Examples of different approaches to the solution of this problem
are the early research on Disturbance Analysis Systems (DASs) and the more
recent developments in the area of Automated Emergency Operating Procedures
(AEOPs).

The early efforts in the field of automated plant process analysis were
hindered by the scarcity of efficient modeling tools suited for self-
contained representation of typical target processes. For example, the
"cause-consequence tree" methodology used in the DAS prototypes produced
"event-oriented" models [1]. This type of representation required a
separate model development effort for each individual event sequence for
which a diagnostic capability was desired. The resulting models often
lacked flexibility, in that even relatively small deviations of an actual
occurrence from a modeled event sequence could compromise the capability by
the model and the associated diagnostic software to match and recognize the
sequence. The outcome of this scarcity of appropriate tools was a
premature reduction of research efforts in this area, and a redirection of
research by major organizations, like the Electric Power Research
Institute, towards the development of computer based models for a
"symptomatic", rather than diagnosis-based, treatment of plant operational
transients [2].

A discrete modeling approach, called the Logic Flowgraph Methodology (LFM)
and designed to overcome some of these difficulties, has been developed and

----

(*) The Author's principal affiliation is with the Lawrence
     Livermore National Laboratory of the University of
     California.

tested at the University of California, Los Angeles. This approach is
neither "event-" nor "symptom-oriented", but can rather be defined as
"process-oriented". As will be discussed in some detail below, the LFM
models can be used in conjuction with the LFM "inference engine" to perform
automated reliability analysis of a plant process [3,4], or to derive in
real time both process diagnostic and process recovery recommendations for
plant operators [3-5].

Described here are the essential features of the LFM discrete simulation
environment. With the description of that environment as background, the
discussion then focuses on the architecture and the procedures of an
"intelligent interface" program which is being developed to greatly
facilitate the LFM user in the construction of nuclear plant process models
based on the LFM modeling primitives and rules. The "intelligent
interface" program that is the object of the discussed research has been
named PROLGRAF-B (for PROcess Logic flowGRAPH Builder) and produces as its
output an "intelligent database" structure containing and representing all
the LFM "objects" necessary to describe the modeled process.

LFM FEATURE SUMMARY AND EXAMPLE

The Logic Flowgraph Methodology (LFM) was developed as a means of
constructing computer implementable models of complex plant processes, such
as those taking place in nuclear power or chemical plants. The LFM models
take the form of directed graphs, with relations of causality and
conditional switching actions represented respectively by "causality edges"
and "conditional edges" that connect network nodes and special operators.
Of these, the first represent important process variables and parameters,
and the latter represent the different types of possible causal or
sequential interactions among them.

The specific usefulness of LFM as a discrete simulation tool derives from
its direct and intuitive applicability to the modeling of causality driven
processes, and from the capabilities offered by the LFM computer-based
"inference engine". This was developed as an integral part of the
methodology, and allows a user to analyze the synthetic LFM models on-line,
e.g. for the direct real-time monitoring and automatic diagnosis of a
process, or off-line, e.g. for the construction of reliability or risk
models.

The reader interested in the details and the applications the LFM
simulation framework will find detailed explanations and discussions in [3
-5]. For the purpose of illustrating some of the fundamental concepts and
modeling "primitives", or "objects", used in LFM we can refer to the simple
LFM graph depicted in Figure 1. The graph shows that between variable 1
(V1) and variable 2 (V2) there normally exists a causal relation of direct
proportionality in terms of possible deviations from rated steady-state
values. This is indicated by the "default" (i.e., shown as unconditioned)
+1 gain value in the directed "multiple gain box" (MGB) between V1 and V2.
However, when a condition of moderate positive variation from steady-state
occurs for variable 3, that is, V3 = +1, the causal link between V1 and V2
is no longer valid, as shown by the 0 gain value that is pointed at in the
same MGB by the "true" outcome of the condition test "is V3 = +1 ?". This
test is in turn represented by the diamond box ("test box" or TB in the LFM
denomination conventions) along the conditional edge from V3 to the gain
box.

USES OF LFM AND COMPARISON WITH AI EXPERT SYSTEMS

As explained above, the computer executable interpretation of the LFM
models makes them especially well suited for the analysis of complex
processes and the diagnosis of unexpected situations that may occur in such

processes. The LFM inference engine is capable of automatically developing
from the LFM graph models, without any help from the user, complete logic
tree analyses of a condition occurring in the modeled process. The only
input needed by the LFM inference engine is the specification of a "top
event", i.e. the event for which an analysis or explanation is required.
If any other conditions and events existing in the process are observable
(i.e. can be detected), they can also be fed as input to the inference
engine. The analytical procedure will use these recorded observations to
appropriately restrict the "search space" in the search for the "root
cause" events and conditions that can be logically deduced as the ones
determining the top event. To work with a simple example of these concepts
we can refer again to Fig. 1. If we ask the LFM code to determine for what
conditions we can have the top event "V2 = 0" (meaning that the variable is
at its unperturbed steady-state) and we do not provide any further
information, the code will give us back the following answer:

"V1 = 0" .or. ("V1 = <any value>" .and. "V3 = +1")

However, if we have observed that V3 has not changed from its steady state
value (i.e., V3 = 0) and we provide this information to the analytical
inference engine, or if the code itself receives direct input to that
effect from a sensor measuring the variable V3, then the deduced root event
will simply be:

"V1 = 0"

The LFM models and software perform functions analogous to those typically
expected from diagnostic "expert systems" developed in the AI (artificial
intelligence) domain, but are also suited for the off-line development of
full fault and success trees for system reliability and risk analysis
applications. With respect to most expert system "shells" and frameworks
LFM presents some notable differences. The LFM representation paradigm is
based on synthetic graphic models which contain in _implicit_ form a
combination of "semantic network" and "if-then rule" knowledge
representations, whereas most expert systems utilize plainly stated "IF-
THEN" rules which are analytic and "verbose" in nature.

Another important feature of the LFM software is its ability to
automatically handle feedback and feedforward loops. This feature allows
the construction of logic trees that have already been checked and adapted
to satisfy all the logic consistency rules that must be obeyed in the
presence of process loops.

It should be also noted that, unlike many AI expert systems, which are
developed from "heuristic" or empirical knowledge of the application
domain, LFM is best suited for the modeling of a given process based on
knowledge that, although qualitatively expressed, is "deep", that is,
rooted in the "cause - effect" reasoning principle.

OBJECT ORIENTED DATABASE REPRESENTATION OF LFM MODELS

The LFM inference engine uses LFM process models as input and generates
logic trees as output. These explain how specific events, defined by the
analyst as the focus of his/her analysis, or actually occurring in the
plant process and thus the subjects of a diagnostic real-time analysis,
have been produced by basic process conditions. The LFM input models
constitute the knowledge base upon which the inference engine operates to
generate its deductions. This knowledge base must be made available to the
inference engine in a form suitable for analysis and deduction by the
latter. The object of the PROLGRAF-B model builder is thus to generate a
database which contains the process deep knowledge of the LFM process
models.

A natural form of representing LFM models is by "object oriented" database
generation. In this type of representation the individual elements
appearing in an LFM model are categorized as belonging to one of a finite
set of LFM object classes. For example, LFM continuous variable nodes
(CVNs) constitute a class of objects, whereas multiple gain boxes (MGBs)
constitute another class. Each class is characterized by a set of
"attributes" or "dimensions" which in turn take on specific "values" in the
instanciation of a specific object within a certain class. This concept is
illustrated by Table I. The table shows the CVN objects which appear in
the LFM model of Fig. 1, their dimensions, and the values taken by these
for that specific instance of inter-object relations. The end product of
the PROLGRAF-B model building procedures is thus an object database
constructed in format similar to that shown in Table I. This constitutes a
convenient means of making available to the LFM inference engine the
process deep-knowledge base required for its cause-effect back-tracing and
reasoning.

PROLGRAF-B MODEL-BUILDING PROCEDURES

The PROLGRAF-B model building procedures consist essentially of a series of
user-guided transformations by which the process knowledge "contained" in
the process piping and instrumentation diagram (P&ID), functional design
specifications, and in the analyst's own understanding, is eventually
molded into the appropriate LFM object database form. The codification of
these procedures in the PROLGRAF-B intelligent interface program will
greatly facilitate the task of a user who wants to construct an LFM process
model and wants to reduce the amount of effort required to become familiar
with the detailed application of the LFM modeling rules and syntax.

The prototype version of the procedures which are being developed to
demonstrate the feasibility of the concept uses piping and instrumentation
diagrams (P&IDs) as the starting point of its "reasoning", and queries the
user through an organized series of question and answer interactions. Four
main stages of program / user interaction are required to operate the
knowledge transformation from the assumed sources of process knowledge to
the LFM object oriented database. These stages are:

- Process and Model Primary Object Identification

- Causality Flow Identification

- Causality Transfer Function and Interaction Specification

- Causality Fault and Conditioning Specification

The PROLGRAF-B interaction with the user follows the reasoning procedure
outlined in the following sections, which illustrate each of the just
listed knowledge transformation stages. Each step in the procedure is
consecutively numbered. Individual steps and/or major blocks of steps are
illustrated by examples making reference to a simple "test case" system
with feedback control, which is depicted in Figure 2 in P&ID format.

PROCESS AND MODEL PRIMARY OBJECT IDENTIFICATION

The identification of the primary "objects" in a given process requires the
user of PROLGRAF-B to recognize and list the fundamental elements appearing
in the P&ID representation of that process. PROLGRAF-B will invite the
user to complete the following steps:

Step 1- List principal components in P&ID.

From examination of the system P&ID the user will identify the principal
system components.

```
Test Case ex.: Pump             - P
               Control Valve - CV
               Flow Sensor    - FS
               Controller     - C
```

Step 2- List "component connectors" linking P&ID components (e.g., pipes,
        wires, pneumatic lines).

The user will also identify the principal elements connecting the
components which have been identified in Step 1.

```
Test Case ex.: Pipe 1 (connects Pump to Control Valve)       - p1
               Pipe 2 (connects Control Valve to Sensor)      - p2
               Wire 1 (connects Sensor to Controller)         - w1
               Wire 2 (connects Controller to Control Valve) - w2
```

Step 2 completes the identification of the primary process objects -- that
is, components and component connectors -- upon which the analyst's
attention has to focus during the LFM model construction. These objects
will not appear in the LFM model itself, but are used as guidance to
identify the LFM model primary objects -- that is, the LFM variable nodes
(VNs) -- as explained in the following procedural step which the user will
be asked to complete by the model builder program:

Step 3- Identify essential parameters characterizing functionality of
        components identified in Step 1.

For each component a list of parameters that can be used to characterize
its performance is provided. The user is invited to identify in the list
those parameters that in his opinion are the best minimal set that can be
used to model the target process at the desired level of detail. The
"minimal set" should be chosen in such a way as to minimize the number of
parameters used to characterize a given component, without suppressing the
representation of physical relations in the process through which faults
and disturbances may manifest themselves.

For a typical pump the "menu" of standard parameters suggested by the
model-builder software could be : <speed>, <input pressure>, <output
pressure>, <head>, <flowrate>; for a controller could be: <set point>,
<input signal>, <output signal>; and so on for each type of catalogued
component. If unsatisfied by the provided menu, the user would be able to
add to the list other parameters he deems necessary to characterize a
component. Once chosen by the analyst, component parameters are considered
by PROLGRAF-B as newly defined LFM "variable node" objects. At the end of
this step the knowledge transformation process has thus progressed to the
stage depicted in Figure 3. This figure shows and lists the parameters
that an analyst could typically choose to describe the "test case" system.
The picture shown in the figure is an hybrid representation in which the
principal P&ID objects (i.e., components and connectors) appear together
with the LFM variable node objects that have been chosen to represent them.
Note that in this example only two of the possible pump characterizing
variables have been selected, namely the pump speed (Ps) and the pump
flowrate (Pf).

CAUSALITY FLOW IDENTIFICATION

Step 4- Recognize and identify the direct cause-effect relations between
        parameters within each component.

Limiting his attention to direct influences of cause and effect within
individual components, the user identifies which variables drive and which
variables are driven in all the identifiable parameter-to-parameter causal
relations.

Test Case ex.: "pump speed Ps drives pump flowrate Pf";
               "control valve command CVc drives control valve position
               CVp"; etc.

Step 5- Recognize and identify the cause-effect relations between
        parameters linked by connectors across component boundaries.

To guide the user through this step, the model-builder program utilizes
the component connectors identified in Step 2 as a means for pre-
identification of possible cause-effect links between pairs of variable
nodes (i.e., parameters) associated with separate components.

Test Case ex.: Pipe 1 (p1) connects "pump" to "control valve", thus cause-
               effect links are possible between the parameters Ps (pump
               speed) and Pf (pump flowrate) on one side, and CVf (control
               valve flowrate), CVp (control valve position) and CVc
               (control valve command) on the other. The user can then
               readily determine that the p1 connector establishes a mutual
               (i.e., bi-directional) causal influence between Pf and CVf.

With reference to the test case example, Figures 4 and 5 illustrate the
corresponding Steps 4 and 5 respectively, whereas Figure 6 shows the
incomplete LFM model that is obtained at the end of Step 5.

CAUSALITY TRANSFER FUNCTION AND INTERACTION SPECIFICATION

The system model that is obtained at the end of Step 5 is not a "complete"
LFM model in that it only shows which cause-and-effect mechanisms exist in
a given process, but does not describe their actual nature. For example,
it shows that in the test case the pump speed influences the pump flowrate,
but it doesn't specify how. It also fails to show if, for those variable
nodes that are influenced by more than one input parameter, any special
non-linear interaction effect take place among the input variable, so that
the influenced "output" parameter is determined by other than just a simple
linear superposition effect. For example, by looking at the model in Fig.
6 one cannot tell (except by guess) how the combined action of the causal
inputs from both the pump flowrate (Pf) and control valve position (CVp)
variable nodes determine the state of the control valve flowrate CVf. In
order to complete the model a few additional model building steps have to
be executed, as described hereinafter.

Step 6- For each variable node in the model, define a vector of discrete
        values sufficient to completely characterize the state of the
        associated variable.

In the course of our most recent research on LFM model based process
knowledge representation, we have developed a more general form of variable
state representation and causal linkage than the one previously used (and
which we have partially described in our summary introduction to the LFM
modeling technique). In the generalized LFM, process variable value ranges
are no longer represented across the board by five continuous "discretized"

states (0, +1, -1, +10, -10: indicating the normal state, small deviations
in the positive or negative direction, and large deviations in the positive
or negative direction, respectively),as described in Refs. [3] and [4]. An
ad-hoc number of states can instead be defined in the updated LFM, each
state being represented by an integer increased by unity over the previous
state (for deviations increasing in the positive direction from the " 0 ",
i.e. normal, state), or decreased by unity (for deviations increasing in
the negative direction from the " 0 " state). In essence, each variable V
is represented by a vector

$$V_i = \{ -N, -(N - 1), \ldots, 0, \ldots, M-1, M \}$$

containing M + N + 1 states (N "negative deviate states", the " 0 "
reference steady state and M "positive deviate states").

Test Case ex.: CVf_i = { -1, 0, 1 }   (low, normal, high);
$\qquad\qquad\quad CVp_i$ = { -1, 0, 1 }   (close more, hold steady, open more);
$\qquad\qquad\quad$ etc.

Step 7- For all the causal relations identified in Steps 4 and 5 define
        "causality transfer functions" (CTFs) to determine how mutual
        influences among process variable states can propagate in the
        system that is being modeled.

CTFs are a generalization of the earlier mentioned MGB (multiple gain box)
LFM "objects". They represent the mapping of the states of a variable
which is upstream in a causality-linked pair, into the states of the
downstream variable. In short mathematical notation, given that U_i is the
state vector of the upstream parameter and D_j is the state vector of the
downstream one, the transfer function of the causality link between the two
is a multiplier vector TF_{ij} such that for each allowed mapping $U_i \to D_j$:

$$D_j = TF_{ij} \cdot U_i$$

This generalization allows a straightforward representation of process non-
linearities, by allowing the gain multiplier to be a function of the
upstream parameter state value. When this is not needed the vector TF_{ij}
becomes a constant and the CTF can be defined and used in the same way as
MGBs were in the earlier LFM modeling conventions.

In Step 7 the user also decides if the number of variable nodes in the
model may be reduced by merging into one node those variables that may not
be necessary to represent separately.

Step 8- Identify "multiple interaction nodes" (MINs), that is, LFM
        variables directly influenced by more than one other variable, and
        select or define an appropriate "input interaction operator" (IIO)
        to describe the interaction.

When more than one causal input into a node exists, the analyst has to
determine if the causal effects merging into the node influence the
corresponding variable in a linearly independent way (e.g., by
superposition of effects), or if a particular non linear interaction may
exist there. The LFM modeling primitives include standard non-linear input
operators such as the "causative-corrective input operator" (CCIO), which
can be used to show corrective actions taking place at the controlled
variable nodes of feedback and feedforward loops. Non standard
interactions are described by "non-linear input operators" (NIOs). These
are multi-dimensional matrices (the number of dimensions depending on the
number of interacting node inputs) which allow the states of the vector
representing a downstream variable which is influenced·by a non linear

interaction to be defined as a function of the combination of the vector
states of the influencing variables.

Test Case ex.: Completion of Steps 6, 7 and 8 transforms the incomplete LFM
model of Fig. 6 into the completed representation shown in
Fig. 7. Note that in this simple model it was sufficient to
represent transfer functions with constant "gains" : +1 for
direct proportionality between variations of causally linked
variables, -1 for inverse proportionality (as required in
the feedback relation between the measured flowrate, FSm,
and the controller output, Co). Note also that a standard
LFM CCIO was used at the input of CVf to show that the
control valve position can exert a corrective action on the
causal effects produced by pump speed and flowrate.
Finally, note that two variable pairs were "reduced" into
one variable, namely FSm - Ci (into FSm) and CVc - CVp (into
CVp).

CONCLUSIONS AND CONSIDERATIONS FOR CONTINUING DEVELOPMENT

The development of the PROLGRAF model builder is still in its initial
stages. The early indications collected in our research, however, appear
positive in indicating that most of the research objectives can be reached.

The activities planned for the next future include a complete formalization
and software implementation of the procedural steps that were outlined and
discussed above. The development and verification of a "stage one"
PROLGRAF-B prototype concept via step-by-step model-building procedures
applied to the simple valve and pump system with feedback discussed above
will be continued. Completion of this "stage one" prototype will
essentially allow a semiautomated construction of process models in LFM
format, which will effectively express the functional cause-effect and
physical relationships in the plant process of interest.

In the second stage of development of PROLGRAF-B the model building
procedures will be extended to cover the conditioning effect of process
faults on the normal process causality relations. This constitutes the
last model building stage (Causality Fault and Conditioning Specification)
of the four which were earlier identified and listed in this paper. It is
presently envisioned that the development of PROLGRAF-B procedures for this
stage will require a preliminary classification of faults into: a)
component internal faults, and b) faults in component connectors. The
classification process will also account for the specific type and nature
of the component or connector being examined, and a suitable database of
"typical faults" will be used to query and guide the user of the model
builder to the identification and specification of the fault types that are
applicable to the individual elements of the plant model being constructed.

The product that the PROLGRAF-B model builder is being designed to generate
is complete logic-flowgraph models of plant processes. Such models, as
briefly discussed in this paper and documented in Refs. 3-5, allow
automated on-line or off-line interactive reliability and diagnostic
analyses to be performed via existing LFM inference software on plant
processes of interest to the analyst.

ACKNOWLEDGEMENTS

This work was sponsored by the U. S. Department of Energy under Contract
No. DE-FG03-86-SF 16610 with the Department of Mechanical, Aerospace and
Nuclear Engineering of the University of California, Los Angeles.

414

TABLE I : Continuous variable nodes in test-case example

DIMENSIONS   VALUES

Identifier	V1	V2	V3
Causal inputs	none	1 from MGB	none
Causal outputs	1 into MGB	none	none
Condtnl. outputs	none	none	1 into TB

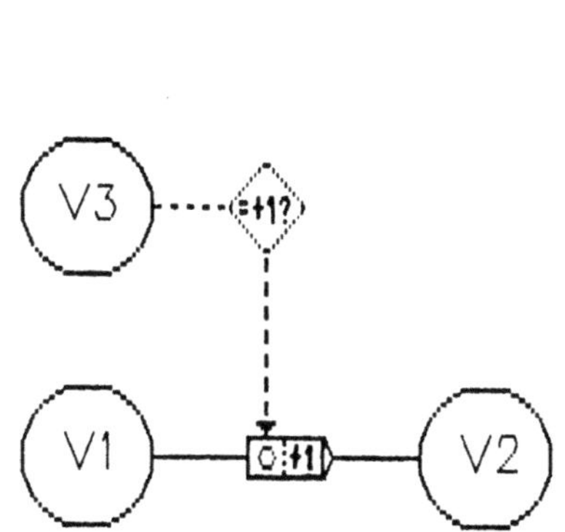

Fig. 1 : Example of LFM model

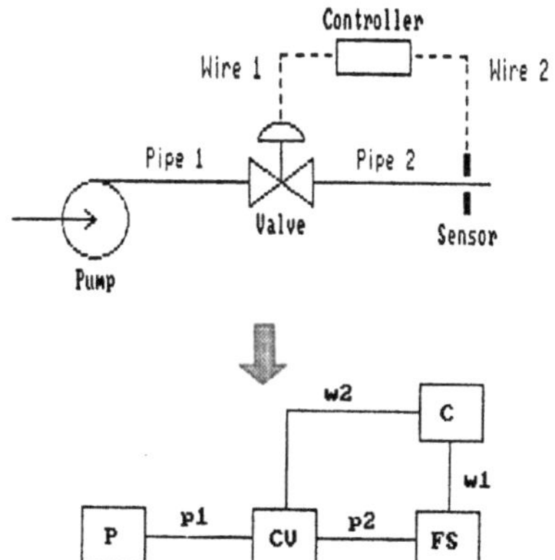

Fig.2 : Identification of plant
system elements from P&ID

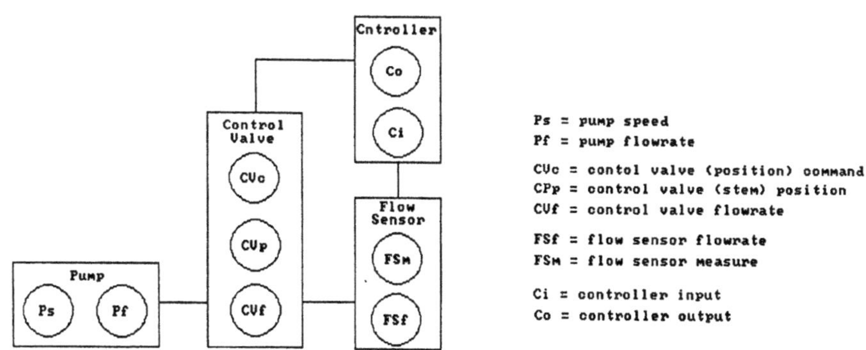

Fig. 3 : Identification of essential process parameters

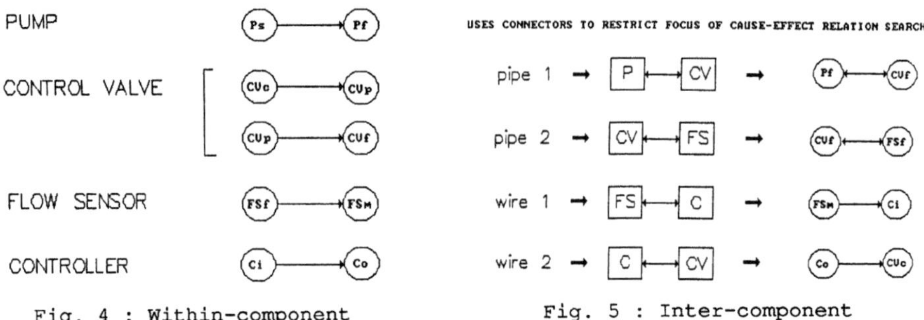

Fig. 4 : Within-component
causal relations

Fig. 5 : Inter-component
causal relations

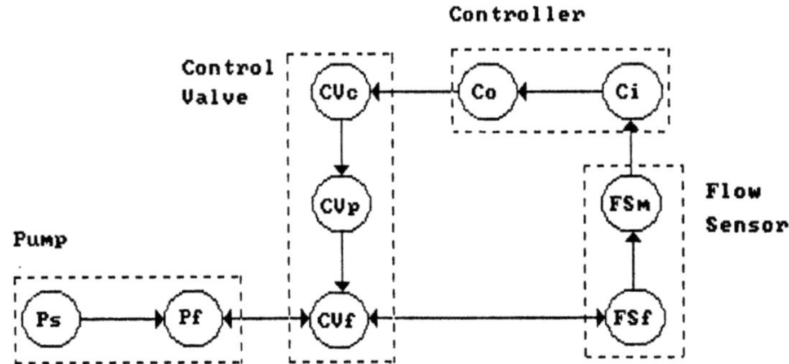

Fig. 6 : Incomplete LFM model

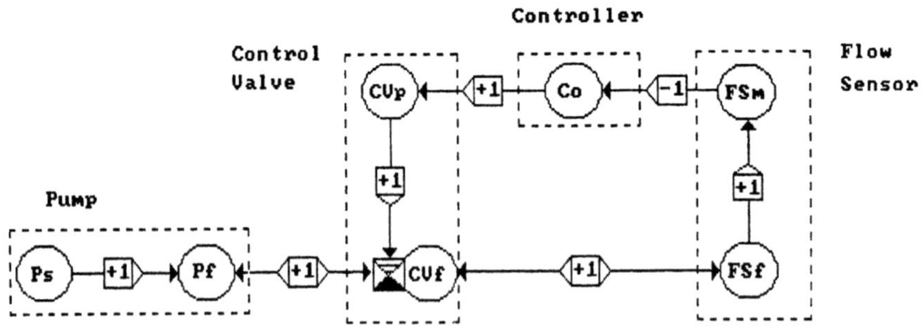

Fig. 7 : Completed LFM model

REFERENCES

1. A.B. Long: "Technical Assessment of Disturbance Analysis Systems",
   Nuclear Safety, Vol. 21, No. 1, 1980
2. D.G. Cain: "Review of Trends in Computerized Systems for Operator
   Support", Nuclear Safety, Vol. 27, No. 4, 1986
3. S. Guarro, D. Okrent: "The Logic Flowgraph: A New Approach to Process
   Failure Modeling and Diagnosis for Disturbance Analysis Applications",
   UCLA-ENG 8507, Jan. 1986
4. S.B. Guarro, D. Okrent: "The Logic Flowgraph: a New Approach to Process
   Failure Modeling and Diagnosis for Disturbance Analysis Applications",
   Nuclear Technology, Vol. 67, Dec. 1984
5. S.B. Guarro, D. Okrent: "Use of Logic Flowgraph Models in a Computer
   Aided Process Analysis and Management System", Proc. of the 1985
   International Topical Meeting on Computer Applications for Nuclear Power
   Plant Operation and Control, Tri-Cities (Pasco), WA, Sep. 8-12, 1985

CHAPTER 8

PLANT CONTROL SYSTEMS

APPLICATION OF AN AI METHOD TO OPTIMAL REACTOR CONTROL PROBLEMS

Yoshikuni Shinohara

Japan Atomic Energy Research Institute
Tokai-mura, Ibaraki-ken, 319-11 Japan

INTRODUCTION

Since the development of mathematical theories and methods of optimization in the fifties such as the dynamic programming method by Bellman and the maximum principle by Pontryagin, a large number of analytical and numerical studies have been made on various optimal reactor control problems. One of the problems which were studied by many authors is optimal reactor shutdown control involving reactor poisoning, especially xenon poisoning.[1-6]

However, most of the problems were studied for the reactor models which were much simplified since mathematical treatment of complex dynamical system, especially nonlinear systems, is generally very difficult except for some types of linear optimal control problems such as the linear regulator problem. Therefore, it is not always possible to apply directly the results of optimal control laws or control programs obtained for such simplified models to more complex actual reactor processes.

In order to bridge the gap between the theoretical study for the simplified model and its practical application to the control of an actual reactor system, it is necessary to make use of various types of knowledges about the problem which fill the gap. For this purpose, the artificial intelligence techniques such as knowledge-based expert control technique[7] may provide a practical method for constructing heuristically a near-optimal or a practically optimal control system by flexible use of various engineering and operational knowledges about the system.

There are various types of expert systems and programming tools for system development which are available on the market and can be used with mini- or micro-computers. However, most of commercially available expert system tools have been developed for open-loop consultation use in which final decision must be made by the users through interactive dialogue with the expert systems. In the application of an expert system technique to closed-loop automatic process control, automatic decision making and real-time execution functions are required.

The development of the simple method of knowledge-based control described in this paper was originally motivated in the study of obstacle avoidance control of a redundant multi-joint robotic manipulator and is

still in its initial stage. This method was tentatively applied also to
the problems of optimal shutdown control involving reactor poisoning in
a high flux thermal reactor to test its applicability to various type of
control problems.

REQUIREMENTS TAKEN INTO ACCOUNT

In the development of the present method of knowledge-based control,
the following requirements were taken into account.

- existence and uniqueness of solution
- consistency of rules
- real-time execution
- structural simplicity
- compactness of system size
- easiness in system management

The knowledge base must be organized in such a way that as the result
of program execution, one and only one control output is always generated
at each control cycle. If the knowledge base is not organized well, there
may occur the cases where no control output is generated.

All the data in the knowledge base must be consistent so as not to
generate inconsistent and contradictory control outputs. If there are
inconsistent data in the knowledge base, it may happen that control outputs oscillate or are not generated.

When a system must be controlled with a short control time interval,
of the order of seconds or milliseconds for instance, the problem of real-
time execution becomes very important. In this case, we must limit the
number of knowledge data in the knowledge base.

One of the advantages of applying the knowledge-based method to
control problems is that the control logic may be modified flexibly only
by modifying the knowledge data and not by modifying the computer program
itself. To maximize this advantage, the structure of the knowledge base
including the form of knowledge representation must be simple and easily
understandable.

Since it was planned to implement the present method of knowledge-
based control on a micro-computer, it was necessary that the size of the
program must be compact enough to be used in it even when the system is
applied to a variety of reactor control problems.

To facilitate operation, maintenance and improvement of the system,
the system should be designed so that it can be managed easily. The
structural simplicity and the compactness of the system are important
factors for this purpose.

A KNOWLEDGE-BASED OPTIMAL CONTROL METHOD

Basic Concept

To explain the basic concept guiding the construction of the present
knowledge-based control method, let us assume that the dynamics of the
system to be controlled can be expressed as follows using the state space
concept.

$$dx/dt = f(x, u),$$

where x is the state vector and u is the control vector, and the associated constraints are described by

$$x \in Xc, \quad u \in Uc,$$

where Xc and Uc signify the constraints for x and y, respectively.

If we have complete knowledge of the system, that is to say, if the function f(x,u) and the associated constraints are completely known and the state vector can be precisely estimated, and if we can find an exactly optimal control law for an optimization problem for the system, it can be expressed by

$$u = F(x), \quad x \in Xc, \quad u \in Uc,$$

as a direct mapping F from the state space X onto the control space U as illustrated conceptually in Fig.1.

Now, we assume that we cannot find the optimal control law u = F(x) for the given system but the optimal control law for a simplified model of reduced dimension which carries the essential characteristics of the original system, as a mapping G from Y onto V.

$$v = G(y), \quad y \in Yc, \quad v \in Vc.$$

Here, y and v are the state and control vectors of reduced dimensions with the corresponding constraints Yc and Vc, respectively.

If we can assume that the essential features of this optimal control for the simplified model are maintained also for the original system, there is a possibility that we can construct a knowledge base in which near-optimal control rules are derived from the knowledge of the essence of the optimal control law for the simplified model and other pertinent knowledges about the original system. And we expect that some near-optimal solutions may be obtained by appropriate combination of these knowledges, which may be expressed symbolically as

$$u = K(G(y), x, Xc, Uc, C),$$

where K(.) signifies the combination of the knowleges listed in the parenthesis and C signifies supplementary knowledge about the given system such as the dynamic characteristics. This is illustrated conceptually in Fig.1.

In this case, there arises a sort of uncertainties or fuzziness caused by the gap between the simplified model and actual system. Uncertainty may also arise if the system state cannnot be estimated with sufficient accuracy. And some measures must be taken to deal with the fuzziness using various practical knowledges about the system behaviour. This situation is also illustrated conceptually in Fig.1. The fuzzy boundaries plotted by the dotted lines signify the fuzziness due to inaccurate estimate of the system state and incomplete knowledge of the optimal control law for the given system, respectively.

<u>System Organization</u>

In the present system, a rule-based model of IF-THEN type is used for knowledge representation because it has the simplest and the most intuitive form of expressing cause-consequence relationships and, therefore, most suitable for control application among various types of knowledge representation. The basic structure of the rule-based model used in the present method is similar to that which was applied in the expert system for a reactor accident diagnosis, but with much simplification.[8]

In order to simplify further the description of rules, the knowledge
base or the rule base is divided into four groups, namely, basic control,
process dynamics, constraint and adjustment rules, as is shown in Fig.2.

The basic control rules include qualitative and quantitative control
rules. A qualitative control rule is an expression like "if x lies in Xr,
then raise the reactor power". A quantitative control rule is a statement
like "if x lies in Xm, then bring the neutron flux level to its maximum
permissible value". An appropriate combination of the basic control rules
generate a candidate control vector.

The dynamics rule is a numerical model of the dynamics of the given
system and is used to predict the value of the state and control vectors
using the current candidate control vector.

The constraint rules are simply the statements of constraints imposed
on the state and control vectors.

The adjustment rules are used to make necessary modification of the
value of the candidate control when the predicted state and/or control
vectors of the system violate the constraints imposed on them. The use of
the adjustment rules makes the expression of the basic control rules much
easier because it becomes unnecessary that a candidate control derived
from a chain of the basic control rules should satisfy at the same time
the constraints imposed on the state vector.

At each control time step, the system state is assumed to be estimated from directly measurable process variables using a state vector
estimator. Using the current estimated state vector, a candidate control
vector is searched by chaining the basic control rules. Prediction of the
state and control vectors are then made using the dynamics rule and the
current candidate control vector. The candidate control vector and the
predicted state and control vectors are then checked for their validity.
If they are valid the candidate control vector is executed, otherwise it
is modified so that they become valid by repeating the above process of
prediction and thereafter using the adjustment rules.

The knowledge manager, which includes a simple inference engine for
the basic control rules, controls the above mentioned process of generating a valid control vector at each control cycle.

APPLICATION TO OPTIMAL REACTOR CONTROL PROBLEMS

The method described above was implemented on a microcomputer and was
applied to obtain near-optimal solutions to the optimal reactor shutdown
control problems involving reactor poisoning as test examples. Computer
simulation study was performed for the following two problems using point
reactor models, for simplified models of which optimal solutions have been
obtained theoretically or numerically.

Minimal Time Problem Involving Only Xenon Effect

The first example is the minimal time shutdown control problem in
which only xenon effect is considered as reactor poison. The problem in
this example is to find the minimal time control program or control law
which gives the specified value of xenon peak after reactor shutdown,
given the initial state of the system. The solution of this problem for a
simplified two-dimensional model, in which the iodine and xenon concentra-

tions in the reactor are the state variables and the neutron flux is the
control variable, has been obtained theoretically using the Pontryagin's
maximum principle. The optimal control law can be expressed using the two-
dimensional xenon-iodine state or phase plane.

The basic control rules for a reactor model of higher dimension are
constructed using the knowledge of the optimal control law expressed in
the xenon-iodine state plane and typical patterns of the time program of
neutron flux variation as well as the knowledge obtained from the computer
simulation of the given system.

The constraints rules include those for maximal and minimal values of
neutron flux during control phase, maximal speed of neutron flux variation, minimal value of excess reactivity and others such as the maximal
speed of control rod movement which have minor effects on the near-optimal
solutions.

The dynamics rule is a reactor dynamics model with temperature reactivity feedback effect and constraints on the rate of change of neutron
flux and the excess reactivity. In this model, the neutron flux is also a
state variable.

The total number of rules including the adjustment rules which are
based on the knowledge about the control dynamics of the system is about
fourty depending upon how the rules are expressed.

In order to check the validity of the method, it was first applied to
the case in which the temperature reactivity feedback effect was neglected
and the constraint for the maximum speed of neutron flux variation was set
very high so that the reactor dynamics is nearly the same as that of the
simplied two-dimensional model. An example of the results of simulation of
the optimal control is shown in Fig.3. The results obtained by the present
method is practically the same as that obtained theoretically. The results
for other cases are also fairly close to those obtained numerically using
an optimum seeking method. The differences in the values of control time
are within about three percent.

<u>Minimal Poisoning Problem Involving Both Xenon and Samarium Effects</u>

The second example is the problem of optimal reactor shutdown control
for minimizing the reactor poisoning at a specified time after shutdown.
In this case both xenon and samarium poisonings are taken into consideration. This problem has never been solved theoretically even for a simplified process model in which the neutron flux is taken as the control
variable and other reactivity effects such as temperature effects are
neglected, but it was studied numerically using the statistical optimum
seeking methods.[9]

From the results of computer simulation study, the knowledge about
the behaviour of the optimal control was gained qualitatively. Therefore,
the basic control rules are constructed using the knowledge about the
features of optimal control programs and the corresponding behaviours in
the xenon-iodine and the samarium-promethium state planes as well as the
knowledge about the dynamic behaviour of the reactor model of higher
dimension which includes temperature reactivity feedback effect and constraints on the rate of change of neutron flux and the excess reactivity.

Other rules are constructed in a similar manner as in the previous
example and the total number of rules are nearly twice as much as that of
the previous case when each rule is expressed in a simple form. The number
of the rules may be reduced to some extent if more information is included

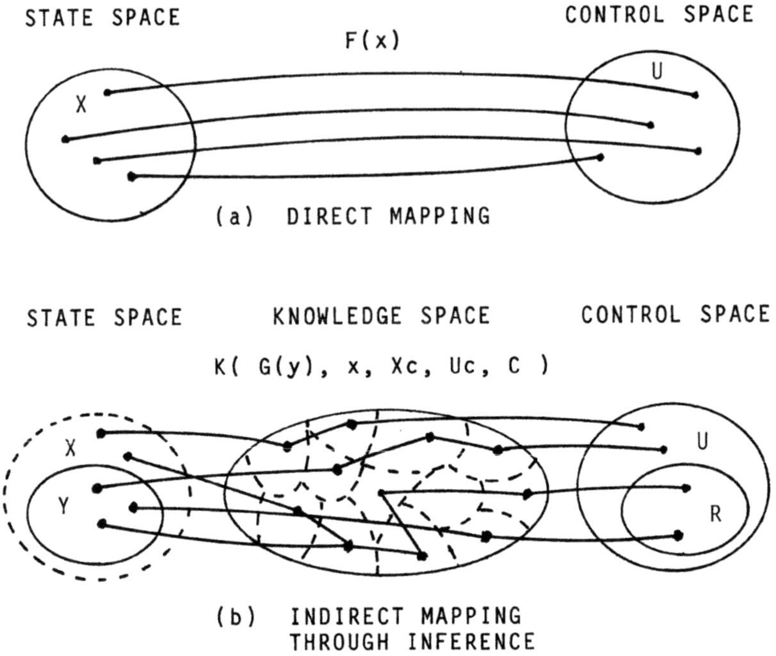

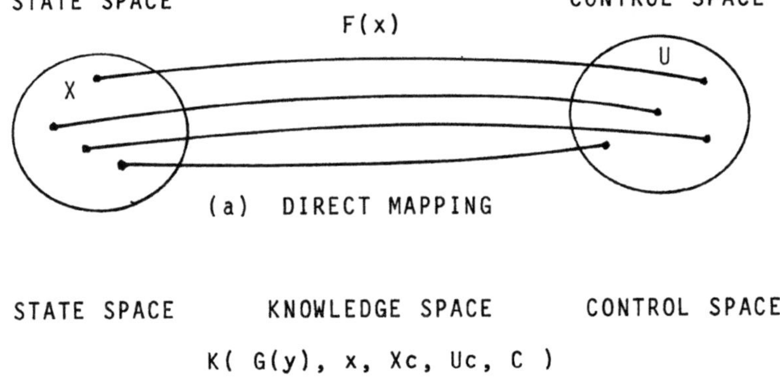

Fig. 1. Conceptual illustration of the knowledge-based control
(a) Optimal control law (assumed to be unknown)
(b) Knowledge-based near-optimal control using rules

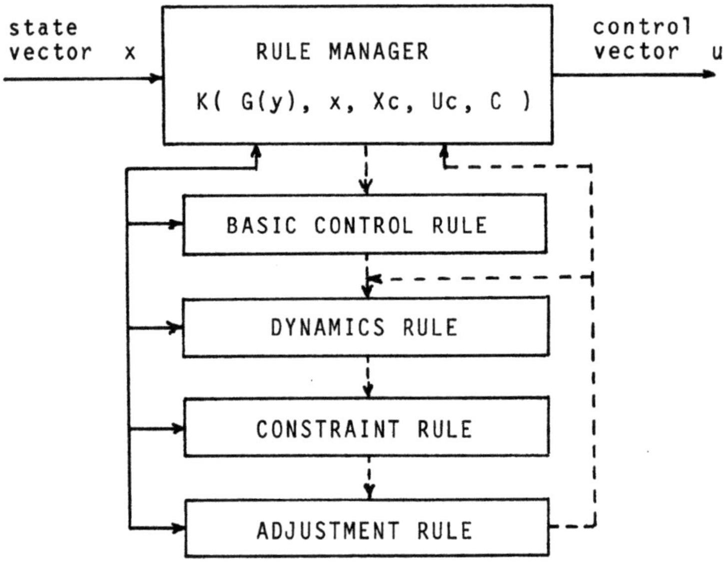

Fig. 2. Grouped structure of the knowledge base

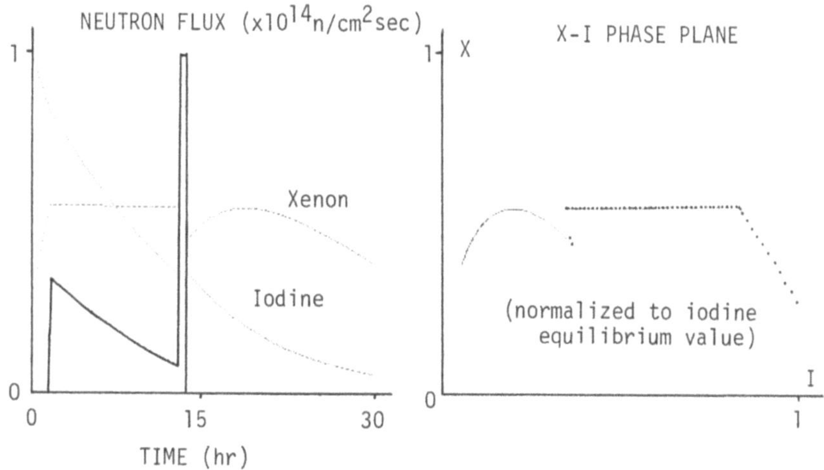

Fig. 3. Example of near-optimal control for the minimal time problem

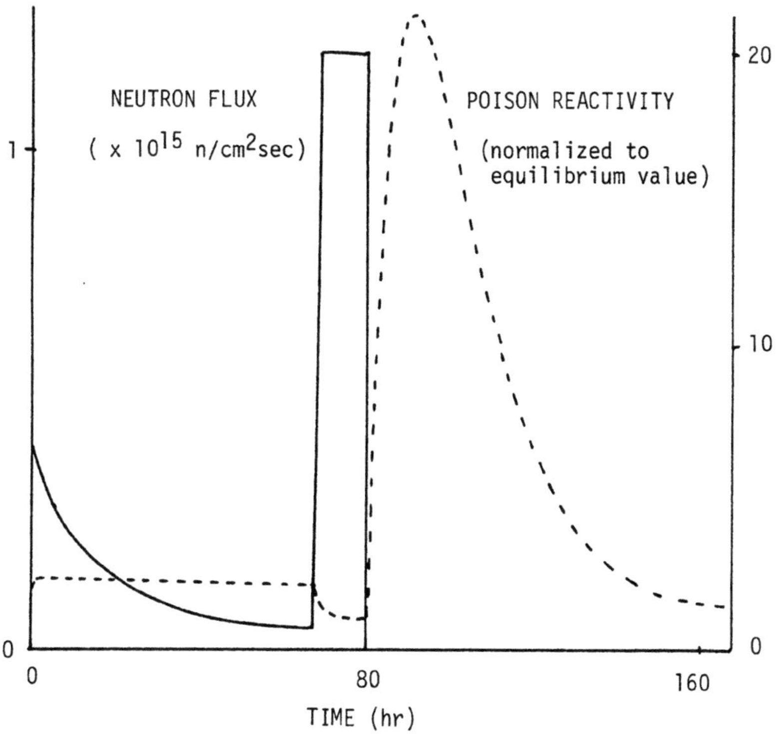

Fig. 4. Example of near-optimal control for the minimal poisoning problem

in each rule expression at the cost of loosing the structural simplicity.

Again, in order to check the validity of the method, it was applied
to the case in which the temperature reactivity feedback effect was neglected and the constraint for the maximum speed of neutron flux variation
was set sufficiently high. An example from the results of computer simulation is shown in Fig.4. The results for other cases are also sufficiently
close to those obtained numerically. The differnces in the values of
reactor poisoning at the specified time after shutdown are within about
five percent.

CONCLUDING REMARKS

The simple method fo knowledge-based control described in this paper
can be applied to control a system near-optimally using the knowledge of
the properties of optimal control for a simplified model of the actual
systems as far as it is possible to assume that the essential properties
of the optimal solutions are kept also in the actual system. Of course,
the more the gap between the behaviour of an actual system and its simplified model becomes, the more difficult it will become to construct rules
which may give near-optimal solutions.

It is planned to apply the present method to the near-optimal control
problems of xenon oscillation, for which a study was recently reported on
the application of an expert system.[10]

REFERENCES

1. Y. Shinohara and J. Valat, Optimalisation de l'empoisonnement xenon
   par minimalisation du pic xenon, C. R. Acad. Sci. Paris, 259:1623-
   1626 (1964)
2. Z. R. Rosztoczzy and L. E. Weaver, Optimum Reactor Shutdown Program
   for Minimum Xenon Buildup, Nucl. Sci. Eng., 20:318-323 (1964)
3. M. Ash, Application of Dynamic Programming to Optimal Shutdown Control, Nucl. Sci. Eng., 244:77-86 (1968)
4. J. J. Roberts and H. P. Smith, Jr., Time-Optimal Solutions to the
   Reactivity-Xenon Shutdown Problem, Nucl. Sci. Eng.,22:470-478 (1965)
5. J. J. Roberts et al., Experimental Application of the Time-Optimal
   Xenon Shutdown Program, Nucl. Sci. Eng., 27:573-580 (1967)
6. Y. Shinohara, Optimal Xenon Control and Related Problems, in: "Application of on-line computers to nuclear reactors," OECD/NEA (1968)
7. K. J. Astrom et al., Expert Control, Automatica, 22:276-286 (1986)
8. Y. Yoshisa et al., Development and Verification of an Accident Diagnostic System for Nuclear Power Plant by Using a Simulator,in: "Proc.
   of ANS International Topical Meeting on Computer Applications for
   Nuclear Power Plant Operation and Control," ANS (1985)
9. M. Kitamura and Y. Shinohara, Study of the Optimal Xenon-Samarium
   Control Problem by a Statistical Search Method, JAERI-M 5712 (1974)
10. D. Bollacasa et al., XIPP - An Expert System for the Control of Axial
    Xenon Oscillations, in: "Proc. of ANS International Topical Meeting on
    Computer Applications for Nuclear Power Plant Operation and Control,"
    ANS (1985)

KNOWLEDGE-BASED SYSTEMS AND INTERACTIVE GRAPHICS FOR REACTOR

CONTROL USING THE AUTOMATED REASONING TOOL(ART) SYSTEM

Magdi Ragheb[*], Bruce Clayton[**] and Paul Davies[**]

[*]Department of Nuclear Engineering
University of Illinois at Urbana-Champaign
103 S. Goodwin Ave., Urbana, Illinois 61801

[**]Inference Corporation, 5300 W. Century Blvd.
Los Angeles, California 90045

ABSTRACT

The use of Knowledge-Based systems and advanced graphics concepts are described using the Automated Reasoning Tool(ART) for a model nuclear plant system. Through the use of asynchronous graphic input/output, the user is allowed to communicate through a graphical display to a Production-Rule Analysis System modelling the plant while its rules are actively being fired. The user changes the status of system components by pointing at them on the system configuration display with a mouse cursor and clicking one of the buttons on the mouse. The Production-Rule Analysis System accepts the new input and immediately displays its diagnosis of the system state and any associated recommendations as to the appropriate course of action. This approach offers a distinct advantage over typing the components statuses in response to queries by a conventional Production-Rule Analysis system. Moreover, two effective ways of communication between man and machine are combined.

INTRODUCTION

The justification for this approach is that the recent developments of highly automated systems and complex systems consisting of a large number of components is modifying an operator's role into that of a supervisory controller who is on top of the hierarchical control structure.[1] In this situation, his activities are settled on a predominantly cognitive level. This consists in monitoring the execution of specified tasks, intervention in case of abnormalities, and planning of future strategies. In spite of the relief from low level tasks, now performed by the control system, the demand on the operator are increased, psychologically and mentally in case of faulty situations, since in this case, large amounts of information must be processed. Erroneous actions may cause severe consequences, or even the loss of the plant. Interactive graphics and Knowledge-Based systems[2] are two new technologies that have reached a state which makes them available for practical applications and usage as operators aids .Their integration will be demonstrated using a model problem consisting of a water supply system bringing coolant water into a nuclear power plant. [3]

MODEL DESCRIPTION

The system shown in Fig. 1 is used in the analysis. It represents a nuclear reactor's water supply system.[4] The coolant is pumped by two pumps connected in parallel. The coolant flow is directed to the reactor core through a valve. A Fault-Tree is first constructed

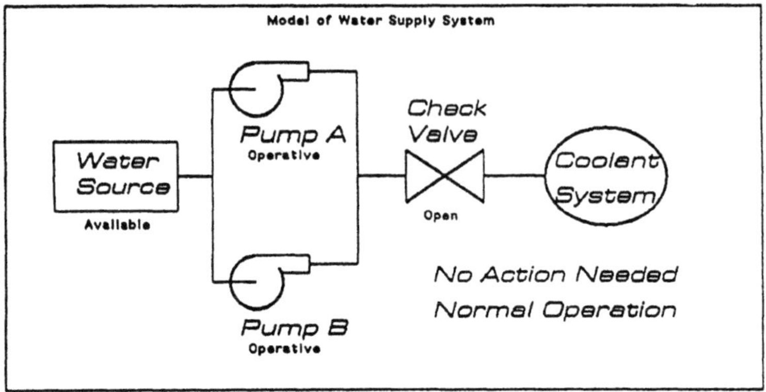

FIGURE 1:   This is the actual interactive display used by the program for asynchronous input/output. The small status labels under each major component of the diagram can be altered by touching them with the mouse cursor.

Table 1 A forward-chaining interactive session where the system is diagnosed as being in an accident status.

Please inform me whether the water-source is
available, undetermined, or unavailable.
AVAILABLE

Please inform me whether the check-valve is
open, undetermined, or closed.
OPEN

Please inform me whether pump-a is
operative or inoperative.
INOPERATIVE

Please inform me whether pump-b is
operative or inoperative.
INOPERATIVE

The recommended course of action is
DECLARE-EMERGENCY.

The system status parameters are
System status is ACCIDENT-STATUS
Water source is AVAILABLE
Check-valve is   OPEN
Pump-a is   INOPERATIVE
Pump-b is   INOPERATIVE
Flow is   IMPOSSIBLE
Pumping is   UNAVAILABLE

to describe the fault state of the system.[5] The complement of the Boolean expression of the top event provides a description of the system in its normal operational state. A Goal-Tree can then be constructed and serve as a logical representation of a model for the plant. The logical gates constituting the Goal-Tree are translated into a Knowledge Base in the form of if/then rules.[6] This Knowledge-Base is searched by an Inference Engine in either the forward-chaining or the backward-chaining mode.

Tables 1 to 3 show standard interactive forward-chaining sessions between an operator and a Production-Rule System. Input by the operator is by typing responses to the system's inquiries. In the first case the system infers from the operator's inputs that a recommended course of action is to declare an emergency, and it summarises the facts on which this recommendation is suggested. Those facts are based on operator's input in the querying process such as Pump-a is INOPERATIVE, and internal state variables inferred by the system such as Flow is IMPOSSIBLE. The second case infers an UNDETERMINED reactor state, and advances a SHUT-DOWN-REACTOR recommendation. The third case infers a NORMAL-OPERATION state, and recommends: NO-ACTION-NEEDED. This approach is very useful as is as an operator's aid. Its usefulness is more apparent as the complexity of the system at hand increases. It involves, however a considerable amount of typing on the videoscreen used by the operator in response to the system's inquiries. The question arises on whether it is possible to improve the interactive process by using graphical displays, and replacing the need for typing an answer by the process of pointing a finger, or moving a mouse cursor on the screen.

Table 2 A forward-chaining interactive session where the system is diagnosed to be in the undetermined state.

Please inform me whether the water-source is available, undetermined, or unavailable.
UNDETERMINED

Please inform me whether the check-valve is open, undetermined, or closed.
OPEN

Please inform me whether pump-a is operative or inoperative.
OPERATIVE

Please inform me whether pump-b is operative or inoperative.
OPERATIVE

The recommended course of action is
SHUT-DOWN-REACTOR.

The system status parameters are
System status is UNDETERMINED
Water source is UNDETERMINED
Check-valve is OPEN
Pump-a is   OPERATIVE
Pump-b is   OPERATIVE
Flow is   POSSIBLE
Pumping is   AVAILABLE

This is in fact possible. Figures 1 to 5 now display interactive sessions by the user where the input to the Production-Rule System is through the mouse cursor. No typing is needed as previously shown in Tables 1 to 3., and the physical display of the components being analyzed provides better communication between the operator and the Production-Rule System running in the background. The visual display can be further supplemented with a description of the logical model of operation of the system in the form of the Goal-Tree from which the Knowledge-Base has been constructed.[7]

Table 3   A forward -chaining interactive session where the system is diagnosed to be in the normal operational mode.

```
Please inform me whether the water-source is
available, undetermined, or unavailable.
AVAILABLE

Please inform me whether the check-valve is
open, undetermined, or closed.
OPEN

Please inform me whether pump-a is
operative or inoperative.
OPERATIVE

Please inform me whether pump-b is
operative or inoperative.
OPERATIVE

The recommended course of action is
NO-ACTION-NEEDED.

The system status parameters are
System status is NORMAL-OPERATION
Water source is  AVAILABLE
Check-valve is   OPEN
Pump-a is        OPERATIVE
Pump-b is        OPERATIVE
Flow is          POSSIBLE
Pumping is       AVAILABLE
```

In Fig.2 the user has used the mouse to indicate that Pump-A is currently "Inoperative." Since there is another pump available, the model-based system infers that no action is needed on the basis that the system is in the normal operational state. The overall system status, as well as the status of the individual components is continuously accessible to the operator through a simple glance at the videoscreen. In Fig. 3, if the operator indicates that the second pump is also inoperative the state of the system is inferred to be in the accident state and a recommendation is given to declare an emergency. This is accompanied by an audible beeping alarm, as well as a visual alarm consisting of several screen flutters. If the user notifies the program that the status of the check-valve is undetermined, this does not modify the accident status and the recommendation of the system, as shown in Fig. 4. The emergency state is terminated if the user notifies the program that Pump-A has returned to the operational state, as shown in Fig. 5. The program changes its recommendation to reflect the fact that the position of the valve is still undetermined. The program will revert to the normal status as soon that it is informed that the operator determines that the valve is in the "open " state.

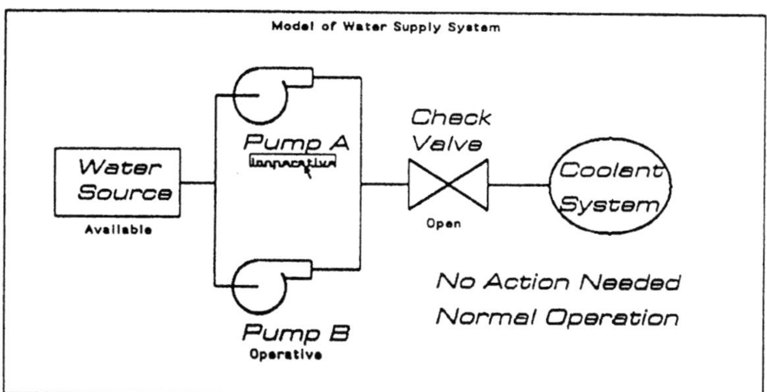

FIGURE 2: Here the user has used the mouse to indicate that Pump-A in currently "inoperative." Since there is another pump available, this new information has no impact on the program's recommendation to the operator.

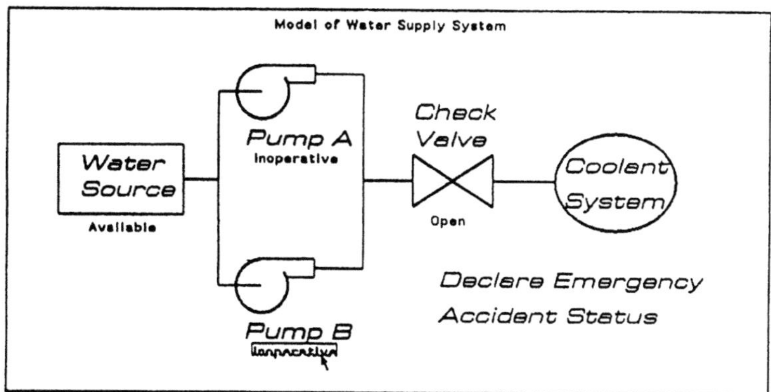

FIGURE 3: When the user indicates that the second pump is also inoperative, however, the system recommends that the operator declare an emergency.

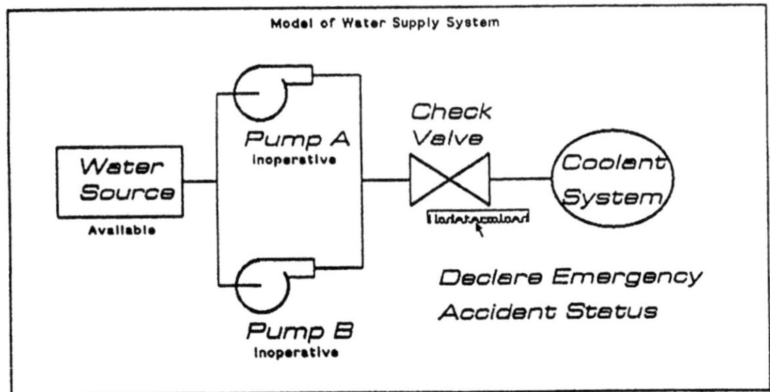

FIGURE 4: If the user now notifies the program that the status of the check valve is uncertain, the program does not change its recommendation. The the water-supply system is still known to be in "accident" status, even if some components are in an "undetermined" state.

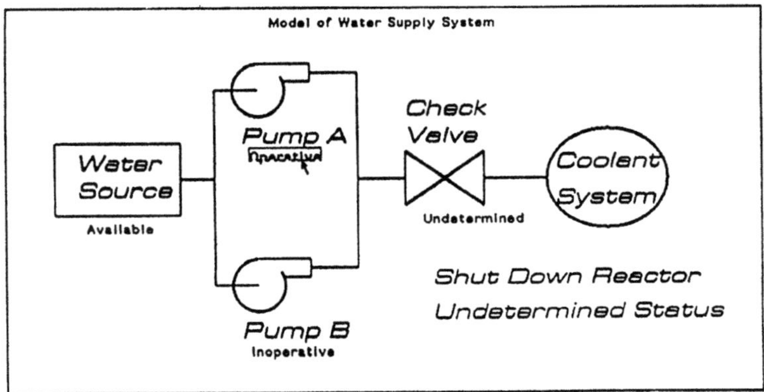

FIGURE 5: The emergency is terminated when the user notifies the program that Pump-A has been repaired. The program changes its recommendation to reflect the fact that the position of the check valve is still unknown. The program will revert to "normal" status as soon as the user determines that the check valve is indeed open.

PROGRAMMATIC ASPECTS

The structure of the rules, definitions and schemata respectively, are here shown:

```
;;; ***********************************************************************
;;; PUMPING AVAILABILITY RULES
;;;
;;; The next two rules watch the status of Pump-A and Pump-B to determine
;;; whether or not water can be pumped through the system.

(defrule pumping-rule-1
        (or
        (pump-A operative)
        (pump-b operative))
  =>
        (assert (pumping available))
  )

(defrule pumping-rule-2
        (pump-a inoperative)
        (pump-b inoperative)
  =>
        (assert (pumping unavailable))
  )
```

```
(defrelation system-status                ;The SYSTEM-STATUS relation
  (?current-status)                        ;has one argument
  "Contains the current status of the system.
      Values may be:
           Normal-operation
           Undetermined
           Accident-status")              ;optional documentation.
```

```
            (DEFSCHEMA ELLIPSE-4
                (INSTANCE-OF ELLIPSE)
                (THICKNESS 2)
                (FILL CLEAR)
                (RADII (70 56))
                (ICON-EXTENT (609 113 753 229))
                (TRANSLATE (680 170))
                (ALU 7)
                (CONTAINED-IN-ICON MODEL))
```

The system was built on an Explorer Lisp machine. Several features existing in the ART system were used:

1. Forward-chaining rules to perform routine inferences and various procedural activities such as animating the interface graphics.

2. Facts, as statements of system component status.

3. Schemata, as a basis for reasoning about graphics objects. These are loaded from a file that is separate from the Production-Rule system.

4. Logical Dependency of the derived facts and their antecedents. If the status of a component is suddenly altered, the Production-Rule system must be capable of retracting and re-evaluating all decisions made downstream at this point.

5. Asynchronous graphic input/output to let the user communicate with the Production-Rule system while its rules are continously being fired in the background.

RECOMMENDATIONS AND CONCLUSIONS

The future application of such methodology will includes the need for experimentation for gathering insights into operators behavior when interacting with Knowledge-Base systems. It must be determined whether there is an improvement in performancee associated with the use of the system, and whether operators can acquire expertise by using such systems. Such systems need to be developed by the users themselves, which requires both an educational effort and an effort to develop automatic programming methods.The future direction of research in this area needs to include the possible use of such systems for procedural aids in control tasks where the system can trace the operator's behavior in the execution of a specific task and then infer if he is on the right track or not. To be able to achieve this, the system should have knowledge about both the plant procedures and the operator's behavior, and would entail major developments in the existing state of knowledge.

REFERENCES

1. K. T. Kimmel, "Knowledge-Based Systems and Interactive Graphics for Process Control Tasks," in: Human Decision Making and Manual Control, H. P. Willumeit,Ed., Elsevier Science Publishers, pp. 277-288 (1986).

2. M. Ragheb and D. Gvillo, "Symbolic Simulation of Engineering Systems on a Supercomputer,"Applications of Artificial Intelligence III, SPIE Vol. 635, p. 368 (1986).

3. M. Ragheb and L. Tsoukalas, "Monitoring the Performance of Devices using a Coupled Probability-Possibility Method," to appear in Int. J. of Expert Systems, Res. and Appl., (1987).

4. M. Ragheb, "HAL-1987," Directory of Applications Software for Cray Supercomputers," Cray Research, Mendota Heights, Minnessota (1987).

5. A. Kuhlman, "Introduction to Safety Science," Springer -Verlag, New York, p. 65 (1986).

6. B. G. Buchanan and E. H. Shortliffe, "Rule-Based Expert Systems," Addisson-Wesley, Reading Masschussets (1984).

7. N. Rasmussen, Department of Nuclear Engineering, Massachussets Institute of Technology, Private Communication.

PROMISES IN INTELLIGENT PLANT CONTROL SYSTEMS

Pedro J. Otaduy

Oak Ridge National Laboratory[*]
Post Office Box X
Oak Ridge, Tennessee 37831-6008, U.S.A.

ABSTRACT

The control system is the brain of a power plant. The traditional goal of control systems has been productivity. However, in nuclear power plants the potential for disaster requires safety to be the dominant concern, and the worldwide political climate demands trustworthiness for nuclear power plants. To keep nuclear generation as a viable option for power in the future, trust is the essential critical goal which encompasses all others.

In most of today's nuclear plants the control system is a hybrid of analog, digital, and human components that focuses on productivity and operates under the protective umbrella of an independent engineered safety system. Operation of the plant is complex, and frequent challenges to the safety system occur which impact on their trustworthiness.

Advances in nuclear reactor design, computer sciences, and control theory, and in related technological areas such as electronics and communications as well as in data storage, retrieval, display, and analysis have opened a promise for control systems with more acceptable human brain-like capabilities to pursue the required goals.

This paper elaborates on the promise of futuristic nuclear power plants with intelligent control systems and addresses design requirements and implementation approaches.

INTRODUCTION

There is a new generation of reactor designs with inherent safety features for which the need for an engineered safety system and even for a reactor control system is debatable. These new designs stem from the original Swedish PIUS light water reactor concept. They are represented in the United States by General Electric's multi-modular sodium-cooled PRISM and Rockwell's gas-cooled SAFR concepts. These designs provide an ample margin of safety using low-power densities and rely on the laws of

---

[*]Operated by Martin Marietta Energy Systems, Inc., for the U.S. Department of Energy under Contract No. DE-AC05-84OR21400.

physics to provide automatic shutdown and decay heat removal. Inherency
not only reduces the number of initiating events with the possibility of
resulting in core damage but also minimizes the need for intervention--
whether by the operator, the control system, or the safety system--and
relaxes the time constraints for corrective action when needed.

Methodologies for risk assessment, reliability analysis, and
determination of the time margin for action based on plant configuration,
sequence of event trees, and confusion matrices are well understood and
accepted by the nuclear industry. It is apparent that these methodologies
will be applied to new nuclear plants. As a result, new nuclear plants
will have a wealth of valuable information that could and should be
incorporated into their control system strategy.

The U.S. nuclear industry is far behind in plant automation and the
use of computers. The role of computers has been limited so far to data
acquisition, data logging and display, and administrative control
activities such as security and area access control. It is well accepted
that computerized signal acquisition facilitates monitoring, display, and
recording of process signals during operation as well as during testing of
reactors and their components. Pioneering work in both nuclear reactor
automation and digital control was done at ORNL's 100-MW(th) High Flux
Isotope Reactor (HFIR). The level of automation implemented with analog
components in the HFIR when it was built in 1968 [1] is still high
compared to standard practices in U.S. plants, and it has been a major
contributor to the HFIR's excellent availability record. The path to
digital reactor control was shown by J. B. Bullock [2], when in 1972 he
demonstrated computer control of the HFIR during a full fuel cycle. Since
then the only computer control implementation in a U.S. nuclear reactor is
the one for reactivity control at the MITR-II research reactor [3].

Canada has been leading the world in the automation and computer
control of commercial power plants since November 1970 when they first
started their 250-MW(e) BLW Gentilly power plant [4]. By 1973 there were
another five digitally controlled plants [four 500-MW(e) CANDUs and one
125-MW(e) PHW)]. All of the CANDU reactors that have followed are
computer controlled. The number of functions performed by the computer
has been increasing steadily, and in the latest one the safety system is
computerized also. The highly automated digital control systems of the
CANDU reactors have been shown to be very reliable and have contributed to
the fact that these reactors have the world's best capacity factor record.

In 1973 the typical CANDU computer system consisted of two identical
16-bit computers with 32 Kbytes of random access memory (RAM), 1.5 Mbytes
of fixed disk, and 2.3 Mbytes of removable hard disk storage operating in
parallel. This computer system controlled reactor power, flow in six
coolant loops, and coolant level in two steam drums. In addition, the
computer handled the refuelling machine, alarms, CRT displays, and data-
logging operations. Note that today it is not unusual to find in
engineers' desks and in their homes, personal computers that exceed the
above computing resources.

The slow acceptance of computers in the nuclear industry may have
been due to the significant time gap between plant design and startup, to
inflexible designs, to safety concerns, to regulatory concerns, to a
general distrust of software developed when the use of computers was
focused in simulation software, to unreliable computer hardware, or to who
knows what else and a little of everything together. It is obvious,
though, that as the industry is exposed to varied applications of computer
technology, whether transcendental or mundane, capability rather than
acceptability becomes the issue.

THE CASE FOR COMPUTERIZED AUTOMATION

The complexity of present nuclear power plants is mostly artificial and results from the use of obsolete technology inadequate for the extraordinary challenge posed by scale factors as well as safety concerns. When safe and reliable operation of a device similar to an industrial-size coffee pot turns into a nightmare, we must conclude that the tools being used to design and operate it are inadequate.

Much of the present complexity is due to the impossibility of integrating design into operation. The lack of automation forces the plant designer to project the design rationale into planes of operation involving the human, planes whose nature does not warrant faithful reproduction of the original at operation time. For instance, if design features require certain actions to occur when specific plant states are reached, the designer must devise means to alert the operator, give him good hints so he can identify the situation readily, and provide him with a set of instructions on the actions to take. This results in the creation of alarm panels and books of procedures documenting the meaning of the alarm, its causes, and actions to follow. The procedures, though, constitute a downgraded redesign of the desired actions forced by the need to map them into the plane of human capability. Likewise, alarms-- usually hardwired to the display panels--often result in alarm avalanches that, although predictable in pattern and sequence by the designer, require intelligence and clear thinking from the operator in addition to good understanding of the system and the procedure collection. The inevitable mismatch between original design actions and actual action at operating time, in addition to the possibility of diagnosis and/or execution error on the human part, makes it necessary to have a human brain in the loop to fill the gap and to make necessary corrections.

Much of the intelligence supposedly needed to operate any complex system is not or should not be necessary. The designer's intent and understanding of the system should be incorporated in the control logic of an automatic system. This will eliminate the need for less knowledgeable persons to act on their own perceptions of the workings of the system and of the designer's rationale. Digital control techniques should effectively embed the designer's knowledge into the control system for perpetuity.

Computerized control systems have invaluable advantages to analog systems mostly related to flexibility and ease of modification. Computerized controls can adapt to design changes; can correct errors; and can accommodate omissions during the design, testing, and commissioning phases. During operation computerized controls can easily be fine-tuned to improve plant performance, modified to incorporate administrative changes, and adapted to component failures and system reconfiguration. No longer should we have plants that when built reflect the status of technology a decade earlier. All during the plant construction period as well as its operating lifetime, computerized controls will allow for timely incorporation of technological advances. One of the research fields with a continuing promise for evolutionary development of tools and techniques of interest to the nuclear power community is that of artificial intelligence (AI).

THE CASE FOR AI

AI is a discipline with a computer science component that has benefitted tremendously from the revolutionary democratization of computer power experienced in this decade. The main contribution of AI has been

that of symbolic processing with its associated logic programming
capabilities. This development is to computer programming what feedback
was to control technology.

Intelligence is related to the capability to create and manipulate
symbols with a purpose. Normally the purpose is to convey thought,
desire, and knowledge. The creation of all kinds of symbols and languages
for specific purposes is one of the signs of human intelligence. Some of
the languages created are specific to the development of computer
programs. Until recently, computer programs could perform only those
tasks amenable to numerical computation. With symbolic processing
languages, it is possible for a program to augment and/or modify another
program and/or itself as it runs, much in the manner of human adaptation
and learning, without losing any of its numeric programming capabilities.
One can see then how symbolic processing would permit the incorporation of
qualitative descriptions of components and systems, planning strategies,
and rules of thumb to control algorithms, thus unifying continuous and
discontinuous control algorithms.

AI research areas with direct impact on control systems are those
involving object-oriented programming (especially for simulation),
learning (adaptive control), and--soon to be disowned by its parents--
expert systems.

Object-oriented programming is a form of programming in which the
programmer first creates a set of class entities and objects. Classes
specify the names of private sets of parameters and procedures
characteristic of objects of that class. Objects are specific instances
of a class. Each object has its own private set of values for the class
parameters. Once the objects are created, the programmer concentrates on
specifying how objects should interact to accomplish the desired goals.
Interaction between objects occurs by message passing. A message consists
of a directive and data. When an object receives a message, the message
is unified with the corresponding procedure in the class to which the
object belongs in the object's private parameter value environment. New
classes can be generated out of existing ones by inheritance mechanisms.
This form of programming results in a highly flexible, reusable, and
easily validated and tested program. For instance, a variety of
controller objects could be defined and be phased in and out of the
simulation while it runs by merely sending a message with the name of the
control object to the object handling simulation configuration.

When applied to control systems, AI learning techniques permit the
development and improvement of control algorithms even when the
characteristics of the controlled system are not known. The search for a
control algorithm, for instance, can be guided by heuristics relating
patterns found when comparing observed and expected system behaviors.

Expert systems (ES) are a special class of knowledge-based systems in
which the knowledge is represented primarily in the form of rules. All
knowledge-based systems are driven by inference engines that are
independent of the knowledge base. The usefulness of expert systems for
power plants comes from the versatility this independence provides.
Expert systems are presently being applied to monitoring plant components
and diagnosis of vibrations and alarm systems. The applicability of
current ES technology to process control, though, depends on the
capability of the inference engine to hybridize its operation, that is to
coexist and communicate with process controllers capable of handling real-
time data, since the treatment of time in inference engines is still in
the research stage.

The role of AI in computerized nuclear plants can be seen as that of
a logical software glue-it-all, providing configuration flexibility and
explanation generation of plant operation. Its impact will depend heavily
on the existence of a quality plant database.

DESIGN REQUIREMENTS

Design should be done using standardized advanced CAE/CAD tools.
These tools should be able to generate computer-readable, transportable
descriptions of their databases. The availability of a complete and sound
plant knowledge base is crucial to the success of intelligent automation.

The control and safety systems should be integral parts of the design
process.

Documentation for every piece of equipment in the nuclear plant
should be comprehensive and be available in computer-readable form in
standard formats. To this effect, the "N" stamp of approval should be
extended to component documentation requirements.

Computer hardware should be fault tolerant and should be implemented
at least in pairs to provide physical redundancy.

Computer software should be highly modular in object-oriented
fashion. Each module should constitute a software object amenable to
being manipulated by other objects. Each object should contain in its
associated database information on its functionality and means of
calibration, testing, simulation, and validation. To this effect
calibration, validation, testing, and simulation programs should be
incorporated as software objects. Most objects, such as sensor objects,
pipe objects, pump objects, controller objects, and other component
objects, would be generated automatically from data in the design
database. Objects such as simulators, validators, and diagnosticians will
come from accepted engineering bags of tricks. Libraries of standardized
validated software objects should be created for the general use of the
nuclear community.

DOE and the nuclear industry should consider making real-time plant
data available to the research community through national networks such as
the one under consideration by the Federal Coordinating Council for
Science, Engineering, and Technology.

IMPLEMENTATION APPROACHES

Regulatory bodies and industry research institutions should encourage
technological advances and their implementation in nuclear plants. The
licensing process should be neither an impediment to the incorporation of
technological advances in nuclear plants nor a justification for the
nuclear industry.

Computer control technology will be introduced into existing nuclear
plants primarily by attrition of analog modules, new CRT-based displays,
new data recorders with digital output characteristics, and new operator
aids. Nonnuclear or safety-related systems will most likely be
computerized in the not too distant future.

A logical way for the industry to try some of the new technology will
be to develop pilot applications to be interfaced on-line to plant safety
parameter display systems (SPDS). It is hoped that in the near future the

HFIR will resume operation and live process signals will be available to the research community. It is imperative, though, that (in order to be able to compete with foreign government-owned nuclear industries) the U.S. Government follow the recommendations of the DOE task force [5] and establish a central facility to support the needs of the nuclear community for design and testing of advanced control systems for automation of nuclear plants.

Future nuclear plants, by the time they are built, should be fully computerized from design to decommission. Following is a glimpse of the plant of the future from today's perspective.

THE NUCLEAR POWER PLANT OF THE FUTURE

The plant of the future will be computer controlled, fully automated, modular, and inherently safe. It will be designed using engineering workstations with standardized CAE/CAD tools. All design data for every component and system in the nuclear plant will be available in computer-readable form. The design database will also contain functional and qualitative descriptions of plant systems as well as results of risk assessment studies. The plant database will be augmented during plant operation by automatic data logging of process and control action events into optical storage devices. The existence of such a comprehensive database will permit evolutionary development of new methodologies for intelligent selection of control strategies, and decision making, and for data retrieval, display, and analysis during the plant's full lifetime.

The plant sensors will be connected to the control computers by means of fiber optic cables, which will reduce the need for containment penetrations, eliminate signal contamination during the transmission process, and be compatible with future optical computers. Sensors will be microprocessor based and will be configured as autonomous objects with knowledge of their limitations, calibration and maintenance requirements, time constants, alternative means to provide a failed signal as well as their role in the system to which they are linked. Sensors will in general digitize their output and keep signal statistics and performance data internally; some sensors, however, due to their placement in environs pernicious to the electronics such as high radiation fields, will require digitalization at a distance from the sensing point. Plant sensors will supply information on the sensed signal and other parameters (such as trends, normality or abnormality, and statistics) only on demand by their local controller and when necessary (such as when the signal changes significantly or abnormal conditions are detected). The plant sensors, with their microprocessors, high-speed fiber optic connections, and local controllers, will constitute an autonomous distributed computer network in which each sensor is a node reporting to a higher level node that validates its assigned signals with physical and analytical redundancy checks and provides the central controller with verified signal data and instrumentation status information. It should be noted that, according to statistics compiled by the IEEE, in today's plants, 95% of shutdowns result from spurious alarms [7].

The control system will consist of two sets of computerized consoles with equal capabilities for monitoring and control operating in parallel. Only one of the consoles will have its control capabilities active, while the other is in stand-by as a backup. The backup console can be used as a crystal ball (CB) for operator training and use by research scholars, utility management, and the general public. The CB will be limited to monitoring the present, analyzing the past, and predicting the near future status of the plant, while the main control console will control the

operation of the plant. The CB, by providing the concerned observer with
the same access to all possibly desirable information on the operation of
the plant as the main control console, will be an effective tool to
promote trust in nuclear plants.

The control computers will be of fault-tolerant design; their
hardware will be configured as a distributed network of processors, and
their software will be based on object-oriented representations of all
plant components and systems. A hierarchical structure with distributed
intelligence will be implemented integrating data acquisition, display,
simulation, control, diagnosis, risk assessment, and expert system
technologies. Plant operation will be fully automated. The plant control
system will continuously monitor plant normality and will keep track of
margins to limits and be aware of safety system functionality to avoid
challenges to the safety system. Normality determination during steady
state operation and transients will be based on validated signal data and
intelligent use of past history, design functionality and equipment
status, and identification of the goal state.

The role of the operator will be that of a high-level supervisor and
task specifier. The operator will always have the capacity to ascertain
the correctness of any computer assertion or action by issuing queries
about the controller rationale and by using the display features to
examine the data.

The operator will be able to interact simultaneously with the monitor
and controller modules while receiving advice from expert systems and
mathematical models of the plant running faster than real time. The form
of operator interaction with the plant will be mainly symbolic through the
computerized console. Operator control directives will be subject to the
same validation scrutiny by an expert system as those directives generated
by the computerized automatic controller. The control system will
maneuver the plant along safe optimal paths to meet the goals set by the
operator.

Information available at both consoles will include design and
operational data on all plant systems and components. Design data will
comprise blueprints, functional schematics, data sheets, operational
characteristics, risk assessment data, performance limits, and maintenance
requirements. Operational data will include the present, past, and
projected what-if status of process parameters and equipment [signals,
valve positions, maintenance and repair, etc.], as well as past and
present audio and video information pertaining to activities inside the
containment and in other plant areas. The projection of what-if and
future status from present trends will be performed by mathematical and
heuristic models of components and systems. These models will be
dynamically configured by an expert system to produce computationally
efficient simulations for accurate near-term prediction of the parameter
of interest. Often this is accomplished by treating all systems other
than the one of interest as lumped parameter systems. The heuristic
models will provide qualitative predictions of the kind an expert operator
would provide.

Both consoles will feature multiple high-resolution screens capable
of intelligent display of computer-generated graphics and text as well as
video, simultaneously in a different screen, in the same screen/different
window, or in the same screen/same window. Some displays will be
presented automatically by the control computer as needed. The operator
will use both voice and electronic pointers such as mouse, light pen, or
touch-sensitive devices to interact with the displays and control the flow
of information. The operator will be able to request the display of plant

schematics, overlay a live display of signal readings of interest on the
schematic, and trigger a faster than real-time simulation of its dynamics.
By selecting a specific device, system, or parameter, the operator will be
presented with a menu. Choices for a plant parameter could be its present
value, history, and trend, whether it is measured, processed, or generated
from others and whether it can be verified by redundancy information.
Choices for a device could be a zoom-in new schematic describing it in
greater detail, a display of a live video picture of the device with
normal or infrared light, a description of its inputs and outputs, its
functionality as part of the system to which it belongs or in relation to
other equipment or systems, its maintenance history, or a simulation of
its operation.

All unmanned equipment areas of the plant, especially the
containment, will be with multiple video cameras with stereo sound and
infrared viewing capabilities. The containment cameras will transmit
their signals by microwave to a static receptor and will propel themselves
along tracks on the walls so that every piece of equipment can be
monitored. The video system will have motion detection capabilities in
both the infrared and normal modes, and automatic as well as operator-
controlled positioning, zooming, focusing, iris opening, and component
lighting. The audio system will have background filtering,
directionality, and focusing characteristics. Both video and audio
systems will automatically convey information to the control computer
whenever triggered by events in the containment.

References

1. F. T. Binford and E. N. Cramer, Ed., The High Flux Isotope Reactor,
    Oak Ridge National Laboratory Report ORNL-3572, (May 1964).

2. J. B. Bullock, G. R. Owens, and W. H. Sides, Jr., Reactor On-Line
    Computer Development at the HFIR, Oak Ridge National Laboratory Report
    ORNL-TM-3679 (February 1972).

3. J. A. Bernard and Asok Ray, "Experimental Evaluation of Digital
    Control Schemes for Nuclear Reactors," 1983 IEEE Trans.

4. W. R. Whittall and E. M. Hinchley, "Canadian Experience in Computer
    Control of Nuclear Plants," Power Plant Dynamics, Control and Testing
    - An Application Symposium, October 8-10, 1973, Knoxville, Tennessee.

5. A. D. Alley, S. Ball, P. Guadio, J. Hopps, E. Kane, R. Meyer, D.
    Nagel, E. Purvis, G. Scharpf, B. Sun, J. Sursock, and C. Tzanos,
    Advanced Control Test Operation - Task Team Report, Prepared for the
    U. S. Department of Energy.

6. K. E. Ball, "Mini-Perspective in Safety Controls," InTech, p. 47
    (July 1986).

USING AN INTEGRATION DEVELOPMENT METHODOLOGY TO PRODUCE A REACTOR

CONTROL SYSTEM FOR THE ADVANCED TEST REACTOR AT THE INEL[a]

Robert L. Smith and John M. Svoboda

Idaho National Engineering Laboratory
EG&G Idaho, Inc.
Idaho Falls, ID 83415

ABSTRACT

A new design methodology is being used at the INEL to produce a distributed reactor console display system and a simulator for the ATR. The
methodology integrates off-the-shelf software and hardware components
into turnkey custom systems that meet the specific needs of the ATR. The
objective of the methodology is to deliver usable systems quickly and at
very low relative costs.

INTRODUCTION

The development of real time process control systems continues to be
one of the shortcomings of traditional development methodologies. The
interface with real world hardware is very complex (inflexible), and the
computing resources are pushed to their limits. Designing a flexible
system that will adjust to the changing process environment and provide
the user full access to the information it produces is a real challenge.
Using an integration approach to the development process provides the
needed design margins to satisfy changes in requirements during the
development period and in the future.

At the Advanced Test Reactor (ATR), located in the Test Reactor Area
(TRA) at the Idaho National Engineering Laboratory (INEL), a major
upgrade of the control room and simulator is underway. The control room
is over 20 years old, and the ATR is scheduled to operate another
20 years. The upgrade is necessary since the vendors of the present control room equipment are no longer providing spare parts and maintenance
is becoming an increasing problem. We do not want spare parts availability and maintenance to affect the availability of the reactor. This
upgrade also gives us the opportunity to implement some improvements to
the control system recommended by a human factors study.

Traditionally, retrofits and upgrades of this type result in long,
complicated development cycles. But, through use of integration development methodology, we are progressing rapidly with a state-of-the-art,

a. Work supported by the U.S. Department of Energy, Assistant Secretary,
Defense Programs, under DOE Contract No. DE-AC07-76ID01570.

high quality, full function real time process control system that exceeds
all design requirements.

In this paper, first the methodology being used is described by con-
trasting it with traditional methodologies. A description of the ATR
system is then presented as an example of this methodology.

TRADITIONAL METHODOLOGIES

Research and experience with traditional development methodologies
have revealed problems in delivering a system that will truly satisfy the
real requirements. The most widely used development technique, structured analysis, has been tried by most engineering organizations but is
only actively used by 10 percent of them. The reason most do not use
these techniques is that the search for requirements is a very involved
process that requires a number of iterations and a large amount of manual
labor, especially when complex, real time distributed systems are needed.
Most users are looking for the easy way out of this dilemma and turn to
promises of tools and fourth-generation languages. These promises will
most likely be fulfilled, but how long it will take and at what cost is
not known.

Conventional development methodologies fall short in environments
where change is required. The specification freeze (signing off on the
specification) removes or restricts flexibility until the system is completed. If the views and needs of the user change after development
starts, it is too late, or a costly restart is needed. Real world environments are constantly changing because of new discoveries and enhancements that are brought out during the development process.

INTEGRATION METHODOLOGY

Faced with the requirement to produce real time computer systems
within budget and schedule, and knowing that traditional methodologies
are not effective, led us to develop the methodology now used. This
methodology provides benefits for all involved in the system development
by allowing each team member (operators, process engineers, hardware/
software engineers, training, safety, quality, managers) to best use
their expertise. The user is provided an environment that includes
training, documentation, and a variety of support personnel so that
interfaces can be prototyped on the actual hardware to be used. The
developers are able to concentrate on the system architecture that links
the user environments together, using standard vendor-independent tools.
Custom development is minimized by using application-specific packaged
software and hardware. Traditional requirement barriers between users
and developers do not exist.

The talents of support personnel, such as human factors, safety,
quality and graphic arts, are utilized to guide the user toward more
functional interfaces. Using these experts as the system is being developed instead of for after-the-fact evaluations allows the design to proceed very rapidly. The final product is very comprehensible and includes
the experience of experts from the computer industry as well as the
nuclear industry. The willingness of the support personnel to become an
integral part of the design process is overwhelming.

Every vendor claims to provide an integrated solution. Since it is
unlikely that any one vendor can stay of top of all the technology areas

needed to support our applications, a variety of hardware and software
vendor's products are integrated into the final system. Thus, we are in
a position to capitalize on the leading-edge technologies of a variety of
vendors, instead of trying to duplicate them. As new products become
available on the open market, they can be integrated into the application
environment. Emerging standards with multiple vendors supporting them
allow us to take advantage of their collective technologies.

Hardware and software tools and packages are selected that provide
the majority of the functions required for the system. By integrating
these functions into a custom application-oriented system, the real
requirements are satisfied. Any custom software that must be written is
isolated and very carefully evaluated. If new hardware or software products become available, they can be integrated into the system just as
easily. We look for products that eliminate our need for costly custom
software.

A particularly strong feature of this design methodology is the reduction in required support. By limiting or eliminating custom developments, maintenance is drastically reduced. Support is primarily provided
by independent software and hardware vendors that have a vested interest
in the success of their products and also have the training and experience specific to their product area. The industry is better able to
adjust to the rapidly changing technology then we are.

The focus of the development effort is to provide environments for a
variety of users, thus avoiding the keyhole affect of the loudest or most
dominant personality driving the development. We are no longer constrained by vendor-specific items, such as operating system and CPU model
numbers. The design is determined by software and hardware elements that
satisfy a particular application domain specific to user needs. The ability to integrate a wide variety of these elements allows us to build
applications that have a custom look without the initial development cost
or continuing maintenance cost. If a package does not perform as needed,
it can be replaced with one that does.

INTRODUCTION TO THE ATR UPGRADE

Integration development methodology was chosen to produce the ATR
systems because it promises to produce a fully functional system, within
a limited budget, that will meet ATR program needs for the next 20 years.
The final system will be fully documented, and training on the subsystems
is available now. The ATR upgrade consists of a redundant control room
display system and a simulator, to be implemented in three phases:

1. Upgrading the ATR control room by providing a redundant display
    system capable of displaying relevant reactor kinetics and
    thermal-hydraulics information,

2. Immediate construction of a new ATR control room simulation
    facility capable of providing operator training for normal
    reactor operation scenarios, and

3. Future upgrading of the simulator to provide operator training
    for ATR reactor kinetics and thermal-hydraulic accident scenarios.

Phase I

The ATR control room display system (CDS) provides for six display
stations, two redundant host computers, and two communication networks

(see Figure 1). The display stations support the rapid display of multiple high-resolution screens that contain information on plant status and parameters. The two host computers provide a centralized data storage and retrieval system that supports the data requirements of the six display stations. The two communications networks provide a redundant communication medium for the redundant display stations, the host computers, and the plant computers that is capable of high-speed data transfer and physical isolation between the CDS system components.

The six display stations are based on standard, off-the-shelf CAD CAM work station technology. The specified work stations have a 32-bit word, operate at two million instructions per second (MIPS), support the UNIX operating system, provide 1280 by 1024 pixel color graphics, and allow for a variety of operator input devices. By selecting off-the-shelf work station technology that is supported by several vendors, we are not locked into one vendor's perception of what the nuclear industry's display requirements will be for the next 20 years. In addition, as ATR program requirements specify higher performance and greater resolution, we are free to go out for competitive bid on the open market to procure state-of-the-art work station technology.

The work station concept excels in fulfilling the ATR requirement for real time graphic display of plant data. There are a variety of user configurable real time graphic display packages available. The ATR project selected interactive packages that support the display of data in over 40 formats, with update rates of less than 1 s. Windows, menus, and icons will also be used to provide the user interface. These interfaces are configured by non-programmer users (graphic artists, operations personnel) instead of programmers. In fact, the building of a new display is as simple as describing the input data source, describing the display, and running the display.

The two host computers are also based on standard, off-the-shelf work stations. The requirements for the hosts dictated a higher performance processor with minimal display capability. Two 4-MIPS, 32-bit processors with monochrome graphics of the same architectural family as the display stations were procured. These machines were supplied by the same vendor as the display stations; yet this was not a hard and fast constraint, as several vendors support the 4-MIPS, 32-bit performance. Again, as ATR program requirements specify higher performance, we are free to go on the open market for competitive bid and are not locked into one vendor for the expected life of the system.

Multiple computer systems such as the ATR CDS require a centralized system configuration responsibility. This configuration may involve system components such as sequential events recorders, display stations, plant computers, data acquisition systems, and simulation computers, as is the case with the ATR CDS. This responsibility resides in the CDS hosts, and a standard, off-the-shelf relational data base has been selected to perform the function. Upon system initialization, the data base generates configuration files that are compatible with the initialization requirements of the different system components. These initialization files are then loaded into each system component through the communications network.

The data base provides a comprehensive set of interactive menus that prompt the user through schema definition, data entry, query screen creation, menu design, report generation, and security setup. Using a fully functional relational data base provides a powerful configuration tool that frees the user from the drudgery of custom programming and allows him to get on with the job of configuring the data base for the needs of

448

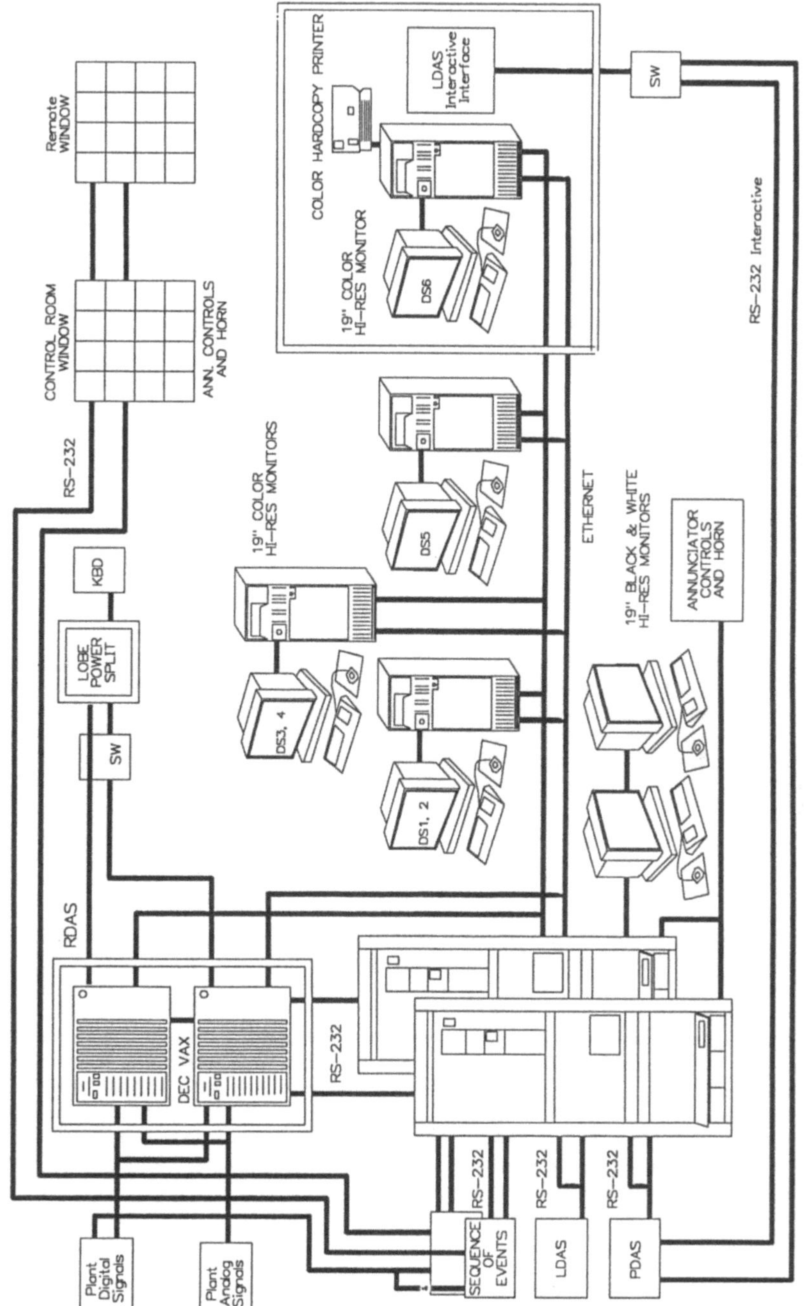

Figure 1. Control room display system block diagram.

a particular application. It also is used by the process engineers to
provide an interface to the system that meets their requirements. They
develop the set of screens, menus, and reports that are needed to view
the process data.

The data base is not used to process the real time data, but it does
provide the description of the data format that is used by every other
component in the system. The storage of real time plant data for historical display is also done by a real time task.

For redundant communication between the display stations, the host
computers, and the plant computers, industry standard local area networks
(LANs) were selected. This selection ensures our ability to expand the
CDS capability far into the future, as a large number of suppliers will
traditionally support well accepted standards.

In addition to providing a communication mechanism, the LAN plays a
vital role in meeting the CDS requirement for isolation between redundant
system components. The isolation is due in whole to the selection of a
LAN that supports a fiber optic transmission medium. The fiber optics
provide grounding isolation, reduced noise, and the elimination of high-
energy fault propagation often associated with metallic conductors. The
single failure possibility has been removed from this system by the use
of passive optical elements in the fiber optic transmission medium.

Phase II

Phase II of the control room upgrade involves the construction of a
simulator that will provide for ATR reactor operator training. This
simulator must be able to train the operator for normal reactor operating scenarios and also accommodate future expansion, during phase III,
into the area of full thermal-hydraulic and reactor kinetics accident
training. The problem is how to implement a limited thermal-hydraulics
level of simulation that is upgradable to a full thermal-hydraulics simulation at a later date with minimal impact on the construction of the
CDS and the Phase II simulator.

The Phase II approach consists of duplicating the CDS portion of the
control room upgrade without the plant computer attached to the LAN. A
simulation computer that simulates the function of the plant computer,
the reactor kinetics, and limited thermal-hydraulics will replace the
plant computer. The same 4-MIPS, 32-bit word work station as the host
was selected to perform this function.

An ATR reactor kinetics model has been under development and in use
at the INEL for the last 20 years. This model was written in Fortran and
runs on a 16-bit minicomputer. As part of this upgrade, the reactor
kinetics were modified, limited thermal-hydraulics were added, and the
code was converted to run under a different operating system. The modification of existing ATR kinetics model software has allowed quick and
efficient turnaround of the major portion of the so called "custom" code
that could not be avoided.

Phase III

During Phase III, the simulator will be upgraded to full thermal-
hydraulics simulator status. Our goal is to accomplish this with minimal
impact on the existing simulator. The limited thermal-hydraulics model of
the Phase II simulator will be upgraded and implemented on a high-
performance computer model. Thermal-hydraulics information from the
Phase III simulation computer will be exchanged with information from the

control logic and reactor kinetics model of the Phase II simulation computer.

The upgraded thermal-hydraulics code will be implemented by configuring RELAP5 for the ATR reactor. RELAP5 is a general purpose thermal-hydraulics code that was developed at INEL. This code can be configured for different pressurized water reactors for a variety of purposes. We shall configure it for ATR thermal-hydraulic accident scenarios. The use of RELAP5 for the thermal-hydraulics simulation will give the operator in training a state-of-the-art reactor accident training tool tailored for the ATR reactor at the expense of configuring RELAP5 for the ATR reactor and a new operating system.

The ATR version of RELAP5 requires more computational power than the work stations that are selected for the CDS and the Phase II simulation computer are capable of. We will go out on competitive bid for an off-the-shelf "super mini" capable of running this version of RELAP5 in real time. This machine will be directly coupled to the Phase II simulation computer via another LAN of the same type used by the rest of the system. Thermal-hydraulic variables will be shared with reactor kinetics and reactor control logic variables.

SUMMARY

A new design methodology is being used at the INEL to produce a distributed reactor console display system and a simulator for the ATR. The methodology integrates off-the-shelf software and hardware components into turnkey custom systems that meet the specific needs of the ATR. The objective of the methodology is to deliver usable systems quickly and at very low relative costs.

Taking advantage of available technology allows more development time to be spent on important long-term features like user interfaces, reliability, packaging, training, configuration management, documentation, and future expansion and less on risky custom hardware and software developments. We are better able to use the talents of specialists in the fields of human factors, reactor operations, graphic arts, quality, and training. Traditional methodologies rely only on hardware and software engineers to provide these functions for the industry they are targeting their products for.

LINGUISTIC CONTROL OF A NUCLEAR POWER PLANT

Joseph J. Feeley and James C. Johnson

Department of Electrical Engineering
University of Idaho
Moscow, Idaho 83843

ABSTRACT

A multivariable linguistic controller based on fuzzy set
theory is discussed and its application to a pressurized water
nuclear power plant control is illustrated by computer
simulation. The nonlinear power plant simulation model has
nine states, two control inputs, one disturbance input, and
two outputs. Although relatively simple, the model captures
the essential coupled nonlinear plant dynamics and is
convenient to use for control system studies. The use of an
adaptive version of the controller is also demonstrated by
computer simulation.

INTRODUCTION

This paper describes the application of fuzzy set theory
to the design of a linguistic controller for a nuclear power
plant. The automatic control of nuclear power plants has been
discussed and debated for some time and a variety of control
systems have been proposed, designed, and analyzed. The
difficulty of the control problem lies in the nonlinear,
dynamic, coupled behavior of the many plant variables and the
need for robust, reliable control. Good control may require
the measurement of many output variables, from reactor power
to generator load, and the control of many input variables,
from control rod position to turbine steam flow, in a plant-
wide, integrated control system. A number of plant control
system designs have been suggested based on modern
multivariable design methods. A far larger number of designs
have been based on clever application of classical control
system design techniques. A major difficulty in both modern
and classical design methods, however, is the need for an
accurate, yet relatively simple, mathematical model of the
plant on which to base the design of the control system. The
objective of this paper is to describe a control system design
method new to nuclear power that does not require an explicit
mathematical model of the plant.

The design method discussed here relies on fuzzy set theory and allows the control system design to be based on a linguistic, rather than a mathematical, model of the plant. The linguistic model is constructed from logical statements about system performance. The model statements do not have to describe every possible system response to every possible combination of inputs and may even include somewhat contradictory statements. In addition, the controller's initial plant model may be modified adaptively during plant operation to further improve controlled plant performance.

FUZZY CONTROL THEORY

Most design methods for automatic control systems rely on the designer's ability to construct an accurate low order mathematical model of the process to be controlled. This model is then used in any of a variety of design procedures to generate a control algorithm[1]. During automatically controlled plant operation the controller output is based on the current plant state, an expression of desired plant performance, and a dynamic mathematical model of plant behavior. During manual operation, on the other hand, trained plant operators, without benefit of mathematical models and control computers and with only approximate knowledge of the plant state, are often capable of meeting or exceeding the performance of automatic control systems. Through experience and training the operator is aparently able to infer an implicit plant model in linguistic form and devise an appropriate control strategy. It was this ability of human operators to devise real-time control strategies that led Procyk and Mamdami[2] to apply Zadeh's[3] fuzzy set theory to the design of automatic controllers based on linguistic models.

<u>Fuzzy Set Theory</u>

Fuzzy set theory is a generalization of conventional set theory and, therefore, includes conventional set theory as a special case. A few of the basic definitions and concepts of fuzzy set theory, and their uses in linguistic control, are reviewed below.

A conventional, or "crisp", set is defined, loosly speaking, as a collection of elements. In conventional set theory, given the set X and a subset of X called C, an element of X, call it x, is either in C or not in C. Fuzzy set theory is a generalization of a conventional set theory in the sense that it allows for partial membership in a set. That is, given a crisp set X a fuzzy subset F consists of the ordered pairs $\{x, u_F(x)\}$ where $u_F(x)$ is the membership function defining the degree of membership of x in F.

The membership function can take on any real value, but is usually normalized to the closed interval [0,1], where 0 indicates nonmembership, 1 indicates full membership and intermediate values indicate the degree of partial membership in the subset. Each element of a fuzzy subset (or, more simply, set) has an associated membership function so that the fuzzy set F of the support set X may be written as

$$F = \{x, u_F(x)\} \qquad x \in X$$

A useful tool in mapping one fuzzy set into another is the fuzzy relation which is analagous to the concept of a function in crisp analysis. A fuzzy relation R from a fuzzy set X to a fuzzy set Y, i.e. R maps X into Y, is itself a fuzzy subset of the outer product space of X and Y. R is characterized by the ordered pairs $\{(x,y), u_R(x,y)\}$ where $u_R(x,y)$ is a bivariate membership function quantifying the membership of (x,y) in R. The value of each $u_R(x_i, y_j)$ is determined by the minimum of $u_F(x_i)$ and $u_F(y_j)$ for each i and j in R. The extension of this concept from two to higher dimensions is straightforward.

Fuzzy relations in different product spaces may be combined using the compositional rule of inference. There are a number of different composition rules but, for the purpose of illustration only the max-min rule is discussed here. The max-min composition states that if R is a fuzzy relation from X to Y and S is a fuzzy relation from Y to Z, then the max-min composition of R and S is a fuzzy relation (often called a relational matrix) denoted by

$$R \# S = \max_j \{ \min_i [\ u_R(x_j, y_i), u_S(x_i, y_j) \ ] \ \}$$

Operationally, the max-min composition is similar to that of ordinary matrix multiplication where the min operation replaces multiplication and the max opereation replaces addition.

The three concepts discussed above, the fuzzy set, the fuzzy relation, and the max-min composition may be combined to process a rule expressed in the form of a logical "IF-THEN" statement such as

IF (fuzzy input variable "Average Primary Coolant Temperature" takes on value "A LITTLE LOW")
THEN (fuzzy output variable "Voltage to Control Rod Actuator" should be "INCREASED A LITTLE").

Specifically, the statement is translated into a relational matrix using the fuzzy relation operation and the output, for a specific input, is inferred using the max-min composition.

When, as is usually the case, multiple rules are specified, they are cast in the form of a series of "IF-THEN-ELSE" statements. The collective effect of all the rules is obtained by "summing" the relational matrices generated by each rule using the logical "max" operation. The net result of the above operations is to translate a series of rules in "IF-THEN-ELSE" format into a relational matrix that provides a mapping from every possible combination of fuzzy inputs to a fuzzy output. Because of the fuzzy nature of the input variables, the rule base may be incomplete, i.e. not every input combination need be considered explictly. Also, the rules may even be contradictory and the most reasonable output will still be inferred (except in certain pathological cases).

Linguistic Control

     The four major operations of a linguistic controller are
fuzzification of the inputs; creation of the rule base or, its
equivalent, the rule base relational matrix; inference of the
output fuzzy set; and defuzzification of the output.

     Fuzzification is the process of transforming the measured
values (numbers) of an input variable into realiztions of the
corresponding fuzzy variable. This is essentially a
quantization procedure where the range of the input variable
is subdivided into into a smaller number of fuzzy values
representing the universe of discourse of the fuzzy variable.

     The designer specifies the rule base in the form of a set
of IF-THEN-ELSE statements that specify what the controller
output should be in response to a given set of controller
inputs. The rules are based on the desired performance of the
system under specified conditions. As noted above, because of
the fuzzy nature of the variables and the logical blending of
the rules, reasonable outputs may be inferred from an
incomplete, or even contradictory, set of rules. For real-
time control applications it is important to note that the
relational matrix may be constructed from the rule base off-
line and prior to operation of the controller, allowing
extremely fast on-line controller operation.

     The contoller output inferred using the max-min
composition described above is itself a fuzzy variable and
must be "defuzzified", i.e. transformed into a real number,
before it can be used by a typical control system actuator. A
number of different procedures are available for
defuzzification and their use is somewhat application
dependent. The simplest method is to chose the output value
with the largest membership grade. This method will present
problems only when the rule base contains contradictory rules,
as evidenced by a double peak in the membership function; or
when there is no clear cut output decision, in which case the
membership function has a broad peak. These somewhat
pathological cases seldom appear in practice, but may be
handled with little additional difficulty when they do. In
this study the center-of-area method, which generates an
output located at a point analogous to the center of gravity
of the output membership function, was used.

Adaptive Linguistic Control

     Adaptive linguistic controllers are linguistic
controllers that modify their relational matrix (i.e., their
rule base) during plant operation. In this study a' priori,
or conventionally obtained, knowledge of system response times
was used to determine which rule was responsible for the
undesireable behavior. This rule is then iteritively modified
to improve system performance. The rate of rule modification
is determined by trial and error and unanswered questions of
convergence and stability remain to be investigated. If
system response times cannot be separated unambiguously, the
responsibilty for undesireable performance can be spread over
a number of rules in a fuzzy manner. Numerical studies have
also shown that good adaptive control can be achieved by

modifying the rule base matrix directly, instead of modifying
the rule itself. This obviates the need for recomputing the
relational matrix during each adaption step and makes on-line
adaptive control feasible.

NUCLEAR POWER PLANT SIMULATION MODEL

A mathematical model of a pressurized water reactor
nuclear power plant was used in computer simulation studies of
the effectiveness of the adaptive linguistic controller. The
model is based on nine coupled nonlinear differential
equations that describe the primary and secondary coolant
systems in sufficient detail for control system analysis. The
nine system state variables are neutron density, one effective
group of delayed neutron emitters, control rod position, fuel
pin temperature, hot leg temperature, cold leg temperature,
steam generator secondary side saturation temperature, steam
flow control valve position, and turbine speed. The two plant
variables assumed available for measurement and use as control
system inputs are primary coolant system average temperature,
and turbine speed. It is also assumed that the controller is
able to manipulate two plant input variables through constant
speed motors. The two controlller output signals are the
voltages applied to control rod and steam flow control valve
actuators. The power removed through the turbine is treated as
a disturbance input. While the model contains many
simplifications and lumpings of distributed variables,
experience has shown that it captures the essence of the
nonlinear multivariable control problem, while still being
relatively easy to use. The simulation model is programmed in
a compiled version of BASIC and runs in nearly real-time on an
IBM PC/AT.

CONTROLLER DESIGN

Simulation studies have shown that satisfactory control
can be obtained using a two-input/two-output controller. The
plant outputs (controller inputs) used for control are average
primary coolant temperature (T_{AV}) and turbine speed (S_T). The
controller also computes and uses the rates of change of these
variables . The plant inputs (controller outputs) are on/off
signals to the control rod actuators and the steam flow
control valve actuator.

Each controller input and output is mapped into a five
state universe of discourse [-2,-1,0,+1,+2] allowing fuzzy set
elements of, respectively, "negative-big", "negative- small",
"zero", "positive-small", and "positive-big". The controller
uses two relational matrices, one for the control rod actuator
signal and one for the steam flow control valve signal. Each
matrix is of dimension 5x5x5x5, with the four coordinate
directions corresponding to T_{AV}, dT_{AV}/dt, S_T, and dS_T/dt.
During operation the fuzzy values of these four fuzzy input
variables determine indices in each of the relational matrices
pointing to the fuzzy values of each of the two corresponding
fuzzy output variables. The output variables are defuzzified
using a center-of-area algorithm.

RESULTS

Although the model was not intended to simulate any
particular power plant, model parameters were selected so that
both transient and steady state performance were typical of a
small (50 Mw) pressurized water reactor plant. Performance of
the linguistic controller during both tracking and regulating
modes was investigated.

Responses of average coolant temperature and turbine
speed to step setpoint changes are shown in Figs. 1 and 2.
Fig. 1 shows relatively smooth, rapid, and overdamped response
of T_{AV} to sequential step changes in setpoint of
-5 F and +10 F, while S_T was held constant. The changes were
acomplished with well coordinated movements of the control
rods and the steam flow control valve. Fig. 2 shows the
response of T_{AV} and S_T to simultaneous setpoint changes of -5
F and +5 rev/sec. Again, well behaved and relatively prompt
response was obtained with reasonable control effort.
The fuzzy controller relational matrices required to produce
these responses were generated in a few learning cycles from a
sparse initial rule base and were used successfully for a wide
variety of transients.

The disturbance rejection properties of the controller
are shown in Fig. 3. In this transient the controller
attempts to maintain T_{AV} and S_T at their setpoint values of
500 F and 30 rev/sec in spite of random 2.5 Mw deviations in
turbine load from its average value of 45 Mw. The controller
was able to maintain T_{AV} within 1 F of its desired value of
500 F and S_T within 1 rev/sec of its desired value of 30
rev/sec. These deviations were less than half the open loop
deviations and were obtained with only moderate control
effort.

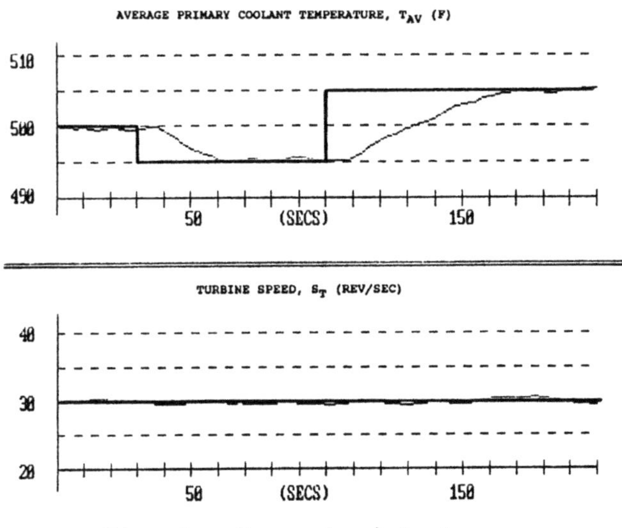

Fig. 1. T_{AV} setpoint change.

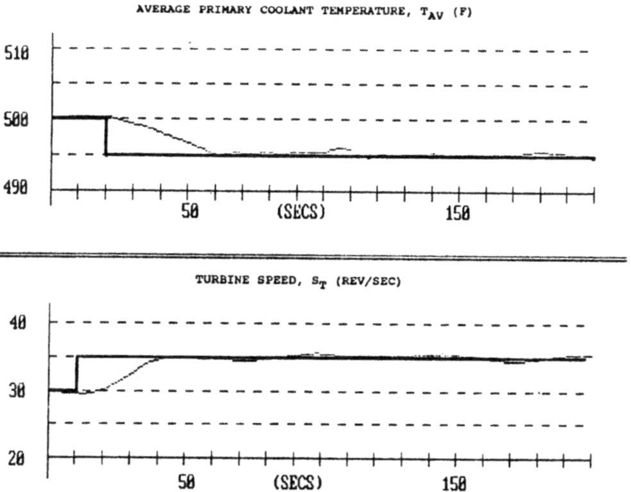

Fig. 2.   Simultaneous T_{AV} and S_T setpoint changes.

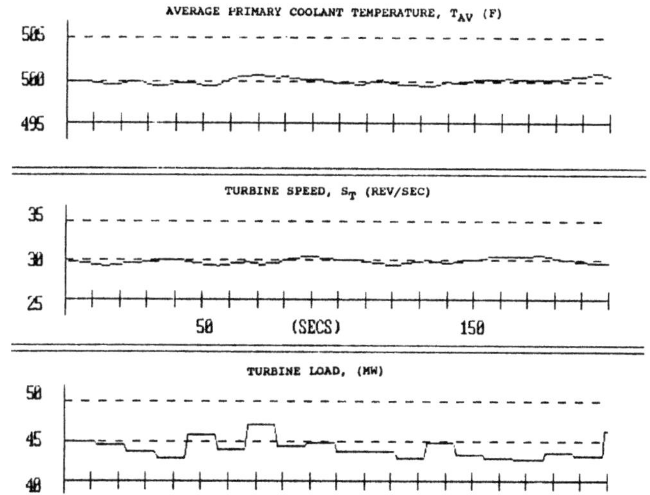

Fig. 3.   Rejection of turbine load disturbance.

CONCLUSIONS

The ability of an adaptive multivariable linguistic
controller to reject disturbances and track setpoint changes
in a simplified nuclear power plant model has been
demonstrated by computer simulation. The design procedure for
the nonadaptive controller is well-founded in fuzzy set theory
and is based on a linguistic model of plant performance. The
linguistic model is formulated as a set of IF-THEN-ELSE
statements relating the plant state to the plant inputs and is
not nearly as precise or detailed as the conventional
mathematical models used in multivariable control system
design. The adaptive version of the controller was developed,
for the special case considered here, through trial and error
computer simulation studies. Serious questions remain open
regarding stability, optimization, and synthesis of adaptive
linguistic controllers and have not been addressed in this
study.

REFERENCES

1. G. H. Hostetter, C. J. Savant, Jr., and R. T. Stefani,
"Design of Feedback Control Systems," Holt, Rinehart and
Winston, New York (1982)

2. T. Procyk and E. Mamdani, A Linguistic Self-Organizing
Process Controller, _Int. Journal of Automatic Control_, Vol. 6,
1978, pp. 15-30

3. L. Zadeh, Outline of a New Approach to the Analysis of
Complex Systems and Decision Processes, _IEEE Trans. on
Systems, Man, and Cybernetics_, Vol. SMC-3, 1973, pp 28-44

TIME-OPTIMAL CONTROL OF REACTOR POWER

John A. Bernard

MIT Nuclear Reactor Laboratory
Massachusetts Institute of Technology
Cambridge, MA 02139

ABSTRACT

Control laws that permit adjustments in reactor power to be made in
minimum time and without overshoot have been formulated and demonstrated.
These control laws, which are derived from the standard and alternate dynamic period equations, are closed-form expressions of general applicability. These laws were deduced by noting that if a system is subject to one or
more operating constraints, then the time-optimal response is to move the
system along those constraints. Given that nuclear reactors are subject to
limitations on the allowed reactor period, a time-optimal control law would
step the period from infinity to the minimum allowed value, hold the period
at that value for the duration of the transient, and then step the period
back to infinity. The change in reactor power would therefore be accomplished in minimum time. This particular response can be achieved by
exploiting the balance between the prompt and delayed neutron populations.
Specifically, the reactor period can be made to change essentially as a step
by initiating the transient with a very high rate of change of reactivity
(prompt effect) and then cutting back on that rate as the delayed effects
build in. The resulting control laws are superior to other forms of time-
optimal control because they are general-purpose, closed-form expressions
that are both mathematically tractable and readily implemented. Moreover,
these laws include provisions for the use of feedback. The results of simulation studies and actual experiments on the 5 MWt MIT Research Reactor in
which these time-optimal control laws were used successfully to adjust the
reactor power are presented.

INTRODUCTION

The Massachusetts Institute of Technology (MIT) and the Sandia National
Laboratories (SNL) are engaged in a joint program to develop and demonstrate
experimentally techniques for the rapid adjustment of a reactor's neutronic
power. It is intended that these methods be used for the transient control
of reactor-powered spacecraft. The specifications imposed on these control
techniques are therefore quite rigid. First, it must be possible to raise
the reactor power by many orders of magnitude both in a few seconds and
without overshoot. Second, the control methodology must be sufficiently
flexible to permit use under a variety of initial conditions. Third, it
should be possible to implement the controller using actuators such as step-

ping motors that are of low weight. Fourth, the control approach should permit continuous operation, including power cycling.

One of the early results of this joint MIT-SNL effort was the elucidation of closed-loop control laws for the time-optimal control of power in reactors that are both subject to a limitation on the allowed reactor period and which can be described by point kinetics. These laws are an extension of the 'Reactivity Constraint Approach', which is a general purpose, supervisory technique developed at MIT to insure that no challenge would be made to a reactor's safety system as the result of any automatic control action [1]. Both the relation of these laws for time-optimal control to the 'Reactivity Constraint Approach' and the results of the initial experimental trials of these laws have been previously reported [2]. The objectives of this paper are (1) to describe the method employed to derive and implement these time-optimal control laws and (2) to report the most recent results from the ongoing experimental program in which the performance of these laws is being evaluated.

TIME-OPTIMAL CONTROL

In general, previous efforts to obtain a minimum time response for reactor power have entailed assembling a set of equations that accurately models a reactor, specifying performance criteria, and applying the conditions for optimality such as those elucidated by the mathematician Pontryagin. So doing yields a set of partial differential equations with split boundary conditions that must be solved iteratively using complex numerical methods. Often it is necessary to linearize the model equations in order to make the mathematics tractable. The result is a sequence of control signals that are valid for a specific reactor model and one particular set of initial conditions. These signals are then applied in an open-loop fashion. Provisions do not exist for the continuous feedback of the power level or period and, should there be an appreciable change in the plant conditions, this predetermined sequence of signals will not produce the desired result. Moreover, both the computational equipment necessary to implement the technique and the time required for the performance of the calculations is often substantial.

The minimum time control laws reported here were obtained by using a very different and novel approach to the problem of time-optimal control. Specifically, it was observed that if a system is subject to one or more constraints, then the time-optimal trajectory will move the system along the most limiting of those constraints. This observation can be used to great advantage. Namely, rather than solve the system's describing equations subject to both the constraint and a performance index, it is more direct to define the physical conditions that correspond to system movement along that limiting constraint. Relative to control of a nuclear reactor, typical constraints are the allowed reactor period, the available rate of reactivity change, and permitted rates of change of temperature. The time-optimal control laws described here are for a reactor subject to a limitation on reactor period. As such, these laws are most applicable to the operation of reactor powered spacecraft. However, they may also prove valuable in the operation of commercial reactors because they can be used to determine the control mechanism velocity necessary to produce virtually any desired power profile. For example, they could be used to adjust reactor power so that temperature changes occur at a uniform rate and thermal stresses are thereby minimized. Also, these laws may be applicable whenever the power is being raised from the level at which criticality was attained to the 'point-of-adding-heat' (i.e., the power level at which feedback effects become appreciable). This span is usually many orders of magnitude and operating procedures often dictate that the power be raised on a constant period.

462

CONTROL APPROACH

The standard approach to a time-optimal, or for that matter, most control problems is to represent the system as a set of first order differential equations. Often, these are linearized about a specific operating point. That route is not followed here. Rather, the system's describing equations are combined through a process of differentiation and substitution so that the constrained parameter is related directly to the system variables. Although cumbersome, this approach has two major advantages. First, it delineates the physical relation between the constrained parameter and the state variables that determine the system's dynamic response. Second, no compromises are necessary in the description of the system. Thus, for example, non-linearities and time-delayed behavior are preserved.

Relative to reactor control, the constrained parameter is taken here to be the period and the state variables are the prompt and delayed neutron populations. Limiting the discussion to space-independent kinetics, the latter are described by the neutron and precursor kinetics equations respectively. These may be combined to yield the dynamic period equation. This relation gives the instantaneous reactor period as a function of the reactivity and the rate of change of reactivity. It may be written in either of two forms [3,4]. The first formulation, which is termed 'standard', is:

$$\tau(t) = \frac{\overline{\beta}-\rho(t) + \ell^{*}[\frac{\dot{\omega}(t)}{\omega(t)} + \omega(t) + \lambda_e(t) - \frac{\dot{\lambda}_e(t)}{\lambda_e(t)}]}{\dot{\rho}(t) + \lambda_e(t)\rho(t) + \frac{\dot{\lambda}_e(t)}{\lambda_e(t)}(\overline{\beta}-\rho(t))} \tag{1}$$

where the standard, effective, multi-group decay parameter, which is a measure of the relative weighting of the various delayed neutron precursor groups, is defined as:

$$\lambda_e(t) \equiv \Sigma\lambda_i C_i(t)/\Sigma C_i(t) \tag{2}$$

and other symbols are defined as:

$\tau(t)$ is the instantaneous or dynamic reactor period,

$\overline{\beta}$ is the effective delayed neutron fraction,

$\rho(t)$ is the net reactivity,

ℓ^{*} is the prompt neutron lifetime,

$\omega(t)$ is the inverse of the dynamic reactor period,

$\dot{\omega}(t)$ is the rate of change of the inverse of the dynamic reactor period,

$\dot{\rho}(t)$ is the rate of change of the net reactivity,

$\lambda_e(t)$ is the effective, multi-group decay parameter,

$\dot{\lambda}_e(t)$ is the rate of change of the effective, multi-group decay parameter,

λ_i is the decay constant of the <u>ith</u> precursor group,

$C_i(t)$ is the concentration of the <u>ith</u> precursor group, and

$\overline{\beta}_i$ is the effective yield of the <u>ith</u> precursor group.

The other formulation of the dynamic period equation, which is termed 'alternate', is:

$$\tau(t) = \frac{\overline{\beta} - \rho(t) + \ell^* [\frac{\dot{\omega}(t)}{\omega(t)} + \omega(t) + \lambda_e'(t)]}{\dot{\rho}(t) + \lambda_e'(t)\rho(t) + \Sigma\overline{\beta}_i(\lambda_i - \lambda_e'(t))} \qquad (3)$$

where the alternate, effective, multi-group decay parameter is defined as:

$$\lambda_e'(t) \equiv \Sigma\lambda_i^2 C_i(t)/\Sigma\lambda_i C_i(t) \qquad (4)$$

Note that the effective, multi-group decay parameters are time-dependent quantities. Their values will vary during transients because the relative concentrations of the various delayed neutron precursor groups change depending on the rate at which power is being raised or lowered. Short-lived precursors dominate during increases and long-lived ones during decreases.

The prompt neutron lifetime is small and terms containing ω can therefore be deleted from both (1) and (3). The resulting equations are general relations that accurately predict the instantaneous reactor period associated with any reactivity pattern provided that the promp critical value is not approached. The terms in the denominators of these modified versions of the standard and alternate dynamic period equations are of special interest to the design of controllers for reactor power. Specifically, if the sum of these terms can be made zero, then the period will be driven to infinity and the change of power in a reactor halted. The physical significance of these terms may be determined by identifying their point of origin in the derivation. This has been done and, considering only the standard equation, it was found that the rate of change of reactivity $(\dot{\rho})$ corresponds to the effect of a changing prompt neutron population while the quantities $\lambda_e \rho$ and $(\dot{\lambda}_e/\lambda_e)(\overline{\beta}-\rho)$ correspond respectively to a changing delayed neutron precursor population and to a changing distribution of delayed neutron precursors within the defined groups. Given that precursor populations are a function of the power history and can not be altered on demand, it is apparent that if an immediate change in the period is desired, an adjustment should be made in the rate of change of reactivity rather than in the reactivity itself. This observation is central to the development of the minimum time control laws.

CONTROL LAWS FOR MINIMUM TIME RESPONSE

Adjustments of reactor power may be accomplished in minimum time if it is possible to change the reactor period in a step-like manner. For example, the period is initially at infinity and the rate of change of power is zero. The period is then stepped to its minimum allowed value and held at that value until the desired power level is attained. The period is then stepped back to infinity. This particular response can be achieved by initiating the transient with a very high rate of change of reactivity (prompt effect) and then cutting back on that rate as the delayed effects build in.

Having obtained the dynamic period equation and knowing the desired time-profile of the reactor period, it becomes a simple matter to obtain the time optimal control laws. The terms in the standard and alternate dynamic period equations are rearranged so that the control signal, which is the rate of change of reactivity, is expressed as a function of the period. So doing yields:

$$\dot{\rho}(t) = (\overline{\beta}-\rho(t))\omega(t) - \lambda_e(t)\rho(t) - (\dot{\lambda}_e(t)/\lambda_e(t))(\overline{\beta}-\rho(t)) +$$

$$\ell^*\dot{\omega}(t) + \ell^*[(\omega(t))^2 + \lambda_e(t)\omega(t) - (\dot{\lambda}_e(t)/\lambda_e(t))\omega(t)] \tag{5}$$

and

$$\dot{\rho}(t) = (\overline{\beta}-\rho(t))\omega(t) - \lambda_e'(t)\rho(t) - \Sigma\overline{\beta}_i(\lambda_i-\lambda_e'(t)) +$$

$$\tag{6}$$

$$\ell^*\dot{\omega}(t) + \ell^*[(\omega(t))^2 + \lambda_e'(t)\omega(t)]$$

where symbols are as previously defined except that $\omega(t)$ is now the inverse of the minimum allowed reactor period and $\dot{\rho}(t)$ is the rate of change of reactivity necessary to achieve that period. Both the terms ρ and $\dot{\rho}$ include thermal feedback effects.

Several aspects of these control laws warrant discussion. First, it may seem strange to write the control signal as a function of the controlled parameter. However, because a time-optimal response has been specified, the desired behavior of the controlled parameter is known. The objective of the control law is therefore to determine the value of the control signal necessary to generate that behavior. Second, the rate of change of reactivity, rather than the reactivity itself, is being used as the control signal. This is logical because, as is evident from the derivation of the dynamic period equation, a changing reactivity corresponds to the effect of the prompt neutrons on the transient and the prompt effect is being used to drive the transient. Third, algorithms based on the alternate control law are easier to implement than those obtained from the standard version because the latter has a term containing the derivative of its effective, multi-group decay parameter and that term is noisy when evaluated in real time. Fourth, the treatment of the term $\ell^*\dot{\omega}(t)$ requires special consideration. Ideally, the change in $\omega(t)$ should occur as a step. However, the presence of the quantity $\ell^*\dot{\omega}(t)$ shows that this is not physically attainable. Precise numerical studies of both control laws have shown that the immediate effect of a step change in $\dot{\rho}(t)$ is a corresponding step change in $\omega(t)$. As a result, the quantity $\omega(t)$ readjusts on a rapid exponential to its specified value. Relative to the control laws, the quantity $\dot{\omega}(t)$ is determined from the expression:

$$\dot{\omega}(t) = [\omega(t) - 1/\tau]/[n(\Delta t)] \tag{7}$$

where τ is the measured reactor period, Δt is the sample interval, and n is the number of sample intervals permitted for $\omega(t)$ to attain its specified value. Given that the quantity $n(\Delta t)$ can be made small, the readjustment of $\omega(t)$ will occur so rapidly (a hundredth of a second or less) as to be essentially a step.

IMPLEMENTATION OF TIME-OPTIMAL CONTROL LAWS

Implementation of both the standard and the alternate control laws has been readily accomplished both by simulation and in real-time on the 5 MWt MIT Research Reactor using the following procedure. First, the quantity $\omega(t)$ is taken to be the inverse of the minimum allowed reactor period. Second, the reactivity, decay parameter, and actual period are either

measured or calculated and each term in the control law is evaluated.
Third, the control law is solved to obtain the rate of change of reactivity
necessary to achieve the specified period. Fourth, using previously obtained knowledge of the differential reactivity worth of the control mechanism,
the required velocity of the control device is determined. The control
mechanism is then moved at this rate. This process is continued until the
reactor power attains the desired level. The quantity $\omega(t)$ is then set to
zero (i.e., infinite period) and the control laws solved for the rate of
change of reactivity necessary to hold the power constant. Feedback effects
are automatically accommodated. For example, if the coolant temperature
were to rise, the measured value of the reactivity would decrease thereby
causing the rate of change of reactivity necessary to maintain the specified
period to increase. A more detailed description of the implementation process is given in [2].

Equations 5 and 6 are suitable for use with any reactor period including
ones that are as short as a few tenths of a second. Also, these control
laws are completely general and can be used to treat reactivity insertions
that are in excess of the prompt critical value. Further information is
given in [5].

DEMONSTRATION OF MINIMUM TIME CONTROL LAWS

Figures 1 and 2 are from a simulation study in which reactor power was
raised from 1 kW to 1 GW, a factor of one million, in 4.15 seconds using the
alternate law. The model selected was that of a thermal reactor with a
prompt neutron lifetime of 100 microseconds and an effective delayed neutron
fraction of 0.00786 $\Delta K/K$. Reactivity feedback effects were excluded for
this particular simulation. The specified constant period was 0.30
seconds. Figure 1 shows the resulting power and reactivity profiles. Note
that no overshoot occurred. The power was leveled in the presence of substantial positive reactivity by making the rate of change of reactivity
negative. That is, the prompt neutron population was intentionally decreased in order to balance exactly the continuing rise in the delayed neutrons. Figure 2 illustrates the behavior of the terms that dominated the
control law during most of the transient. (<u>Note</u>: Not shown is the $\ell^*\dot{\omega}$ term
which, as discussed in [2, 5], is of large magnitude whenever the value of
the specified period is changed.) Note the discontinuous changes in the
control signal that correspond to the moments at which the period was
stepped from infinity to 0.30 seconds and from 0.30 seconds to infinity.
Also note that the required rate of change of reactivity was a maximum at
the outset of the transient and was then gradually reduced as the terms
representing the delayed effects grew in magnitude.

Figure 3 illustrates the use of these time-optimal control laws for an
actual power increase of 500 kW on the 5 MWt MIT Research Reactor. This
increase was performed under closed-loop, digital control using a variable
speed motor to drive the reactor's regulating rod. A previously described
experimental protocol [6] was used during the course of this and other on-
line trials of these control laws. The data shown are from a trial in which
the specified period was 300 seconds. The period was stepped to this value
by initially requiring a rate of reactivity insertion of about 3.3 millibeta
per second. Then, as the delayed terms grew in magnitude, the rate of
change of reactivity was reduced. On reaching full power, the period
stepped back to infinity by making the rate of change of reactivity negative. The power increase then stopped. Given that the reactor period was
maintained constant at 300 seconds for the entire transient, the power
change was accomplished in the minimum time possible for a reactor limited
to a 300 second period. As proof, note that the initial and final power
levels were 1030 and 1500 kW. Hence, from the definition of reactor period,

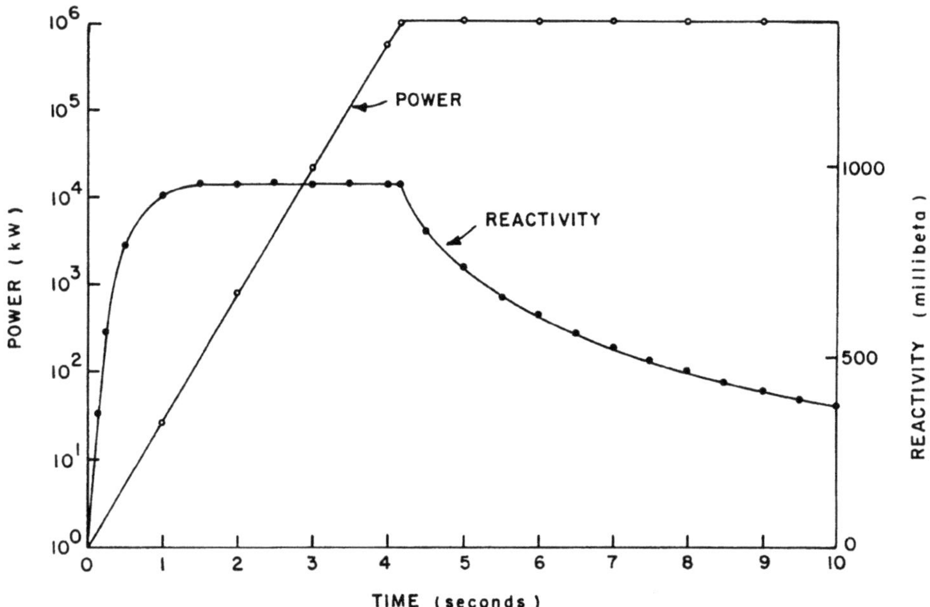

Figure 1. Power Increase of Six Orders of Magnitude Using Alternate 'Constant Period' Control Law with a Specified Period of 0.30 Seconds — Simulation Study, 23 Dec. 86

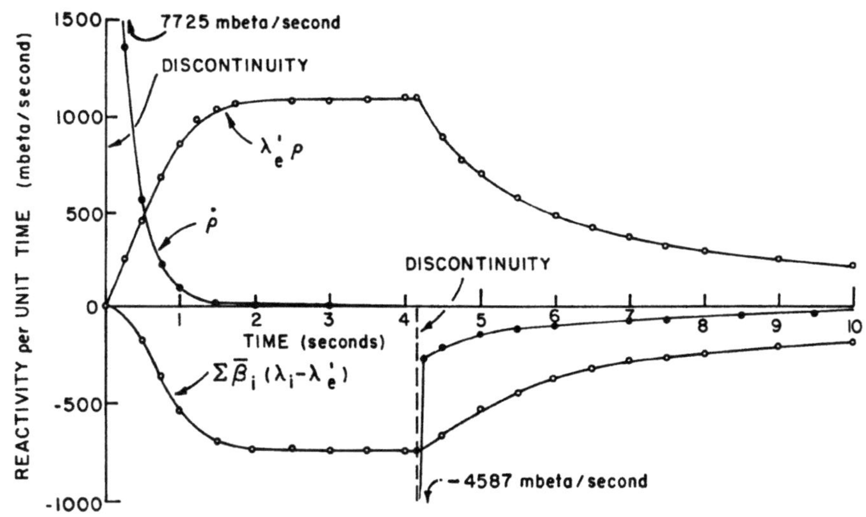

Figure 2. Component Terms of Alternate 'Constant Period' Control Law for Study Shown in Figure 1

the minimum possible transit time was $\tau \ln(P_F/P_O)$ or 113 seconds. This is in excellent agreement with the 114 second figure obtained from the experiment. Figure 3 shows the power and reactivity profiles. Note that there was no power overshoot. The reactivity is that of the control mechanism. Note that there is a rapid rate of change of reactivity at both the outset and termination of the transient. This occurs because the prompt effect (the $\dot{\rho}$ term) is being used to drive the period from infinity to 300 seconds and from 300 seconds to infinity respectively. The rate of change of reactivity decreased during the interval 10-114 seconds because the delayed contributions to the period were increasing. However, in contrast to the simulation study, it did not approach zero. This difference occurred because, on the real reactor, the temperature of the primary coolant was rising and the controller automatically added reactivity in order to compensate for this effect.

Figure 4 is from an experimental study in which one of the MIT Research Reactor's shim blades was used on the actuator. The reactivity worth of the shim blades is roughly an order of magnitude greater than that of the regulating rod. The controller was programmed so that the alternate time-optimal law would repeatedly raise and lower the power by 1 MW while on a period of ± 100 seconds as appropriate. As shown, the performance of the control law was excellent.

Additional experimental studies are planned for both the MIT Research Reactor and the Sandia National Laboratories' Annular Core Research Reactor. The latter, a modified TRIGA known as the ACRR, uses beryllium-uranium oxide fuel. It will be used to verify the performance of these laws on periods of a few tenths of a second.

ASSESSMENT OF TIME-OPTIMAL CONTROL LAWS

When compared to other techniques for achieving a time-optimal response, it is apparent that the control laws reported herein represent a superior means of achieving a minimum time response. First, they are general purpose closed-form expressions suitable for use on any reactor that can be described by point kinetics. Second, they are mathematically tractable. Third, they are exact. No approximations were made in their derivations. They rigorously describe the non-linear dynamics of a nuclear reactor. Fourth, they have been demonstrated on an actual reactor in real-time. Fifth, these laws are readily implemented because the desired control mechanism velocity (as opposed to merely position) is specified. Sixth, the resulting power trajectory is readily determined because the period is known in advance. This should facilitate coordination of heat removal systems with energy production. Seventh, corrections are automatically made for reactivity effects such as those associated with changing coolant temperatures and xenon. Eighth, both the reactor period and the power trajectory are known. Hence, there are two quantities available for feedback of the conventional type in which the product of a gain and an error between the actual and specified parameters is used to improve system response. Ninth, these laws are quite flexible. Assuming that there is no limitation on the available differential reactivity worth, they can be used to obtain the control mechanism velocity necessary to produce any desired profile of the neutronic power. Finally, it should be recognized that these 'time-optimal' control laws are consistent with the Pontryagin approach. Specifically, if the Pontryagin condition for time-optimal control is applied to a system that is subject to a constraint, then the solution will be to move the system along the trajectory established by that constraint. These control laws do exactly that. Only, they are closed-form expressions amenable to feedback and not, as is the result with the Pontryagin approach, a sequence of open-loop signals valid for only one particular set of boundary conditions.

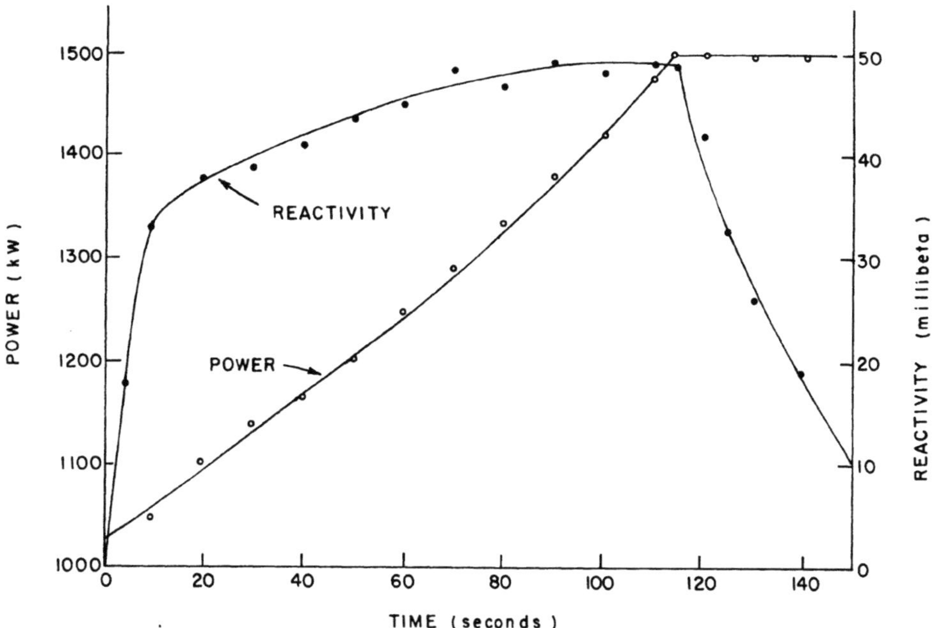

Figure 3: Power Increase of 500 kW Using Alternate 'Constant Period' Control Law with a Specified Period of 300 Seconds — Experimental Data, 4 May 87

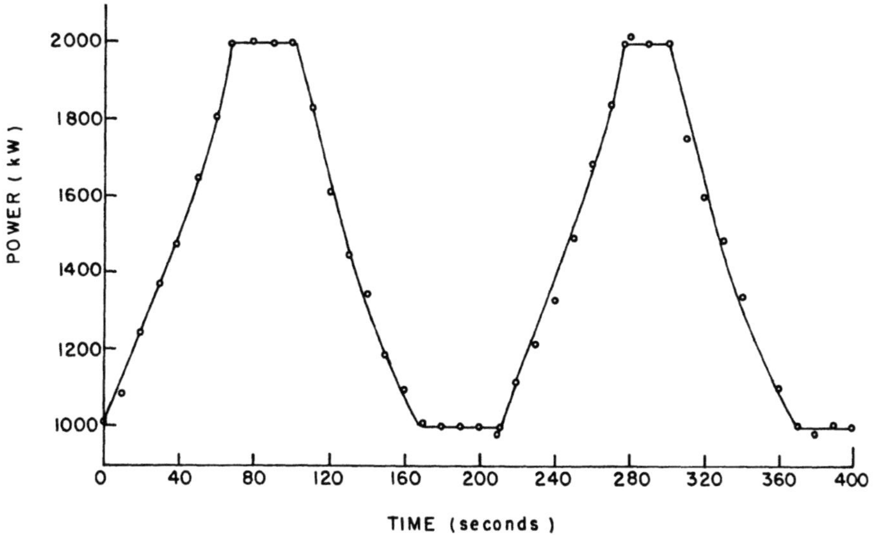

Figure 4: Power Cycling on ±100 Second Period Using Alternate 'Constant Period' Control Law — Experimental Data, 20 July 87

SUMMARY

A new methodology for the elucidation of time-optimal control laws has been described. This approach is suited for systems that are subject to one or more limiting constraints. The system's describing equations are combined so that the control signal is a function of the constrained parameter. The time-optimal profile of the constrained parameter is then determined. In general, this profile will be step-like. Substitution of the time-optimal profile for the constrained parameter then results in the value of the control signal necessary to achieve a minimum time response. This technique was successfully applied to a nuclear reactor and the resulting time-optimal control laws were verified by both simulation and actual experiment on the 5 MWt MIT Research Reactor.

ACKNOWLEDGEMENTS

The contributions of Professors Allan F. Henry, David D. Lanning, and John E. Meyer of the MIT Department of Nuclear Engineering and of Mr. Kwan S. Kwok and Mr. Paul T. Menadier of the MIT Nuclear Reactor Laboratory are acknowledged. Appreciation is also expressed to Mr. Frank V. Thome and Mr. Frank J. Wyant of the Sandia National Laboratories. Also, the assistance of Mr. Ara Sanentz in preparing this paper is noted. Computational equipment was provided by the Charles Stark Draper Laboratory. The research itself was supported by a grant from the U.S. Department of Energy via the Sandia National Laboratories.

REFERENCES

1. J. A. Bernard, D. D. Lanning, and A. Ray, Use of Reactivity Constraints for the Automatic Control of Reactor Power, IEEE Trans. Nucl. Sci., NS-32(1):1036-1040 (1985).

2. J. A. Bernard, K. S. Kwok, D. D. Lanning, A. F. Henry, and J. E. Meyer, Application of the Reactivity Constraint Approach to the Transient Control of Spacecraft Reactors, Proc. Fourth Symp. Space Nuclear Power Systems, Albuquerque, NM (Jan. 1987).

3. J. A. Bernard, Derivation of the Standard Dynamic Period Equation, Trans. Am. Nucl. Soc., 55 (1987).

4. J. A. Bernard, A. F. Henry, and D. D. Lanning, The Design and Experimental Evaluation of a Closed-Loop Digital Controller Based on an Alternate Formulation of the Dynamic Period Equation, Proc. Topical Mtg. Am. Nucl. Soc. Reactor Physics and Safety, Saratoga Springs, NY, 610-621 (Sept. 1986).

5. J. A. Bernard, Evaluation of 'Constant Period' Control Laws for the Time-Optimal Control of Reactor Power, IEEE Trans. Nucl. Sci., NS-35(1) (1988).

6. J. A. Bernard, K. S. Kwok, R. S. Ornedo, D. D. Lanning, and J. H. Hopps, The Application of Digital Technology to the Control of Reactor Power: A Review of the MIT Reactor Experiments, Proc. Sixth Power Plant Dynamics, Control, and Testing Symp., Knoxville, TN (April 1986).

A COMPUTER CONTROL SYSTEM FOR A RESEARCH REACTOR

Kevan C. Crawford and Gary M. Sandquist

University of Utah
Salt Lake City, Utah

Introduction

Monitoring the complex prossesses within a nuclear reactor facility places a tremendous burden on the reactor operators. The difficulty of operation varies widely depending on the reactor application. While control details vary from one application to another, problems with the man/machine interface are manifest in all applications. It has long been recognized that extensive computerized automation and control would reduce many of these problems. [1] In recent years, computers have been placed in many new facets of reactor control and management. [2]

Most reactor applications until now, have not required computer control of core output. Commercial reactors are generally operated at a constant power output to provide baseline power. However, if commercial reactor cores are to become load following over a wide range, then centralized digital computer control will be required to make the entire facility respond as a single unit to continual changes in power demand. Navy and research reactors are much smaller and simpler and are operated at constant power levels as required, without concern for the number of operators required to operate the facility. For navy reactors, centralized digital computer control may provide space savings and reduced personnel requirements. Computer control will offer research reactors versatility to efficiently change a system to develop new ideas. The operation of any reactor facility would be enhanced by a controller that does not panic and is continually monitoring all facility parameters. Eventually, very sophisticated computer control systems may be developed which will sense operational problems, diagnose the problem, and depending on the severity of the problem, immediately activate safety systems or consult with operators before taking action.

Another application of nuclear reactors is the spaced-based reactor currently being developed. [3] For this application, current manual control methods are unacceptable because of extreme volume and weight limitations as well as the apparent requirement of no human interaction. Past space applications have included low power reactors that were run at a constant power. However, waste heat disposal problems dramatically increase as the power demand increases and power requirements may change from low levels to extremely high levels and back again in very short periods of time, making load following essential. In this case, centralized digital computer control is a necessity.

Centralized digital computer control of reactor power and safety systems, however, has not been readily accepted by all members of the nuclear industry. Based on experience with older electronic components and designs, skeptics have questioned the reliability of increasingly more complicated electronic circuitry. However, as electronics have progressed away from vacuum tubes and mechanical relays, the probability of failure has been greatly reduced. Still, others seem more concerned with the reliability of the software. But as long as the software has been tested and approved under all foreseeable circumstances, just the same as standard and emergency procedures for current control systems, the software should not fail.

Some facilities have decided to implement limited automated control of absorber rods. A Canadian commercial power application uses distributed controllers to manage the flux profile to prevent localized "hot spots" from occurring. [4] This application, however, does not control power output and does not have input from the scram systems. Another application of computer control is the distributed digital computer control of the transient rod of the Annular Core Research Reactor at Sandia Laboratories. [5] This project utilized the transient rod position as feedback to a processor which controlled rod withdrawal. Rod withdrawal schemes were designed and manually entered into the processor prior to a power pulse. During execution of the pulse, the processor controls rod withdrawal rates. The limitations of the system include control on a single rod, no power feedback control and no scram capability.

The Nuclear Engineering Laboratory at the University of Utah has developed a centralized digital computer control system for the 100 Kilowatt TRIGA research reactor located on campus. While the research reactor application is relatively simple compared to other applications, it is also ideal for the initial development and testing of automated control.

An International Business Machines (IBM) Series/1 minicomputer monitors and displays physical parameters, executes operator rod movement and scram commands, performs safety limit checks to ensure safe operation of the reactor, and utilizes linearized inverse reactor kinetics equations for steady state power control. Reactor control also stores current values of reactor physical parameters on hard disk. Additional software systems have been developed to perform rod reactivity and thermal power calibrations, control tank water demineralization and cooling, manage laboratory security, and control pneumatic sample transfer. The system employs no distributed controllers and has direct control of absorber rod movements and rod magnet scrams. Software and hardware checks are built into the system as well as manual overrides on rod movement and control rod magnet current. The control system is retrieved from the system disk where the channel calibration settings are retained. With strict access control, the disk settings are far less likely to be accidentally or intentionally destroyed than easily accessible hardware calibration points. This method of system storage also provides for the capability to develop automatic software self-correction during operation. The present control system allows for manual updates of the calibration points during regular rod reactivity, thermocouple, and thermal power calibrations required by the UUNEL technical specifications.

This paper is primarily concerned with the application of elementary control concepts to digital computer control of a nuclear reactor. These concepts include data acquisition and display, data/setpoint comparison, and scram decisions. While application of these concepts seems trivial, available documented evidence of a complete application currently does not exist. Therefore, the primary contribution of this work is to provide, for the first time, a working hardware and software design that combines elementary control techniques and reactor control philosophy to improve reactor output control. In addition, automatic control techniques are presented and investigated. With this design, more sophisticated software control techniques for reactor output may be developed which will significantly reduce human interaction with the control system in some applications, and completely eliminate human interaction with the control system in other applications.

<u>Reactor Control Theory</u>

The control of nuclear reactors has always followed conservative time-tested philosophies. Nuclear reactor control is perhaps unique because it is a combination of quite different classical operational philosophies known as the conventional power station control philosophy and the airplane control philosophy. [6] The emphasis of the control theory of this work is the automatic control philosophy which is implemented using the classical operational philosophies as a foundation.

The conventional power station control philosophy is characterized by the disconnection of expensive equipment and shutdown of the entire plant upon detection of a significant component failure. Under the airplane control philosophy philosophy, the occurrence of a significant component failure should not require the shutdown of any or all safety systems which are necessary for continued operation.

The software features a design which incorporates self-checking modules. Since the operating system is multitasking, the system was divided into modules of similar function. This allows for faster program development as well as provides a method for software self-checking. The monitoring modules sense physical parameters detected by their hardware counterparts and compare the values to setpoints. These modules make the scram decisions and pass the information to the Rod Control Module. The Rod Control Module will receive the scram conditions and either continue normal operation or cut the control rod magnet current. Additionally, the Rod Control Module senses conditions on the control rod drives as well as interacts with the operator by means of a control panel to provide manual initiation of rod movements. The hardware counterparts of the Rod Movement Module have redundancy as well as a "fail safe" design. The movement of a rod requires that a relay switch to the "off" position. This type of design provides a dynamic brake for the rod drive, but more importantly, does not allow unpredictable results when a rod drive relay fails. Since solid state relays fail in the "on" position a relay failure would result in the inability to move the rod. Since rod movement occurs during a minute fraction of the operating time, the use of solid state relays which fail in the "on" position is preferred to the conventional relay system which would fail in the "off" position and result in the uncontrolled drive of an absorber rod. If the reactor operator observes that a rod does not respond to drive commands, then the operator may use another rod drive to safely control the reactor.

The Master Module serves as the system supervisor. If the Master Module does not receive a completion code from any one of the other modules, the Master Module assumes the system has ceased to function properly and will then provide immediate and direct scram capabilities.

Since the operational philosophies have a higher priority than automatic control, they determine when automatic control may be utilized. Unlike the operational philosophies, automatic control is not concerned with component failures. The object of this philosophy is the movement of absorber rods to produce a requested power. When the current power level is compared to the requested power level, an error may be detected which will require the movement of a control rod to correct the error.

The automatic control philosophy adds another dimension to the classical operational philosophies. Using this philosophy, normal operator interactions with the computer would be limited only to run parameters such as power, period, and run time. This eliminates subjective judgements, objective judgements under pressure, and distractions to the operator and insures the reactor will be operated in a safe and controlled manner as well as providing reproducible operations.

Before a control scheme can be designed, the control system must be defined. A block diagram for the control system is shown in Figure 1. The first transfer function of the forward loop transfer function is the operation performed on the error by the computer controller to produce the output signal. The reactivity is input into the second transfer function of the forward loop, the reactor function. The output is the actual power and serves as the feedback signal which is compared directly to the requested power.

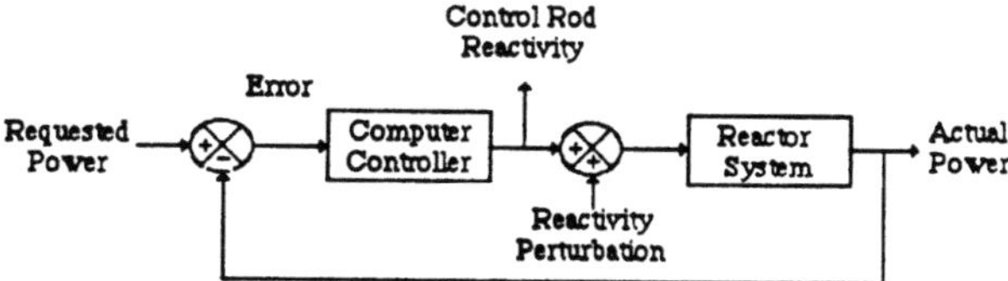

Figure 1. Feedback Control Diagram.

Since the computer controller responds to reactivity perturbations generated in the nuclear reactor, characteristics of the nuclear system must first be modelled. Exhaustive analyses of this function exist. [7] When the equation that governs nuclear reactor kinetics was derived, the time dependent and space independent neutron and energy balance equations were used. The equations for a research reactor include prompt neutron effects, the effects of six delayed groups of neutrons, a source term, and temperature effects on the system reactivity. [8,9]

Reactivity may be inserted into a reactor through a number of mechanisms. Some of the mechanisms, such as fission product buildup and moderator and coolant effects, may be eliminated from consideration because of their long time response with respect to the computer sampling interval. The void coefficient of reactivity and effects from chemical shims will not be considered because they are not present in most research reactor systems. The reactivity can then be described by the summation of fuel temperature effects and absorber rod movement.

However, while the behavior of the nuclear system is the foundation for understanding and developing a practical control scheme, the computer controller is the focus of this work. From the computer controller's point of view, the input variable is the power and the output variable is the reactivity. The computer has the ability to perform the comparison of the actual power to the requested power to produce an error signal. Therefore, the comparison will be performed inside the computer transfer function.

If the computer were to have an exact model of the reactor, it would maintain running accounts of the delayed neutron and prompt neutron concentrations as well as the temperatures of all influential regions. However, because it is impractically for the computer to know all the neutron concentrations, an approximation may be assumed. The computer will estimate the state of criticality based on the data collected at the last two sampling intervals, point 0 and point 1 of Figure 2. Knowing the condition of criticality, the computer may take no action and leave the reactor in a supercritical condition or the computer may return the reactor to exactly critical and leave the reactor at an elevated power level or the computer may wish to return the power to the requested level by reversing the perturbation and making the reactor subcritical.

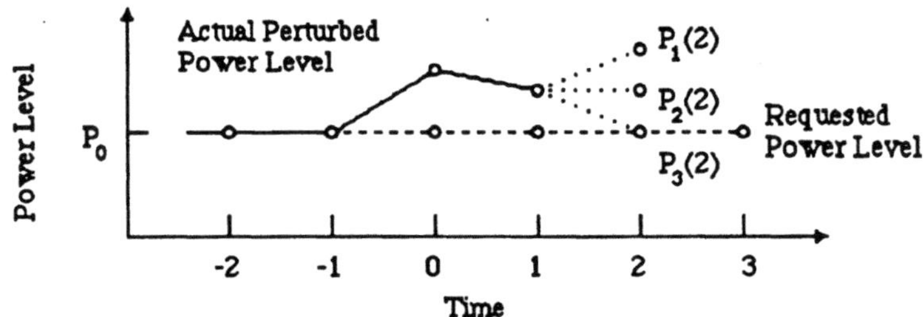

Figure 2.   General Power Perturbation.

The resulting reactivity correction from the perturbation shown in Figure 2 is described by the following equation.

$$\Delta\rho_c = \frac{1}{P_0}\{P_0 - P(1) + \frac{P_0 - 2P(1) + P(0)}{\Lambda\tau} +$$

$$\sum_i \alpha_i^*[(P(0) - P(1))(1 - e^{-k_i^*\tau})(1 - \frac{1}{k_i^*\tau}(1 - e^{-k_i^*\tau})) +$$

$$(P(1) - P_0)(e^{-k_i^*\tau} - \frac{1}{k_i^*\tau}(1 - e^{-k_i^*\tau})) + (P_0 - P(0))(1 - e^{-k_i^*\tau})^2]\} . \quad (1)$$

If the time required to collect all data is much longer than the time required to calculate the next control rod reactivity change, then the complexity of equation (1) does not decrease the system performance. For the computer controller at UUNEL, the calculation time is insignificant in comparison to the time required for sampling, filtering, and converting all analog and digital inputs. However, if the sampling time is reduced through improved interfacing, then the possibility exists where the computation time may become the major source of performance degradation. To reduce the computation time, the power at sampling point 0 may be assumed to be at the requested power as demonstrated in Figure 3. This assumption sacrifices the accuracy of the value of the reactivity between sampling points 0 and 1, but will significantly reduce the computation time. Equation (1) then reduces to the following equation.

$$\Delta\rho_c = \frac{P_0 - P(1)}{P_0}\{1 + \frac{2}{\Lambda\tau} + \sum_i \alpha_i^*[1 - \frac{1 - e^{-k_i^*\tau}}{k_i^*\tau}(2 - e^{-k_i^*\tau})]\}\ . \tag{2}$$

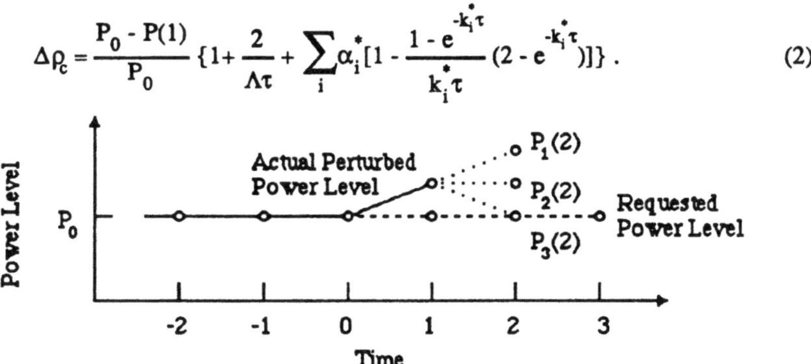

Figure 3. Simple Power Perturbation.

If the control scheme includes the ability to change power levels during operation, then the possibility may exist for the computer to place the reactor in a prompt critical condition. This would be caused by large power errors, the inability of the rod drives to move large amounts of reactivity, and the time delay of the reactor response. The problem may be reduced by increasing the rate of reactivity movement or by not allowing large power errors to appear to the computer transfer function. If the rod drive mechanisms are adequate for the range of reactor applications, then software change is the simplest solution and will require the comparison of the reactor period to a safe limiting value.

Controlling the positive reactor period by using the reactivity as the variable to be compared and controlled requires many computational and accounting procedures. Therefore, to keep computer computations to a minimum, a more direct method may be used. The ratio of $P(2)$ to $P(1)$ over a constant sampling interval should not exceed a value corresponding to the minimum period requested by the operator. If a particular minimum positive period is set at run time or prior to run time, the computer will convert the requested minimum allowable period to a ratio as follows:

$$\frac{P'(2)}{P'(1)} = \exp\ [\frac{\Delta t}{T_s}]\ . \tag{3}$$

After each sampling cycle, the computer will determine the ratio of $P_0(2)$ to $P(1)$. If the ratio exceeds the ratio of equation (3), then the value of $P_0(2)$ must be temporarily changed by using the following equation.

$$P(2) = \frac{P'(2)}{P'(1)}\ P(1)\ .$$

The reactor control equation (1) may then be used to calculate the reactivity change necessary to attain the desired power level without exceeding the requested minimum reactor period.

Very small reactivity changes also present a problem. As the automatic control scheme stabilizes the actual power at the level of the requested power, reactivity change requests will become very small. The computer will attempt to make the change request based on differential reactivity worth curves for the control rod and data collected from the rod position indicator circuits. The computer, however, will compute rod positions distributed with a characteristic deviation around an average value. If the deviation becomes relatively large in comparison to the requested change in rod position, then the computer will not be able to detect the change of rod position. In this case, the two solutions currently available is for the computer to make an estimate of rod position change based on the average value of the rod position and the rate of rod withdrawal or to allow for a larger margin of power error before the controller takes action. Another solution not currently available is to redesign the rod position indicator circuits to collect data with a higher precision.

Another problem to consider is the delay in the control rod movement generated by the rod drive motor. If the delay is approximately the same magnitude as the sampling interval, then the delay may cause instabilities in the system. While the control rod drive response time has not been measured, it appears to be less than one tenth the sampling interval. However, this problem should be reconsidered when reducing the sampling interval.

The last problem considered here is the ability of the system to compensate for spurious electronic transients. There are two basic approaches to suppress transients: the hardware approach and the software approach. If the hardware is unable to suppress the transient, then the digital filters in the software will act on the transient. First, an average data value is calculated from several raw data points acquired during the sampling cycle. The average value is weighted with the previous running value. While this method increases the response time of the instrumentation, an unusual data point will have very little effect on control variables.

It is important to remember that the control scheme presented is appropriate for research reactors, particularly the TRIGA reactor at the University of Utah. Other reactor applications may contain design or operational features which may include the effects of additional physical parameters. Power reactors may wish to include coolant temperature and void coefficients of reactivity. The requested power level for a commercial power reactor could be derived from the generator load, providing a load following capability which currently does not exist in these reactors. Pulsing reactors may wish to include a second order fuel temperature coefficient of reactivity as well as ramp reactivity insertion. This scheme only accounts for six delayed group of neutrons and the a first order fuel temperature coefficient of reactivity.

The experimental facilities developed for this project provide the foundation for further development of sophisticated control schemes. Using the control equations that approximate the physical system, control systems may be developed which will, in some cases, place the operator another step away from the reactor system, and in other cases, may completely eliminate the need for human operators.

Evaluation

Operating the reactor without previously testing the control system is unacceptable. However, the complete behavior of the automatic power level control will remain unknown until the entire system is installed. Therefore, the automatic control philosophies presented should be evaluated in a manner similar to the evaluation of the operational control philosophies. Experiments could be designed to simulate the complete system or could be designed to test portions of the system while operating the reactor from the old console.

Two major items should be addressed by the automatic control philosophy evaluation. First, the derivative term in equations (1) and (2) may cause the control scheme to overreact if the power signal contains too much noise. Since the signal noise is affected by the sensors, interfaces and digital filters, the evaluation should determine if the hardware and software are compatible with the control scheme. Additionally, the evaluation should also determine if the closed loop feedback control scheme is stable and how the system reacts in time, and should provide a comparison of error measurement to the uncontrolled system.

An experiment was designed and performed to examine the system hardware and software compatibility with the control schemes. While operating the reactor near maximum power where the fuel temperature is affected by the power level, ten grams of cadmium placed in the central irradiator of the core was oscillated with an amplitude of one inch at periods ranging from 2 seconds to 20 seconds to induce reactivity perturbations. A program was activated on the control computer which read the core power output from an ion chamber located near the core. The Weiner digital filter was used to average 150 data points and provide a discretely sampled power level to an automatic control subroutine. [10] The subroutine calculated the control rod reactivity change required to bring the power to a requested power level by the next sampling interval. Both equations (1) and (2) were substituted into the subroutine. The output of the open loop control scheme experiment was the current suggested control rod reactivity. The results indicated that the data was sufficiently smooth for the two control methods presented to accurately predict the reactivity perturbation in magnitude and phase with respect to the power oscillation.

The control rod position for all of the periods tested on both control equations appears to be bounded around an average value corresponding to the unperturbed critical condition.

These observations were performed over periods lasting approximately 90 seconds.

The only method of testing a closed loop control scheme prior to full installation of the system is to mathematically model the system on the computer. The simulation starts with the reactor power at the requested power. The reactor behavior is modelled using the differential reactor kinetics equations with an additional term is added to describe a reactivity perturbation from an external source. These differential equations are used to calculate the difference in neutron populations and fuel temperature for very small time steps. The difference in each value is added to the values of the previous time step to determine the new value.

The external reactivity perturbation is changed for each small time step within a sampling interval, to simulate a smooth reactivity change. The external reactivity follows a sine function with the amplitude and period determined before each run. The sine function was chosen because any shape reactivity perturbation may be simulated with a Fourier series.

At the end of a sampling interval, the average value for the preceding interval is used in the control equation to determine the magnitude and direction of the control rod reactivity change to be executed before the next sampling interval begins. After the control reactivity has been incremented, the program returns to the stepping procedure used to simulate the reactor.

Before the closed loop computer simulation was attempted, an open loop simulation of the first test was conducted to compare the computer simulation with the actual reactor system. The simulations routines were verified to quite accurately describe the the reactor response.

The next series of numerical experiments was designed to evaluate three groups of reactor parameters under the conditions of external perturbation in the closed loop feedback control system. Long lived delayed neutron and fuel temperature effects were tested with an external perturbation period of 10 seconds. Short lived delayed neutrons were tested with an oscillation period of 0.10 second and prompt neutrons were tested with an oscillation period of 0.001 second. Two or three reasonable sampling intervals were used for each test. Sampling intervals ranged from 0.001 of the perturbation period to 0.1 of the perturbation period. Four control schemes, including equations (1) and (2) and their versions without delayed neutron and temperature effects were compared.

Figures 4, 5, 6, and 7 are representative of the results of these experiments. All control schemes were shown to be stable under all of the conditions tested, by observing the power and reactivity behavior of at least 50 reactivity perturbation periods. The abbreviated control schemes performed similarly to their unabridged versions for sampling intervals less than one second. The automatic control schemes reduced the mean error by better than 50 percent for 10 second perturbation periods and 75 percent for 0.1 second perturbation periods.

In comparing the control schemes, two items should be considered, the mean power error, or the square root of the mean squared signal error (RMS), and the time needed to return the system to equilibrium after the initial perturbation.

The performance of equation (1) is graphically illustrated in Figure 8. These data are representative of the other three control equations. Directly comparing equilibrium RMS values, equation (2) consistently obtained RMS values 15 percent lower than the RMS values of equation (1) for all of the conditions tested. However, when comparing the time response to the initial perturbation by observing the integral of the mean error, equation (1) returns the system to equilibrium faster than equation (2). The integral over the first five sampling intervals is smaller for equation (1) than for equation (2).

Unusual power behavior for control equation (2) is observed to begin between reactivity oscillation periods of 1.0 second and 0.1 second. The unusual behavior is shown in Figure 7. While this behavior may be undesirable, it does not indicate instability or larger mean power error. This behavior continues in shorter reactivity periods for equation (2).

While the experiments were designed to test the performance range of several automatic control schemes, most observable research reactor perturbations are either step reactivity changes from experiment movements or long period oscillations from coolant temperature changes. For both reactivity insertion formats, it may be concluded that the automatic control equations (1) and (2) perform similarly. Equation (2) performs slightly better than equation (1) by producing lower RMS signal errors at equilibrium for oscillating reactivity perturbations. Equation (1) performs slightly better than equation (2) by bringing the reactor system to equilibrium faster for the initial oscillating reactivity perturbation. Therefore, another criterion may be used to compare the control schemes.

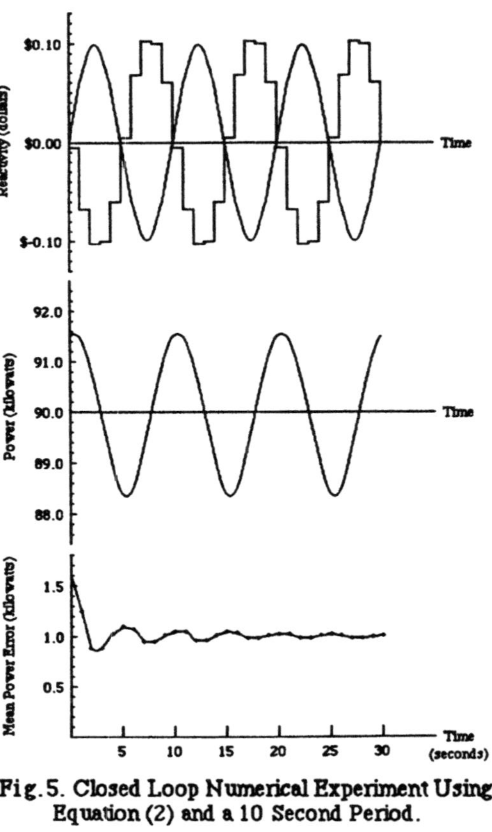

Fig.4. Closed Loop Numerical Experiment Using Equation (1) and a 10 Second Period.

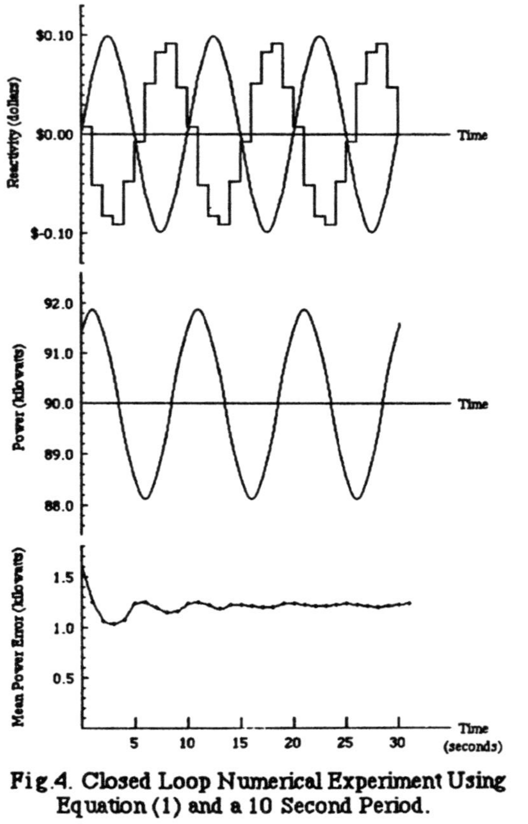

Fig.5. Closed Loop Numerical Experiment Using Equation (2) and a 10 Second Period.

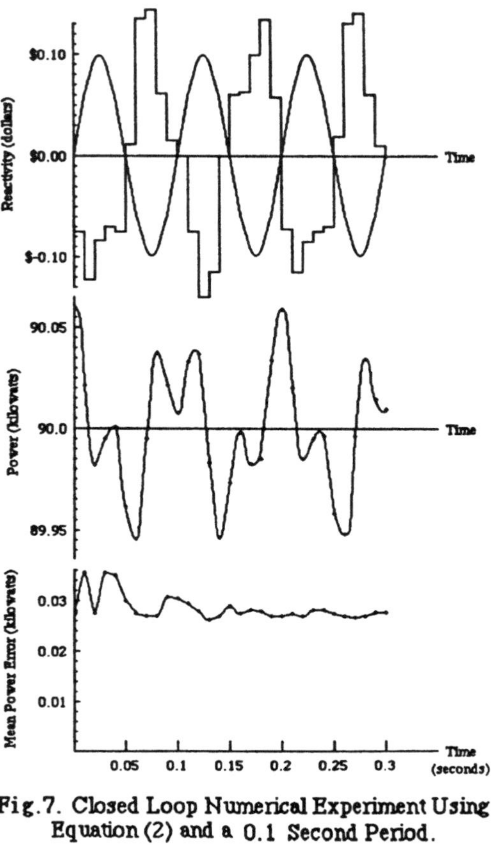

Fig.6. Closed Loop Numerical Experiment Using
Equation (1) and a 0.1 Second Period.

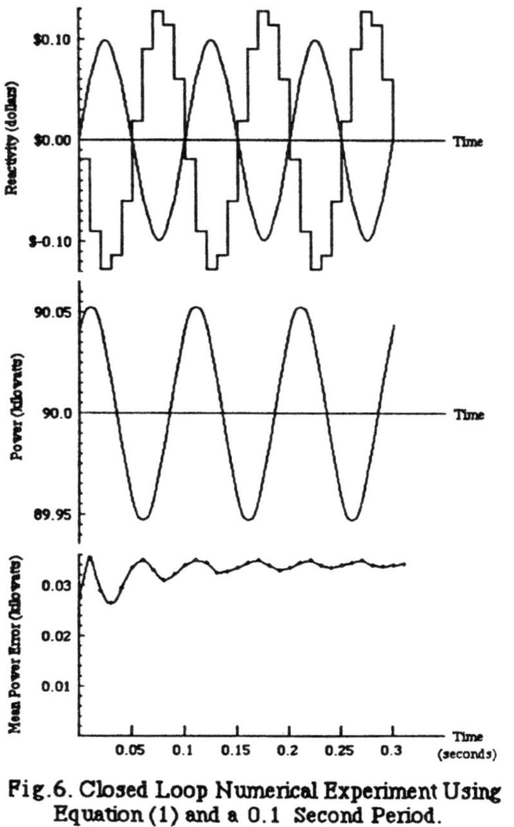

Fig.7. Closed Loop Numerical Experiment Using
Equation (2) and a 0.1 Second Period.

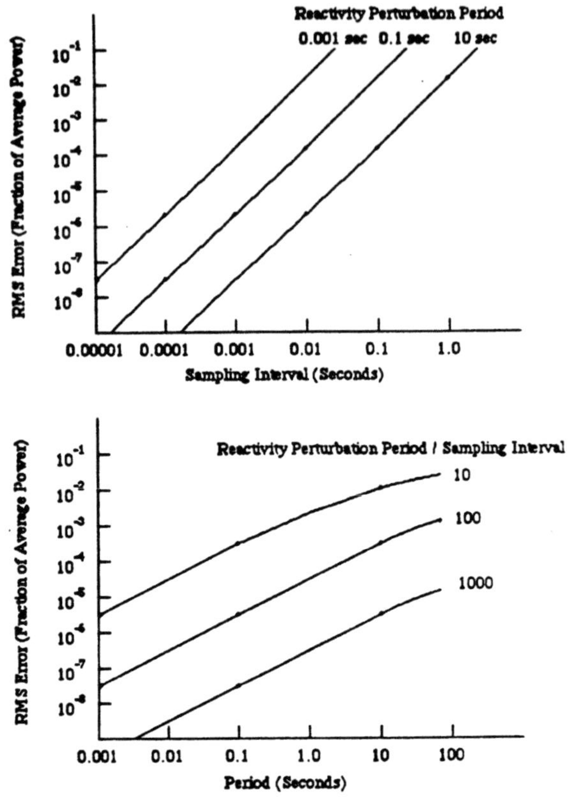

Figure 8. RMS Error Performance Evaluation for Equation (1).

Figure 8 demonstrates the value of short sampling intervals. If control rod reactivity calculations, such as equation (1), require a significant percentage of the sampling interval time, then the abbreviated versions of the control equations, such as equation (2), are desirable. However, if reactivity calculations are an insignificant portion of the sampling interval, then perhaps the most efficient method of reducing the power error is to acquire faster hardware or to streamline data collection techniques.

<u>References</u>

1. Oakes, L. C., "Automation of Reactor Control and Safety Systems at ORNL," <u>Nuclear Safety</u>, Vol. 11, No. 2, March-April 1970, pp. 115-118.
2. Shukla, J. N., and Iubelt, J. A., "Computer Based Systems in Boiling Water Reactors," <u>Comput. Soc.</u>, Vol. 10, Nos. 3-4, Spring-Summer 1980, pp. 21-26.
3. Marshall, E., "DOE's Way-Out Reactors," <u>Science</u>, Vol. 231, p. 1357, March 1986.
4. Hinchley, E., and Kugler, G., "On-line Control of the CANDU-PHW Power Distribution," Report No. 5045, Atomic Energy of Canada Ltd., March 1975.
5. Boldt, K. R., Sullivan, W. H., and Kefauver, H. L., "Description and Characterization of the ACRR's Programmable Transient Rod Withdrawal Mode," <u>Seventh Biennial U.S. TRIGA Users' Conference</u>, General Atomic Company, San Diego, March 1980.
6. Schultz, M. A., <u>Control of Nuclear Reactors and Power Plants</u>, Second Edition, McGraw-Hill Book Co., Inc., New York, 1961.
7. Hetrick, D. L., <u>Dynamics of Nuclear Reactors</u>, University of Chicago Press, Chicago, Ill., 1971.
8. Sandquist, G. M., "General Solution of Reactor Kinetic Equations for One

Group of Delayed Neutrons and Time Dependent Reactivities," <u>Journal of Nuclear Energy</u>, Vol. 26, p. 213, 1972.

9. Bell, G. I., and Glasstone, S., <u>Nuclear Reactor Theory</u>, Van Nostrand Reinhold Co., New York, 1970.

10. Brown, R. G., <u>Introduction to Random Signal Analysis and Kalman Filtering</u>, John Wiley and Sons Inc., New York, 1983.

STRUCTURE OF CONTROLLERS FOR THE DIRECT DIGITAL CONTROL OF REACTOR POWER:

LICENSING IMPLICATIONS

John A. Bernard

MIT Nuclear Reactor Laboratory
Massachusetts Institute of Technology
Cambridge, MA 02139

ABSTRACT

The structure of controllers for the direct digital control of power in
a nuclear reactor is examined and features possibly important to licensing
are identified. This was done by first reviewing the human approach to process control and then discussing designs whereby digital systems could be
made to emulate each of the functions that a human operator now performs.
Features identified as relevant to licensing include the separation of the
safety and control systems, the use of supervisory algorithms that incorporate the concept of 'feasibility of control', signal validation, verification of signal implementation, and the use of either fault-tolerant or
redundant hardware. Examples supporting the above findings are drawn from
the MIT experience in first designing and then licensing a digital, closed-
loop controller for the 5 MWt MIT Research Reactor.

INTRODUCTION

Digital technology is increasingly being recognized as a means of improving the reliability, efficiency, and safety of industrial processes.
The nuclear industry, which historically has been slow to embrace this
approach, is now pursuing the digital option with vigor [1,2]. There are
many reasons for this trend. First, nuclear plants are increasingly becoming the major source on a given power grid and must therefore load-follow.
Maintenance of thermal limits and power distributions in the presence of the
resulting complex rod patterns necessitates the use of real-time digital
controls. Second, improvements in the structure of software, the advent of
fault-tolerant hardware, and stunning increases in computer memory have made
practical the implementation of distributed digital control systems. Third,
current planning envisions the construction of clusters of medium-sized,
modular reactors. Closed-loop control will be a necessity if the cost of
operating such facilities is to be contained. Lastly, the increased use of
digital technology would improve the man-machine interface by allowing
licensed operators to monitor the overall plant without having to manipulate
it as well. Canada, and now Japan, are using this approach to advantage.

While the potential benefits of digital technology are now recognized,
the necessary guidance on the application of digital systems to plant operation and control is, in large measure, yet to be developed. This paper

addresses that question. Specifically, it examines the approach utilized by
humans to formulate control decisions and then suggests techniques whereby
closed-loop digital systems could be structured in order to achieve a similar or greater degree of reliability. As such, this paper draws heavily on
the results of an on-going program at the Massachusetts Institute of Technology (MIT) to develop and apply advanced instrumentation and control
methodologies to nuclear reactors. Among this program's achievements are
techniques for signal validation using the parity space approach, a supervisory controller for reactor power, methods for the on-line reconfiguration
of control laws, and a rule-based controller [3-6]. Each of these technologies has been successfully demonstrated on the 5 MWt MIT Research Reactor
(MITR-II). The supervisory controller, which is designated as the MIT-CSDL
Non-Linear Digital Controller or NLDC, is of particular interest to the subject of this paper because it was licensed by the U.S. Nuclear Regulatory
Commission for general use on the MITR-II in April 1985.

The specific objectives of this paper are (1) to describe the approach
that human operators follow in formulating control decisions, (2) to examine
the status of digital control techniques with emphasis on the extent to
which the human approach is emulated, (3) to suggest a possible generic
structure for a digital controller, and (4) to discuss the acceptance of
such systems in terms of regulatory criteria. The basic premise is that
closed-loop digital controllers may be licensable if they are structured so
as to provide each of the functions that human operators now perform when
using manual control.

HUMAN APPROACH TO PROCESS CONTROL

A study was recently completed in which the licensed operators of the 5
MWt MIT Research Reactor were asked to describe the process by which they
adjusted the reactor's power [6]. These findings and the results of other
more extensive studies such as those by Sheridan [7] and Kelly [8] suggest
that the human approach to process control entails four subtasks. These are
planning, prediction, implementation, and assessment. Figure One is a schematic of the overall process. The planning phase involves determination of
the desired system response. This is accomplished by first noting the operational objectives and then determining the desired plant state and the most
efficient means for achieving it given the confines of approved procedures.
Digital technology including methods for procedure prompting and automatic
searches for technical specification requirements is currently available to
assist the operators of nuclear plants with the planning process. Closed-
loop digital controllers, as discussed here, would not supplant the
operator's role in planning. Rather, they would be used to improve his or
her ability to implement a particular strategy.

The predictive subtask is perhaps the most important because an operator is capable of controlling a complex plant in proportion to his or her
ability to anticipate the effect that any of the available control options
will have on the plant's behavior. Operators form concise mental models of
the plant's dynamics and then use these models to preplan control actions.
These models are composites of the operator's training including theoretical
understanding, drills, and actual experience. Using such a model and knowing both the plant's current status and any information on trends, the operator estimates the future behavior of the system. This predicted response
is then compared to the desired one. Adjustments to the control signal are
made only if it appears that the response of the plant will not be as
projected. As a result, operators generally achieve excellent control with
only a few manipulations of the control mechanisms.

It is illuminating to examine the nature and use made of the concise

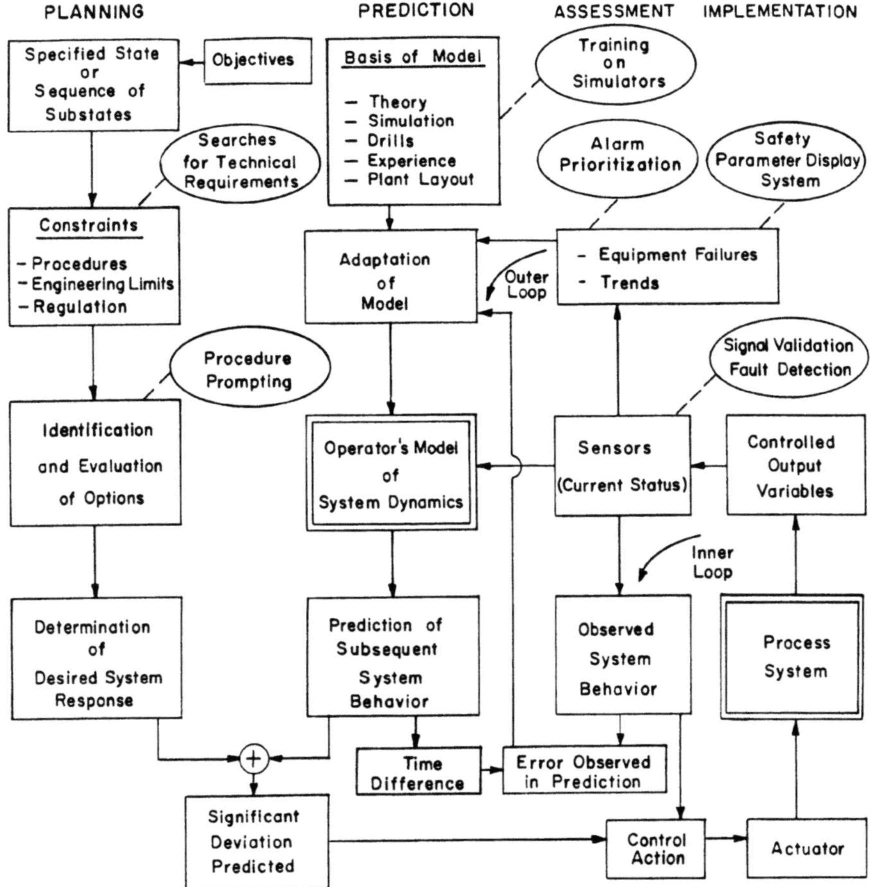

Figure 1. Model of Human Approach to Process Control

mental models. Operators do not predict the exact future state of a plant. Rather, they identify those conditions under which they will be able to achieve their objectives. For example, it was found from the study of the MIT Research Reactor's operators that a model might consist of the realization that insertion of one of the reactor's control mechanisms at or below a certain position will be sufficient to terminate all transients for which the rate of rise of power was within a specific bound. Models are therefore used both to project plant response and to define limiting conditions for operation. This realization is crucial to the structuring of digital controllers for safety-constrained processes such as nuclear reactors.

Implementation of the specified control action often requires the simultaneous application of signals to several plant subsystems. This process is often entirely automated.

The assessment phase of the control process is the most complex of the four subtasks. It is characterized by inner and outer feedback loops. The inner loop is the simplest. Feedback from the controlled variable is used to ascertain that the required control action is being implemented. The

outer loop fulfills two functions. First, it provides information for use
with the model. Data on plant trends and on equipment availability are preferred to mere status information because the former facilitate prediction.
Second, this outer loop permits operators to evaluate their understanding of
the plant. Specifically, if their estimates of the controlled parameters
are in error, then their model may require revision.

In addition to identifying the functional aspects of an operator's job,
it is important to categorize the manner in which he or she performs that
job. Observations of the licensed MITR-II operators showed that they generally relied on only a few sensors which they scanned no more than once every
few seconds. They verified the correctness of their control decisions by
comparing these sensor readings with their expectations of the plant's
response. Changes to the control signal were made only in response to
genuine needs. For example, effects such as high frequency noise were
ignored. In general, the more skilled the operator, the more he or she
functioned by recognizing the appropriate patterns of actions and responses.
Analytic skills were normally only used if the operator was forced to deal
with situations that were not frequently encountered. In summary, operators
rely on pattern recognition techniques rather than on numerical or analytic
skills. Additional information is given in [9].

ASSESSMENT OF DIGITAL TECHNOLOGY FOR CONTROL

Figure Two shows four schematics that represent successive developments
in the technology of closed-loop controllers. Uppermost is a manual system.
The loop is closed only in the sense that the operator adjusts the control
device in response to the measured value of the output. Next is a simple
feedback system typical of many of the analog (and some digital) controllers
that are now in service. The output of the process is compared to a
reference value and the resulting error transmitted to the controller which
in turn drives the actuator so that the plant output is forced to the specified value. Such systems are referred to as P-I-D controllers because the
control signal is most effective if it is proportional to both the derivative and the integral of the error as well as to the error itself. The
design of such systems requires that both the controller and the plant be
represented using transfer functions. These functions are in fact models
which are used to predict response to a step change in demand. The P-I-D
approach reduces the sensitivity of the plant to distrubances and relieves
the operator of the tedium of continuously adjusting the actuator. Its disadvantage is that the technique is applicable only to linear time-invariant
systems having a single input and output. The third schematic is an
application of state analysis. Each variable in both the controller and the
plant is fed back via its own gain coefficient. The result is a flexible
design that permits controller response to be optimized in terms of a performance index. Also, systems with multiple inputs and outputs can be
treated. This technique is dependent upon the existance of accurate plant
models. State analysis is widely used, especially in the aerospace industry. It is, however, restricted to linear systems. The fourth schematic
depicts model-algorithmic control which is a recently developed method
applicable to non-linear systems. The output of a mathematical model and
that of the actual plant are compared and the difference is used to produce
a control signal that drives the plant to the desired state.

This brief examination of digital closed-loop controllers shows that
the trend is for an increased use of detailed models as a means of improving
accuracy, optimizing response, and accommodating non-linear systems. One
aspect of the predictive portion of the control process is therefore being
addressed. However, if the overall performance of a digital controller is
to equal that of a human operator then the other functions that a human per-

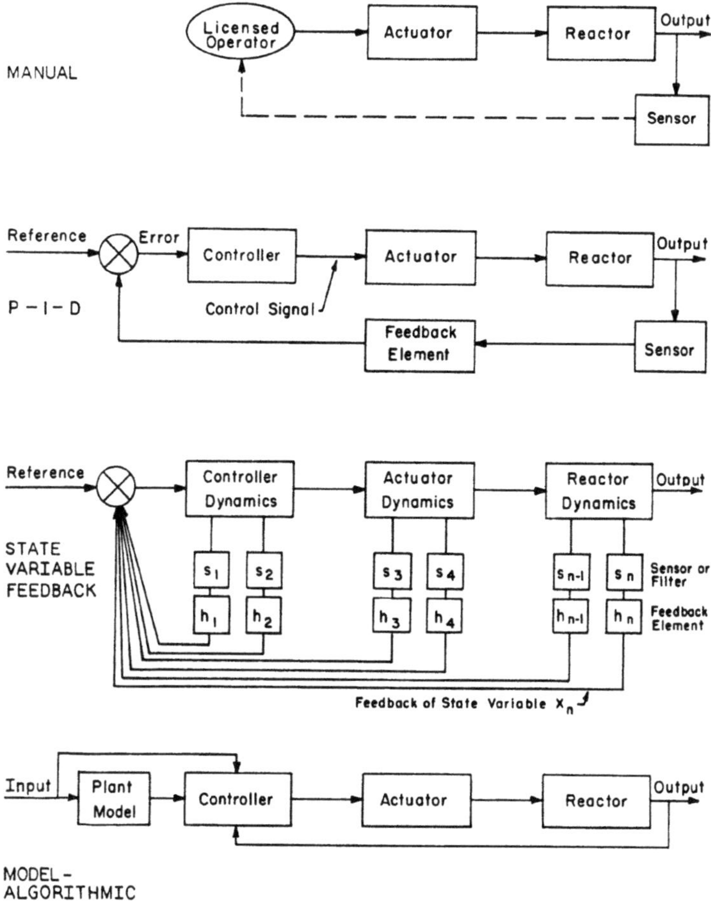

Figure 2. Successive Developments in the Technology
of Closed-Loop Digital Controllers

forms must also be emulated. These include detection of abnormalities by
checking for patterns of consistency, the recognition of limiting conditions
within which control is always feasible, and the assessment of the effects
of a given control action. Also, system reliability must be considered.

Patterns of Consistency

The capability to identify abnormal responses is crucial to achieving
proper control. Operators accomplish this by repeatedly checking for consistency. Are all the sensor readings for a given parameter in agreement?
Is the value of each parameter within the expected range for the given plant
condition? Is the plant's dynamic response as expected? Is the change consistent with the signal that was applied? The need to develop efficient
digital methods for pattern recognition is the subject of much research and
recent advances in both artificial intelligence and parallel processing
architectures are encouraging. Also, as shown in Figure One, digital technologies now exist for certain pattern recognition tasks including signal
validation, alarm prioritization, and safety parameter displays [1,2].

<u>Feasibility of Control</u>

The control algorithm should fulfill two major functions. First, it should specify the desired plant trajectory and, if the actual state of the plant differs from the specified one, generate a feedback signal that reduces the error. Second, it should both define the envelope of conditions under which it will be possible to halt the transient and preclude operation beyond that envelope. It is this latter function in which many proposed algorithms are deficient. The origin of the problem may be that the two functions are often combined and hence the latter one goes unrecognized. For example, many physical processes can be adequately described as second order linear systems. The parameters associated with such a system may be preset so that the degree of overshoot (including no overshoot) can be specified exactly. Hence, for this case, prediction of the desired system trajectory and identification of the conditions under which a transient can be halted are synonymous. However, if a system is of higher order or non-linear or time-delayed, then this will not be true and it will be necessary to define separately the conditions for no overshoots. Relative to nuclear reactors, algorithms which can adjust the power while guaranteeing the avoidance of overshoots are defined as having the property of 'feasibility of control' [4]. This issue is only now being recognized within the control community [10].

<u>Assessment</u>

Plant and component models that can be run faster than real time should be employed to assess plant response. These models could be used to discern between incipient casualties and slowly occurring natural changes. For example, is a gradual loss of condenser vacuum the result of an increasing ambient temperature or an impending air ejector failure? Also these models would provide a method for testing and evaluating proposed control actions prior to actual implementation. Currently, the only use made of mathematical models in digital closed-loop controllers is to determine the desired system trajectory. Real time models of most plant components do exist [11]. Active research is needed to unify these models, to identify methods for their calibration, and to integrate their use within the controller.

<u>Reliability</u>

The equipment used to implement a digital controller for a safety-constrained system should be highly reliable. This can be accomplished through the use of either a single fault-tolerant processor or redundant units operated in parallel. Also, techniques for the on-line reconfiguration of equipment may be of use in increasing overall reliability [5]. As is the case with signal validation, reliability is an achievable goal. Another issue that concerns reliability is the calibration of the mathematical models that have become an integral part of most digital controllers. The validity of these models should be verified at regular intervals.

CONTROLLER STRUCTURE

Figure Three shows the structure of the MIT-CSDL Non-Linear Digital Controller or NLDC. As mentioned earlier, this controller has been licensed for general use on the MIT Research Reactor. Moreover, the basis of this approval was the design of the controller and not a limitation on the reactivity worth of the associated actuator. The NLDC is designed in terms of general principles and may be used to control the neutronic power of any reactor that can be described by space-independent kinetics. This includes research, test, small to mid-size commercial, and spacecraft reactors. Hence, the structure of this digital controller and the experience being

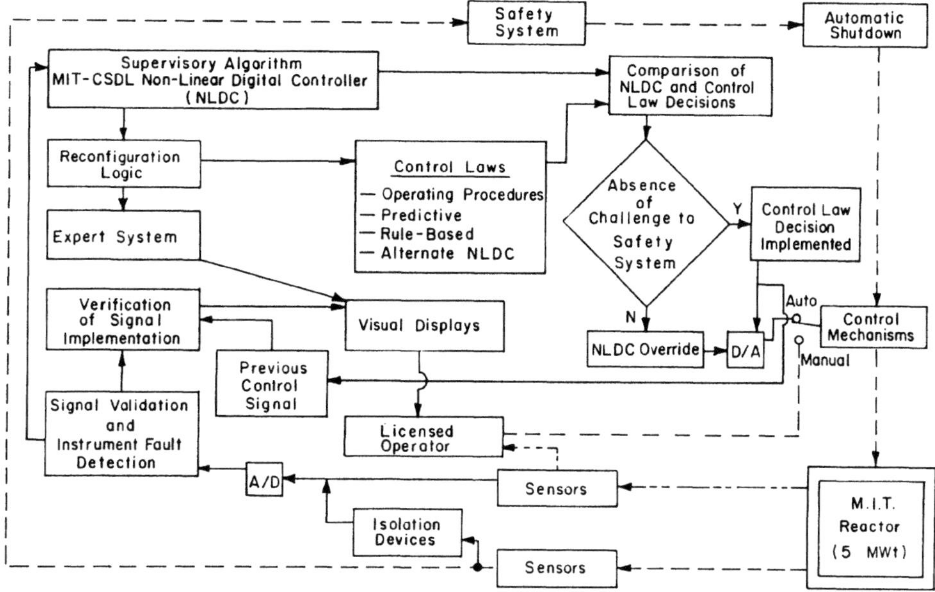

Figure 3. Controller Structure

gained from its use may be relevant to the establishment of standards concerning the application of digital technology to the operation of nuclear reactors. The NLDC's principal structural features are as follows:

Separation of Safety and Control Systems

The nuclear safety system is separate from the closed-loop controller. The work 'separate' is defined as meaning that the output of an instrument used in the safety system must not be influenced by interaction with the control system. Thus, if an instrument is common to both systems, its signal must be passed through an isolation device, such as an optical transformer, to preclude any possibility of feedback from the control system. The purpose of keeping the two systems separate is to insure that the capability of the safety system to perform its intended function will never be compromised.

Signal Validation and Instrument Fault Detection

All sensor information is processed by signal validation and fault detection routines. There are several methods for accomplishing this. The simplest is to verify that each reading is within the range expected for a given plant condition. A more sophisticated approach is to identify the largest consistent subset of signals and reject any that is not a member of that set. A further refinement is to incorporate a real-time system model that generates an analytic value for the measured parameter. Sensor readings are then checked for consistency both with each other and with the calculated value. The NLDC uses this latter method as part of a numerical technique known as the 'parity space approach' [3]. Instrument fault checks are also performed and the weighting factor for each sensor is adjusted in proportion to the frequency with which its readings are judged to be valid. Thus, reliance on a failing sensor is gradually reduced thereby assuring a 'bumpless' transition when complete failure actually occurs.

Supervisory Algorithm and Plant-Specific Control Laws

The controller consists of the supervisory algorithm, a bank of control laws, and a reconfiguration logic. The supervisory algorithm establishes the limiting conditions within which control will always be feasible and thereby guarantees that there will not be a challenge to the safety system as the result of any automatic control action. It does this by requiring a change in the _present_ value of the control signal if conditions are such that an overshoot could occur at some _future_ time. The NLDC uses the 'reactivity constraint approach' which is a technique suitable for reactors governed by a limitation on power level [4]. A different method would be required for a pulsed reactor in which total energy production is limiting or for a large commercial reactor in which both power level and thermal limits are of concern. The control law bank consists of laws suited for each operational mode. The reconfiguration logic selects the one that is most applicable to the current plant condition [5]. The decision of the selected control law and that of the supervisory algorithm are then compared and the more conservative is implemented. This approach has the advantage that it combines a general-purpose, supervisory algorithm that precludes challenges to the safety system with a plant specific control law.

Expert System

Also part of the controller are the rudiments of an expert system. This currently provides, at a very basic level, information to the operator on the rationale for the reconfiguration logic's choice of control law. This portion of the controller will eventually be expanded to provide a means of assessing control law performance and identifying plant trends.

Verification of Signal Implementation

The implementation of the control signal should be verified because there could be a failure in either the computer interface or in the actuators. Verification should be performed in the broadest possible manner. For example, if the controller's decision was to reduce the reactivity, then it should be determined both that a control device was inserted and that the period did actually lengthen.

REGULATORY CONSIDERATIONS

The safety evaluation supporting the general use of the MIT CSDL Non-Linear Digital Controller or NLDC on the MIT Research Reactor addressed the issue of whether or not an unreviewed safety question would be created. Details of that evaluation are given in [12].

An unreviewed safety question is considered to exist if the probability of occurrence or the consequences of an accident or malfunction of equipment important to safety previously evaluated in the facility's safety analysis report is increased, or the possibility for an accident or malfunction of a different type than any evaluated previously is created, or the margin of safety as defined in the basis of any technical specification is reduced. The MITR-II had, as part of its original license, approval to use automatic controllers subject to a limitation on the reactivity worth of the associated absorber. The absence of challenges to the safety system was assured by restrictions on the actuator rather than by requirements on the design of the controller. General use of the NLDC would reverse this philosophy because the absence of challenges to the safety system would be assured by incorporating the concept of 'feasibility of control' in the controller's design. Hence, both this technical specification and its basis required revision. No other unreviewed safety questions were identified.

The basic argument contained in the safety evaluation was that (1) the
safety and control systems would remain separate and the capability of the
former would therefore be unchanged, (2) the functions of the control system
would be unchanged and hence there was no possibility of a new type of accident nor would the consequences of an existing (potential) accident be
increased, and (3) the probability of occurrence of any analyzed (potential)
accident would not be increased. Evidence for this last portion of the
argument was that the NLDC had been successfully tested via simulation,
ex-core mockups, and extensive on-line trials under the original technical
specification. The new technical specification that now governs general use
of the NLDC contains five stipulations. These are (1) that the controller
incorporate the principle of 'feasibility of control', (2) that each proposed control law be the subject of a safety analysis, (3) that the safety
system be separate, (4) that should a limiting condition (in this case too
short a period) be approached, control revert to manual, and (5) that the
operability of the period trip be tested at approved intervals. There is no
limitation on the reactivity worth of the associated absorbers. This
digital controller has now been used for several hundred power increases and
decreases of 20% or more. No challenge to the safety system, not even an
incipient one, has ever occurred.

Relative to the licensing of digital controllers for other reactors, it
is apparent that both the separation of the safety system and the use of a
supervisory algorithm that incorporates the concept of 'feasibility of
control' are important considerations. Also relevant are signal validation,
the verification of signal implementation, and the use of either fault-
tolerant or redundant hardware because, if properly maintained, each reduces
the probability of a malfunction and thereby facilitates demonstration that
an unreviewed safety question does not exist. In addition, a multi-tiered
structure in which a supervisory algorithm reviews the decisions of plant-
specific control laws may offer certain generic advantages. Specifically,
the supervisory concepts are based on general principles and the software
required for their implementation is concise. Hence, it should be possible
to standardize licensing for this portion of the controller. Moreover,
given that the proposed actions of the plant-specific control laws are subject to review by the supervisory tier, the specific details of these laws
should not be of concern to the overall licensing of the controller.
Detailed scrutiny of these laws and their implementing software may not be
essential. Rather it should only be necessary to show that the software
that implements these laws can not create a fault that disables the overall
control program. Hence, the use of a multi-tiered control structure may
offer the possibility of standardized, regulatory approval while at the same
time allowing for individual plant differences.

SUMMARY

A possible methodology for the identification of criteria for the
licensing of closed-loop, digital controllers was suggested. Specifically,
it was proposed that such controllers will be licensable if they are structured to provide each of the functions now performed by human operators
using manual control. The human approach to process control and the status
of digital technology were then reviewed and structural features important
to the design of digital controllers identified. These included the separation of the safety and control systems, the incorporation of the concept of
'feasibility of control', signal validation, verification of signal implementation, and the use of fault-tolerant or redundant hardware. Also it was
argued that a multi-tiered controller that combined a supervisory algorithm
with plant specific laws might offer the possibility of standardized licensing. Experience obtained on the 5 MWt Research Reactor suggests that, if
properly structured and maintained, digital controllers can be used both

safely and to great advantage. The extension of this technology to large
light water reactors should improve operation by increasing reliability,
reducing cost, and enhancing the man-machine interface.

AKNOWLEDGEMENTS

Special appreciation is extended to Professor David D. Lanning and Mr.
Lincoln Clark, Jr. for their many contributions to the design and safety
evaluation of the MIT-CSDL Non-Linear Digital Controller. Appreciation is
expressed to Carolyn Hinds, Ara Sanentz, and Leonard Andexler for their
assistance in manuscript preparation. This research was supported by the
U.S. Department of Energy under Contract No. DE-AC02-86NE37962-A000.

REFERENCES

1. *Proc. Intl. Topical Mtg. Computer Applications Nucl. Power Plant Opera-
    tion and Control*, American Nuclear Society, Pasco, WA (Sept. 1985).

2. *Proc. Electric Power Research Institute Seminar*, Expert System Applica-
    tions in Power Plants, Boston, MA (May 1987).

3. A. Ray, M. Desai, and J. Deyst, Fault Detection and Isolation in a
    Nuclear Reactor, *J. Energy*, 7(1):79-85 (1983).

4. J. A. Bernard, A. F. Henry, and D. D. Lanning, Application of the
    'Reactivity Constraint Approach' to Automatic Reactor Control, *Nuc.
    Sci. Eng.*, 98(2) (1988).

5. R. S. Ornedo, J. A. Bernard, D. D. Lanning, and J. H. Hopps, Design and
    Experimental Evaluation of an Automatically Reconfigurable Control-
    ler for Process Plants, *Am. Control Conf.*, Minneapolis, MN, 3:1662-
    1668 (1987).

6. J. A. Bernard, The Construction and Use of a Knowledge Base in the
    Real-Time Control of Research Reactor Power, *Proc. Sixth Power Plant
    Dynamics, Control, and Testing Symp.*, Knoxville, TN (April 1986).

7. T. B. Sheridan and W. R. Ferrell, "Man-Machine Systems: Information,
    Control, and Decision Models of Human Performance," MIT Press,
    Cambridge, MA (1974).

8. C. R. Kelley, "Manual and Automatic Control," John Wiley and Sons, New
    York, NY (1968).

9. J. A. Bernard, R. S. Ornedo, and D. D. Lanning, Human Approach to Pro-
    cess Control and the Role of Digital Technology, *Am. Control Conf.*,
    Minneapolis, MN, 2:934-940 (1987).

10. S. E. Phillips and D. E. Seborg, Conditions that Guarantee No Over-
    shoot for Linear Systems, *Am. Control Conf.*, Minneapolis, MN, 1:628-
    636 (1987).

11. S. P. Kao and J. E. Meyer, A Plant-Computer-Based Pressurized Water
    Reactor Primary System Thermal Hydraulic Model, *Trans. Am. Nucl.
    Soc.*, 49:469-470 (1985).

12. J. A. Bernard and D. D. Lanning, Issues in the Closed-Loop Digital
    Control of Reactor Power: The MIT Experience, *IEEE Trans. Nucl.
    Sci.*, NS-33(1):992-997 (1986).

492

CHAPTER 9

EQUIPMENT DIAGNOSTICS